AF644425

PENTAGON'S

GLOCLIMATOPAEDIA

3. Emissions Trading & Carbon Management

PENTAGON'S

GLOCLIMATOPAEDIA

3. Emissions Trading & Carbon Management

Dr. A.N. Sarkar

PENTAGON PRESS

Pentagon's Gloclimatopaedia / Dr. A.N. Sarkar

ISBN 978-81-8274-548-3 (Set)

First Edition: 2012

Copyright © Reserved

All rights reserved. No part of this publication may be reproduced, Stored in a retrieval system, or transmitted, in any form or by any means, electronic, mechanical, photocopying, recording, or otherwise, without first obtaining written permission of the copyright owner.

Published by

PENTAGON PRESS
206, Peacock Lane, Shahpur Jat,
New Delhi-110049
Phones: 011-64706243, 26491568
Telefax: 011-26490600
email: rajan@pentagonpress.in
website: www.pentagonpress.in

Printed at
Syndicate Binder, Noida.

Preface

Climate change basically implies a long-term significant change in the "average weather" that a given region experiences. It involves changes in the variability or average state of the atmosphere over durations ranging from decades to millions of years. These changes can be caused by dynamic process on earth, external abiotic and biotic forces, and more prominently by human activities. Global climate change (GCC), is widely believed to be caused by the Greenhouse Gases (GHG). The Intergovernmental Panel on Climate Change reported for the first time in 1990 that greenhouse gas (GHG) concentrations in the global atmosphere were rising as a result of human GHG emissions, principally from fossil-fuel (coal, petroleum and natural gas) burning. Common GHGs include water vapor, carbon dioxide, methane, nitrous oxides, chlorofluorocarbons, hydrofluorocarbons, perfluorocarbons, sulfur hexafluoride, ozone, and aerosols. Natural processes and human activities emit GHG. The accumulation of GHG in the atmosphere regulates the earth's temperature causing global warming and other interrelated effects. Since Global climate change is inextricably linked with the enhanced build-up of greenhouse gases, emissions trading in the form of carbon credits or CERs is opening up a new vista of trade opportunities with concomitant prospect for reduction of emissions—both by the developed and developing nations of the world. Various national and international programmes undertaken by the government and voluntarily or by the non-government agencies have positively impacted the progressive reduction of emissions only to some insignificant extent in many parts of the world.

Kyoto Protocol

The Kyoto Protocol, international treaty formulated in 1999, but came into force in 2005, binds most developed nations to a 'cap and trade' system for the six major greenhouse gases. Emission quotas in the shape of firmed-up commitments were agreed by each participating country, with the intention of reducing their overall emissions by 5.2% of their 1990 levels by the end of 2012.

Under the Treaty, for the 5-year compliance period from 2008 until 2012 nations that emit less than their quota will be able to sell emissions credits to nations that exceed their quota. It is also possible under the convention of the protocol for developed countries within the trading scheme to sponsor carbon projects that provide a reduction in greenhouse gas emissions in other countries, as a way of generating tradeable carbon credits. The Protocol allows this through Clean Development Mechanism (CDM) and Joint Implementation (JI) projects, in order to provide flexible mechanisms to aid regulated entities in meeting their compliance with their caps. The UNFCCC validates all CDM projects to ensure that they create genuine additional savings and that there is no carbon leakage.

Emissions Trading

Emissions trading is an administrative approach used to control pollution by providing economic incentives for achieving reductions in the emissions of pollutants. The term "emissions trading" refers to "a market where, for specified pollutants, parties can buy or sell allowances or permits for emissions, or credits for reductions in emissions."Emissions trading might happen in a region, country or internationally. In a carbon market, tradable units are either allocated by a regulatory body, or produced by GHG emission reduction projects. The former system often involves a "cap-and-trade" scheme, where a regulator caps the participants' allowable emission levels, then issues tradable units equal to the cap. Participants who emit less GHG than allowed can sell their left-over allowances at market price. The emission cap produces the environmental benefit and might be lowered over time. Project-based transactions involve buyers purchasing emission credits from a project that can "demonstrate GHG emission reductions compared with what would have happened otherwise." There is also an expanding retail segment that sells emission reductions to individuals and companies wanting to voluntarily offset their "carbon emission footprints." According to the World Bank's 2007 report on carbon markets: "Reports of increased interest of banks, credit card issuers, private equity funds and others in this segment suggest that it could grow exponentially if only there were a credible, voluntary standard for such assets".

Carbon Credits

Carbon credits correspond to a determined tradable quantity of greenhouse gas (GHG) emissions. They are essentially a permit to emit one metric tonne of CO_2 into the atmosphere. The goal is to achieve a zero net increase in GHG emissions, because each tonne of increased emissions is 'offset' by project-based GHG reductions. They are a key component of national and international emissions-trading schemes. Credits can be exchanged between businesses or individuals, or bought and sold in international markets at the prevailing market price. Carbon credits create a potential market for reducing greenhouse emissions by giving a monetary value to emissions, in effect making emissions a commodity. Credits can be used to finance carbon reduction schemes between trading partners and around the world. They are normally quoted in metric tonnes of carbon dioxide equivalent, and are used to offset emissions from any process that emits GHGs (such as the combustion of fossil fuels), whether in industry, transportation or the household. Carbon credits are used in signatory countries to the Kyoto Protocol to meet emission reduction targets announced annually through international conventions. Carbon credits are created when GHG emissions are reduced below some business-as-usual baseline. This baseline allowance is set by governing bodies such as the United Nations (through the UN Framework Convention on Climate Change) or, in North America, the Chicago Climate Exchange. The basic requirements are that emission reductions are measurable and last at least five years.

Trend in Carbon Market

Carbon emissions-trading has been steadily increasing in recent years. According to the *World Bank's Carbon Finance Unit*, 374 million metric tonnes of carbon dioxide equivalent (tCO_2e) were exchanged through projects in 2005, a 240% increase relative to 2004 (110 $mtCO_2e$) which was itself a 41% increase relative to 2003 (78 $mtCO_2e$).In terms of dollars, the World Bank has estimated that the size of the carbon market was 11 billion USD in 2005, 30 billion USD in 2006, and 64 billion in 2007. The Marrakesh Accords of the Kyoto protocol defined the international trading mechanisms and registries needed to support trading between countries, with allowance trading now occurring between European countries and Asian countries. Various proposals are being investigated to see how these systems might be linked across markets, with the *International Carbon Action Partnership (ICAP)* as an international body to help co-ordinate this.

Trading of Carbon Credits and India's Perspectives

India, along with other developing nations, is at an advantage as it can implement approved CDM projects for the purposes of trading CERs. One third of the total CDM projects registered with UNFCCC are from India. India's carbon credits' trading is expected to reach $100 billion by 2010. In 2007, a total of 160 new projects were registered with UNFCCC. As a consequence, Indian industry managed to generate over 27 million carbon credits. Indian projects receive further impetus by way of investments and finance from developed nations who are potential buyers of CERs. The Multi-Commodity Exchange of India Ltd entered into an alliance with the Chicago Climate Exchange in 2005 to introduce carbon credit trading in India. This association has integrated Indian markets with their global counterparts to cover risks associated with futures trading of carbon credits and ensuring best prices. CDM projects, mostly in key sectors such as manufacturing, energy, agriculture, mining and mineral production, would thus result in providing a boost to the Indian economy. The major CDM projects are located in Rajasthan, Andhra Pradesh, Maharashtra Karnataka, Himachal Pradesh and Punjab.

Exchange-based Trading of Carbon Credits

The role of an exchange is to facilitate emissions trading but also to guarantee each trade. In order to fulfil this role, access to the exchange floor is restricted to its members/seat-holders, who can act as buyers or sellers. In order to trade Carbon credits on an exchange like the New York Mercantile Exchange (NYMEX), ParisBourse, the International Petroleum Exchange, or the Sydney Futures Exchange, one must go through a member of the exchange. Currently there are at least six exchanges trading in carbon allowances: the Chicago Climate Exchange, European Climate Exchange, Nord Pool, PowerNext, Multi Commodity Exchange and National Commodity and Derivatives Exchange. Recently, NordPool listed a contract to trade offsets generated by CDM called Certified Emission Reductions (CERs). Many companies now engage in emissions abatement, offsetting, and sequestration programs to generate credits that can be sold on one of the exchanges.

Headquartered in the financial capital of India, Mumbai, MCX (www.mcxindia.com) is a demutualised nationwide electronic Multi-commodity futures exchange set up by Financial Technologies with permanent recognition from Government of India for facilitating online trading, clearing and settlement operations for futures market across the country. The exchange started operations in November 2003 and presently is the top commodity exchange in the country with 70% market share for quarter ended June 30, 2007 of the total commodity exchange industry in the country. Among the top ten commodity derivatives exchanges in the world, MCX ranks among the top three bullion, energy and copper exchanges in the world in terms of contracts traded. Apart from being accredited with ISO 9001:2000 for quality management, it is the world's first and only multi-commodity exchange to have achieved ISO 27001:2005 certification, the global benchmark for information security management systems.

Accounting and Taxation Aspects of Carbon Trading

The emergence of the opportunity of revenue generation by taking up structured Clean Development Mechanism (CDM) projects has given a new dimension to Accounting and Taxation of the transacted Carbon credits through exchanges/ derivative markets. India's carbon credits' trading is expected to reach $100 billion by 2010. In 2007, a total of 160 new projects were registered with UNFCCC. As a consequence, Indian industry managed to generate over 27 million carbon credits. Indian projects receive further impetus by way of investments and finance from developed nations who are potential buyers of CERs. The Multi-Commodity Exchange of India Ltd entered into an alliance with the Chicago Climate Exchange in 2005 to introduce carbon credit trading in India. This association has integrated Indian markets with their global counterparts to cover risks associated with futures trading of carbon credits and ensuring best prices. The major CDM projects in India are located in Rajasthan, Andhra Pradesh, Maharashtra Karnataka, Himachal Pradesh and Punjab. Trading in CERs, although at a nascent stage, has resulted in huge foreign exchange earnings for Indian suppliers.

As CERs are capital assets, tax liability should ideally be admitted under the head Capital Gain, and claim for concessional rate of taxation should also be made if credit is held for more than 36 months immediately preceding the date of transfer. This gives an opportunity to take a decision about timings of sale of such credits, keeping a balance between cash-flow needs, interest factor and difference in rate of tax between long-term and short-term holdings. As there would be no cost of acquisition for self-generated CER credits, section 55(2) of the Income Tax Act will come into operation, and total sale consideration will be liable for Capital Gains Tax (long-term/short-term) according to the period of holding. In Indian circumstances, if sale of CER credits happen to overseas buyers, of the property held overseas, such sale, though sale of 'goods', will not attract any sales tax.

Emissions Trading Modalities: Cap and Trade System

Emissions-trading is an administrative approach used to control environmental pollution by providing economic incentives for achieving reductions in the emissions of pollutants. Kyoto is a 'cap and trade' system that imposes national caps on the emissions of Annex I countries. On average, this cap requires countries to reduce their emissions 5.2% below their 1990 baseline over the 2008 to 2012 period. The first step in setting up a cap-and-trade program is deciding which greenhouse gases and emissions sources are covered and who is responsible for holding allowances. Some sectors that might be included under the cap are electric power, manufacturing, transportation, or fossil fuel use. For a proper understanding on the modalities of Emission trading a good understanding of Carbon-economics encompassing a number of elements of business accounting and finance are required, such as the economic modeling of demand and supply of carbon credits and allowances, forward and spot pricing, risk management business value, capital allocation and the (possible) International Financial Reporting Standards (IFRS) directives for accounting for carbon and related transactions. In addition, taxation issues of direct carbon taxes, value-added (VAT) and goods and services (GST) taxes, and transfer pricing implications of carbon trading also need to be considered.

Potential Barriers in Emissions Trading

Emissions trading strategy should utilize financial market best practice and spread market exposure over the year for trading a net long or net short position. This exposes the firm to an average price traded value over the year and not a single one off price. The larger the short or long exposure the closer the position should be actively managed by financially experienced personnel.

Options, floating price forwards and swaps can also be used to ensure an average annual price exposure and minimize transaction costs. An effective and active trading strategy will help ensure market trends are tracked and market timing of transactions are effective. There are few barriers to cap and trade emissions trading so long as markets evolved especially across Europe. Liquidity has improved and one common market across Europe has evolved for emissions which is more akin to financial markets like foreign exchange than flow based commodity markets where there are still massive barriers to the creation of an effective pan European traded market.

Certification and Verification

Since emission reductions used to generate credits under article 12 require considerably more scrutiny, a certification function is necessary to ensure that only certified emission reduction become part of the allowance system. Certified allowances should be treated in the same way as all other allowances. The certification process is one practical way to help ensure that the trading system runs smoothly. While the ultimate authority for certification should be the Conference of the Parties, the operational authority for certification can, and should, be delegated to subordinate bodies specifically designed to fulfil that function. Some certification authority could be delegated to specific governmental units within participating nations or communities of participating nations or even to private certification entities, providing that certain preconditions were met. While certification is sufficient for transfer of a credit, use of a credit to fulfil part of an assigned amount obligation would also require verification. Certification on the other hand, would provide assurance that a specific emission reduction or carbon absorption would be forthcoming from the project, verification would provide the assurance that these expectations had de facto materialized.

Compliance and Enforcement

Multiple commitment periods offer significant opportunities for enforcing compliance. The principal tools for enforcing compliance include declaring non-compliant Parties ineligible for trading and reducing assigned amounts in subsequent commitment periods. Currently, the Protocol establishes that there will be negotiations to fix assigned amounts in subsequent commitment periods, but it has perhaps not been generally recognized how important that task is in promoting compliance within the first commitment period. Transparency is another tool for ensuring compliance, since much of the domestic law presumes that enforcement will be based upon public opinion and normative pressures to comply. Transparency is an important complement to enforcement because it makes violations apparent, and the fear of detection promotes compliance. Transparency systems require the disclosure of basic information regarding obligations, actual emissions and trading activity, in order to allow judgements about compliance status. To date, transparency systems have relied heavily on self-reported data, and have worked best when data have been made available to the public. Such data need to be collected and verified at the international level, which will require the use of international public databases of self-reported and verified information, under the responsibility of the authorized body. Numerous issues concerning carbon trading regimes have yet to be worked out, but could significantly affect the various parties involved in trading. These issues include:

- Whether countries' control regimes will be compatible with trading;
- The effect of restrictions on permit availability or demand;
- Impacts of international transfer payments;
- Measurement and reporting of emissions, sinks, and costs; and
- Accountability and enforceability.

Conducting Carbon Emissions Inventory

For an organization, the term carbon footprint or greenhouse gas inventory includes the carbon emissions and other greenhouse gas (GHG) emissions generated directly from the organization's activities or use of fuels, and also indirectly from the use of electricity and from the use and disposal of materials, products and services. By measuring its carbon footprint or GHG inventory, the organization can manage and reduce emissions over time, and also use it for disclosure to stakeholders or for marketing and corporate social responsibility (CSR) purposes. After calculating the carbon footprint, it is then possible for the organization to take active steps to manage the emissions.

Carbon Market: The Regulatory Instruments

The emerging global carbon market is growing exponentially as states and countries around the world use the power of the "market" to control greenhouse gas emissions. In the US, a voluntary market has sprung up to help provide a method to track these emission reductions for concerned individuals and businesses. Carbon offset and renewable energy credits typically represent investments in alternative energy enterprises, energy efficiency technology development and forestry- or agricultural-related (carbon sequestration) projects. International and Federal legislation has been enacted to deal with GCC issues. The Montreal Protocol was originally signed in 1987 and substantially amended in 1990 and 1992. The Montreal Protocol governs compounds that deplete ozone in the stratosphere—chlorofluorocarbons (CFCs), halons, carbon tetrachloride, and methyl chloroform. The Protocol provided that these compounds were to be phased out by 2000 (2005 for methyl chloroform).

India as a Potential Supplier of CERs

Over 100 million tonnes of CERs are potentially available globally through 2012 The portfolio of 78 projects approved by the NCDMA to date has the potential to generate approximately 117 million tonnes of CERs (cumulative) until the end of 2012, according to the project developers' estimates. India continues to be a leader in both small and large scale projects. Two HFC-23 destruction projects and one PFC reduction project from a large aluminium smelter have been approved. In addition, projects related to renewable energy and energy efficiency and fuel switch have the potential to generate another 17and 37.4 million tonnes of CERs until 2012, respectively. India, being one of the leading generators of CERs through CDM, has a large scope in emissions trading. Analysts forecast that its trading in carbon credits would touch US$ 100 billion by 2010.

Currently, the total registered CDM projects are more than 300, almost 1/3rd of the total CDM projects registered with the UNFCCC. The total issued CERs with India as a host country till now stand at 34,101,315 (around 34 million), again around 1/3rd of the total CERs issued by the UNFCCC. In value terms (INR), it could be running into thousands. Various industries that have scope of generation of CERs; viz. Agriculture, Energy (renewable and non-renewable sources), Manufacturing, Fugitive emissions from fuels (solid, oil and gas), Metal production, Mining and mineral production, Chemicals, Afforestation and reforestation. Further, there has been a surge in number of registered projects in India. In 2007, a total of 160 new projects were registered with the UNFCCC indicating that more than half of all registered projects in India happened last year. It is expected that with increasing awareness this would go further up in the future. The number of expected annual CERs in India is hovering around 28 million and considering that each of these CERs is sold for around 15 euros, on an average, the expected value is going to be around Rs 2,500 crore.

Financial Accounting Issues of CDM Credits in India

The contribution of India to the cumulative global CO_2 emissions from 1980 to 2003 is only 3.11% and yet, India is one of the major players in the global market on the supply side of CERs. Indian companies have started getting credit of CERs and some of them have also entered into sale arrangement with buyers in the international market. As this is a new concept, it has given rise to interesting financial accounting dimensions. Issues that were duly considerer are (i) how to account for expenditure on CDM projects, (ii) whether or not to account for self-generated CERs held with registry, (iii) if credits are to be accounted, at what point of time these should be recognized in books of accounts and at what value, and (iv) how to account for sale consideration of CERs and its disclosure in accounts and notes. Answers to these questions are found within existing pronouncements of ICAI as well as Schedule VI requirements.

The Emerging International Carbon Market

The overall value of the global carbon market was estimated at over $10 billion in 2005. In the first quarter of 2006 alone, the value of carbon market transactions reached $7.5 billion. This growth is driven by high prices in the EU ETS market for allowances. EUAs worth $8.2 billion were traded in 2005. This corresponded to 322 million tCO_2e and represented almost a 40-fold increase over the previous years' volumes. Recent findings indicate that EU governments had been conservative and may have over-allocated allowances to installations for the EU ETS phase I (2005-07). This resulted in a volatile market when first year reports were made public in May 2006. It is now widely anticipated that there will be much tighter compliance limits in phase II (2008-12). This would result in strong demand for both EU allowances and CERs. In addition to the EU ETS, several EU governments have established national programs to purchase carbon credits to meet their Kyoto Protocol compliance targets. Demand from the Japanese private sector for carbon credits remains strong. The Japanese Government has announced new plans to purchase CERs, adding further to demand. Other demand drivers include the establishment of a regional greenhouse gas initiative in the northeastern United States, the continuing operation of the New South Wales carbon market in Australia and the Chicago Climate Exchange. The supply of carbon credits has improved, particularly in Asia. The PRC and India are now leading suppliers of CERs.

Political leaders from major industrialized countries and global business leaders from a range of industrial sectors have affirmed that market-based approaches and an efficient and effective "carbon market" are key elements of a successful global climate regime. These approaches will create a market value for greenhouse gas (GHG) emissions and a price signal to guide action by countries, businesses and other entities. The G8 Gleneagles Plan of Action on climate change, clean energy and sustainable development specifically flagged "tradable certificates and trading of credits" and "project-based and voluntary offset mechanisms" as elements of a carbon market that need further policy development. The World Economic Forum Roundtable, in a statement released just prior to the G-8 Summit in Germany, urged governments to "establish a long term, market-based policy framework extending out to 2030...." and "define greenhouse gas emissions rights through a cap and trade system or other market-based mechanisms."

Looking Beyond 2012

If the carbon market is to play a significant role in helping to achieve the deep GHG emission reductions required over the next 20 years, decision-makers will need to consider how best to broaden and deepen the reach of that market. The broader the reach of the carbon market – through greater participation of countries and/or coverage of key sources and sectors and/or companies – the more efficient and effective it will be in minimizing costs and mobilizing investment. Broad coverage can also address concerns about leakage of emissions-intensive economic activity between countries as well as competitiveness more generally. Broadening and deepening the market will likely require a more flexible range of options to meet widely differing national circumstances and needs. The criteria for the years after 2012 is how to broaden participation in the carbon market in developing countries in a way that minimizes cost uncertainties and promotes economic growth and investments in low-carbon technology. In particular, the engagement of less advanced developing countries in the carbon market needs to be supported. The CDM will likely continue to play an important role and may expand its scope to address policy-based crediting and a broader range of projects. Efforts are underway to strengthen and streamline the CDM. A strengthened CDM can help build practical experience and confidence with market-based approaches across the developing world and position the CDM to contribute beyond 2012.

The Voluntary Carbon Markets

The voluntary carbon market (VCM) has developed independently of government targets and policies and is a place where anybody, from businesses, to NGOs, to individuals can participate in the business of offsetting. Carbon credits are also created in the voluntary market, but unlike the compliance market where credits are tradeable under the Kyoto flexible mechanisms, credits in the voluntary market are generally non-fungible—they are not tradeable between schemes. There is a legally-binding voluntary market where parties can set self-imposed, legally binding greenhouse gas emissions reductions targets: an example of this is the Chicago Climate Exchange. In voluntary carbon markets, activities that reduce GHGs produce verified emission reductions (VERs) that can be sold to companies or individuals wishing to voluntarily reduce their carbon footprints. GHG emission reduction projects developed under the Kyoto Protocol's Clean Development Mechanism (CDM) have been highly successful in reducing emissions and generating Certified Emission Reductions (CERs), which are then purchased by governments and organizations in Europe and Japan to help meet their emission reduction targets.

The current voluntary offset market is seen as potentially undercutting the compliance market with cheaper offsets that are not clearly additional and sending the wrong price signals. Since the public and the media often do not distinguish between the compliance and the voluntary market, there is also a risk of damaging the reputation of compliance markets. To secure quality and transparency in the voluntary market, it is argued that voluntary offset standards should closely follow CDM procedures and apply them to VERs (e.g. the CDM approach to additionality, the documentation of reductions, and the monitoring and verification processes). Standards that share this viewpoint include VER+ and the Voluntary Offset Standard (VOS).

Voluntary Carbon Standards

The Voluntary Carbon Standard has been developed by The Climate Group and the International Emissions Trading Association (IETA). The

Voluntary Carbon Standard aims to ensure that all voluntary emission reductions projects that want to trade in VCUs are independently verified to meet specific criteria and that these will represent "real, quantifiable, additional and permanent project-based emission reductions." The VCS will provide protocols and criteria to certification entities and project developers on the specifications for creating, verifying, and registering VCUs. The VCS has created a registry managed by the Bank of New York which is used to register, transfer and retire VCUs from the market. There are two elements to standardized offset crediting: (1) streamlining the estimation of project baselines by using standard assumptions and emission factors; and (2) deciding the "additionality" of projects using standard eligibility criteria. The basic requirements for developing standardized baselines have been well-documented. Standardized approaches work best where projects and their baseline alternatives are fairly homogeneous; where similar technologies and practices are in use across wide geographic areas; and where performance data on those technologies and practices are readily available. Even where certain types of projects are highly suited to standardized methods, most carbon offset programs have relied on at least some project-specific data in order to accurately estimate baseline emissions. Because standardization involves balancing policy objectives related to cost, equity, and environmental integrity, institutional processes should be transparent and accountable. Efforts should focus on project categories with high suitability for standardization and the greatest potential contribution to sustainable development, such as renewable energy, industrial energy efficiency and process emissions, and efficient vehicles.

Carbon Credit Offsetting and its Standardization

A 'Carbon Credit' becomes a 'Carbon Offset' when the value of the credit is invested in projects that reduce emissions. Tradable GHG reductions are often purchased to offset or "neutralize" a company's carbon dioxide emissions, and are therefore commonly referred to as "carbon offsets." Carbon offsets are a unique commodity. In essence, they are a claim or property right to a specific, verified reduction in GHG emissions. Carbon offsets originate from projects or activities (i.e., "offset projects") explicitly designed to reduce GHG emissions. In purchasing and retiring an offset, a company in effect makes the claim that it has foregone reducing its own emissions, but is instead paying someone else to achieve the same net result. Carbon offsets are intangible, and so their integrity depends on assurances about how well they meet certain defining criteria.

Offsetting is a way for individuals and companies to compensate their own greenhouse gas emissions by investing in projects that remove carbon from the atmosphere or reduce or eliminate GHG emissions compared to the conventional methods. By offsetting emissions level for an activity (e.g. a flight), the idea is that the activity becomes "carbon neutral". Carbon offsets are generated by renewable energy, efficiency projects and other projects that avoid the release of CO_2 or other greenhouse gasses. This rise of offsets has been accompanied by greater 'Carbon literacy' amongst companies and consumers, with hard questions being asked, not just of the offset providers, but also of the relevance of offsets in the wider carbon emissions solution/climate change challenge.

Carbon Footprint

A carbon footprint is "the total set of GHG (greenhouse gas) emissions caused directly and indirectly by an individual, organization, event or product" (UK Carbon Trust, 2008). An individual, nation, or organization's carbon footprint is measured by undertaking a GHG emissions assessment. Once the size of a carbon footprint is known, a strategy can be devised to reduce it. Carbon offsets, or the mitigation of carbon emissions through the development of alternative projects such as solar or wind energy or reforestation, represent one way of managing a carbon footprint. The concept and name of the carbon footprint originates from the ecological footprint discussion. The carbon footprint is a subset of the ecological footprint.

Achieving Carbon Neutrality

'Carbon neutral', also called 'Carbon neutrality' is a term used to describe the action of organizations, businesses and individuals taking action to remove as much carbon dioxide from the atmosphere as each put in to it. Being carbon neutral, or having a net zero carbon footprint, refers to achieving net zero carbon emissions by balancing a measured amount of carbon released with an equivalent amount sequestered or offset. The carbon neutral concept may be extended to include other greenhouse gases (GHG) measured in terms of their carbon dioxide equivalence—the impact a GHG has on the atmosphere expressed in the equivalent amount of CO_2. The term climate neutral is used to reflect the fact that it is not just carbon dioxide (CO_2), that is driving climate change, even if it is the most abundant, but also encompasses other greenhouse gases regulated by the Kyoto Protocol, namely: methane (CH_4), nitrous oxide (N_2O), hydrofluorocarbons (HFC), perfluorocarbons (PFC), and sulphur hexafluoride (SF6). The overall goal of carbon neutrality is to achieve a 'Zero-carbon footprint'. This practice is often called 'carbon offset' or 'offsetting'. Being carbon neutral involves calculating your total climate-damaging carbon emissions, reducing them where possible, and then balancing your remaining emissions, often by purchasing a carbon offset: paying to plant new trees or investing in "green" technologies such as solar and wind power.

Carbon neutrality offers individuals, businesses and other institutions the opportunity to take the personal responsibility for the global warming implications of their lifestyle. In response to a rapid increase in public awareness about global warming in the last 12 months, many corporations are focusing on how they can manage their carbon impacts. Many organizations in India are aiming to define their carbon footprint, to reduce this footprint by reducing the emissions from the business and, increasingly, to reduce their impact further by purchasing 'green energy' (from renewable sources) or carbon offsets. Some companies, wishing to go further, are becoming 'carbon neutral', where they completely offset their carbon footprint to theoretically have zero carbon impact on the environment. These companies can capture the Indian market by playing the significant role as the retail seller as well as buyer of the carbon credit.

Abatement of GHG Emissions: The Challenge of the 21st Century

The challenge of the 21st century will be for human society to regain a healthy relationship with our 'living' earth. The signs are clear that right now we are failing in that task. Unless we change our ways, it is likely that comfortable civilization as we know it will disappear. Climate change is not just an environmental problem—it is a societal problem. The simple truth is that humanity is changing the atmosphere—every day we add around 70 million tonnes of invisible carbon dioxide, the very gas that we know controls the temperature of the planet. Climate change is a critical threat to the continued existence of our civilization. Reduction of emissions will be the hallmark of carbon responsible companies and the use of a transparent carbon philanthropy mechanic, supported by clear cost metrics, to reflect the voluntary financial commitments of companies, either through support for charities and/or internal emissions reduction programmes, will deliver tangible benefits that investors, employees and consumers can identify. The use of a financial mechanic for reporting carbon will be helpful preparation for a regulatory environment, support the important link between carbon efficiency and cost efficiency, thereby bringing carbon into business operations as an environmental and cost metric.

Carbon Literacy

Carbon literacy embraces the following attributes: (a) Possessing a general knowledge or awareness of the concepts, causes, and the effects of atmospheric pollution or greenhouse gases on the environment, especially in relation to the climate of earth as in climate change or global warming. (b) To have the skills necessary to limit or reduce personal and/or collective impact, i.e. carbon footprint, on the environment through more efficient and wise use of available resources and energy. (c) To have the knowledge and skills necessary to participate in personal carbon trading systems and platforms which focus on financial incentives and investments in new methods, systems, and technologies that limit or reduce greenhouse gas pollution, as in carbon capturing or carbon sequestering, and the research and development of alternative energy sources.No single person is expected to understand every detail about all of the fundamental climate science literacy concepts. Full comprehension of these interconnected concepts will require a systems-thinking approach, meaning the ability to understand complex interconnections among all of the components of the climate system. Moreover, as climate science progresses and as efforts to educate the people about climate's influence on them and their influence on the climate system mature, public understanding will continue to grow

World civilization is dependent on fossil fuels and is starting to recognize the consequences. However we are still predominantly carbon illiterate. Carbon dioxide, the primary greenhouse gas caused by burning fossil fuels, is invisible to all and unknown to many and there are few economic penalties for its release. The key question is how to break this vicious circle and create action. Consumers are becoming more aware but it is unrealistic to expect major change until carbon is properly priced in the economy which requires government action. Lead Energy/ Power generating companies have the ability to break the cycle by helping raise consumer awareness, challenging government and looking for business opportunities in a carbon-constrained world. There is clearly a need for more research to be done in understanding what exactly encourages people to reduce their emissions; on the extent to which the practice of offsetting has an effect on such behaviour; and on how much it can play a role in educating people about climate change in order to meeting Corporate Social Responsibility (CSR) obligations, to generate goodwill, or to attract the growing number of customers by environmental action.

Carbon Management and Corporate Social Responsibility

More and more companies are now demonstrating their 'carbon consciousness by seeking out inventive and broad viewed carbon offsets. One approach that is becoming more and more desirable is to invest in smaller scale, high quality projects that demonstrate social or biodiversity benefits in addition to carbon mitigation. Some organizations look to lower the impact of their business on the environment by including projects that have a strong social focus, such as including disadvantaged people in the completion of the carbon projects themselves or by combining agro-forestry projects that open up Fair Trade commercial opportunities to the farmers.

It's this emphasis on complementary benefits that cements carbon offsetting into the CSR arena and provides another area of advantageous marketing for the more savvy company. It provides reassurance that your investment is providing a wide range of conscientiousness benefits in addition to compensating for the greenhouse gases you emit. Whatever offsetting portfolio one opt for, one should be clear on how it fits with the CSR strategy and be confident of both the footprint calculations and the quality of the carbon credits. The overall success of a carbon management plan relies on the co-operation and involvement of everyone within an organization including senior management. This commitment is essential to driving such a complex and serious subject and showing staff that they are serious about the initiative. Genuine management commitment will ensure that the necessary resources are devoted and employees are more likely to become involved. Carbon management can be time and resource intensive; so without management support, direction and an understanding of the issue, it can be very difficult to make real change.

Broad Framework of International Carbon Market

Possible elements of a framework that would enable to provide a broad institutional and operational mechanism for an international carbon market could include:

- binding fixed emission limits for industrialized countries; for industrialized countries unable to agree to the above, binding fixed or dynamic emission limits for some sectors in some regional groupings – or possibly economy-wide binding dynamic emission limits;
- binding transnational sectoral emission limits (fixed or dynamic) for some key sectors represented by multinational "operators" (e.g., cement, steel, aluminium); individually customized non-binding baselines for specific sectors to generate credits, while attracting investment in key sectors of developing countries; and a project-based crediting mechanism to provide coverage of emission reduction and sink enhancement activities not already covered by other market-based mechanisms.
- While an international framework is essential, the ability of the international carbon market is to minimize costs
- The International Carbon Bank and Exchange provide a platform that enables individual and corporate clients to keep track of greenhouse gases in a secure environment. Clients can use their account at the ICBE to establish baselines, introduce emission reductions, and acquire offsets in real-time.

Carbon Risk Management

Different accounting treatments can be adopted based upon the purpose for which the allowances are held. In broad terms allowances, as marketable instruments, could be held for: (a) Hedging of current or future period emissions exposures as a result of taking on a cap; or (b) Speculative purposes to benefit from expected changes in the market price of the allowance. Shareholders require, and so management should provide, clarity over whether the allowances are being held for hedging or speculative purposes as the risk profile of the company, and its financial results, are directly affected by these choices. To illustrate this point an example has been set out below. In the example the company has entered into a forward sale of emission allowances for delivery in 2007. The diagram is based on a hypothetical transaction. Under the UK Scheme it would be expected that the number of allowances that a company expects to hold for each compliance period would be less than the forecast emissions profile. The question to be answered is whether the sale has been entered into for hedging or trading purposes. For a company to be able to demonstrate that it is buying allowances for hedging purposes it needs to be able to demonstrate that the allowances are required to meet a future emissions exposure or are within their planned hedging policy. Key elements of the operating and control framework to enable a company to support the transaction and provide additional evidence are: (a) A carbon risk management policy; and (b) The availability and credibility of management information to support the implementation of that policy.

International Carbon Finance for Adaptation

Carbon finance is a new branch of Environmental finance. Carbon finance explores the financial implications of living in a carbon-constrained world, a world in which emissions of carbon dioxide and other greenhouse gases (GHGs) carry a price. Financial risks and opportunities impact corporate balance sheets, and market-based instruments are capable of transferring environmental risk and achieving environmental objectives. Issues regarding climate change and GHG emissions must be addressed as part of strategic management decision-making. The general term is applied to investments in GHG emission reduction projects and the creation (origination) of financial instruments that are tradable on the carbon market. At the International Level, the Carbon Finance Unit serves as IFC's in-house resource for all carbon finance-related issues, providing services directly to buyers and sellers. The Unit leads in the development of new products, and serves as an advisor on a variety of products and services to support private sector participation in the evolving carbon market.

The World Bank Group has been active in scaling up the activities towards a low-carbon economy by developing the carbon market (the Bank currently manages 9 carbon funds with a total investment of $2 billion), utilizing GEF (Global Environment Facility) resources, and mobilizing private sector capital. Since its beginning in 1991, the GEF has provided $6.2 billion in grants and generated over $20 billion in co-financing from other sources to support over 1,800 projects that produce global environmental benefits in 140 developing countries and countries with economies in transition. Further improvements in World Bank and other International Financial Institutions' financial instruments include scale-up of umbrella carbon facilities, ESMAP (Energy Sector Management Assistance Program) window for clean energy project development, MIGA (Multi Investment Guarantee Agency) insurance for carbon credits, expanded use of contingent grants and guarantees, bundling programmatic Clean Development Mechanism, the European Bank for Reconstruction and Development's Energy efficiency facility, the Inter-American Development Bank's Sustainable energy initiative and the Asian Development Bank's Asia Pacific Energy Efficiency Fund.

According to International Finance Corporation the term "carbon fund" or "carbon vehicle" refers to a collective investment scheme that receives money from investors and uses this money to buy carbon credits from, or invest into greenhouse gas (GHG) emissions reduction projects, generally CDM and JI. After a certain defined period, the carbon fund will then give investors carbon credits and/or cash in return. The emission reduction projects can be sourced from different developers, countries, and technologies. Carbon funds can be initiated and managed by development banks, governmental agencies, private financial institutions, or consultancies. In terms of investors, ICF identifies the following types of shareholder structures: government(s), private firm(s), or a mix of both (multi-shareholder).

Carbon funds have become an important player in the carbon markets over the past few years and have experienced considerable growth both in terms of the number of vehicles and the amount of capital raised. However, the amount of capital secured or targeted by funds created over the last two years appears to be slowing down. Indeed, the lack of clarity on the post 2012 climate policy framework at international and domestic levels, increasing competition on the buyers' side, and the decline in the number of large and low risk Clean Development Mechanism (CDM) projects create a challenging and uncertain environment for new entrants in the carbon finance community. The future of the carbon fund market is complex and uncertain. However, faced with increasing challenges, whether carbon vehicles will come out as winners or losers will mainly depend on their investment strategy and their flexibility to adapt.

Climate Change, Energy Security and Sustainable Development Issues

The Ministry of Power (BEE, 2007), Government of India, brought out a detailed status position paper on "Addressing Energy Security and Climate Change" in October, 2007 in a Report Form, which will bring about integration and various initiatives institutionalized through specialized institutions, such as the Ministry of New and Renewable Energy, the Bureau of Energy Efficiency, and the Technology Information, Forecasting and Assessment Council, with specific mandates to promote climate friendly technologies. The National Environment Policy, 2006, provides the basis for the integration of environmental considerations in the policies of various sectors. The Policy Statement for Abatement of Pollution, 1992, stressed the prevention of pollution at the source based on the "polluter pays" principle. The Forest Policy, 1988, highlights environmental protection through preservation and restoration of the ecological balance. Sustainable development has become part of all climate change policy discussions at the global level, particularly due to adoption of Agenda 21 and the various Conventions resulting from the UNCED-1992. The generally accepted and used definition as given by the Brundtland Commission in 1983 is 'development that meets the needs of the present without compromising the ability of future generations to meet their own needs'.

Greenhouse Gas (GHG) Emissions Reduction Strategy

India emits the fifth most carbon of any country in the world. At 253 million metric tons, only the U.S., China, Russia, and Japan surpassed its level of carbon emissions in 1998. Carbon emissions have grown nine-fold over the past forty years. Going forward, carbon emissions are forecast to grow 3.2% per annum until 2020. To put this in perspective, carbon emissions levels are estimated to increase by 3.9% for China and by 1.3% for the United States. India is a non-Annex I country under the United Nations Framework Convention on Climate Change, and as such, is not required to drastically reduce its carbon emissions. Climate change and other sustainable development policies are often but not always synergistic. However, sustainable development has become an integrating concept embracing economic, social and environmental issues. Three critical components in promoting sustainable development are economic growth, social equity and environmental sustainability. Switching to more sustainable development paths can make a major contribution to climate change mitigation, but implementation may require overcoming multiple barriers. Climate change and other sustainable development policies often benefit each other, though not always. There is a growing understanding of the possibilities to choose and implement mitigation options in several sectors to create synergies and avoid conflicts with other aspects of sustainable development. Making development more sustainable by changing development paths can make a major contribution to climate change mitigation, but implementation may require resources to overcome multiple barriers.

Energy Carbon Nexus

Energy consumption and climate change are highly correlated. Leading enterprises increasingly understand the links between greenhouse gas (GHG) emissions and energy consumption. To put it simply, if we reduce energy use, we will reduce GHG emissions. Companies need short-term solutions to optimize energy usage and better understand their GHG emissions. In the long term, they want to optimize the usage and financial impact of energy and reduce GHG emissions. Energy is a key contributor to carbon and other GHG emissions. Recent energy prices have seen extreme volatility because of supply-and-demand imbalances and financial speculation. Most energy experts anticipate a return to higher energy pricing when the current economic downturn ends. The ability to manage the acquisition, use, and cost of energy will be critical to combating climate change and to promoting sustainable, profitable growth.Companies and other organizations are being challenged to create an effective, fiscally responsible, and sustainable response. Commercial and industrial businesses around the world are under pressure to cut energy usage and spending, go "green," and take measures to address climate change. The drivers for change come from many sources such as those from: (i) Consumers who place increasing value on good environmental stewardship and reflect it in their buying preferences; (ii) Retailers who demand their suppliers embrace their own particular brand of green strategy; (iii) Governments that globally mandate change; (iv) Investors who rightly demand to know the green credentials of the companies in which they are investing.

Energy for Sustainable Development

Energy is a critical instrument for Socio-economic and sustainable development. Energy is central to sustainable development and poverty reduction efforts since it affects all aspects of development—social, economic, and environmental—including livelihoods, access to water, agricultural productivity, health, population levels, education, and gender-related issues. And yet, some 2.4 billion people in developing countries lack modern fuels for cooking and heating and approximately 1.6 billion people do not have access to electricity. The current energy infrastructure is unsustainable, vulnerable to natural disasters, and woefully insufficient in many developing countries.

The Government of India should therefore provide incentives, within legal and policy frameworks, for public, private, and community level action towards greater fuel efficiency and the use of appropriate technologies. Countries with such incentive systems can serve as examples of utility based incentives for Renewable Energy Technologies (RETs), tax incentives, fuel displacement levies and feed-in laws, extension of loans to renewable energy promoters and consumers shifting to RETs, market mechanisms, and pricing policies such as net metering facilities and tax-exemption on public bus purchases. Such incentives have the potential of encouraging private sector players to promote the use of renewable energy, including wind, solar, and hydropower. Developed countries should further international cooperation to transfer renewable energy technologies and capacity building to developing countries.

The Government of India should also implement existing commitments with relevance to energy for sustainable development, including those made in Agenda 21, the WSSD, and the Johannesburg Plan of Implementation. It is critical for countries to implement fully and expeditiously the UN Framework Convention on Climate Change, its Kyoto Protocol, and related climate change agreements, policies, and partnerships. Government should involve civil society in multi-stakeholder participatory processes during the design, implementation, and monitoring of energy programmes.

Harmonizing Global Energy Markets and Energy Security

Approaches to ensuring energy supply security in the 21st century should differ from past approaches that concentrated on oil substitution. Besides sustainable growth challenges, new approaches need to tackle the new energy security issues raised by market liberalization. The enhanced role of markets is tied closely to the process of globalization. Globalization, which is still gaining momentum, has encouraged competition and strengthened markets and regional and international trade, particularly for crude oil and oil products, natural gas, and energy services. Globalization is bringing new opportunities for energy security, such as better access to markets and services and the transfer of technologies that are helping to reduce the cost of energy exploration and expand proven reserves. International trade in energy resources (including Emissions trading) and services is vital for energy security.

Fulfilling the commitments as agreed and at the schedules approved would greatly affect the use of energy resources and could compromise global economic progress. There is a large gap between the commitments and the means for implementation. For instance, emissions reduction targets agreed upon by negotiators were not necessarily implemented by legislators or other policymakers. Implementation of such targets is hindered not only by cost but also by the need to maintain energy security. All indications are that fossil fuels will continue to dominate global energy resources for at least the first decades of the 21st century. Moreover, the demand for energy services will continue to increase. Most of the growth will be in developing countries, which can ill afford the high cost of containment measures. It is therefore essential to find means to contain energy-related emissions without compromising energy security.

A Futuristic Vision of Emission Trading: A Roadmap for Copenhagen Summit

The foregone paragraphs highlight the various specific natures of emerging problems that come in the way to abate climate change with a view to suggest probable solutions through integrated and holistic approaches of emissions reduction to attain carbon neutrality and sustainable energy development with the help carefully designed local as well as global initiatives and inter-governmental cooperation. An attempt has also been made to clearly bring out as to what extent the various approaches and strategies suggested are implementable with appropriate institutional arrangement and governing principles. The various limitations, especially in respect of global norms for emissions reduction commitments for compliance of Kyoto Protocols and observing Global accounting Standards for Emissions trading of Carbon credits have been briefly touched upon with desired lines of actions, as detailed in the following paragraphs:

Given the backdrop of the accumulated knowledge on the scientific reasons for global climate change and research findings emanated therefrom on the various adverse impacts of these changes on ecology, human and animal life etc. it necessitates a proper documentation and analysis of these events to guide the future options, strategies and long- term policies leading to formulating goal-oriented abatement of the declining climate change trend and their stabilization in order to save the planet from further ecological devastation. Global review of the scientific literature on the subject, their careful scrutiny and analysis should stimulate the modern civil society at large to have deeper appreciation and understanding of the interrelationship between the desirability of reversing the processes of climate change as well as building simultaneously the concept as well as Action–oriented programmes of 'Sustainable Energy Development' to bear upon its impact on the reversal of the former processes, especially at a time when "Global warming" and "Greenhouse effects" have been quite alarming along with Global Energy and Economic Crisis.

Emissions reduction through the flexible Clean Development Mechanism (CDM)-based country, regional and global Models initiated and developed through concerted efforts by various International Organizations and Financial Institutions can go a long way to progressively reduce emissions of Greenhouse gases (GHGs) to make the planet a safer place to live in: Sustainable Energy Development can serve as a catalyst to this value-chain. Notable examples of such global initiatives with prospect of developing into Inter-country and inter-regional programmes for replication and adaptation are: (i) European Union Emission Trading Scheme (EU ETS), (ii) UK Emissions Trading Scheme (UK-ETS), (iii) Indo-German Energy Program (IGEN), (iv) CleanAir Canada, (v) Japan's Voluntary Emissions Trading Scheme (JVETS), (vi) International Greenhouse Partnerships (IGP) Program, (vii) UNIDO Knowledge Network for Industrial Technology Transfer, (viii) US-Japan Agreement on Cooperation in Environmental Protection, (1975) , (ix) United Nations Economic Commission for Europe—Long Range Trans-boundary Air Pollution (LRTAP), (x) South Asian Regional Initiatives (SARI), (xi) Global Environment Facility (GEF), (xii) World Bank Pilot Programme for Climate Resilience (PPCR), (xii) Mexican Multilateral Climate Change Fund (MCCF), (xii) Bi- and Multilateral Carbon Auction Levy Funding under IPREA, (xiv) Germany's BMU Funded International Climate Initiative (ICI), (xv) UNEP's Carbon Neutral Network, (xvi) Society of Building Science Educators (SBSE) etc.

The role of an exchange is to facilitate emissions trading but also to guarantee each trade. In order to trade Carbon credits on an exchange like the New York Mercantile Exchange (NYMEX), ParisBourse, the International Petroleum Exchange, or the Sydney Futures Exchange, one must go through a member of the exchange. Currently, there are at least six exchanges trading in carbon allowances: the Chicago Climate Exchange, European Climate Exchange, Nord Pool, PowerNext, Multi Commodity Exchange and National Commodity and Derivatives Exchange. Recently, NordPool listed a contract to trade offsets generated by CDM carbon projects called Certified Emission Reductions (CERs). Many companies are now engaged in emissions abatement, offsetting, and sequestration programs to generate credits that can be sold on one of the exchanges. The emergence of the opportunity of revenue generation by taking up structured Clean Development Mechanism (CDM) projects has given a new dimension to Accounting and Taxation of the transacted Carbon credits through exchanges/ derivative markets. For instance, India's carbon credits' trading is expected to reach $100 billion by 2010. In 2007, a total of 160 new projects were registered with UNFCCC. But, such transactions through Exchanges should be subject to maintaining Global accounting standards with regulatory and legal safeguards for accountability, transparency and compliance with stipulated disclosure norms. For a proper understanding on the modalities of Emission trading a good understanding of Carbon-economics encompassing a number of elements of business accounting and finance are required, such as the economic modeling of demand and supply of carbon credits and allowances, forward and spot pricing, risk management business value, capital allocation and the (possible) International Financial Reporting Standards (IFRS) directives for accounting for carbon and related transactions. In addition, taxation issues of direct carbon taxes, value-added (VAT) and goods and services (GST) taxes, and transfer pricing implications of carbon trading also need to be considered.

Harmonizing emerging global markets and energy security in one hand and continue with the comprehensive strategies of sustainable development without slowing down of global economy with lesser dependence of conventional sources of energy on the other hand will indeed be a gigantic task, given the fact of extensive uses of fossil-fuels (i.e. major polluters through GHG emissions) as major feedstock to sustain the present pace of industrial growth and future energy demands placed especially by the G-8 Nations and/ or BRIC (Brazil, Russia, India and China) Nations. Sustainable development has become part of all climate change policy discussions at the global level, particularly due to adoption of Agenda 21 and the various Conventions resulting from the UNCED-1992. The generally accepted and used definition as given by the Brundtland Commission is 'development that meets the needs of the present without compromising the ability of future generations to meet their own needs'. Approaches to ensuring energy supply security in the 21st century should therefore differ from past approaches that concentrated on conventional energy (predominantly fossil-fuels) substitution. Besides sustainable growth challenges, new approaches would need to tackle the new energy security issues raised by market liberalization. The enhanced role of markets is tied closely to the process of globalization. Globalization, which is still gaining momentum, has encouraged competition and strengthened markets and regional and international trade, particularly for crude oil and oil products, natural gas, and energy services. Globalization is bringing new opportunities for energy security, such as better access to markets and services and the transfer of technologies that are helping to reduce the cost of energy exploration and expand proven reserves. International trade involving energy resources (including Emissions trading) and services therefore becomes vital for energy security.

The idea of sustainable development is gaining acceptance and momentum on the official level as well as among the public. Sustainable development demands environmental preservation. Energy production and utilization, particularly in the case of fossil fuels, can be major sources of environmental degradation. These detrimental environmental impacts have a direct bearing on the future of energy—in terms of fuels and the extent of their use—and on energy security. The United Nations Framework Convention on Climate Change, adopted at the Rio Earth Summit in 1992 and the Kyoto Protocol, signed by more than 160 countries in 1997 (UNFCC, 1997), call for major reductions of greenhouse gas emissions, which are caused mainly by energy use.

In the G-8 Meeting of the Major Economic Forum (MEF) held on 9th July, 2009 in L'Aquila, Italy, the leaders of the largest economies agreed to a goal of achieving at least a 50% reduction in global greenhouse gas emissions by 2050. As global emissions have risen by around a fifth since 1990, a later baseline would represent a higher final emissions goal. It was recognized that the 80% emissions reduction goal would require robust mid-term reductions, although no intermediate targets were set. Equally important will be taking a holistic review of the basis and criteria being used in the 'Cap & Trade' mechanism of Emissions Trading. A rational and balanced approach for emissions reduction commitments by a Nation should ideally be the use of quantum of emissions generated in absolute term (i.e. primary criteria) over to computing on per capita (i.e. secondary criteria) basis. Secondly, the Voluntary Carbon Market should be equally 'Kyoto Protocol Compliant' by bringing in a more stringent 'regulatory regime' to fix accountability in terms of both domestic and international Tax and Tariff control mechanism. But, this perception does not suo motto go ahead sans implicit, inherent and practical challenges. In recent G-8 Summit, what was being asked and debated upon is that Global emissions be less than half their 1990 levels by 2050; having peaked before 2020. Since emissions from the developing countries are on the whole lower than those of the developed economy—and will need to be continued to rise in short terms as they maintain economic growth and address poverty—it has been proposed that the developed countries cut emissions at least by 80 % relative to 1990 by 2050, with major steps towards this goal over the next decade. Developing countries on their part will also need to significantly slowing and peaking emissions reduction growth in the coming decade. To substantiate this point, attention may be drawn to a New Report from the "Breaking the Climate Deadlock" Project involving a Strategic Partnership between 'A' Country and the Climate Group, showing how major reductions even by 2020 are achievable if we focus actions on certain key technologies (viz. eco-technologies involving mainly renewable), deploy policies that have been proven to work (viz. appropriate technologies) and invest now in developing those future technologies (viz. Carbon capture storage, coal-bed methane, geothermal, ocean hydrates etc.) that will take time to mature to operationalize.

The year 2009 is a crucial year for climate change. At a pivotal UNCCC Global Summit on Climate Change in Copenhagen by the end of 2009, Governments of the world need to agree the continuation of the global climate treaty, building on and strengthening the Kyoto Protocol. The G8 leaders meeting in Italy this July have availed the opportunity to set the course for the Copenhagen conference to succeed. In doing this they realized the following advantages of action against climate change:

- Action now will start to clean up our climate act, keeping the option open for the planet to avoid catastrophic climate change.
- Action now will set the world on a sustainable energy path, which in the short run can help stabilize the economy, and in a few decades can provide ample energy for the whole planet.

Action now can help develop a low-carbon economy that helps avoiding millions of climate refugees and massive cost for later adaptation to climate impacts. But a more immediate and perhaps bigger battle on Climate change to take a cognizance of this crucial aspect looms large in Copenhagen Global Climate Summit in coming December. According to 2004-UNFCCC Global Convention, developed countries must periodically negotiate with global partners on mitigation commitments to avoid what is termed as 'dangerous anthropogenic interference' with its climate systems. Fulfilling the commitments, as agreed and at the schedules approved, would greatly affect the use of energy resources and could compromise global economic progress. This is so because there is seemingly a large hiatus between the commitments and the means for implementation. For example, targets agreed upon by negotiators were not necessarily implemented by legislators or other policymakers in the past decade. Implementation of such targets is hindered not only by cost but also by the need or domestic economic compulsions to maintain both energy security and the projected pace of growth over a planned period of time.

To sum-up the impressions drawn from these experiences, all indications are that fossil fuels will continue to dominate global energy scene for at least the first couple of decades of the 21st century. Moreover, the demand for energy services will continue to increase. Most of the growth will be in developing countries, which can ill afford the high cost of containment measures. It is therefore essential to find means to contain energy-related emissions without compromising energy security in a cost-effective, participatory, sustainable and eco-friendly manner. As an environmental safeguard, growing importance will have to be given on efficient 'Carbon Management' with programme-based 'complementarity' and 'adaptability' amply demonstrated through some of the Nation-wide 'Carbon Neutrality', 'Carbon Foot-printing/ Offsetting' and 'Carbon Literacy 'Programme initiatives pioneered in a Mission-Mode manner by some of the developed economies of the world; namely the USA, UK, Canada, Germany, Japan and Australia. These learnings and experiences should be extrapolated to and shared by the developing economies of the world in the sprit of building synergy and evolving a unified 'Carbon-Neutral Global Society.'

IPCC has reported that a plethora of human and industrial activities have raised the greenhouse gas (i.e. emissions) concentrations level at an alarming rate globally. The Kyoto Protocol has, only to very limited extent, succeeded through appropriate programme interventions in restricting the developed Nations (G-8 Nations) to progressively cut on emissions through, among others, 'Cap & Trade System'. Trading of Carbon credits in the form of CERs has provided a growing International market to the mutual advantage of Annexure-I and Non-Annexure-I nations, particularly through CDM and to an insignificant degree Joint ventures. Exchange-based trading of Carbon credits has opened up a new vista of trade opportunities in the global carbon market.

For developing a sound marketing system with built-in efficiency, speed of transactions, transparency in reporting and revenue recognition etc. there is a need to comply with the Global Accounting Standards, Tax Planning, access to Multi-commodity Exchange Market, certification, verification and enforcement procedures. Carbon neutrality and Carbon literacy will pay a significant and pivotal role in promoting and expanding carbon market in future. Business Models for CERs transactions need to be developed in India particularly in the energy, manufacturing/ processing and transport sector. Voluntary carbon market with particular focus on Carbon Offsetting will substantially help in emission reduction entitlements for Annexure-I country such as India in future for which there will be a need to develop voluntary carbon market standards.

The possibility of costly disruption from rapid climate change either globally or locally, calls for greater attention to carbon management in the decision making process. Business entities need to consider issues such as trading in carbon allowances (or permits), investment in low-CO_2 emission technologies, counting the costs of carbon regularity compliance and passing on the increased cost of carbon regulation to consumers through higher prices. Consumers need to consider if, given a choice, they are willing to pay a higher price for CO_2 neutral products and services so as to play their part in reducing CO_2 emissions. Such considerations require information for informed decision making, thus the need for strategic management accounting techniques and measures is quite evident and forthcoming for achieving sustained gains from carbon neutrality vision for future.

In view of this projected futuristic vision there seems to be a clear need to have a comprehensive knowledge and understanding about the vast complexity and implicit linkage between emerging global concerns over the adverse impact of emissions on environment in one hand and the urgency to curb emissions within limits through proactive interventions and extended Global cooperation on the other hand. Several program initiatives on progressive emissions reduction recently taken at the International, Regional as well as Country and inter-Country level (i.e. partnership) have paved the way to evolve an integrated and holistic approach to the impending global problem to seek pragmatic solution. The experiences gained and shared in this endeavour so far, however, seem to fall apart in the absence of consolidation of information as a 'Centralized Global Repository' that can be used as a 'Vision

Document' to guide future strategic policies. A careful scanning and screening of Global literature on Emission Trading, widely being used as a potential means to explore new business opportunities to partly resolve environmental and concomitant climate change problems, has convinced the author to attempt a Book on the theme 'Emissions Trading and Carbon Management' at a time when needed most. To complement the effort further, it was also thought prudent to prepare and append with the Book an exhaustive! Glossary of Technical Terms related to 'Emission Trading' and allied discipline areas to help develop a better comprehension and clearer understanding of the subject for need-based application, referencing, interpretation and adaptation by the international workers and communities at large. The Book is divided into the following five chapters:

Chapter 1: Emissions of Green House Gases (GHG): The Characteristics and Environmental Impacts
Chapter 2: Emissions Trading and Accounting of Carbon Credits
Chapter 3: Global Development Initiative in Abatement of Climate Change and Promoting Emissions Trading
Chapter 4: Regulatory Regime and International Carbon Finance
Chapter 5: Carbon Management: Carbon Neutrality and Carbon Literacy

The book has provided immensely valuable and a plethora of updated information on the state of global climate change principally resultant of emissions, the compositions and characteristics of Greenhouse gases and their impact particularly on environment and human life, modus operandi of emissions trading and accounting of Carbon credits, global initiatives and cooperation in abatement of emissions, regulatory and governance aspects of carbon credit flows, Carbon management and International Carbon finance, Carbon neutrality, carbon footprints and last but not the least Carbon literacy to help build a Carbon-neutral Global community.

The author feels that the book will be useful to the readers at large, social scientists, research scholars and dedicated workers in the field of ecology, environment, economics, industries and energy sector.

Dr. A.N. Sarkar

Contents

CHAPTER 1

Emissions of Green House Gases (GHGs): The Characteristics and Environmental Impacts

Global Climate Change

Climate change refers to long-term fluctuations in temperature, precipitation, wind, and other elements of the Earth's climate system. Natural processes such as solar-irradiance variations, variations in the Earth's orbital parameters, and volcanic activity can produce variations in climate. The climate system can also be influenced by changes in the concentration of various gases in the atmosphere, which affect the Earth's absorption of radiation. The Earth naturally absorbs and reflects incoming solar radiation and emits longer wavelength terrestrial (thermal) radiation back into space. On average, the absorbed solar radiation is balanced by the outgoing terrestrial radiation emitted to space. A portion of this terrestrial radiation, though, is itself absorbed by gases in the atmosphere. The energy from this absorbed terrestrial radiation warms the Earth's surface and atmosphere, creating what is known as the "natural greenhouse effect." Without the natural heat-trapping properties of these atmospheric gases, the average surface temperature of the Earth would be about 33°C lower (IPCC, 2001). Climate change basically implies any long-term significant change in the "average weather" that a given region experiences. It involves changes in the variability or average state of the atmosphere over durations ranging from decades to millions of years. These changes can be caused by dynamic process on earth, external forces including variations in sunlight intensity, and more recently by human activities (http://www.bibalex.org/Libraries/Presentation/Static/Climate_Change.pdf).

Global climate change (GCC), is widely believed to be caused by the Greenhouse Gases (GHG).

The United Nations Intergovernmental Panel on Climate Change (IPCC) constructed several emission trajectories of GHG needed to stabilize global temperatures and climate change impacts. The IPCC predicted that the range of global mean temperature change from 1990 to 2100 could range from 1.1°C to 6.4°C.Regardless of analytical methodology, global average temperature and sea level are expected to rise under all scenarios (IPCC, 2007). The IPCC has reported, "The balance of evidence suggests a discernible human influence on global climate." It is certain that human activities have significantly increased the amount of gases in the atmosphere that contribute to this effect. While it is unclear exactly what the impacts of a rapidly warming planet will be, it is clear that there will be significant changes. There are many gases that contribute to the greenhouse effect – some directly and others indirectly. The most important of these gases have been identified by IPCC, and focused upon by the international community as the emissions that should be reduced to curb the "enhanced greenhouse effect."

The Intergovernmental Panel on Climate Change (IPCC) has been of the view that warming of the climate system is unequivocal, citing evidence of increases in global average air and ocean temperatures, widespread melting of snow and ice, and rising global average sea level. Furthermore, most of the observed increase in global average temperatures is very likely due to the unprecedented increases in atmospheric concentrations of global greenhouse gases caused by human activities. Strong, immediate and sustained action is essential to avert severe environmental, health, economic and security impacts (ICAP, 2007).

Greenhouse Gas (GHG) Emissions

Although the Earth's atmosphere consists mainly of oxygen and nitrogen, neither plays a significant role in enhancing the greenhouse effect because both are essentially transparent to terrestrial radiation. The greenhouse effect is primarily a function of the concentration of water vapor, carbon dioxide, and other trace gases in the atmosphere that absorb the terrestrial radiation leaving the surface of the Earth (IPCC, 1996). Changes in the atmospheric concentrations of these greenhouse gases can alter the balance of energy transfers between the atmosphere, space, land, and the oceans. A gauge of these changes is called radiative forcing, which is a simple measure of changes in the energy available to the Earth-atmosphere system (IPCC, 1996). Holding everything else constant, increases in greenhouse gas concentrations in the atmosphere will produce positive radiative forcing (i.e., a net increase in the absorption of energy by the Earth). Climate change can be driven by changes in the atmospheric concentrations of a number of radiatively active gases and aerosols. There is clear evidence that human activities have affected concentrations, distributions and life cycles of these gases (IPCC, 1996). Naturally occurring greenhouse gases include water vapor, carbon dioxide (CO_2), methane (CH_4), nitrous oxide (N_2O), and ozone (O_3). Several classes of halogenated substances that contain fluorine, chlorine, or bromine are also greenhouse gases, but they are, for the most part, solely a product of industrial activities.

Chlorofluorocarbons (CFCs) and hydrochlorofluorocarbons (HCFCs) are halocarbons that contain chlorine, while halocarbons that contain bromine are referred to as bromofluorocarbons (i.e., halons). Because CFCs, HCFCs, and halons are stratospheric ozone depleting substances, they are covered under the Montreal Protocol on Substances that Deplete the Ozone Layer. The UNFCCC defers to this earlier international treaty; consequently these gases are not included in national greenhouse gas inventories. Some other fluorine containing halogenated substances—hydrofluorocarbons (HFCs), perfluorocarbons (PFCs), and sulfur hexafluoride (SF6)—do not deplete stratospheric ozone but are potent greenhouse gases. These latter substances are addressed by the UNFCCC and accounted for in national greenhouse gas inventories.

There are also several gases that, although they do not have a commonly agreed upon direct radiative forcing effect, do influence the global radiation budget. These tropospheric gases—referred to as ambient air pollutants—include carbon monoxide (CO), nitrogen dioxide (NO_2), sulfur dioxide (SO_2), and tropospheric (ground level) ozone (O_3). Tropospheric ozone is formed by two precursor pollutants, volatile organic compounds (VOCs) and nitrogen oxides (NOx) in the presence of ultraviolet light (sunlight). Aerosols—extremely small particles or liquid droplets—often composed of sulfur compounds, carbonaceous combustion products, crustal materials and other human induced pollutants—can affect the absorptive characteristics of the atmosphere. However, the level of scientific understanding of aerosols is still very low (IPCC, 2001).

Carbon dioxide, methane, and nitrous oxide are continuously emitted to and removed from the atmosphere by natural processes on Earth. Anthropogenic activities, however, can cause additional quantities of these and other greenhouse gases to be emitted or sequestered, thereby changing their global average atmospheric concentrations. Natural activities such as respiration by plants or animals and seasonal cycles of plant growth and decay are examples of processes that only cycle carbon or nitrogen between the atmosphere and organic biomass. Such processes—except when directly or indirectly perturbed out of equilibrium by anthropogenic activities—generally do not alter average atmospheric greenhouse gas concentrations over decadal timeframes. Climatic changes resulting from anthropogenic activities, however, could have positive or negative feedback effects on these natural systems.

Atmospheric concentrations of these gases, along with their rates of growth and atmospheric lifetimes, are presented in Table 1.

Table 1: Global Atmospheric Concentration (ppm unless otherwise specified), Rate of Concentration Change (ppb/year) and Atmospheric Lifetime (years) of Selected Greenhouse Gases

Atmosphere Variable	CO_2	CH_4	N_2O	SF_6[a]	CF_4[a]
Pre-industrial atmospheric concentration	278	0.700	0.270	0	40
Atmospheric concentration (1998)	365	1.745	0.314	4.2	80
Rate of concentration change[b]	1.5c	0.007c	0.0008	0.24	1.0
Atmospheric Lifetime	50-200[d]	12[e]	114[e]	3,200	>50,000

[a]Concentrations in parts per trillion (ppt) and rate of concentration change in ppt/year.
[b]Rate is calculated over the period 1990 to 1999.
[c]Rate has fluctuated between 0.9 and 2.8 ppm per year for CO_2 and between 0 and 0.013 ppm per year for CH_4 over the period 1990 to 1999.
[d]No single lifetime can be defined for CO_2 because of the differentiates of uptake by different removal processes.
[e]This lifetime has been defined as an "adjustment time" that takes into account the indirect effect of the gas on its own residence time.
Source: IPCC (2001).

The Intergovernmental Panel on Climate Change (IPCC, 1990) reported for the first time in 1990 that greenhouse gas (ghg) concentrations in the global atmosphere were rising as a result of human ghg emissions, principally from fossil-fuel burning. The historic record of ghg concentrations in the atmosphere shows between 200 and 300 per million by volume (ppmv). Since industrialization, the impact of human fossil fuel burning has raised ghg concentration in the atmosphere by 30%. Scientists have found that the four most important variable greenhouse gases, whose atmospheric concentrations can be influenced by human activities, are carbon dioxide (CO_2), methane (CH_4), nitrous oxide (N_2O), and chlorofluorocarbons (CFCs).

Historically, CO_2 has been the most important, but those other atmospheric trace gases are also radioactively active, in that they can affect Earth's heat budget and thereby contribute to a greenhouse warming of the lower atmosphere. The principal greenhouse gases that enter the atmosphere because of human activities are: Carbon dioxide, methane, nitrous oxide and fluorinated gases. Common GHG include water vapor, carbon dioxide, methane, nitrous oxides, chlorofluorocarbons, hydrofluorocarbons, perfluorocarbons, sulfur hexafluoride, ozone, and aerosols. Natural processes and human activities emit GHG. The accumulation of GHG in the atmosphere regulates the earth's temperature. Hydrofluorocarbons (HFCs) are synthetic man-made chemicals that are used as a substitute for CFCs for automobile air conditioners and refrigerants. Perfluorocarbons (PFCs) have stable molecular structures and do not break down though the chemical processes in the lower atmosphere. Two common PFCs are tetrafluoromethane and hexafluoroethane. Concentrations of tetrafluoromethane in the atmosphere are over 70 parts per trillion (ppt).

U.S. emissions during 2006 decreased by 1.1 percent overall from the previous year, according to the national greenhouse gas inventory released on 15 April, 2008 by EPA (EPA, 2006). Total emissions of the six main greenhouse gases in 2006 were equivalent to 7,054.2 million metric tons of carbon dioxide. These gases include carbon dioxide, methane, nitrous oxide, hydrofluorocarbons, perfluorocarbons and sulfur hexafluoride. The report indicates that overall emissions have grown by 14.7 percent from 1990 to 2006, while the U.S. economy has grown by 59 percent over the same period. The decrease in emissions in 2006 was due primarily to a decrease in carbon dioxide emissions associated with fuel and electricity consumption. The following factors were primary contributors to this decrease:

- Compared to 2005, 2006 had warmer winter conditions, which decreased consumption of heating fuels, as well as cooler summer conditions, which reduced demand for electricity;
- Restraint on fuel consumption caused by rising fuel prices, primarily in the transportation sector; and
- Increased use of natural gas and renewables in the electric power sector.

Extending the traditional input-output model to account for the environmental impacts of production processes reveals the channels by which environmental burdens are transmitted throughout the economy. In particular, the environmental input-output approach is a useful technique for quantifying the changes in the levels of greenhouse emissions caused by changes in the final demand for production activities. The input-output model can also be used to determine the changes in the relative composition of greenhouse gas emissions due to exogenous inflows. In this paper we describe a method for evaluating how the exogenous changes in sectorial demand, such as changes in private consumption, public consumption, investment and exports, affect the relative contribution of the six major greenhouse gases regulated by the Kyoto Protocol to total greenhouse emissions. The empirical application is for Spain, and the economic and environmental data are for the year 2000. Results show that there are significant differences in the effects of different sectors on the composition of greenhouse emissions. Therefore, the final impact on the relative contribution of pollutants will basically depend on the activity that receives the exogenous shock in final demand, because there are considerable differences in the way, and the extent to which, individual activities affect the relative composition of greenhouse gas emissions (Butner-http://www.recercat.net/handle/2072/1750).

There is a natural balance between the energy coming into our atmosphere from the sun and the energy escaping from the earth back into space. A simplified description of this energy balance depicts the energy that comes in as relatively short wavelength UV radiation and the energy that goes out as longer wavelength IR radiation. Much of the high-energy radiant energy from

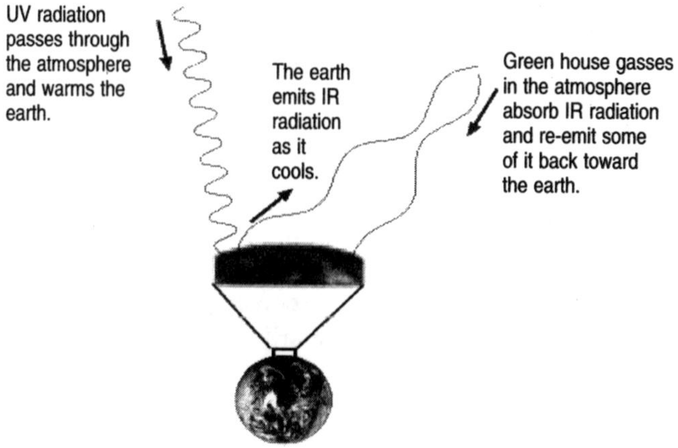

Fig. 1: The Greenhouse Effect: Some of the ultraviolet light that comes from the sun passes through the earth's atmosphere and warms the earth. As the earth cools, it emits infrared radiation. Some of this infrared radiation is absorbed by the greenhouse gases. When the gas particles re-emit the IR radiation, a portion of it is sent back toward the earth. Thus, the greenhouse gases trap some of the energy that would have otherwise escaped. This trapped energy leads to temperatures on the earth that is higher than they would be without the greenhouse gases.

the sun is absorbed in the upper atmosphere, but some of the ultraviolet radiant energy passes through. The lower-energy end of the ultraviolet portion of the light spectrum reaches the earth's surface and warms it. As the earth cools down again, it releases energy as infrared radiation (Figure 1) (http://www.mpcfaculty.net/mark_bishop/greenhouse.htm)

The greenhouse gases absorb some of the IR energy released as the earth cools. When they re-emit it, some of the energy is sent back toward the earth, rather than out into space. If the greenhouse gases were not there to trap some of this IR energy, the earth would be about 33°C (60°F) colder than it is. The gases are called greenhouse gases because their effect is like the glass panels in a greenhouse. Radiant energy passes through the panels to enter the greenhouse, but the panels slow down the passage of heat energy leaving the greenhouse.

Increases in the levels of the greenhouse gases trap too much of the escaping energy. It is this situation that scientists believe is causing global warming. Table 2 compares today's levels of five greenhouse gases with their levels 100 years ago. The concentrations are described in parts per million (ppm) or parts per billion (ppb). Parts per million is the equivalent of milligrams of substance per kilogram total. For example, 360 ppm CO_2 in the atmosphere means that each kilogram of atmosphere contains 360 mg of CO_2. Parts per billion (ppb) is the same as micrograms of substance per kilogram total. (There are a billion, or 109, micrograms per kilogram.) For example, 310 ppb N_2O in the atmosphere means that there are 310 mg of N_2O per kilogram of atmosphere. The table also lists some of the human, or anthropogenic, sources of these gases.

Table 2: Greenhouse Gases Concentration Levels

Greenhouse Gas	*Concentration 1800s*	*Concentration 1990s*	*Percent Increase*	*Anthropogenic Sources*
CO_2	280 ppm	360 ppm	29%	burning fossil fuels; deforestation
CH_4	0.95 ppm	1.7 ppm	79%	agriculture; fuel leakage
CFCs	0	0.7 ppb	—	refrigerants
N_2O	250 ppb	310 ppb	24%	agriculture; combustion
O_3	15 ppb	20-30 ppb	33-100%	urban pollution

Because carbon dioxide plays a major role in global warming, many of the proposed solutions to our problem are aimed at decreasing the levels of carbon dioxide in the atmosphere. This might be done in three general ways. *One approach* would be to stop the emission of carbon dioxide into the atmosphere. Because the burning of fossil fuels accounts for about 80% of these emissions, one important focus is to develop alternatives to fossil fuel use. For example, if electricity generated by solar energy could take the place of electricity generated by coal-burning power plants, the release of carbon dioxide into the atmosphere would be greatly reduced.

A *second approach* is to continue generating carbon dioxide, but to trap it before it is released into the air. One suggestion is to liquefy the CO_2 produced at fossil fuel power plants and pump the liquid deep into the ocean. It would remain in the ocean water and not return to the surface for a long time. In the interim, scientists would be developing alternatives to fossil fuels.

A *third approach* to lowering the carbon dioxide levels in the atmosphere would be to remove the CO_2 after it has been released into the air. Because CO_2 mixes throughout the atmosphere, this is generally considered to be too big a task, but at least one suggestion for removing it has been considered seriously: alter the chemistry of the earth's oceans such that they will absorb the excess CO_2. A huge amount of carbon dioxide is constantly dissolving in the ocean from the air and at the same time leaving the ocean and returning to the air. The ocean takes about 100 billion tons of carbon dioxide out of the atmosphere per year, and ultimately returns 98 billion tons of it. The remaining 2 billion tons end up as organic deposits on the sea floor. If the rate of escape from the ocean to the air could be slowed, the net shift of CO_2 from the air to the ocean would increase, and the levels of CO_2 in the air would fall. The goal is to find some way to increase the use of CO_2 in the ocean before it can escape.

Scientific Explanation for Global Warming

The concept of global warming itself is controversial because of the complexity of our global weather system and the difficulty of predicting changes in it. For example, as the concentrations of the greenhouse gases in the atmosphere increase, the temperature of the earth increases, and more water evaporates from the surface of the earth. The effect of this mechanism is difficult to predict. Water itself is a greenhouse gas, so an increase in the gaseous H_2O concentrations in the atmosphere would trap more IR radiation and could lead to an even greater increase in temperature. On the other hand, the increased water vapor also leads to increased clouds. These, especially at low altitudes, can reflect the incoming UV radiation back into space and lead to cooling. Meanwhile, the greenhouse effect of the water molecules in high clouds seems to outweigh the reflective effect of the clouds. The IR energy sent back to Earth from the water molecules in the high clouds is greater than the energy from UV radiation that the high clouds reflect back into space (Drake: http://www.infibeam.com/Books/info/Frances-Drake/Global-Warming-The-Science-of-Climate-Change/0340653027.html).

Another example of the complexity of global warming stems from a relationship between this topic and stratospheric chemistry. If the IR radiation trapped by excess greenhouse gases in the lower atmosphere had been allowed to move into the stratosphere as the natural cycle would have allowed, the gases in the stratosphere would have absorbed some of it and been warmed. Thus, the increase in the greenhouse gases is thought to have caused a cooling of the stratosphere. This cooling may have increased the formation of ice crystals there, and provided a surface on which ozone molecules are destroyed at a greater rate than would otherwise occur. Less ozone allows more UV radiation to reach the earth's surface. This can damage the plants that take CO_2 from the air and lead to an increase in the CO_2 levels in the atmosphere. The extra CO_2 absorbs IR radiation that would have reached the stratosphere, and the cycle repeats (Figure 2).

Scientists are well aware of the complexities of global warming, but they are confident that the recent refinements in measuring changes in the

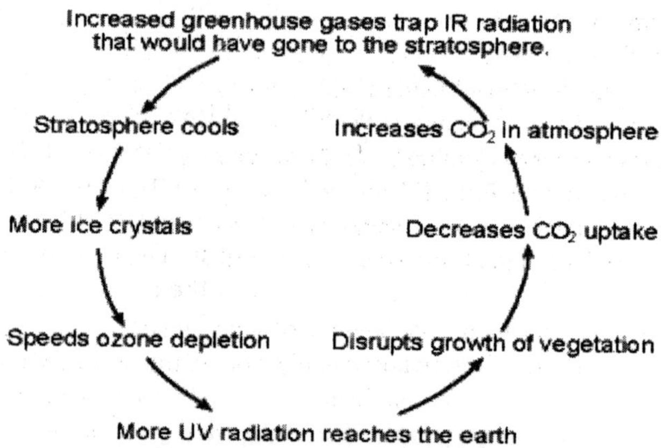

Fig. 2: Greenhouse Gases and Stratospheric Cooling

atmosphere, the new computer models being developed to predict long-range changes in climate, and the new experiments that these developments have inspired will soon combine to yield a detailed understanding of the global environment. It is this understanding that will in turn produce a successful strategy for solving the problems we face.

Greenhouse Effect

The greenhouse effect is the natural warming of the Earth's atmosphere. Solar radiation enters the atmosphere mainly as light, and some of that radiation is absorbed by the Earth's surface then changed to heat that is re-radiated into the atmosphere where it is absorbed by greenhouse gases then re-radiated back to Earth again (http://www.sciencelearn.org.nz/contexts/icy_ecosystems/science_ideas_and_concepts/greenhouse_effect).

The "Greenhouse Effect" is the heating of the Earth due to the presence of greenhouse gases. It is named this way because of a similar effect produced by the glass panes of a greenhouse. Shorter-wavelength solar radiation from the sun passes through Earth's atmosphere, and then is absorbed by the surface of the Earth, causing it to warm. Part of the absorbed energy is then reradiated back to the atmosphere as long wave infrared radiation. Little of this long wave radiation escapes back into space; the radiation cannot pass through the greenhouse gases in the atmosphere. The greenhouse gases selectively transmit the infrared waves, trapping some and allowing some to pass through into space. The greenhouse gases absorb these waves and reemits the waves downward, causing the lower atmosphere to warm (Fig. 3). (www.eb.com:180). In terms of distribution of GHG in Earth's atmosphere, Carbon Dioxide is clearly the majority (Fig. 4).

The greenhouse effect occurs naturally, and without it, the Earth would be –19ºC on average. Total solar radiation entering the atmosphere is 342 Watts per square metre ($W\ m^{-2}$). This is mainly visible light (short wave radiation) but also some ultra-violet radiation and infra-red radiation (heat or long wave radiation). Of the incoming solar radiation:

- 31% is reflected back to space by clouds, aerosols, the Earth's surface and atmosphere;
- about 20% is absorbed by the atmosphere; and
- 49% is absorbed by the Earth's surface then transformed to infra-red radiation (heat) and re-radiated into the atmosphere.

Some heat is lost to space, while the rest is absorbed by greenhouse gases, warming the Earth's atmosphere to an average 14ºC.

The greenhouse effect is a natural process by which some of the radiant heat from the Sun is captured in the lower atmosphere of the Earth, thus maintaining the temperature of the Earth's surface. The gases that help capture the heat, called "greenhouse gases," include water vapor, carbon dioxide,

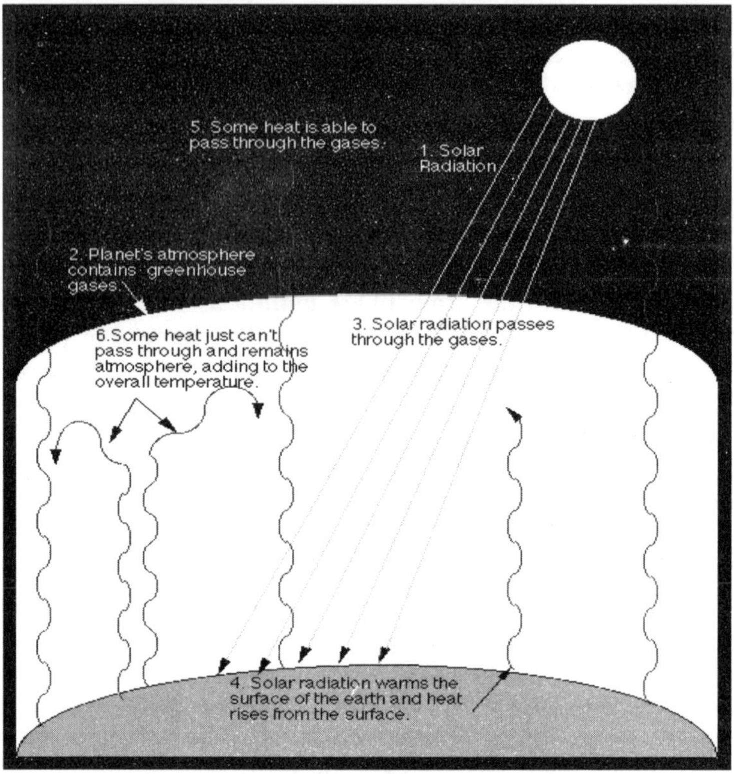

Fig. 3: The Process of Global Warming and how Greenhouse Gases Create the "greenhouse effect"
(www.eecs.umich.edu/mathscience/funexperiments/agesubject/lessons/images/diagrampage.html)

GREENHOUSE GASES

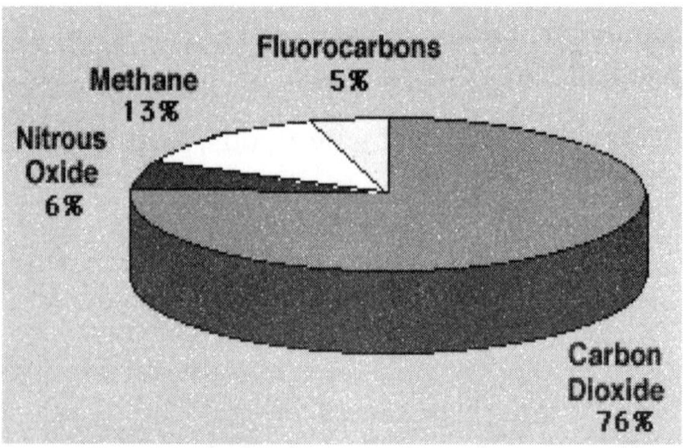

Fig. 4: Distribution of GHG in Earth's Atmosphere: Carbon Dioxide is Clearly the Majority
(www.abcnews.com/sections/us/global106.html)

methane, nitrous oxide, and a variety of manufactured chemicals. Some are emitted from natural sources; others are anthropogenic, resulting from human activities.

Over the past several decades, rising concentrations of greenhouse gases have been detected in the Earth's atmosphere. Although there is not universal agreement within the scientific community on the impacts of increasing concentrations of greenhouse gases, it has been theorized that they may lead to an increase in the average temperature of the Earth's surface. To date, it has been difficult to note such an increase conclusively because of the differences in temperature around the Earth and throughout the year, and because of the difficulty of distinguishing permanent temperature changes from the normal fluctuations of the Earth's climate. In addition, there is not universal agreement among scientists and climatologists on the potential impacts of an increase in the average temperature of the Earth, although it has been hypothesized that it could lead to a variety of changes in the global climate, sea level, agricultural patterns, and ecosystems that could be, on net, detrimental.

The most recent report of the Intergovernmental Panel on Climate Change (IPCC) concluded that: "Our ability to quantify the human influence on global climate is currently limited because the expected signal is still emerging from the noise of natural variability, and because there are uncertainties in key factors. These include the magnitudes and patterns of long-term variability and the time-evolving pattern of forcing by, and response to, changes in concentrations of greenhouse gases and aerosols, and land surface changes. Nevertheless, the balance of evidence suggests that there is a discernable human influence on global climate."

U.S. Greenhouse Gas Emissions

In 2006, total greenhouse gas emissions in the United States were 7,181.4 million metric tons carbon equivalent, according to 2007 estimates published by the Energy Information Administration (EIA). Of this total, 5,825.5 million metric tons, or 82 percent, was due to carbon emissions from the combustion of energy fuels—the focus of this report. By 2004, total U.S. greenhouse gas emissions had risen to 5,923.2 million metric tons carbon equivalent. EIA's *Annual Energy Outlook 2007*—(AEO, 2007) projects that energy-related carbon emissions will reach 42,880 million metric tons in 2030. Projected emissions continue to rise at an average annual rate of 1.1 percent a year from 2004 to 2030, reaching 7,958 million metric tons of carbon emissions in 2030. Because energy-related carbon emissions are a large portion of total greenhouse gas emissions, any efforts to reduce greenhouse gas emissions will likely have a significant impact on the energy sector; however, as discussed later, there are a number of factors outside the domestic energy market that also affect emissions levels (http://www.infoplease.com/ipa/A0004686.html).

To put U.S. emissions in a global perspective, the United States produced about 22 percent of the worldwide energy-related carbon emissions in 2004, which totaled 26.9 billion metric tons, as noted in EIA's *International Energy Outlook 2007* (IEO, 2007). Although continued increases in carbon emissions are expected for the United States and other industrialized countries, much more rapid increases are projected for the developing countries of Asia, the Middle East, Africa, and Central and South America. As a result, global carbon emissions from energy use are expected to increase at an average annual rate of 1.8 percent from 2004 through 2030, reaching 16.7 billion metric tons, to which the United States would contribute about 19 percent.

Global Warming

Global warming is the increase in the average temperature of the Earth's near-surface air and oceans since the mid-20th century and its projected continuation (http://en.wikipedia.org/wiki/Global_warming). Global surface temperature increased 0.74 ± 0.18°C (1.33 ± 0.32°F) during the last century. The Intergovernmental Panel on Climate Change (IPCC) concludes that increasing greenhouse gas concentrations resulting from human activity such as fossil fuel burning and deforestation are responsible for most of the observed temperature increase since the middle of the 20th century. The IPCC also concludes that natural phenomena such as solar variation and volcanoes produced most of the warming from pre-industrial times to 1950 and had a small cooling effect afterward. These basic conclusions have been endorsed by more than 40 scientific societies and academies of science, including all of the national academies of science of the major industrialized countries.

Climate model projections summarized in the latest IPCC report indicate that global surface temperature will probably rise a further 1.1 to 6.4°C (2.0 to 11.5°F) during the twenty-first century. The uncertainty in this estimate arises from the use of models with differing sensitivity to greenhouse gas concentrations and the use of differing estimates of future greenhouse gas emissions. Some other uncertainties include how warming and related changes will vary from region to region around the globe. Most studies focus on the period up to 2100. However, warming is expected to continue beyond 2100 even if emissions stop, because of the large heat capacity of the oceans and the long lifetime of carbon dioxide in the atmosphere.

Increasing global temperature will cause sea levels to rise and will change the amount and pattern of precipitation, probably including expansion of subtropical deserts. The continuing retreat of glaciers, permafrost and sea ice is expected, with the Arctic region being particularly affected. Other likely effects include shrinkage of the Amazon rainforest and Boreal forests, increases in the intensity of extreme weather events, species extinctions and changes in agricultural yields.

Political and public debate continues regarding what actions (if any) to take in response to global warming. The available options are mitigation to reduce further emissions; adaptation to reduce the damage caused by warming;

and, more speculatively, geo-engineering to reverse global warming. Most national governments have signed and ratified the Kyoto Protocol aimed at reducing greenhouse gas emissions.

Global Warming Potential (GWP)

Global Warming Potentials (GWPs) are intended as a quantified measure of the globally averaged relative radiative forcing impacts of a particular greenhouse gas (http://www.epa.gov/climatechange/emissions/downloads/ghg_gwp.pdf). It is defined as the cumulative radiative forcing:

- Both direct and indirect effects
- Integrated over a period of time from the emission of a unit mass of gas relative to some reference gas (IPCC 1996).

Each gas differs in its effectiveness at absorbing heat, the effect it has on other greenhouse gases, and the average time it stays in the atmosphere and consequently it's potential for warming over time. Methane, for instance, is considered to be 23 times more effective at warming than carbon dioxide over 100 years (Table 3). Although methane lasts about 12 years in the atmosphere, whereas carbon dioxide lasts for 50–200 years, methane indirectly affects ozone and water vapour, making it a more effective greenhouse gas. Methane comes from many sources—in New Zealand, many cows and sheep are responsible for high methane production.

Table 3. Global Warming Potential (GWP) for Greenhouse Gases over 100 years

Gas	GWP
Carbon dioxide (CO_2)	1
Methane (CH_4)	23
Nitrous Oxide (N_2O)	296
Hydroflurocarbons	12-1200
Sulphur hexafluoride (SF_6)	222000

It could be noted from the above table that Sulphur hexafluoride (SF_6) has the highest order of global warming potential, followed by Hydroflurocarbons. GWP values allow policy makers to compare the impacts of emissions and reductions of different gases. According to the IPCC, GWPs typically have an uncertainty of roughly about 35 percent, though some GWPs have larger uncertainty than others, especially those in which lifetimes have not yet been ascertained. The following table provides the compiled Statistics on the Global Warming Potentials (GWP) and Atmospheric Lifetimes (Years) used in the Inventory.

Table 4: Global Warming Potentials (GWP) and Atmospheric Lifetimes (Years) used in the Inventory

Gas	Atmospheric Lifetime	100 year GWP[a]	20 year GWP	500 year GWP
Carbon dioxide (CO_2)	50,200	1	1	1
Methane (CH_4)[b]	12±3	21	56	6.5
Nitrous oxide (N_2O)	120	310	280	170
HFC-23	264	11,700	9,100	9,800
HFC-125	32.6	2,800	4,600	920
HFC-134a	14.6	1,300	3,400	420
HFC-143a	48.3	3,800	5,000	1,400
HFC-152a	1.5	140	460	42
HFC-227ea	36.5	2,900	4,300	950
HFC-236fa	209	6,300	5,100	4,700
HFC-4310mee	17.1	1,300	3,000	400
CF_4	50,000	6,500	4,400	10,000
C_2F_6	10,000	9,200	6,200	14,000
C_4F_{13}	2,600	7,000	4,800	10,000
C_6F_{14}	3,200	7,400	5,000	10,700
SF_6	3,200	23,900	16,300	34,900

[a] GWOPs used here are calculated over 100 year time horizon.

[b] The methane GWP includes the direct effects and those indirect effects due to the production of tropospheric ozone and stratospheric water vapor. The indirect effect due to the production of CO_2 is not included.

Source: IPCC (1996)

Contribution of India to Global GHG Emissions Build-up

In recent years, the development planning in India has increasingly incorporated measurable goals for enhancement of human well being, beyond mere expansion of production of goods and services and the consequent growth of per capita income. Many developmental targets are even more ambitious than the UN Millennium Development Goals; several of which are directly or indirectly linked to energy and therefore to GHG emissions. India holds over 1 billion people, i.e. over 16% of global population. Endowed with coal, India's energy system has evolved around coal. India's share in global CO_2 emissions is still very small (Table 5). The contribution of India to the cumulative global CO_2 emissions from 1980 to 2003 is only 3.11%. Thus, historically and at present India's share in the carbon stock in the atmosphere is relatively very small when compared to the population. India's carbon emissions per person are twentieth of those of the US and a tenth of most Western Europe and Japan (Figure 5).

Changes in the Atmospheric Gas Composition

Earth's atmospheric composition has changed throughout its existence.

Table 5: Share of Global GHG Emissions (%)

	1990	2003
United States	23.04	23.06
China	10.41	14.07
Russia[1]	9.67	6.38
Japan	4.54	4.79
India	2.63	4.07
Germany[2]	4.24	3.35
Canada	2.19	2.39
United Kingdom	2.76	2.24
Italy	1.91	1.85
France	1.80	1.63
Rest of World	38.61	36.17

[1]Russia 1990 numbers are for 1992.
[2]Germany 1990 numbers are for 1991.
Source: Based on data from USDOE.

Fig.5: Per capita CO_2 Emissions from Energy for the Year 2003

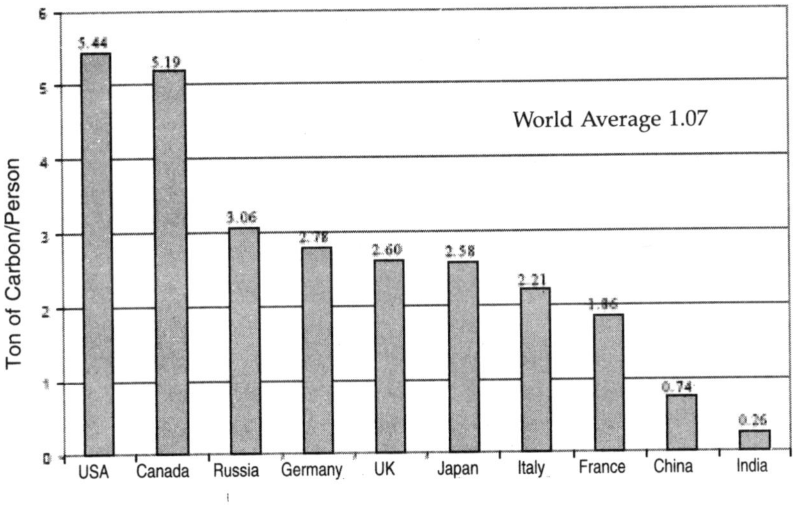

Initially, there was very little oxygen and the atmosphere consisted mainly of carbon dioxide and nitrogen with a little methane, ammonia and sulphur gas. When blue-green algae evolved several billion years ago, they used sunlight, carbon dioxide and water to produce carbohydrates and released oxygen as a by-product (i.e. photosynthesis). Slowly the atmosphere accumulated enough oxygen for oxygen-breathing organisms to survive. Today, Earth's atmosphere is composed of:

- 78% Nitrogen
- 21% Oxygen

- 0.9% Argon
- 0.037% Carbon Dioxide
- Neon, Helium, Methane, Krypton and hydrogen in smaller proportions
- Water vapour in varying amounts

Global Warming vs. the Greenhouse Effect

The Earth goes through natural cycles of temperature increase and decrease due to changes in the Earth's orbit called Milankovitch cycles. However, since 1860, the increase in global temperature by about 0.6°C is considered by the Intergovernmental Panel on Climate Change (IPCC) to result mainly from increased greenhouse gas concentrations. From 1750 to 2001, there was an increase of 31 percent carbon dioxide, 150 percent methane and 16 percent nitrous oxide concentration in the atmosphere. This is due to an increase in the burning of fossil fuel, change in land use and agriculture, and decrease of carbon sinks such as forests through deforestation. To gain a better understanding of the relationship between Earth's atmospheric composition and temperature, scientists take ice cores from Antarctica. They find the age of the ice and the gas composition of the air bubbles trapped in that ice (http://www.niwascience.co.nz/pubs/wa/09-1/ice.htm). Scientists are trying to better understand the influence of the Antarctic seasonal changes on greenhouse trace gases and also the effect of the ozone hole over Antarctica on New Zealand and internationally.

Characteristics of Green House Gases

Atmospheric concentrations of greenhouse gases that absorb infrared light can lead to additional global warming, causing climate change (Figure 6). Past and future emissions of the greenhouse gases carbon dioxide, methane and nitrous oxide and the impact on climate are described on this page (http://www.lenntech.com/greenhouse-effect/greenhouse-gases.htm).

The impact of greenhouse gases on the enhanced greenhouse effect is determined by their residence time in the atmosphere. When atmospheric residence time is greater, the total impact of a greenhouse gas on global warming is larger. Atmospheric residence time represents the average amount of time the molecules of a greenhouse gas exist in air before it is somehow removed. The average residence time of the greenhouse gases carbon dioxide and nitrous oxide is more than a century. Consequentially, these greenhouse gases will impact global warming long after emission cut-backs are achieved. Contrastingly, methane only has a residence time of one decade. The Physico-chemical characteristics of different GHGs are as described below:

Carbon Dioxide

Carbon Dioxide (CO_2) is a colourless, odourless non-flammable gas and is the most prominent Greenhouse gas in Earth's atmosphere. It is recycled

Fig. 6: Integrated Framework of Climate Change by IPCC

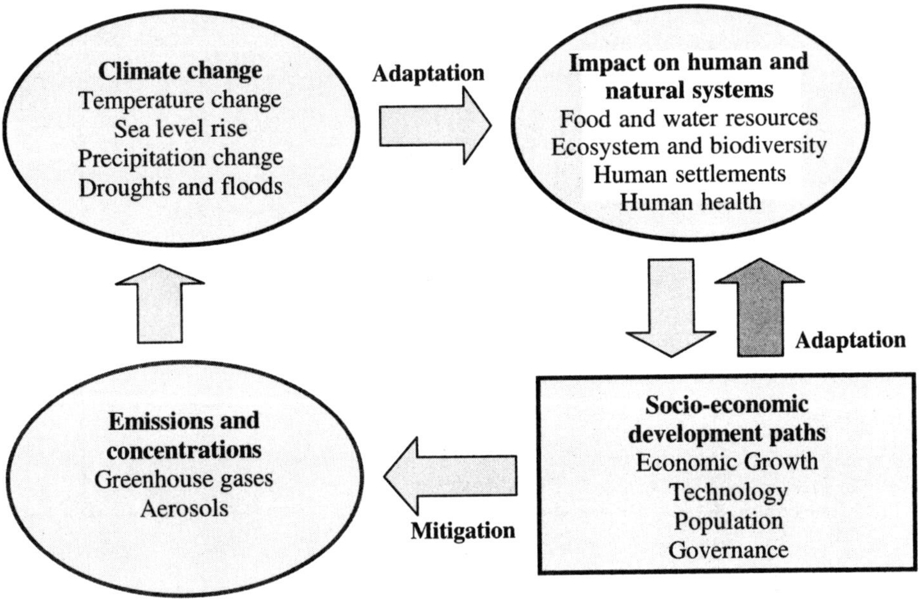

through the atmosphere by the process photosynthesis, which makes human life possible. Photosynthesis is the process of green plants and other organisms transforming light energy into chemical energy. Light Energy is trapped and used to convert carbon dioxide, water, and other minerals into oxygen and energy rich organic compounds. (Encyclopedia Britannica, Volume 25) Carbon Dioxide is emitted into the air as humans exhale, burn fossil fuels for energy, and deforests the planet. Every year humans add over 30 billion tons of carbon dioxide in the atmosphere by these processes, and it is up thirty percent since 1750 (www.envirolink.org/orgs/edf/sitemap.html). An isolated test at Mauna Loa in Hawaii revealed more than a 12% (316 ppm in 1959 to 360 ppm in 1996) increase in mean annual concentration of carbon dioxide. Mauna Loa, located in Hawaii, is the world's largest volcano at 40,000 cubic km and 4,170 meters above sea level (Fig. 7). Ice core samples have also shown a dramatic increase in carbon dioxide levels. Drilling deep into glaciers and polar ice caps and taking out samples of ice, then melting the ice and capturing the gas has shown an increase in carbon dioxide concentrations over the past 100 years.

In 1996, carbon dioxide world emissions increased by 2.8%. The U.S. reported a 3.3% increase in CO_2 concentrations. The U.S. continues to emit CO_2 more than any other country in the world, accounting for 25% of all emissions. The European Union had an increase of 2.2%, much larger than a small increase of 1.1% in 1995. Eastern Europe had a decreasing rate of -2.4%. China's increase in 1996 was 4.7% (http://infoweb.magi.com/~dwalsh/wfsesr.html) Fossil Fuels were created chiefly by the decay of plants from millions of years ago. We use coal, oil and natural gas to generate electricity, heat our homes,

Fig. 7: Ice Core Samples and Samples at Mauna Loa, Hawaii, Reveal Increase CO_2 Concentrations

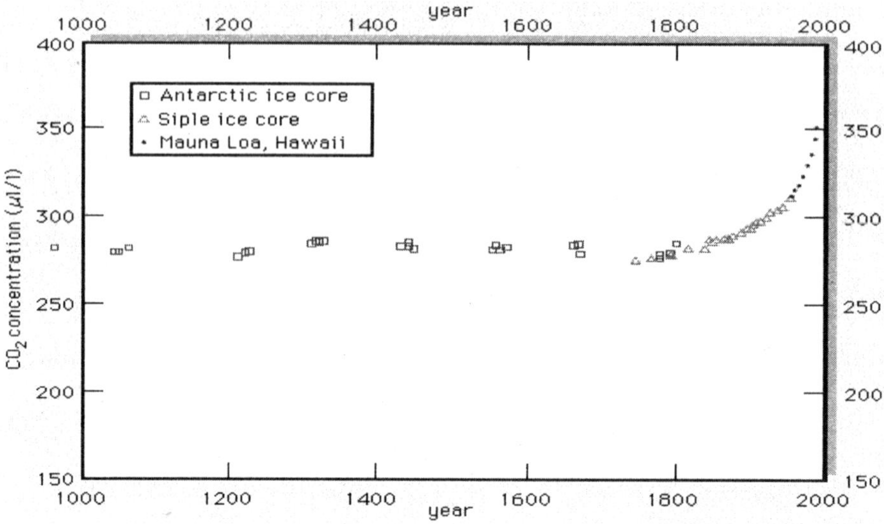

Source: *Encyclopedia Britannica*, Volume 27.

power our factories and run our cars. These fossil fuels contain carbon, and when they are burned, they combine with oxygen, forming carbon dioxide. The two atoms of oxygen add to the total weight. The World Energy Council reported that global carbon dioxide emissions from burning fossil fuels rose 12% between 1990 and 1995. (www.eb.com:180). The increase from developing countries was three times that from developed countries. Middle East carbon dioxide emissions from burning of fossil fuels increased 35%, Africa increased 12%, and Eastern Europe increased rates by 75% from 1990-1995 (Fig. 8).

Carbon dioxide production is not the same for every country. Still, we all need to reduce our greenhouse gas emissions because CO_2 is a trans-boundary air pollutant. This basically means that pollution is not limited to the country where it is emitted. This is the same for all greenhouse gases and air pollutants. Figure 9 shows the atmospheric CO_2 concentration in the past century. CO_2 is the greenhouse gas with the highest anthropogenic emission. The main cause of anthropogenic CO_2 emissions to the atmosphere is fossil fuel combustion. Fossil fuels are oil, coal and natural gas. These are exploited by humans to gain energy.

IPCC Projections

It is clear that a sustainable society will obtain more energy from renewable energy sources than a society based on increasing wealth. This results in lower eventual carbon dioxide emissions in both sustainable IPCC scenarios (Figure 10). The global scenario scores better for carbon dioxide emissions than does the regional approach.

Fig. 8: The Increase of Carbon Dioxide in the Air over the Past few Centuries

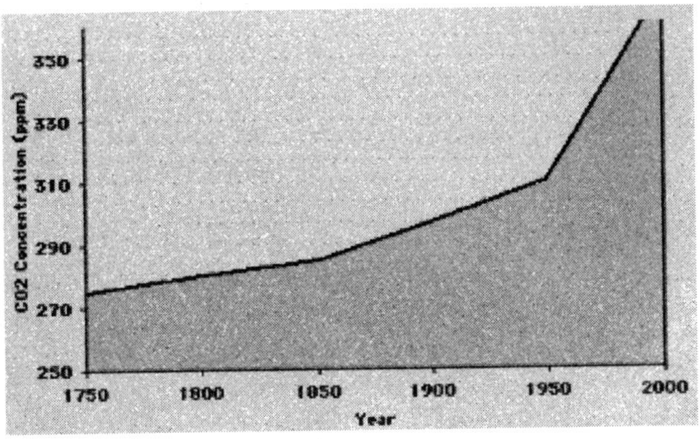

Source: http://infoweb.magi.com/~dwalsh/wfsesr.html.

Fig. 9: Past Global Atmospheric Carbon Dioxide Concentration

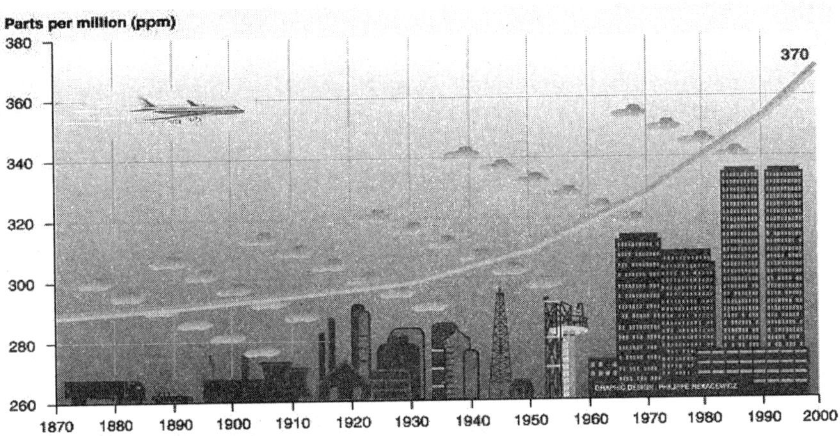

Source: http://infoweb.magi.com/~dwalsh/wfsesr.html

CO_2 is the main contributor to the anthropogenic radioactive forcing because of changes in concentrations from pre-industrial times. Whereas CO_2 emissions are by-and-large attributable to two major sources; energy consumption and land-use change, other emissions arise from many different sources and a large number of sectors and applications. Projections of emissions of CO_2 in 2100 range from more than 40 to less than 6 billion tons of elemental carbon, meaning a range from almost a sevenfold increase to roughly the same emissions level as in 1990.

Methane

Methane is a colourless, odourless, flammable gas. It is formed when plants decay and where there is very little air. It is often called *swamp gas* because it is abundant around water and swamps. Bacteria that breakdown

Fig. 10: Carbon Dioxide Emissions According to the SRES

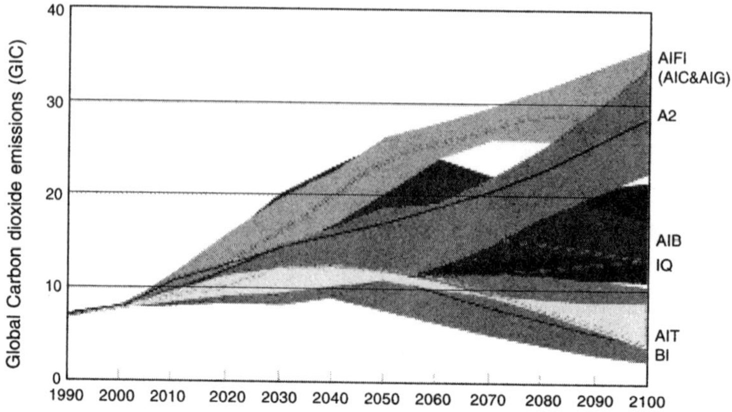

Source: www.envirolink.org/orgs/edf/sitemap.html.

organic matter in wetlands and bacteria that are found in cows, sheep, goats, buffalo, termites, and camels produce methane naturally. Since 1750, methane has doubled, and could double again by 2050 (Fig. 11). Each year we add 350-500 million tons of methane to the air by raising livestock, coal mining, drilling for oil and natural gas, rice cultivation, and garbage sitting in landfills. (www.envirolink.org/orgs/edf/sitemap.html) It stays in the atmosphere for only 10 years, but traps 20 times more heat than carbon dioxide.

Rice cultivation has developed into a large business; farmland has doubled in the past 45 years (www.envirolink.org/orgs/edf/sitemap.html). It feeds 1/3 of the World's population. It grows mostly in flooded fields, where bacteria in waterlogged soil release methane. Livestock such as cows, sheep, goats, camels, buffaloes, and termites release methane as well. There are six different sources of atmospheric methane. In order of importance these are wetlands,

Fig. 11: Methane is on the Rise since 1750

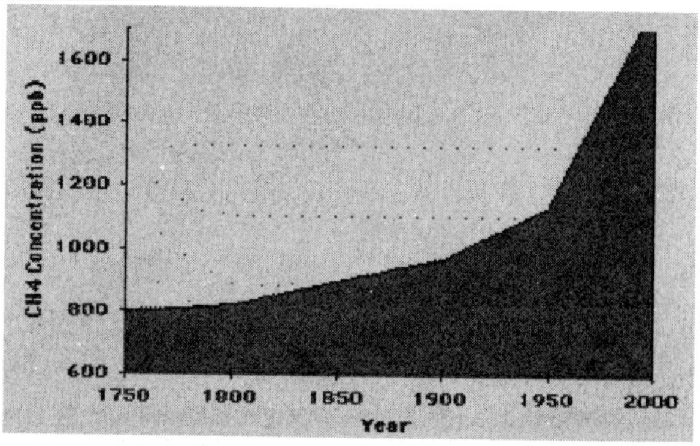

Source: www.envirolink.org/orgs/edf/sitemap.html.

fossil fuels, landfills, ruminant animals, rice paddies and biomass combustion. Methane has a larger global warming potential than carbon dioxide. Still, emissions are lower than those of carbon dioxide. It is estimated that methane produces about one-third the amount of global warming of carbon dioxide.

Past emissions

After the industrial revolution, there was an increase in global atmospheric methane concentrations. In the past 200 years, methane concentrations have increased from approximately 620 to approximately 1700 ppb (Figure 12). The radioactive forcing of methane has also increased.

IPCC projections

Methane (CH_4) emissions are expected to increase most in the two regional scenarios. Increases are most pronounced in the regional wealth scenario where

Fig. 12: Past Emission Levels of Methane (IPCC)

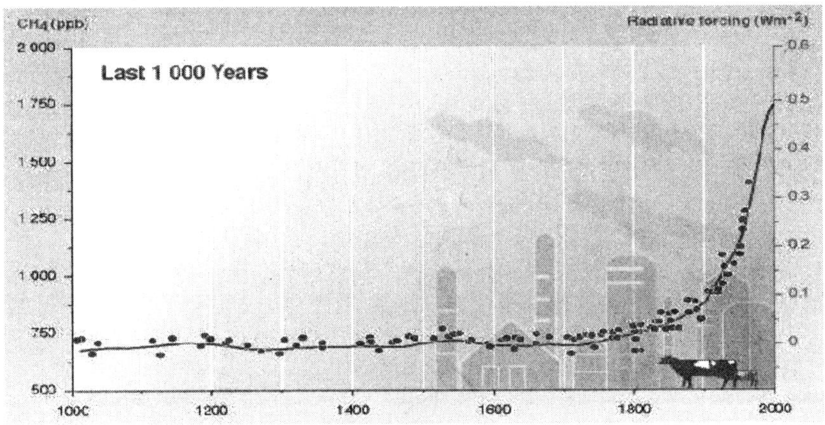

Source: www.envirolink.org/orgs/edf/sitemap.html.

emissions may rise to between 549 and 1069 Mt CH_4 by 2100, compared to 310 Mt CH_4 in 1990. In the globalized scenarios, the CH_4 emissions level off and subsequently decline sooner or later in the 21st century. The National Oceanic and Atmospheric Climate Administration (NOAA) reported that the atmospheric build-up of methane has slowed down greatly (Figure 13). It is claimed that if current trends continue, it may reach zero within a few decades (Dlugokencky *et al.*, 1998). This finding was however not reported on major news broadcasts or websites, for some reason.

Nitrous Oxide

Nitrous oxide is another colourless greenhouse gas; however, it has a sweet odour. It is primarily used as an anesthetic because it deadens pain and for this characteristic is called laughing gas. This gas is released naturally from

Fig. 13: Contrast between Findings of the IPCC and NOAA

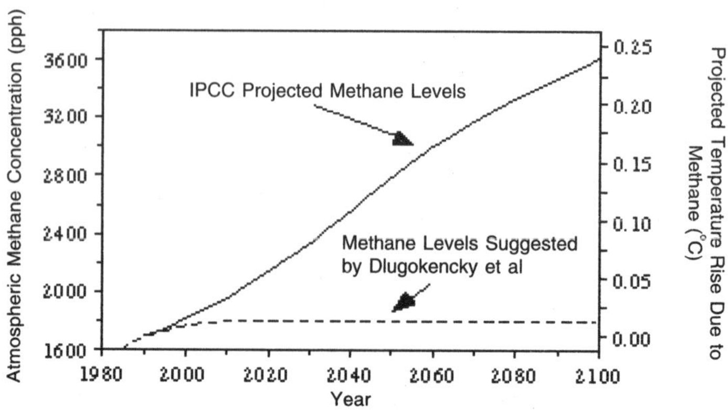

Source: www.envirolink.org/orgs/edf/sitmap.html.

oceans and by bacteria in soils. Nitrous oxide gas rose by more than 15% since 1750 (Fig. 14). Each year we add 7-13 million tons into the atmosphere by using nitrogen based fertilizers, disposing of human and animal waste in sewage treatment plants, automobile exhaust, and other sources not yet identified. It is important to reduce emissions because the nitrous oxide we release today will still be trapped in the atmosphere 100 years from now.

Nitrogen based fertilizer use has doubled in the past 15 years. These fertilizers provide nutrients for crops; however, when they breakdown in the soil, nitrous oxide is released into the atmosphere. In automobiles, nitrous oxide is released at a much lower rate than carbon dioxide, because there is more carbon in gasoline than nitrogen. The greater part of the natural supply of nitrous oxide gas is released by oceans. Processes occurring in soils cause most of the remainder. The gas is a by-product of the biological de-nitrification process in anaerobic environments and of the biological nitrification process

Fig. 14: Nitrous Oxide has been on the Rise since 1750

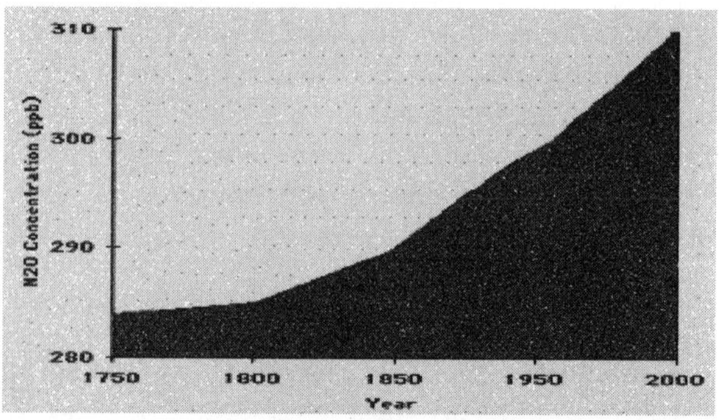

Source: www.envirolink.org/orgs/edf/sitmap.html

Fig. 15: Past Nitrous Oxide Emissions (IPCC)

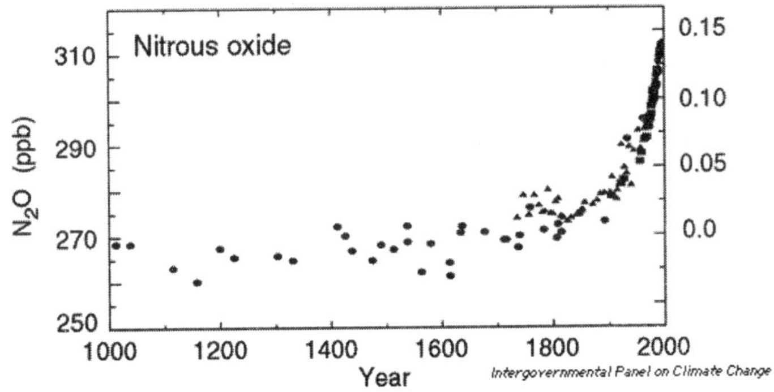

in aerobic environments. About a third of current N_2O emissions are anthropogenic. These anthropogenic emissions stem from agricultural soils; cattle feed lots and the chemical industry.

Past Emissions

The atmospheric concentration of nitrous oxide (N_2O) has increased by 46 ppb (17%) since 1750 and continues to increase (Figure 15). The present N_2O concentration has not been exceeded during at least the past thousand years.

IPCC Projections

Future nitrous oxide (N_2O) emissions are mainly determined by food supply, because these emissions largely stem from soil processes induced by agriculture. Like CH_4, N_2O emissions are generally highest in the regional wealth scenario (Figure 16).

N_2O emissions are generally affected by large uncertainties, because these are mainly caused by bacterial soil processes and are therefore difficult to measure. Extensive research into N_2O sources is still required.

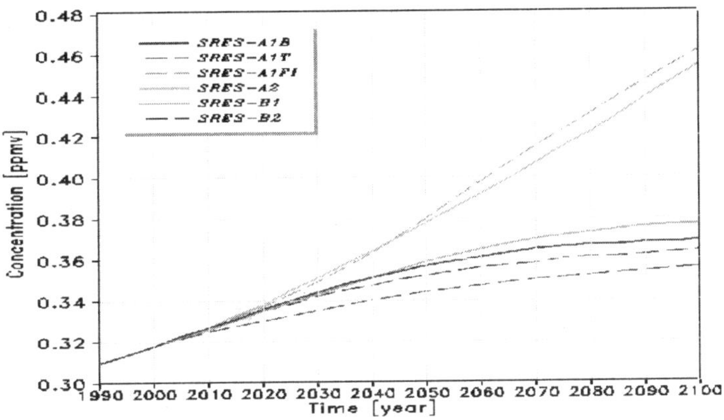

Fig.16: Nitrous Oxide emissions according to the SRES scenarios in a) global, (b) OECD, (c) former Soviet Union, (d) Asia, (e) Africa, (f) Latin America

CFCs: The Super Greenhouse Gas

For nearly eighty years since the Arrhenius paper, climate scientists assumed that CO_2 was the main anthropogenic or manmade greenhouse gas (e.g., SMIC Report, 1971). Since CO_2 does not react with other gases in the atmosphere, the greenhouse effect was largely a problem of solving the physics, thermodynamics and dynamics of climate. This picture changed drastically when it was discovered that there are other man-made gases, which on a per molecule basis could be up to ten thousand times stronger than the CO_2 greenhouse effect (Ramanathan, 1975). Chlorofluorocarbons, or CFCs, used as refrigerants and propellants in deodorizers, drug delivery pumps, etc are some of the strongest of such super greenhouse gases. These are purely synthetic gases. In 1974, Molina and Rowland published a famous paper in Nature (Molina and Rowland, 1974). They proposed that CFC11 and CFC12 (known then as Freon 11 and Freon 12) will build up in the atmosphere including the stratosphere, because of their century or longer life time. According to their theory, UV radiation from the sun will photo-dissociate the CFCs, and the released chlorine atoms will catalytically destroy ozone in the stratosphere.

CFCs have disproportionately large greenhouse effect. There are three important reasons (Ramanathan, 1975): (1) CFCs absorb and emit radiation in the 8–12 mm region. The background atmosphere is quite transparent in this region; i.e., the natural greenhouse blanket is thinnest in the 8–12 mm region, and for this reason this region is called as the atmospheric window. The background water vapour has very little absorption. (2) Next, the quantum mechanical efficiency (also knows as transition probability) of CFCs is about 3–6 times stronger than that due to CO_2. In addition, CFCs have many absorption bands in this region. (3) Lastly, the CFC concentrations are so low (part per billion or less) that their effect increases linearly with their concentration, where as the CO_2 absorption is close to saturation since their concentration is about 300,000 times larger. So it's a lot harder for a CO_2 molecule to enhance the greenhouse effect than CFCs. These three factors combine to make CFCs a super greenhouse gas. Within a period of 10 years after the CFC paper by Ramanathan in 1975, several tens of anthropogenic greenhouse gases were added to the list (e.g., Wang *et al.*, 1976; Ramanathan *et al.*, 1985).

Fluorocarbons

Fluorocarbons are a general term for any group of synthetic organic compounds that contain fluorine and carbon. Many of these compounds, such as chlorofluorocarbons (CFCs), can be easily converted from gas to liquid or liquid to gas. Because of these properties, CFCs can be used in aerosol cans, refrigerators, and air conditioners. Studies in the 1970s showed that when CFCs are emitted into the atmosphere, they break down molecules in the Earth's ozone layer (The World Book). Since then, the use of CFCs has significantly decreased and they are banned from production in the United States. The substitute for CFCs is hydro-fluoro-carbons (HFC's). HFCs do not harm or breakdown the ozone molecule, but they do trap heat in the atmosphere, making it a greenhouse gas, aiding in global warming. HFCs are used in air conditioners and refrigerators. The way to reduce emissions of this gas is to be sure that in both devices the coolant is recycled and all leaks are properly fixed.

Other Greenhouse Gases

CFCs (Chlorofluorocarbons) are gaseous compounds that contain molecules with carbon atoms bonded exclusively to fluorine or chlorine. CFCs are applied for example as refrigerants. These compounds have perhaps the greatest global warming potential among trace gases to induce global warming, because they are very persistent. CFCs absorb infrared radiation in the 8-13 μm wavelengths. Each CFC-molecule has the potential to cause the amount of global warming normally caused by tens of thousands of CO_2 molecules. The chlorofluorocarbons $CFCl_3$ and CF_2Cl_2 have been released into the atmosphere in large quantities in the past and have long residence times. Application of CFCs is now prohibited in most countries. Many countries have agreed to reduce emissions under the 1987 Montreal protocol on substances that deplete the ozone layer. The combined effect of methane, nitrous oxide, ozone and CFCs is now almost as large as that of carbon dioxide. Concentrations of these greenhouse gases are usually summarized as an Effective Carbon Dioxide Concentration.

Emissions from Combustion of Natural Gas

Natural gas is an extremely important source of energy for reducing pollution and maintaining a clean and healthy environment. In addition to being a domestically abundant and secure source of energy, the use of natural gas also offers a number of environmental benefits over other sources of energy, particularly other fossil fuels. Natural gas is the cleanest of all the fossil fuels. Composed primarily of methane, the main products of the combustion of natural gas are carbon dioxide and water vapor, the same compounds we exhale when we breathe. Coal and oil are composed of much more complex molecules, with a higher carbon ratio and higher nitrogen and sulfur contents. This means that when combusted, coal and oil release higher levels of harmful emissions, including a higher ratio of carbon emissions, nitrogen oxides (NO_x), and sulfur dioxide (SO_2). Coal and fuel oil also release ash particles into the environment, substances that do not burn, but instead are carried into the atmosphere and contribute to pollution as can be seen from Table 6 (http://www.naturalgas.org/environment/naturalgas.asp).

The combustion of natural gas, on the other hand, releases very small amounts of sulfur dioxide and nitrogen oxides, virtually no ash or particulate matter, and lower levels of carbon dioxide, carbon monoxide, and other reactive hydrocarbons.

Table 6: Contribution of Fossil Fuels of Emissions level
(Pounds per Billion Btu of Energy Input)

Pollutant	Natural Gas	Oil	Coal
Carbon Dioxide	117,000	164,000	208,000
Carbon Monoxide	40	33	208
Nitrogen Oxides	92	448	457
Sulfur Dioxide	1	1,122	2,591
Particulates	7	84	2,744
Mercury	0.000	0.007	0.016

Source: EIA—Natural Gas Issues and Trends, 1998.

The use of fossil fuels for energy contributes to a number of environmental problems. As the cleanest of the fossil fuels, natural gas can be used in many ways to help reduce the emissions of pollutants into the atmosphere. Burning natural gas in the place of other fossil fuels emits fewer harmful pollutants into the atmosphere, and an increased reliance on natural gas can potentially reduce the emission of many of these most harmful pollutants.

Industrial and Electric Generation of Emissions

Pollutant emissions from the industrial sector and electric utilities contribute greatly to environmental problems in the United States. The use of natural gas to power both industrial boilers and processes and the generation of electricity can significantly improve the emission profiles for these two sectors. Natural gas is becoming an increasingly important fuel in the generation of electricity. As well as providing an efficient, competitively priced fuel for the generation of electricity, the increased use of natural gas allows for the improvement in the emissions profile of the electric generation industry. According to the National Environmental Trust (NET) in their 2002 publication entitled 'Cleaning up Air Pollution from America's Power Plants', power plants in the U.S. account for 67 percent of sulfur dioxide emissions, 40 percent of carbon dioxide emissions, 25 percent of nitrogen oxide emissions, and 34 percent of mercury emissions. Coal fired power plants are the greatest contributors to these types of emissions. In fact, only 3 percent of sulfur dioxide emissions, 5 percent of carbon dioxide emissions, 2 percent of nitrogen oxide emissions, and 1 percent of mercury emissions come from non-coal fired power plants.

Natural gas fired electric generation, and natural gas powered industrial applications, offer a variety of environmental benefits and environmentally friendly uses, including:

- *Fewer Emissions*—Combustion of natural gas, used in the generation of electricity, industrial boilers, and other applications, emits lower levels of NO_x, CO_2, and particulate emissions, and virtually no SO_2 and mercury emissions. Natural gas can be used in place of, or in addition to, other fossil fuels, including coal, oil, or petroleum coke, which emit significantly higher levels of these pollutants.

- *Reduced Sludge*—Coal fired power plants and industrial boilers that use scrubbers to reduce SO_2 emissions levels generate thousands of tons of harmful sludge. Combustion of natural gas emits extremely low levels of SO_2, eliminating the need for scrubbers, and reducing the amounts of sludge associated with power plants and industrial processes.
- *Reburning*—This process involves injecting natural gas into coal or oil fired boilers. The addition of natural gas to the fuel mix can result in NO_x emission reductions of 50 to 70 percent, and SO_2 emission reductions of 20 to 25 percent.
- *Cogeneration*—The production and use of both heat and electricity can increase the energy efficiency of electric generation systems and industrial boilers, which translates to requiring the combustion of less fuel and the emission of fewer pollutants. Natural gas is the preferred choice for new cogeneration applications.
- *Combined Cycle Generation*—Combined cycle generation units generate electricity and capture normally wasted heat energy, using it to generate more electricity. Like cogeneration applications, this increases energy efficiency, uses less fuel, and thus produces fewer emissions. Natural gas fired combined cycle generation units can be up to 60 percent energy efficient, whereas coal and oil generation units are typically only 30 to 35 percent efficient.
- *Fuel Cells*—Natural gas fuel cell technologies are in development for the generation of electricity. Fuel cells are sophisticated devices that use hydrogen to generate electricity, much like a battery. No emissions are involved in the generation of electricity from fuel cells, and natural gas, being a hydrogen rich source of fuel, can be used. Although still under development, widespread use of fuel cells could in the future significantly reduce the emissions associated with the generation of electricity.

Essentially, electric generation and industrial applications that require energy, particularly for heating, use the combustion of fossil fuels for that energy. Because of its clean burning nature, the use of natural gas wherever possible, either in conjunction with other fossil fuels, or instead of them, can help to reduce the emission of harmful pollutants.

Pollution from the Natural Gas Vehicles

The transportation sector (particularly cars, trucks, and buses) is one of the greatest contributors to air pollution in the United States. Emissions from vehicles contribute to smog, low visibility, and various greenhouse gas emissions. According to the Department of Energy (DOE), about half of all air pollution and more than 80 percent of air pollution in cities are produced by cars and trucks in the United States.

Natural gas can be used in the transportation sector to cut down on these high levels of pollution from gasoline and diesel powered cars, trucks, and buses. In fact, according to the EPA, compared to traditional vehicles, vehicles operating on compressed natural gas have reductions in carbon monoxide emissions of 90 to 97 percent, and reductions in carbon dioxide emissions of 25 percent. Nitrogen oxide emissions can be reduced by 35 to 60 percent, and other non-methane hydrocarbon emissions could be reduced by as much as 50 to 75 percent. In addition, because of the relatively simple makeup of natural gas in comparison to traditional vehicle fuels, there are fewer toxic and carcinogenic emissions from natural gas vehicles, and virtually no particulate emissions. Thus, the environmentally friendly attributes of natural gas may be used in the transportation sector to reduce air pollution.

Natural gas is the cleanest of the fossil fuels, and thus its many applications can serve to decrease harmful pollution levels from all sectors, particularly when used together with or replacing other fossil fuels. The natural gas industry itself is also committed to ensuring that the process of producing natural gas is as environmentally sound as possible.

Effects of Global Warming on Environment

Melting of Iceberg

Tens of thousands of years ago, "armadas of ice" crumbled off of the ice sheet covering North America into the Atlantic Ocean. Meanwhile, thousands of miles away on the other side of the continent, icebergs calved off of another ice sheet into the Pacific. Their synchrony—just uncovered by new research—suggests the events might be connected in a long-distance domino effect. The fact that melting at one location may influence ice sheets afar may be useful in understanding the behavior of ice today, according to the study's authors."What it is saying is that these ice sheets are connected," said lead author Ingrid Hendy of the University of Michigan. "If we melt Greenland, we could raise sea level and affect Antarctica. Or, if we melt Greenland, we can affect the tidewater glaciers up in Alaska."

Climatic Impacts of Icebergs—Impacts on Ice Sheets and Sea Level

Apart from local weather effects, such as fog production, icebergs have two main impacts on climate. Iceberg production affects the mass balance of the parent ice sheets, and melting icebergs influence both ocean structure and global sea level.

The Antarctic Ice Sheet has a volume of 28 million cubic km (about 6.7 million cubic miles), which represents 70 percent of the total fresh water (including groundwater) in the world. The mass of the ice sheet is kept in balance by a process of gain and loss—gain from snowfall over the whole ice sheet and ice loss from the melting of ice at the bottom of the ice shelf and from the calving of icebergs from the edges of the ice shelf. The effect of summer runoff and from sublimation off the ice surface is negligible.

Annual snowfall estimates for the Antarctic continent start at 1,000 cubic km (240 cubic miles). If the Antarctic Ice Sheet is in neutral mass balance, the annual rate of loss from melting and iceberg calving must be close to this value; indeed, estimates of iceberg flux do start at this value, though some run much higher. Such apparently large fluxes are still less than the mean flow rate of the Amazon River, which is 5,700 cubic km (about 1,370 cubic miles) per year. In Antarctica the annual loss amounts to only one ten-thousandth of its mass, so the ice sheet is an enormous passive reservoir. However, if losses from iceberg calving and ice-shelf meltings are greater than gains from snowfall, global sea levels will rise.

At present, the size, and even the sign, of the contribution from Antarctica are uncertain. Consequently, Antarctic ice flux has not been included as a term in the sea-level predictions of *Climate Change 2007*, the fourth assessment report of the Intergovernmental Panel on Climate Change (IPCC). What is more certain is that the retreat of glaciers in the Arctic and mountain regions has contributed about 50 percent to current rates of sea-level rise. (The rest is due to the thermal expansion of water as the ocean warms.) An increasing contribution is coming from a retreat of the Greenland Ice Sheet, and part of this contribution is occurring as an iceberg flux (http://www.britannica.com/EBchecked/topic/281212/iceberg/273740/Climatic-impacts-of-icebergs.

Global Warming Impacting Vulnerability of Climate Change

The effects of global warming in the 21st century and beyond are expected

Fig. 17: An iceberg melting near Greenland's Ammassalik Island is just one sign of global warming. A record amount of ice has melted in the Arctic. (By John Mcconnico Associated Press)

Fig. 18: Melting Iceberg

Source: National Geographic-http://shop.nationalgeographic.com/product/203/4879/114.html)

to be disastrous, according to the summary of a scientific report issued on April 6, 2007 by the world's leading panel of climate change scientists. And many of those changes are already under way. Climate Change 2007: Impacts, Adaptation and Vulnerability (http://environment.about.com/od/globalwarming/a/ipcc_report_two.htm), the report from Working Group II of the United Nations' Intergovernmental Panel on Climate Change (IPCC)—which took six years to compile and draws on the research of 2,500 scientists from more than 130 countries—also makes it clear that while poor people worldwide will suffer most from the effects of global warming, no person on Earth will escape its consequences. The effects of global warming will be felt in every region and at all levels of society. The summary report on the effects of global warming is part two of a four-part IPCC report that will be released in stages throughout 2007. The first part, released in February 2007, confirmed with 90 percent certainty that global warming is now unstoppable and humans are to blame for a significant portion of the heat-trapping greenhouse gases that have caused global temperatures to rise dramatically since the middle of the 20th century.

Action Needed to Contain Global Warming

The rising global temperatures may bring some temporary benefits, according to the report, most notably increased food production due to more rainfall and longer growing seasons at mid-to-high latitudes and fewer deaths associated with cold weather. But scientists expect these short-term benefits to be outweighed by increased drought, flooding, water scarcity and hunger in other regions as well as more death and disease worldwide. "It is now clear that we are to blame for the last 50 years of warming, and this is already causing adverse changes to our planet," said Catherine Pearce, climate campaigner for Friends of the Earth International. "Unless we take action to reduce emissions now, far worse is yet to come, condemning millions in the poorest parts of the world to loss of lives, livelihoods and homes. Climate change is no longer just an environmental issue. It is a looming humanitarian catastrophe, threatening ultimately our global security and survival."

Findings of the Report include:

- Projected climate change is likely to affect millions of people who are already vulnerable. Heat waves, floods, storms, fires and droughts will cause increased death, disease and other harm. Global warming is also expected to lead to more deaths due to malnutrition, diseases that cause diarrhea, cardio-respiratory diseases related to higher concentrations of ground-level ozone, and wider distribution of diseases carried by insects, rodents, etc.
- Millions more people are projected to be at risk from coastal flooding due to sea level rise, especially in densely populated and low-lying settlements that already face other challenges, such as hurricanes and tropical storms.
- Approximately 20-30 percent of plant and animal species assessed so far are likely to be at higher risk of extinction if increases in global average temperature exceed 1.5-2.5 degrees Celsius. The global average temperature already has increased by about 0.74 degrees Celsius over pre-industrial levels.
- In the course of the current century, water supplies stored in glaciers and snow cover are projected to decline, reducing water availability in regions supplied by melt-water from major mountain ranges, where more than one-sixth of the world population currently lives. For example, glacial retreat in the Himalayas will disrupt downstream water supplies, which will have implications for billions of people across India, China, Nepal and Bhutan.
- By 2020, 75-250 million people in Africa will be exposed to water scarcity due to climate change.
- During the same period, yields from rain-fed agriculture in some African countries could be reduced by 50 percent.
- Latin America faces a risk of significant biodiversity loss by mid-century as increases in temperature and associated decreases in soil water lead to gradual replacement of tropical forest with savanna in parts of the Amazon region.
- People living on small islands, such as those found in the Caribbean and the Pacific, are particularly vulnerable to rising sea levels, extreme weather, and deteriorating coastal conditions associated with global warming. Climate change is projected to reduce water resources on many small islands and to adversely affect the livelihoods of island communities by undermining fisheries, tourism and other core components of island economies.

There are many environmental problems coming from the increase concentration of greenhouse gases in Earth's atmosphere. As Jeff Rubin of ABC NEWS reported, several signs indicate that we've begun changing Earth's climate: increased water vapor in the atmosphere, glaciers and polar ice caps appear to be melting, floods and droughts are becoming more severe, and sea levels have risen, on average, between 4 and 10 inches since 1990. (www.abc.com/sections/us/global106.html) Experts concur, by 2100, we might get a 2 foot sea level rise, but the catch is, levels might continue to rise 2 or 3 feet per century, for 1000 years (www.abcnews.com/sections/us/global106.html) (Fig 19). These rises in sea level can increase the salinity of freshwater throughout the world, and cause coastal lands to be washed under the ocean. Warmer water and increased humidity may encourage tropical cyclones, and changing wave patterns could produce more tidal waves and strong beach erosion on the coasts.

Feedstock Diversity in Energy Production and Emissions Reduction

Fossil-fuels, chiefly coal, oil and natural gas, now supply most of the world's energy (Fig. 20). Only a small amount comes from renewable sources, which do not release gases that trap heat in the atmosphere. If we could get more of our energy from renewable sources, we could reduce the amount of fossil fuels we burn. By the year 2050, renewable sources could provide forty percent of the energy needed in the world. Use of renewable energy can help both to slow global warming and to reduce air pollution. (www.doc.mmu.ac.uk/aric/gcc/cell.html#pos6)

Fossil fuels also emit greenhouse gases when burned. Coal emits high amounts of greenhouse gases, and the world may be supplied with enough of it to last over 100 years. Oil emits high amounts of greenhouse gases and also

Fig. 19: Flooding from Global Warming may be already happening

Source: www.abcnews.com/sections/us/global106.html.

Fig. 20: This Pie Graph shows the Breakdown of how the World Produces its Energy

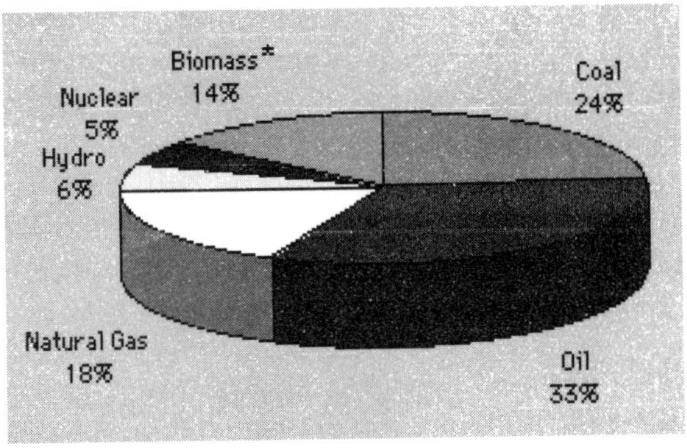

Source: www.envirolink.org/orgs/edf/sitemap.html.

other types of air pollution harmful to the environment. The world's oil supply is also estimated to last over 100 years. Natural Gas is the lowest of all fossil fuels in greenhouse gas emissions; supplies are projected to last over 100 years (Table 7) (www.doc.mmu.ac.uk/aric/gcc/cell.html#pos6).

Table 7: This Table shows what Percentage of CO_2 comes from Oil, Natural Gas, and Coal. For example, in 1996, 44.7% of the world's CO_2 emissions came from the combustion of oil

Country (region)	OIL	Natural Gas	Coal
World	44.7%	18.4%	36.9%
Canada	51.8%	30%	18.2%
United States	45%	21.3%	33.7%
European Union	56.2%	19%	24.8%
China	17.4%	1.1%	81.5%
Japan	64.6%	9.9%	25.5%

Source: http://infoweb.magi.com/~dwalsh/wfsesr.html.

Acid Rains

Acid rain is rain or any other form of precipitation that is unusually acidic. It has harmful effects on plants, aquatic animals, and infrastructure. Acid rain is mostly caused by human emissions of sulfur and nitrogen compounds which react in the atmosphere to produce acids. In recent years, many governments have introduced laws to reduce these emissions (http://www.google.co.in/search?hl=en&q=Acid+Rains&btnG=Google+Search&meta=&aq=f&oq=).

"Acid rain" is a popular term referring to the deposition of wet (rain, snow, sleet, fog and cloud-water, dew) and dry (acidifying particles and gases) acidic components. A more accurate term is "acid deposition". Distilled water, which contains no carbon dioxide, has a neutral pH of 7. Liquids with a pH less than

7 are acidic, and those with a pH greater than 7 are bases. "Clean" or unpolluted rain has a slightly acidic pH of about 5.2, because carbon dioxide and water in the air react together to form carbonic acid, a weak acid (pH 5.6 in distilled water), but unpolluted rain also contains other chemicals (Likens *et al*, 1987).

$$H_2O\ (l) + CO_2\ (g) \rightarrow H_2CO_3\ (aq)$$

Carbonic acid then can ionize in water forming low concentrations of hydronium and carbonate ions:

$$2H_2O\ (l) + H_2CO_3\ (aq)\ CO_3^{2-}\ (aq) + 2\ H_3O^+\ (aq)$$

The term acid rain does not convey the true nature of the problem and therefore scientists use the term "acid depositions". This is because the acid which has formed due to pollution may return to the earth as a solid or a gas and not just as rain. Depending upon the climatic conditions it could also come down as rain, fog, or snow, and in the wet form it is known as "acid precipitation" (http://www.essortment.com/all/acidraineffect_rqmz.htm). The following illustrations (Fig. 21a&b) depict both cases of 'Acid depositions' and 'Acid precipitation'.

Acid rain is a widespread problem found all over the world. It is the result of chemicals from burned fossil fuel mixing with moisture in the atmosphere and falling to the ground as rain, snow, sleet, etc. Acid deposition is a more precise name than acid rain because acid can fall to the ground as rain, snow, sleet, hail, and anything else. It can also combine with dry particles and fall to the ground; therefore it is called dry deposition whereas acid coming down in rain is called wet deposition.

Acid rain forms in the atmosphere from chemicals created by the burning of fossil fuels. When coal and oil are burned they release sulfur dioxide (SO_2) and two nitric oxides, nitric oxide (NO) and nitrogen dioxide (NO_2) designated by the symbol NO_X. These dangerous chemicals come from the smoke that pours from factory chimneys and exhaust from car tail pipes. Once released, sulfur dioxide and nitrogen oxides react with other chemicals in the air, water vapor, and sunlight to produce sulfuric and nitric acids, the acids in acid rain. The more sulfuric and nitric acids present, the higher the acidity of the rain. On a 14 point scale, 7 being neutral, anything less than 7 is considered acidic and anything more is considered alkaline. Battery acid is about 1 acidity and lemon juice is around 2 acidity. Regular rain is between 5 and 6 acidity. Acid rain is somewhere between 2 and 5.5. The worst case of acid rain ever recorded was in Wheeling, West Virginia where the rain had an acidity of 2.2. These levels of acidity seriously harm plants, trees, and all other life (http://www.geocities.com/CapeCanaveral/Hall/9111/DOCS.HTML).

All over the world, acid levels are rising which endangers trees, lakes, streams, drinking water supplies, monuments, and animal life. Basically everything is affected by acid rain. Not only life is affected, but also buildings, roads, even metals on cars and bridges.

Fig 21(A & B): Acid Rains in the forms of 'Acid Depositions' and Acid Precipitations'

When acid rain occurs, it does not immediately affect acidity in lakes and streams. The water dilutes the acid so only over a long period of time can the water become too acidic. In the spring, something called acid shock can happen. Snow, that contains acid, can build up and when it melts all the acid runs into the streams and lakes at one time. When acid levels are too high it can kill small organisms like algae. When the algae die other bigger organisms that eat algae die from starvation. It is a big chain reaction until all life is gone. Some lakes, i.e. the lakes in the Adirondack Mountains of New York State have been stripped of all life because the acid levels are too high. Acidic water that seeps into soil can also kill trees that are nearby. Buildings are taking a toll from acid rain.

The acid corrodes the stones and cement. Now that acid rain is widely acknowledged throughout the world, more and more people are doing something about it. For example: In November 1990, President George Bush signed the Clean Air Act, which had an annual cost of up to $25 billion. It cut sulfur dioxide emissions by 10 million tons a year. It also cut nitrogen oxides by 2 million tons a year. Europe agreed to cut sulfur dioxide emissions by 30% of 1980 levels, and a freeze (not increasing) on nitrogen oxide emissions at 1987 levels. Now all factories are required to be outfitted with scrubbers, which cost $150 million each. They are very expensive to maintain, but they remove 95% of sulfur dioxide after coal is burned. In scrubbers, poisonous gases are sprayed with a mix of water and lime

Sources

Certain industries, as well as emissions from vehicles give rise to increase of sulphur dioxide and nitrogen oxides in the air. These emissions change into sulphates and nitrates under the influence of sunlight and moisture, and get converted into sulphuric acid and nitric acid, which come down as acid rain. Coal generally contains between 2 - 3 % sulphur, and when it is burned, this sulphur is released into the atmosphere. Electric companies and other industries which burn coal cause a lot of emission of sulphur dioxide. Other industries which process raw ore containing sulphides in order to obtain copper, zinc, or nickel also because of an increase in sulphur dioxide levels in the atmosphere.

The major source of emissions of nitrogen oxides into the air is from vehicles and other places where fossil fuels are burned. Forest fires, often caused by man, either deliberately or accidentally, are another source of pollution. Naturally occurring phenomena like volcanic activity, lightning, or organic decay, also gives rise to an increase in atmospheric pollutants but not much can be done about these causes. However over 90% of sulphur dioxide emissions and around 95% of nitrogen oxides released into the air are from man made sources.

The problem of acid rain is not new. It was first noticed during the 17th century, when people observed the effects of industrialization on plants and animals. As far back as in 1872, the Scottish chemist, Angus Robert Smith wrote a book "Air and Rain: The beginnings of Chemical Climatology", in which he used the term "Acid Rain", and the name has stuck. The problem has become severe since the 1960s when fishermen noticed a sharp reduction in the quantity of fish in lakes of North America and Europe. The havoc done by acid rain is not localized in the place where it is caused. The atmospheric emissions may travel for several days and over long distances depending upon wind and climatic conditions, before coming down as acid rain. The problem caused in an industrialized area may therefore result in acid rain in the surrounding forests or lakes, or even further away. It is believed that around 50% of the acid rain that occurs in Canada is due to pollution caused in the United States of America, and the effect of polluting industries in England can be felt in Norway.

Effects of Acid Rain on Plant Life

Both natural vegetation and crops are affected by acid rain. The roots are damaged by acidic rainfall, causing the growth of the plant to be stunted, or even in its death. Nutrients present in the soil, are destroyed by the acidity. Useful micro organisms which release nutrients from decaying organic matter, into the soil are killed off, resulting in less nutrients being available for the plants. The acid rain, falling on the plants damages the waxy layer on the leaves and makes the plant vulnerable to diseases. The cumulative effect means that even if the plant survives it will be very weak and unable to survive climatic conditions like strong winds, heavy rainfall, or a short dry period. Plant germination and reproduction is also inhibited by the effects of acid rain.

Effects of Acid Rain on Aquatic Life

The action of acid rain causes harmful elements like mercury and aluminium to be leached from the soil and rocks and it is then carried into the lakes where aquatic life may be affected. Warning signs have been posted at several lakes, telling about the dangers of eating fish which may have been poisoned by mercury. Just as the soil has a natural ability to neutralize the acidity of rain water, within a certain limit, so also lakes and other water bodies can to a certain extent nullify the effects of acid rain. However as the acidity increases, the natural mechanisms are no longer able to cope. As the water gets more acidic its pH goes down. As the pH reaches 5.5, plankton, certain insects and crustaceans begin to die. At a pH of around 5.0, the fish population begins to die. When the pH drops below 5.0, all the fish have died, and the bottom of the lake lies covered with un-decayed material. Every year during the spring thaw, there is a sudden increase in the acidity of the lakes as frozen acid is suddenly deposited in them. This "Acid Shock" prevents the reproduction of aquatic species, or results in the deaths of the hatchlings.

Effects on Animals and Birds

All living organisms are interdependent on each other. If a lower life form is killed, other species that depended on it will also be affected. Every animal up the food chain will be affected. Animals and birds, like waterfowl or beavers, which depended on the water for food sources or as a habitat, also begin to die. Due to the effects of acid rain, animals which depended on plants for their food also begin to suffer. Tree dwelling birds and animals also begin to languish due to loss of habitat.

Effects on Human Beings

Mankind depends upon plants and animals for food. Due to acid rain the entire fish stocks in certain lakes have been wiped out. The economic livelihood of people who depended on fish and other aquatic life suffers as a result. Eating fish which may have been contaminated by mercury can cause serious health problems. In addition to loss of plant and animal life as food sources, acid rain gets into the food we eat, the water we drink, as well as the air we breathe. Due to this asthmatic people and children are directly affected. Urban drinking water supplies are generally treated to neutralize some of the effects of acid rain and therefore city dwellers may not directly suffer due to acidified drinking water. But out in the rural areas, those depending upon lakes, rivers, and wells will feel the effects of acid rain on their health. The acidic water moving through pipes causes harmful elements like lead and copper to be leached into the water. Aluminium which dissolves more easily in acid rain as compared to pure rainfall has been linked to Alzheimer's disease. The treatment of urban water supplies may not include removal of elements like Aluminium, and so is a serious problem in cities too.

Other Effects

All living things, whether plants or animals, whether living on land or in the water or trees, are affected either directly or indirectly by acid rain. Even buildings, bridges and other structures are affected. In cities, paint from buildings has peeled off and colours of cars have faded due to the effects of acid rain. From the Taj Mahal in India to the Washington Monument, great buildings all over the world have been affected by the acid rainfall which causes corrosion, fracturing, and discoloration in the structures. In Europe, structures like The Acropolis in Greece and Renaissance buildings in Italy, as well as several churches and cathedrals have suffered visible damage. In the Yucatan peninsula in Mexico, and in places in South America, ancient Mayan Pyramids are being destroyed by the acid rain. Temples, murals, and ancient inscriptions which had previously survived for centuries are now showing severe signs of corrosion. Even books, manuscripts, paintings, and sculpture are being affected in museums and libraries, where the ventilation system cannot eliminate the acid particles from the air which circulates in the building. In some parts of Poland, trains are required to run slowly, as the tracks are badly damaged due to corrosion caused by acid rainfall.

Environmental Impact of Acid Rain

Acid deposition - more commonly known as acid rain, results from man-made emissions of sulphur-dioxide and nitrogen-oxides, through the burning of fossil-fuels, principally for energy and transportation. Acid rain has negative effects on the environment in which we live. Since acid rain is a trans-boundary pollution problem, acidic emissions produced by one country can be deposited in another. Sweden and Norway, for example, both receive more than 90% of their sulphur pollution from abroad (http://www.ace.mmu.ac.uk/eae/acid_rain/older/Impacts.html). If large quantities of acid rain are deposited they may have detrimental consequences for wildlife, forests, soils, freshwater and buildings. Acid rain acidifies the soils and waters where it falls, killing off plants and animals. Surface water acidification can lead to a decline in, and loss of, fish populations and other aquatic species including frogs, snails and crayfish. Acid rain affects trees, usually by weakening them through damage

Fig. 22: Effects of Acid Rain on Surface Waters

to their leaves. Certain types of building stone can be dissolved in acid rain.

The ecological effects of acid rain are most clearly seen in the aquatic, or water, environments, such as streams, lakes, and marshes. Acid rain flows into streams, lakes, and marshes after falling on forests, fields, buildings, and roads. Acid rain also falls directly on aquatic habitats. Most lakes and streams have a pH between 6 and 8, although some lakes are naturally acidic even without the effects of acid rain. Acid rain primarily affects sensitive bodies of water, which are located in watersheds whose soils have a limited ability to neutralize acidic compounds (called "buffering capacity"). Lakes and streams become acidic (i.e., the pH value goes down) when the water itself and its surrounding soil cannot buffer the acid rain enough to neutralize it. In areas where buffering capacity is low, acid rain releases aluminum from soils into lakes and streams; aluminum is highly toxic to many species of aquatic organisms (http://www.epa.gov/acidrain/effects/surface_water.html).

Impact of Acid Rain on Lakes and Streams

Many lakes and streams examined in a National Surface Water Survey (NSWS) suffer from chronic acidity, a condition in which water has a constant low pH level. The survey investigated the effects of acidic deposition in over 1,000 lakes larger than 10 acres and in thousands of miles of streams believed to be sensitive to acidification. Of the lakes and streams surveyed, acid rain caused acidity in 75 percent of the acidic lakes and about 50 percent of the acidic streams. Several regions in the U.S. were identified as containing many of the surface waters sensitive to acidification. They include: the Adirondacks and Catskill Mountains in New York state, the mid-Appalachian highlands along the east coast, the upper Midwest, and mountainous areas of the Western United States. In areas like the Northeastern United States, where soil-buffering capacity is poor, some lakes now have a pH value of less than 5. One of the most acidic lakes reported is Little Echo Pond in Franklin, New York. Little Echo Pond has a pH of 4.2.

The acidification problem in both the U.S. and Canada grows in magnitude if "episodic acidification" is taken into account. Episodic acidification refers to brief periods during which pH levels decrease due to runoff from melting snow or heavy downpours. Lakes and streams in many areas throughout the U.S. are sensitive to episodic acidification. In the Mid-Appalachians, the Mid-Atlantic Coastal Plain, and the Adirondack Mountains, many additional lakes and streams become temporarily acidic during storms and spring snowmelt. For example, approximately 70 percent of sensitive lakes in the Adirondacks are at risk of episodic acidification. This amount is over three times the amount of chronically acidic lakes. In the mid-Appalachians, approximately 30 percent of sensitive streams are likely to become acidic during an episode. This level is seven times the number of chronically acidic streams in that area. Episodic acidification can cause "fish kills." Emissions from U.S. sources also contribute to acidic deposition in eastern Canada, where the soil is very similar to the soil of the Adirondack Mountains, and the lakes are consequently extremely vulnerable to chronic acidification problems. The Canadian government has estimated that 14,000 lakes in eastern Canada are acidic.

Acid Rains Affecting Fish and Other Aquatic Organisms

Acid rain causes a cascade of effects that harm or kill individual fish, reduce fish population numbers, completely eliminate fish species from a water body and decrease biodiversity. As acid rain flows through soils in a watershed, aluminum is released from soils into the lakes and streams located in that watershed. So, as pH in a lake or stream decreases, aluminum levels increase. Both low pH and increased aluminum levels are directly toxic to fish. In addition, low pH and increased aluminum levels cause chronic stress that may not kill individual fish, but leads to lower body weight and smaller size and makes fish less able to compete for food and habitat. Some types of plants and animals are able to tolerate acidic waters. Others, however, are acid-sensitive and will be lost as the pH declines. Generally, the young of most species are more sensitive to environmental conditions than adults. At pH 5, most fish eggs cannot hatch. At lower pH levels, some adult fish die. Some acid lakes have no fish. The chart below (Table 8) shows that not all fish, shellfish, or the insects that they eat can tolerate the same amount of acid; for example, frogs can tolerate water that is more acidic (i.e., has a lower pH) than trout (http://www.epa.gov/acidrain/effects/surface_water.html).

Acid Rain Affecting Ecosystems

Together, biological organisms and the environment in which they live are called an ecosystem. The plants and animals living within an ecosystem are highly interdependent. For example, frogs may tolerate relatively high levels of acidity, but if they eat insects like the mayfly, they may be affected because part of their food supply may disappear. Because of the connections between

Table 8: Acid Tolerance Levels of Different Species of Fish

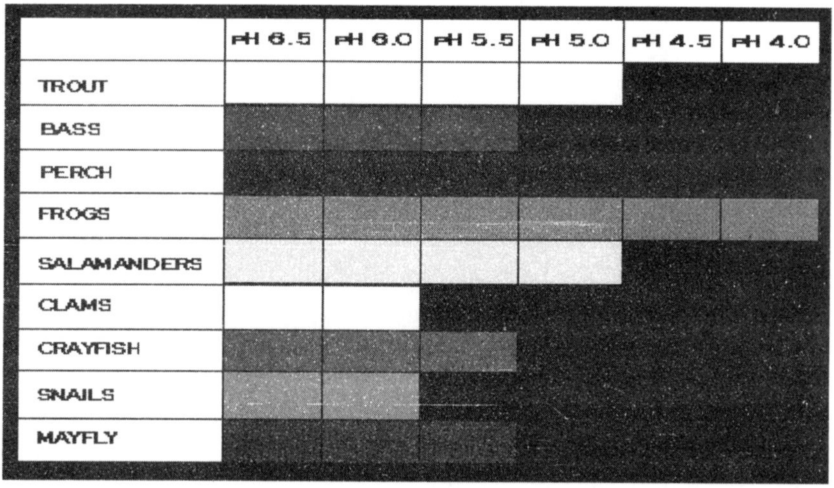

Souecs: http://www.epa.gov/acidrain/effects/surface_water.html

the many fish, plants, and other organisms living in an aquatic ecosystem, changes in pH or aluminum levels affect biodiversity as well. Thus, as lakes and streams become more acidic, the numbers and types of fish and other aquatic plants and animals that live in these waters decrease.

Role of Nitrogen (in Acid Rain) on Other Environmental Problems

The impact of nitrogen on surface waters is also critical. Nitrogen plays a significant role in episodic acidification and new research recognizes the importance of nitrogen in long-term chronic acidification as well. Furthermore, the adverse impact of atmospheric nitrogen deposition on estuaries and near-coastal water bodies is significant. Scientists estimate that 10 to 45 percent of the nitrogen produced by various human activities that reaches estuaries and coastal ecosystems is transported and deposited via the atmosphere. For example, about 30 percent of the nitrogen in the Chesapeake Bay comes from atmospheric deposition. Nitrogen is an important factor in causing eutrophication (oxygen depletion) of water bodies. The symptoms of eutrophication include blooms of algae (both toxic and non-toxic), declines in the health of fish and shellfish, loss of sea-grass beds and coral reefs, and ecological changes in food webs. According to the National Oceanic and Atmospheric Administration (NOAA), these conditions are common in many of our nation's coastal ecosystems. These ecological changes impact human populations by changing the availability of seafood and creating a risk of consuming contaminated fish or shellfish, reducing our ability to use and enjoy our coastal ecosystems, and causing economic impact on people who rely on healthy coastal ecosystems, such as fishermen and those who cater to tourists.

Fig. 23: Effects of Acid Rain - Human Health

Acid rain looks, feels, and tastes just like clean rain. The harm to people from acid rain is not direct. Walking in acid rain, or even swimming in an acid lake, is no more dangerous than walking or swimming in clean water. However, the pollutants that cause acid rain—sulfur dioxide (SO_2) and nitrogen oxides (NO_x)—do damage human health. These gases interact in the atmosphere to form fine sulfate and nitrate particles that can be transported long distances by winds and inhaled deep into people's lungs. Fine particles can also penetrate indoors. Many scientific studies have identified a relationship between elevated levels of fine particles and increased illness and premature death from heart and lung disorders, such as asthma and bronchitis (http://www.epa.gov/acidrain/effects/health.html).

Based on health concerns, SO_2 and NO_x have historically been regulated under the Clean Air Act, including the Acid Rain Program. In the eastern U.S., sulfate aerosols make up about 25 percent of fine particles. By lowering SO_2 and NO_x emissions from power generation, the Acid Rain Program will reduce the levels of fine sulfate and nitrate particles and so reduce the incidence and the severity of these health problems. When fully implemented by the year 2010, the public health benefits of the Acid Rain Program are estimated to be valued at $50 billion annually, due to decreased mortality, hospital admissions, and emergency room visits. Decreases in NO_x emissions are also expected to have a beneficial impact on human health by reducing the nitrogen oxides available to react with volatile organic compounds and form ozone. Ozone impacts on human health include a number of morbidity and mortality risks associated with lung inflammation, including asthma and emphysema.

Ozone Depletion

Ozone depletion describes two distinct, but related observations: a slow, steady decline of about 4% per decade in the total volume of ozone in Earth's stratosphere (ozone layer) since the late 1970s, and a much larger, but seasonal, decrease in stratospheric ozone over Earth's polar regions during the same period. The latter phenomenon is commonly referred to as the ozone hole. In addition to this well-known stratospheric ozone depletion, there are also tropospheric ozone depletion events, which occur near the surface in Polar Regions during spring (http://en.wikipedia.org/wiki/Ozone_depletion).

The detailed mechanism by which the polar ozone holes form is different from that for the mid-latitude thinning, but the most important process in both trends is catalytic destruction of ozone by atomic chlorine and bromine (http://www.atm.ch.cam.ac.uk/tour/part3.html). The main source of these halogen atoms in the stratosphere is photo-dissociation of chlorofluorocarbon (CFC) compounds, commonly called freons, and of bromo-fluorocarbon compounds known as halons. These compounds are transported into the stratosphere after being emitted at the surface. Both ozone depletion mechanisms strengthened as emissions of CFCs and halons increased.

CFCs and other contributory substances are commonly referred to as ozone-depleting substances (ODS) (http://www.scientificamerican.com/article.cfm?id=chlorofluorocarbons-cfcs). Since the ozone layer prevents most harmful UVB wavelengths (270–315 nm) of ultraviolet light (UV light) from passing through the Earth's atmosphere, observed and projected decreases in ozone have generated worldwide concern leading to adoption of the Montreal Protocol banning the production of CFCs and halons as well as related ozone depleting chemicals such as carbon tetrachloride and trichloroethane. It is suspected that a variety of biological consequences such as increases in skin cancer, damage to plants, and reduction of plankton populations in the ocean's photic zone may result from the increased UV exposure due to ozone depletion.

Evidence that human activities affect the ozone layer has been building up over the last 20 years, ever since scientists first suggested that the release of chlorofluorocarbons (CFCs) into the atmosphere could reduce the amount of ozone over our heads. The breakdown products (chlorine compounds) of these gases were detected in the stratosphere. When the ozone hole was detected, it was soon linked to this increase in these chlorine compounds. The loss of ozone was not restricted to the Antarctic - at around the same time the first firm evidence was produced that there had been an ozone decrease over the heavily populated northern mid-latitudes (30-60N). However, unlike the sudden and near total loss of ozone over Antarctica at certain altitudes, the loss of ozone in mid-latitudes is much less and much slower – only a few percentage per year. However, it is a very worrying trend and one which is the subject of intense scientific research at present. Many of these findings have since been reinforced by a variety of internationally supported scientific investigations involving satellites, aircraft, balloons and ground stations, and the implications are still being quantified and assessed.

New Clues to Ozone Depletion

According to the Report published in the *Science Daily* (July 27, 2007) — large quantities of ozone-depleting chemicals have been discovered in the

Antarctic atmosphere by researchers from the University of Leeds, the University of East Anglia, and the British Antarctic Survey (http://www.sciencedaily.com/releases/2007/07/070726104756.htm).

The team of atmospheric chemists carried out an 18-month study of the make-up of the lowest part of the earth's atmosphere on the Brunt Ice Shelf, about 20 km from the Weddell Sea. They found high concentrations of halogens —bromine and iodine oxides—which persist throughout the period when there is sunlight in Antarctica. A big surprise to the science team was the large quantities of iodine oxide, since this chemical has not been detected in the Arctic.

The source of the halogens is natural—sea-salt in the case of bromine, and in the case of iodine, almost certainly bright orange algae that coat the underside of the sea ice around the continent. These halogens cause a substantial depletion in ozone above the ice surface. This affects the so-called oxidizing capacity of the atmosphere - its ability to "clean itself" by removing certain—often man-made—chemical compounds. The iodine oxides also form tiny particles (a few nano-metres in size), which can grow to form ice clouds, with a consequent impact on the local climate.

Scientists now plan to carry out further research to assess what impact this may be having on the local environment. Very recent satellite observations by one of the team, Dr Alfonso Saiz-Lopez, have confirmed that iodine oxides are widespread throughout coastal Antarctica. John Plane, professor of atmospheric chemistry at the University of Leeds, said: "Halogens in the lowest part of the atmosphere have important impacts on ozone depletion, the ability of the atmosphere to remove potentially harmful compounds, and aerosol formation. All these atmospheric phenomena are linked to climate change. We still have to work out what the ramifications of this discovery are. These exciting results also show how important it is to keep exploring the atmosphere - there seems to be plenty more to find out." Using high-tech measuring equipment, a beam of light was projected across the ice shelf and the spectrum of the reflected light analyzed and chemical levels measured. The work was carried out in a new atmospheric observatory at Halley Station, operated by the British Antarctic Survey, and was supported by funding from the U.K.'s Natural Environment Research Council.

Environmental Indicators of Ozone Depletion

Over 60 years ago, chlorofluorocarbons (CFCs) were invented in the United States, and they soon found many uses throughout the world in refrigeration, air conditioning, and other industrial processes. Due to scientific evidence that CFCs and other chemicals destroy ozone in the upper atmosphere, the United States, the country which has traditionally been the largest emitter of CFCs worldwide, is rapidly scaling back the use of these chemicals and phasing out their production (http://www.epa.gov/Ozone/

science/indicat/index.html) (Fig 24-A & B).

The ozone (O_3) layer in the stratosphere protects life on earth from exposure to dangerous levels of ultraviolet light. It does so by filtering out harmful ultraviolet radiation from the sun. When CFCs and other ozone-

Fig. 24-A: Depletion of Ozone Layers

Source: http://www.epa.gov/Ozone/science/indicat/index.html.

Fig. 24-B: US Impact of CFC on the Ozone Layer (Source: USITC)

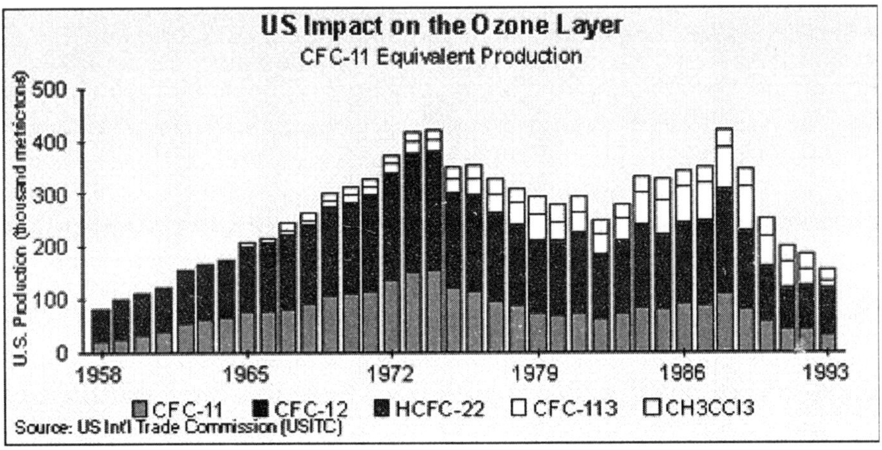

Source: USITC.

degrading chemicals are emitted, they mix with the atmosphere and eventually rise to the stratosphere. There, the chlorine and the bromine they contain catalyze the destruction of ozone. This destruction is occurring at a more rapid rate than ozone can be created through natural processes. The degradation of the ozone layer leads to higher levels of ultraviolet radiation reaching Earth's surface. This in turn can lead to a greater incidence of skin cancer, cataracts, and impaired immune systems, and is expected also to reduce crop yields, diminish the productivity of the oceans, and possibly to contribute to the decline of amphibious populations that is occurring around the world.

The chemicals most responsible for the destruction of the ozone layer are chlorofluorocarbons, carbon tetrachloride, methyl bromide, methyl chloroform, and halons. Chlorofluorocarbons have long been widely used as coolants in refrigerators and air conditioners and as foaming agents, solvents, and aerosol propellants. Carbon tetrachloride and methyl chloroform are solvents used for essential industrial applications. In the United States, carbon tetrachloride is now used almost entirely as a feedstock for the production of chlorofluorocarbons. Hydrogenated CFCs (HCFCs) have many of the same uses as CFCs and are increasingly employed as interim substitutes for CFCs. Long predicted, the degradation of the ozone layer was confirmed when a large hole in the layer over Antarctica was reported in 1985. Smaller but significant stratospheric decreases have been seen over more populated regions of the Earth. Subsequent research established that industrial chemicals are responsible for the observed depletions of ozone over Antarctica and play a major role in global ozone losses.

Human Activities

Chlorine and bromine are emitted to the atmosphere from both natural and human sources. These very stable human-made chemicals are not soluble in water and are not broken down chemically in the lower atmosphere. Thus, they survive long enough to reach the stratosphere. The CFCs and carbon tetrachloride are relatively un-reactive in the lower atmosphere (the troposphere) and move unscathed into the stratosphere where they are decomposed by intense sunlight, releasing chlorine to catalyze the destruction of ozone molecules. Certain ozone-depleting chemicals (HCFC-22 and methyl chloroform) are more reactive in the troposphere and deliver less of their initial chlorine load to the stratosphere. Halons also are generally reactive in the troposphere and deliver only a fraction of their initial load of bromine to the stratosphere, but bromine is 40 times more efficient at destroying ozone than chlorine. Increasing attention is being focused on the ozone-depleting role of methyl bromide, which has three potentially major human sources (soil fumigation, biomass burning, and the exhaust of automobiles using leaded gasoline), in addition to a natural oceanic source. U.S. production of ozone-depleting gases has declined significantly since 1988, and has now reached levels (measured by their ozone depletion potential) comparable to those of

30 years ago. Because of the international agreements to decrease production and ultimately to phase out production of CFCs and halons, total equivalent chlorine (total chlorine and bromine, with adjustments to account for bromine's higher ozone depletion potential) in the troposphere peaked between 1992 and 1994 and has since decreased. Total chlorine abundance in the stratosphere is at or near peak; stratospheric bromine is likely still increasing. Increasing ozone losses are predicted for the remainder of the decade, with gradual recovery by the mid-21st century.

The Ozone Hole

Starting most clearly in the austral (southern hemisphere) spring of 1980, a massive, continental-sized hole—accounting for one-quarter to one-half of the ozone—appeared over the largely unpopulated core of Antarctica. For the past 18 years, this hole has grown in size. Recently, ozone depletion over the South Pole has became so severe that less than one-third of the average October levels (measured from the 1950s through the 1970s) remained on many days of that same month in 1993. The 1995 Antarctic ozone hole opened up before mid-September and was the largest ever recorded that early in the austral spring (http://www.ucsusa.org/global_warming/science_and_impacts/science/the-science-of-ozone.html). The Antarctic ozone hole took most atmospheric chemists by surprise. The original Rowland/Molina theory predicted that the most severe ozone loss would occur relatively high in the stratosphere (above 30 km). In fact, the largest depletions over Antarctica occurred in the middle range between 13 and 21 km.

What atmospheric scientists had failed to factor in was the powerful "polar vortex." The winter stratospheric air over the Antarctic routinely gets colder than the air anywhere else. Because the winter stratospheric air over most of Antarctica is isolated from warmer, lower-altitude air by a strong set of confining winds known as the polar vortex, this air gets so cold that polar stratospheric clouds regularly form between the altitudes of approximately 12 to 25 km. Particle surfaces in these clouds cause chemical reactions that create even more reactive halogen chemicals. Thus, when the first light of austral spring reaches these molecules, they break apart and release massive levels of the chlorine and bromine atoms that drive catalytic destruction cycles. Additional chemical reactions also related to the surface particles in the stratospheric clouds actually lengthen ozone-destroying cycles.

Over the past decade, ground aircraft and satellite-based measurements have shown direct, quantitative relationships between the occurrence of polar stratospheric clouds, the creation of high levels of reactive halogen chemicals, and ozone destruction. In recent years, these halogen-driven destruction cycles have been vigorous enough to destroy almost all of the ozone in the core of the polar vortex—between 14 and 18 km—and much of the ozone several kilometers above and below these altitudes.

Amount of Stratospheric Ozone Loss

When the Antarctic polar vortex breaks up in the late spring, the ozone-poor stratospheric air resulting from the ozone hole spreads northward over populated regions of Australia, New Zealand, and South America. This effect has been strong enough in recent years to cause significantly increased levels of UV-B to fall on portions of these regions. In the northern hemisphere, the arctic polar vortex is much weaker that the Antarctic vortex, so the arctic stratosphere does not get as cold for as long. Polar stratospheric clouds do form sporadically, however. After the resulting air is exposed to sunlight, high levels of reactive halogen chemicals form immediately. These patches of stratospheric air often experience significant ozone depletion, sometimes as they are driven south over populated regions of Europe, Asia, and North America. It is not clear whether current multitude ozone losses are due to polar ozone-depleted air mixing down to lower latitudes, or to increased on-site halogen-catalyzed ozone destruction. This uncertainty is being investigated by ongoing atmospheric measurement and modeling programs. It is abundantly clear, however, that recent years have shown a disturbing level of ozone loss at middle as well as high latitudes.

Starting with ozone levels measured in the late 1960s and early 1970s—well before atmospheric chlorine and bromine levels were high enough to cause significant ozone depletion—satellite and ground-launched ozone measurements have documented northern hemisphere winter/spring downward trends of about 6 percent per decade, and summer/fall downward trends of 3 percent per decade for 1979-1994. Seasonal differences are less pronounced in the southern hemisphere, but yearly averaged downward trends of 4 to 5 percent per decade have been recorded. Only in the tropics (20° N - 20° S) is there no statistically significant loss of ozone.

Expected Ozone Loss in the Future

Atmospheric chlorine and bromine levels are expected to peak around 1998, if current international agreements limiting the production and release of halocarbons are implemented and enforced. This suggests that maximum northern hemisphere midlatitude winter/spring ozone losses of 12 to 13 percent of historic background levels will probably occur in that time frame (about 2.5 percent more loss than in 1994). Summer/fall mid-latitude losses of 6 to 7 percent are also anticipated. Such losses would be accompanied by increases in levels of harmful ultraviolet solar radiation of approximately 15 percent in winter/spring and approximately 8 percent in summer/fall. Because of the long life of many halocarbons, particularly CFCs, it will be a long time before stratospheric ozone levels are completely restored. Current projections envision a significant annual Antarctic ozone hole for several decades into the next century.

Negative Impacts of Ozone Depletion

Effects on Human and Animal Health: Increased penetration of solar UV-B radiation is likely to have high impact on human health with potential risks of eye diseases, skin cancer and infectious diseases.

Effects on Terrestrial Plants: In forests and grasslands, increased radiation is likely to change species composition thus altering the bio-diversity in different ecosystems. It could also affect the plant community indirectly resulting in changes in plant form, secondary metabolism, etc.

Effects on Aquatic Ecosystems: High levels of radiation exposure in tropics and subtropics may affect the distribution of phyto-planktons, which form the foundation of aquatic food webs. It can also cause damage to early development stages of fish, shrimp, crab, amphibians and other animals, the most severe effects being decreased reproductive capacity and impaired larval development.

Effects on Bio-geo-chemical Cycles: Increased solar UV radiation could affect terrestrial and aquatic bio-geo-chemical cycles thus altering both sources and sinks of greenhouse and important trace gases, e.g. carbon dioxide (CO_2), carbon monoxide (CO), carbonyl sulfide (COS), etc. These changes would contribute to biosphere-atmosphere feedbacks responsible for the atmosphere build-up of these greenhouse gases.

Effects on Air Quality: Reduction of stratospheric ozone and increased penetration of UV-B radiation result in higher photo dissociation rates of key trace gases that control the chemical reactivity of the troposphere. This can increase both production and destruction of ozone and related oxidants such as hydrogen peroxide, which are known to have adverse effects on human health, terrestrial plants and outdoor materials.

The ozone layer, therefore, is highly beneficial to plant and animal life on earth filtering out the dangerous part of sun's radiation and allowing only the beneficial part to reach earth. Any disturbance or depletion of this layer would result in an increase of harmful radiation reaching the earth's surface leading to dangerous consequences.

Emissions from Thermal Power Plants

India is the world's fourth largest economy and has a fast growing energy market. India's current power capacity is 30% short of demand. Coal and petroleum are the primary sources of energy. High ash content in Indian coal and inefficient combustion technologies contribute to India's emission of air particulate matter and other trace gases, including gases that are responsible for the greenhouse effect. The USAID Office of Environment, Energy & Enterprise in India promotes clean energy development through efficient energy use and pollution reduction. This organization has sponsored the project, Anthropogenic Emissions from Energy Activities in India, to assess the emissions of greenhouse and other trace atmospheric gases from energy

Fig. 25: Steam and Smoke are Emitted from a Coal-fired Power Station in England

Source: http://science.howstuffworks.com/carbon-capture.htm/printable.

activities in India. The findings of the study are as follows (http://www.enzenglobal.com/pdf_downloads/environmental_impact.pdf):

The main emissions from coal combustion at thermal power plants are carbon dioxide (CO), nitrogen oxides (NO), sulfur oxides (SO), chlorofluorocarbons (CFCs), and air- borne inorganic particles such as fly ash, soot, and other trace gas species. Carbon dioxide, methane, and chlorofluorocarbons are greenhouse gases. These emissions are considered to be responsible for heating up the atmosphere, producing a harmful global environment. Oxides of nitrogen and sulfur play an important role in atmospheric chemistry and are largely responsible for atmospheric acidity. Particulates and black carbon (soot) are of concern, in addition to possible lung tissue irritation resulting from inhalation of soot particles and various organic chemicals that are known carcinogens.

Carbon Dioxide Emissions

Utilities in India burn mostly coal with approximately 10 –30% excess air. The total carbon obtained from analysis is converted to CO after the reaction (combustion) is complete. Total CO emissions for 1997 from all the power plants in India are estimated at 1.1 Teragrams (Tg) per day or 397 Tg per year. Average CO emission per unit of electricity is 1.04 Gig grams (Gg). Technological improvements in efficient combustion of coal can lead to greater production of electricity per unit of coal that will effectively reduce CO emission per unit of electricity. Although the current per capita carbon dioxide (CO) emission in India is only one quarter of the world average and about twenty times less than United State's averages, the growth rate of emissions is very high. Because of this growth, the region is expected to soon become a major contributor of greenhouse gases, such as CO and other air pollutants.

Sulfur Dioxide Emissions

The sulfur content in Indian coal is low compared to United States coal. Acid rain due to sulfur dioxide emissions is presently not of great concern. However, increasing coal use or blending Indian coal with imported coal of higher calorific value (further increasing electricity production) needs to be carefully addressed through viable technological options. Average SO emissions per unit of electricity are 0.0069 Gg. Total SO emissions are estimated to be 7.33 Gg per day or 2.7 Tg per year.

Emissions of Oxides of Nitrogen

Oxidation of nitric oxide (NO) discharged in combustion products forms nitrogen dioxide (NO) in the atmosphere. These oxides of nitrogen are responsible for the formation of photochemical smog. Nitric oxide emission per unit of electricity is estimated as approximately 0.00056 Gg. Total NO emissions are estimated to be 0.5 Gg per day and 0.185 Tg per year. Nitrogen oxides are important chemical species in the atmosphere since they contribute to its acidity; they also act as precursor gases for the formation of tropospheric ozone. Tropospheric ozone is a greenhouse gas responsible for global warming and is also known to have an adverse affect on plants. NO emissions should be kept at a minimum possible level. Lower concentrations of NO lessen the formation of tropospheric ozone even when other precursor gases like carbon monoxide (CO) are present in higher concentrations.

Carbonaceous Material and Black Carbon (soot)

Incomplete and/or inefficient combustion processes of fossil fuel generate soot. A recently conducted Indian Ocean Experiment (INDOEX) suggests that the presence of soot carbon in the atmosphere over the northern Indian Ocean hinders its natural heating processes by about 15%. Enhancement of boundary layer heating can significantly influence regional hydrological cycles and climate. Present calculations show that soot emissions are produced at a rate of 22.0 Gg per year from Indian thermal power plants. Soot emissions in India have not been studied thoroughly so far; these are the first estimates of soot emission from Indian thermal power plants. Appropriate technological intervention to prevent soot carbon emissions may possibly not only reduce the chances of soot escaping into the atmosphere (where it can potentially change the radiation balance), but can also lead to further increases in electricity production.

General Observations

1. Most power plants in India are running at an efficiency rate of 20-30%, which is lower than the efficiency rates of 35-40% in the US. The reason for inefficient combustion should be probed and corrective measures taken to modernize India's plants. These actions will reduce national emissions as well

as increase electricity production.

2. Dispersion and transport of emitted greenhouse gases and other pollutants over the entire Indian region need to be mapped to better understand the impact on the climate, as well as on human, animal, plant, and ecological systems.

3. Further research is needed on population exposure to these polluting gases. Pollutants' effects and power plant density need to be studied.

Analyses of detailed air characteristics and associated meteorological parameters, as well as emissions of greenhouse gases and other pollutants from energy activities in India, will help climate modeling and understanding the impact of these emissions on India's climate, human and plant health, and agriculture.

Impact of Emissions on Air Quality

Rapid industrialization worldwide has triggered off a wave of economic development bringing prosperity and advancement for many nations. However, such activity has not been without degradation of the environment, through the depletion and contamination of existing natural resources and discharge of pollutants into air, water and dumping of solid waste on land. All of these have exerted an enormous load on the environment as a whole damaging the quality of resources available to all existent life forms.

Air Pollution is amongst the most serious global environmental concerns. Apart from incidents and mishaps such as London Fog, Bhopal Gas Tragedy, there are cases that have long-term hazards and need to be addressed urgently. Amongst these are the Ozone Hole problem, Global Warming issue and Acid Rain problem. These concerns however, have led to the formation of specific governing bodies regulating emissions, laws and regulations for controlling these problems (Azmi *et al*, 2008). In the study of air pollution, it is essential to have a complete insight into the nature, behavior and characteristics of a pollutant along with the prevalent meteorological characteristics of the environment into which it is released. This provides the basic data regarding its impact on the environment as well as developing effective measures to check its impairing properties and control the damage that can be caused to nature. Computer techniques are increasingly being used in conjunction with both deterministic and statistic approaches provide powerful tools to study the impact of pollutants on the environment. They also enable prediction of likely outcomes and help to develop environmental safe guards. These models validated with actual field and laboratory data help to verify the authenticity of novel approaches and their compatibility with the Environmental Protection Authority (EPA) criteria. Therefore, air quality models are indispensable tools for evaluating the impact of air pollutants on urban environments and human health.

In a world of global supply chains and rapidly expanding trade, ocean shipping—currently the dominant mode of transport for international cargo—

is becoming an increasingly important source of air pollution and greenhouse gas emissions. Today, ocean-going vessels transport 90 percent of all trade by volume to and from the 25 members of the European Community (EC), and nearly 80 percent by weight of all goods shipped in and out of the United States (EC, 2006, US DOT, 2003). Over the last three decades, activity in the marine shipping sector, as measured in metric ton-kilometers, has grown on average by 5 percent every year. Since emissions from ocean-going vessels have only been moderately controlled, this growth has been accompanied by a commensurate increase in the sector's contribution to local and global air pollution.

Ocean-going vessels contribute significantly to global emissions of nitrogen oxides (NOx), sulfur oxides (SOx), and particulate matter (PM). It is estimated that by 2020, ship emissions contributions to the European Union (EU) NOx and SOx inventories will surpass total emissions generated by all land-based mobile, stationary and other sources in the twenty-five nations (EC 2005). Air quality impacts from ocean-going vessels are especially significant in port cities and nations with extensive coastlines adjacent to shipping corridors. Studies making use of geographic marine activity data have estimated that about 70–80 percent of all ship emissions occur within 400 km (248 miles) of land. Pollutants such as NOx, SOx, and PM have been linked to a variety of adverse public health outcomes, including increased risk of premature death from heart and pulmonary diseases and worsened respiratory disease. Marine emission sources are therefore responsible for a growing share of the public health impacts of exposure to air pollution in many regions. Although ocean-going vessels are among the most efficient modes of freight transport, they also generate substantial quantities of greenhouse gas emissions. Currently, carbon dioxide (CO_2) emissions from the international shipping sector as a whole exceed annual total greenhouse gas emissions from most of the nations listed in the Kyoto protocol as Annex I countries (Kyoto Protocol, 1997).

The goal of the International Council on Clean Transportation (ICCT) is to dramatically reduce conventional pollutant and greenhouse gas emissions from personal, public, and goods transportation in order to improve air quality and human health, and mitigate climate change. The Council is made up of leading regulators and experts from around the world that participate as individuals based on their experience with air quality and transportation issues. The ICCT promotes best practices and comprehensive solutions to improve vehicle emissions and efficiency, increase fuel quality and sustainability of alternative fuels, reduce pollution from the in-use fleet, and curtail emissions from international goods movement (ICCT, 2007). The ICCT recommends that emission mitigation measures should be adopted at all major port facilities and be fully integrated with local and/or regional air quality plans. Each port type has access to a range of implementation mechanisms to reduce emissions from ships at berth. For example, landlord ports can include emission reduction requirements in their lease agreements with tenant operators. Operating ports can directly implement some infrastructure measures.

A comprehensive study was conducted by the Federal Aviation Administration (FAA)/Environmental Protection Agency (EPA) on the air quality to assess the impact of aircraft emissions of carbon monoxide (CO), hydrocarbons (HC), and oxides of nitrogen (NOx) in the vicinity of airports (Yamartino *et al*, 2008). This assessment includes the results of recent modeling and monitoring efforts at Washington National (DCA), Los Angeles International (LAX), Dulles International (IAD), and Lakeland, Florida airports and an updated modeling of aircraft generated pollution at LAX, John F. Kennedy (JFK) and Chicago O'Hare (ORD) airports. The Airport Vicinity Air Pollution (AVAP) model which was designed for use at civil airports was used in this assessment. In addition the results of the application of the military version of the AVAP model the Air Quality Assessment Model (AQAM) are summarized. Both the results of the pollution monitoring analyses in Volume I and the modeling studies in Volume II suggest that: maximum hourly average 60 CO concentrations from aircraft are unlikely to exceed 5 parts per million (ppm) in areas of public exposure and are thus small in comparison to the National Ambient Air Quality Standard of 35 ppm; maximum hourly HC concentrations from aircraft can exceed 0. 25 ppm over an area several times the size of the airport; and while annual average NO_2 concentrations from aircraft are estimated to contribute only 10 to 20 percent of the NAAQS limit level

If rising demands for national and international air travel are to be accommodated, mean air-traffic movements will likely have to rise. With only limited scope for reducing emissions from jet engines, it appears quite possible that airport capacity may at some locations become constrained by air-quality requirements. During takeoff and climb-out, NO_2—the pollutant of principal concern - constitutes some 5-10% of NO_x at the engine nozzle exit, and a fraction equilibrating at several times this downstream. Assessing associated impacts on local air quality is complicated by the action not just of transport processes of the ambient atmosphere, but also of those forced by the passage of the aircraft

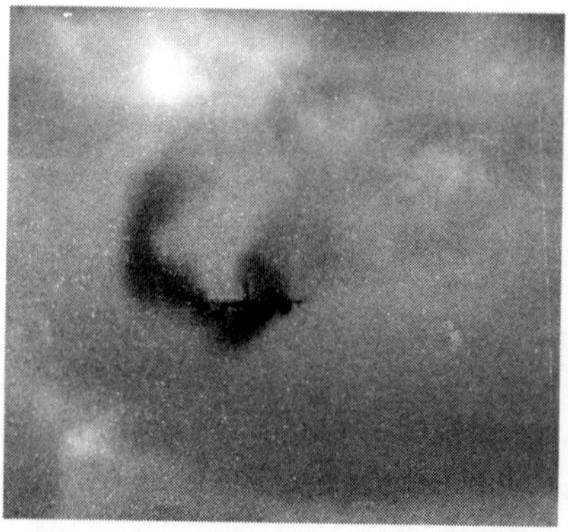

itself. The latter are not well understood, and not represented satisfactorily in current airport air-quality models.

There is increasing concern over the environmental impacts arising from pollutants emitted by aircraft engine exhausts. These emissions can:

- Affect local or regional air quality;
- Affect the earth's climate directly, e.g. carbon dioxide (CO_2) emissions, or indirectly through modification of processes involved in the climate system, e.g. cloud formation;
- Lead to ozone production or destruction (depending on the altitude of the emissions).

Kerosene is the main component of aviation fuel which is used to provide propulsion for current commercial aircraft. Like other fossil fuels, kerosene produces CO_2 and water vapour as the products of complete combustion. As the combustion process is not completely efficient, partial oxidation products such as carbon monoxide (CO) and oxygenated organic compounds are produced and emitted. Depending on the load on the engine and the mixing processes in the combustor, soot can be formed and emitted together with unburnt hydrocarbons.

The emission profiles vary with the phase of the flight. Although the sulphur content of the fuel is reduced during the refining process, aviation fuel still contains a small amount of sulphur and this is converted to sulphur oxides (SO_x) during the combustion process. The temperatures reached in the jet engine combustor result in the formation of nitrogen oxides (NO_x) through high temperature reactions of the nitrogen and oxygen present in the combustion air.

The impact of the aviation industry on local air quality, especially in the vicinity of airports, has long been recognized. The pollutants of concern are the emissions of NO_x, CO_2 hydrocarbons and soot. Emissions limits from aircraft engine exhausts have been defined for these pollutants by the International Civil Aviation Organization (ICAO), although the limits only apply during the Landing Take-Off (LTO) cycle. In addition to aircraft emissions, local air quality is further compromised by pollution from motor vehicles along extensive road networks that provide access to airports (http://www.caa.co.uk/default.aspx?catid=68&pagetype=90&pageid=52).

The other environmental concern for the aviation industry is the contribution of aircraft exhaust emissions to climate change. The most important emissions are considered to be those of CO_2 and NO_x. Subsonic aviation currently contributes between 2-3% of the CO_2 emitted from all fossil-fuel combustion. The emissions of NO_x lead to the production of ozone (O_3) in the upper troposphere. There is also increasing scientific concern about SO_x and soot emissions. The impact of aerosol and cloud formation from these species could potentially have a large but currently uncertain impact on climate change.

Air Quality and Emissions Trading

Emission trading is a market-based alternative to conventional regulation, in which sources facing high costs to control pollution have the flexibility to meet their emission limits by purchasing excess reductions from other sources that can afford to lower their emissions further than federal or state regulations of the USA require. Currently, a trading program is being implemented at the federal level of USA to reduce acid rain. At the state level, California is operating a trading program to assist the Los Angeles area in complying with the federal air quality standard for ozone (Bearden, 1999). Although specific savings are difficult to estimate, initial monitoring data indicate that these programmes likely have reduced overall emissions at lower costs than conventional regulation might have done. Their early success has encouraged the consideration of trading to address other air quality concerns, including regional problems with ozone transport, possible increases in pollution if the electric utility industry is restructured, private sector compliance with EPA's Tier 2 proposal to reduce emissions from new motor vehicles, reducing vehicular pollution through telecommuting, and proposals to reduce greenhouse gas emissions as would be required by the Kyoto Protocol. Whether trading would be effective in addressing these issues depends on program design, implementation, enforcement, and accurate monitoring data to verify that transactions between pollution sources represent actual reductions in emissions.

Conventional air quality regulations place fixed limits on emissions from individual sources and require them to install specific technologies to control pollution. However, compliance may be more economically feasible for some sources than it is for others. An emission trading is a market-based alternative to conventional regulation which permits sources facing high pollution control costs to meet their emission limits by purchasing excess reductions from other sources that can afford to lower their emissions further than federal or state regulations require. Trading has the potential to improve air quality in cases where a pollutant disperses over a broad geographic area and the environmental objective is to control total emissions rather than limit local emissions from individual sources. In the United States, experience with trading programs began in the 1970's when the Environmental Protection Agency (EPA) used its authorities under the Clean Air Act to develop more flexible policies on emission control measures that states could pursue to comply with federal air quality standards. Most trading programs have focused on pollutants generated from large stationary sources that account for a significant share of total emissions, such as electric utility plants and industrial boilers that burn fossil fuels.

For diffuse pollutants, decreasing total emissions over large areas through trading has the potential to improve overall air quality. For example, trading is well suited for pollutants such as nitrogen oxides (NOx), sulfur oxides (SOx), and carbon dioxide (CO_2) because they can drift far from their points of release. However, trading is less suited for pollutants that remain locally concentrated because excess reductions in one area would not improve air quality elsewhere, and localized problems with air quality could arise from trading if sources clustered in one area are permitted to offset substantial increases in pollution with credits or allowances. Whether trading can improve overall air quality depends on many factors related to program design, implementation, and enforcement.

Emissions Trading Programs for Clean Air in USA

While early trading programs were based mostly on credits, allowance programs with emissions caps have become more common. Although exact savings are difficult to estimate, recent experiences with allowance trading under the federal acid rain program and in California's efforts to control ozone in the Los Angeles area indicate that these programs likely have reduced overall emissions at lower costs than conventional regulation. Whether an allowance program to control ozone transport in 12 northeastern and Mid-Atlantic States, which began in May 1999, will achieve similar results depends in part on the cooperation of the states involved. Each program is discussed below.

Federal Acid Rain Program: To combat acid rain, the Clean Air Act Amendments of 1990 established an allowance program to reduce emissions of sulfur dioxide (SO_2). The program capped national emissions at roughly 50% of 1980 levels. Phase I of the program began in 1995 and involves over 400 coal-fired electric utility generating units. Phase 11 begins in 2000 when the number of participating units will increase to over 2,000. EPA reports that actual emissions have been below the established cap each year from 199510 1998 (EPA, 1998). However, emissions have remained below the cap because utilities banked about 31% of total allowances during this time. Consequently, emissions could rise in coming years if utilities decide to use their banked allowances to achieve compliance. As the program enters its second phase, the substantial increase in the number of participating sources also could make the program more difficult to administer and perhaps reduce its effectiveness.

Regional Clean Air Incentives Market: In January 1994, California's South Coast Air Quality Management District introduced the Regional Clean Air Incentives Market (RECLAIM) in the Los Angeles area. RECLAIM is an allowance program aimed at reducing emissions of NOx and SOX. The program's goal is to achieve an 80% reduction in emissions by 2003. Stationary sources that annually emit at least 4 tons of NOx or SOX receive allowances, which are based on past peak production and existing emission requirements. The total number of allowances is equal to the cap on emissions. The state reports that actual emissions have been below the cap each year from 1994 to 1997 (RECLAIM Audit Report for the 1997). However, some critics argue that actual reductions in emissions have been less dramatic than a comparison to the cap suggests because emission allowances were initially high.

Interstate Trading: In 1994, the Ozone Transport Commission, representing 12 mid-Atlantic and northeastern states, the District of Columbia, and EPA, agreed to implement an interstate trading program that caps NOx emissions at less than half of 1990 levels to assist the region in complying with the federal air quality standard for ozone. The first phase of the program began in May 1999 and allows large electric utilities and industrial boilers within the region to trade NOx allowances to meet their emission limits (http://ncseonline.org/nle/crsreports/air/air-27.cfm). The second phase of the program is scheduled to begin in 2003 when the emissions cap will be lowered. EPA will record each trade and track monitoring data on actual emissions. Each participating state will be responsible for allocating allowances to pollution sources and enforcing program requirements within its borders. Since enforcement will be a state role, the effectiveness of the program largely will depend on the combined efforts of the states in carrying out their implementation responsibilities.

Milestones in Auto Emissions Control in USA

Air pollution and cars were first linked in the early 1950's by a California researcher who determined that traffic was to blame for the smoggy skies over Los Angeles. At the time, typical new cars were emitting nearly 13 grams per mile hydrocarbons (HC), 3.6 grams per mile nitrogen oxides (NOx), and 87 grams per mile carbon monoxide (CO). Since then, the Federal Government has set standards to bring down levels of these pollutants, and the auto industry has responded by developing new emission control technologies. The current Federal certification standards for exhaust emissions from cars are 0.25 gram per mile HC, 0.4 gram per mile NOx, and 3.4 grams per mile CO. The standard for evaporative HC emissions is 2 grams per test. The improvements came about in a series of steps (USEPA, 1994):

- 1964 California requires minimal emission control systems on 1966 model cars.
- 1966 Congress requires minimal emission controls on all 1968 and later cars.
- 1970 Congress adopts the first major Clean Air Act, establishes the U.S. Environmental Protection Agency (EPA), and gives the new Agency broad responsibility for regulating motor vehicle pollution. The clean air law calls for 90 percent reductions in automotive emissions. New cars must meet a 0.41 gram per mile HC standard and a 3.4 grams per mile CO standard by 1975; NOx emissions must be reduced to 0.4 gram per mile by 1976. The law also directs EPA to set health-based "National Ambient Air Quality Standards" for six pollutants, all of which are present in auto emissions to some degree.
- 1971 New cars must meet evaporative emission standards for the first time; charcoal canisters to trap gasoline vapors appear.
- 1972 Exhaust gas recirculation (EGR) valves appear as automakers strive to meet NOx standards.

1974 Congress delays the HC and CO standards until 1978 and sets interim standards at the request of the auto industry. Congress adopts the Energy Policy Conservation Act, setting the first fuel economy goals. The Corporate Average Fuel Economy (CAFE) program establishes a phase-in of more stringent fuel economy standards beginning with 1975 models.

1975 The first catalytic converters appear, and with them, unleaded gasoline, in response to HC and CO standards.

1977 Congress amends the Clean Air Act. At the request of automakers, the HC standard is delayed until 1980 and the CO standard until 1981. The NOx standard is delayed until 1981 and is relaxed to 1 gram per mile.

1981 New cars meet the amended Clean Air Act standards for the first time. Sophisticated three-way catalysts with on-board computers and oxygen sensors appear in most new cars.

1983 Inspection and Maintenance (I/M) programs are established in 64 cities nationwide, requiring passenger vehicles to undergo periodic testing for malfunctioning emission control systems.

1985 EPA adopts stringent emission standards for diesel-powered trucks and buses, to take effect in 1991 and 1994.

1989 EPA for the first time sets fuel volatility limits aimed at reducing evaporative emissions.

1990 EPA imposes strict limits on diesel fuel sulfur content to help buses and trucks meet the 1985 emission standards (which become effective in the early 1990's).

Congress amends the Clean Air Act to require further reductions in HC, CO, NOx, and particulate emissions. The amendments also introduce a comprehensive set of programs aimed at reducing pollution from motor vehicles. These include lower tailpipe standards; more stringent emission testing procedures; expanded I/M programs; new vehicle technologies and clean fuels programs; transportation management provisions; and possible regulation of emissions from non-road vehicles.

1991 EPA promulgates lower tailpipe standards for HC and NOx as required by the 1990 Clean Air Act. These standards take effect beginning with 1994 models. 3 of 2 Milestones are in Auto Emissions Control.

1992 Standards setting emission limits for carbon monoxide at cold temperatures (20°F) are established for the first time. Oxygenated gasoline is introduced in cities with high CO levels.

1993 Limits on sulfur content of diesel fuel take effect, enabling catalyst technology to reduce diesel particulate emissions.

1994 Phase-in begins for cleaner vehicle standards and technologies required by the 1990 Clean Air Act.

Emissions Control during the Beijing Olympics

A new study has determined that the air in Beijing during the 2008 Olympics was cleaner than the previous year's, thanks to aggressive efforts by the Chinese government, which could help policymakers adopt pollution-curbing policies to improve air quality (http://in.news.yahoo.com/139/20090726/981/tsc-emission-control-during-beijing-olym.html).

Led by Max Zhang, assistant professor of mechanical and aerospace engineering, Cornell University, the study indicated that such measures as regulating traffic density and encouraging public transportation can have a significant impact on local air quality. As per the 'Mission Approach' the study helped local regulators and policymakers to adopt long-term sustainable emission controls to improve air quality. The study was based on air quality readings before, during and after the Olympics. Leading up to the Olympics, the Chinese government barred more than 300,000 heavy-emission vehicles - mostly trucks—from the roads. The city also implemented rules in which only some people were allowed to drive on certain days based on their license plate numbers. As a result, close to 2 million vehicles were pulled from the roads. The other mandates involved halting construction and decreasing the use of coal in favour of natural gas for electricity.

In 2007 and 2008, the researchers collected air quality data from equipment installed at two elevations on a building in the heart of Beijing. They also tracked emissions from vehicles in different areas of the city by following randomly selected cars and trucks in a minivan equipped with sensitive instruments for detecting carbon particles, including carbon monoxide, carbon dioxide (CO_2) and black carbon, or soot. The researchers concluded that black carbon pollution is significantly greater at ground level than at higher elevations, and diesel trucks are a major source of black carbon emission during the summer in Beijing. The researchers found that car emissions of black carbon were down 33 percent in 2008 compared with their 2007 readings. CO_2 decreased 47 percent, and ultrafine carbon-based particles - those that measure less than 100 nano meters - decreased 78 percent. The sharp drops were most likely due to a new emission standard implemented in Beijing in 2008, in which all new registered vehicles as well as gasoline and diesel fuel engines were required to achieve emissions standards equivalent to European Union regulations. A similar standard was mandated starting in June 2008 for 20,000 buses and 66,000 taxies. The improved fuel quality probably enhanced the performance of engines and catalytic converters, the researchers reported.

CHAPTER 2

Emissions Trading and Accounting of Carbon Credits

Emissions Trading: The Definition and Concept

Emissions trading is an administrative approach used to control pollution by providing economic incentives for achieving reductions in the emissions of pollutants (http://en.wikipedia.org/wiki/Emissions_trading). The term "emissions trading" refers to "a market where, for specified pollutants, parties can buy or sell allowances or permits for emissions, or credits for reductions in emissions" (Canada National Round Table, 2001). Because emissions-trading uses markets to determine how to deal with the problem of pollution, it is often touted as an example of effective free market environmentalism. Emissions trading might happen in a region, country or internationally.

Emissions Trading refers to the creation of surplus emission reductions at certain stacks, vents or similar emissions sources and the use of this surplus to meet or redefine pollution requirements applicable to other emissions sources. This allows one source to increase emissions when another source reduces them, maintaining an overall constant emission level. Facilities that reduce emissions substantially may "bank" their "credits" or sell them to other facilities or industries (http://www.ecomii.com/dictionary/emissions-trading).

Emissions trading concept also allows the existence of a *Tradable-permit System* in which a greenhouse gases emitter (firm or country under obligation to limit its total air pollution emissions to a specified level) can buy/sell permission to emit a certain amount of emissions from/to other emitters (who are below/above their limit). The market price of these permits (called 'pollution credits') reflects the marginal cost of emission reduction and gives an emitter the incentive to install and manage a cost effective pollution control system as an income producing asset (http://www.businessdictionary.com/definition/emissions-trading.htm).

International trade in greenhouse gas emissions is specifically provided for in the 1997 Kyoto Protocol. At the heart of the Protocol is an undertaking by nearly all the industrialized countries to reduce their emissions by a fixed percentage below what they were in 1990. These reductions, which are to be made over the period 2008-2012, will be costly for many countries. Since energy production and use produce the lion's share of greenhouse gas emissions, especially CO_2, the responsibility for achieving emission reductions will fall most heavily on the energy sector. In an effort to mitigate those costs, the framers of the Protocol proposed three innovative "market mechanisms," including emission trading. Central to any future trading regime will be the notion of avoided emissions, negotiable units of account that represent tonnes of CO_2 in the atmosphere. Countries will be able to trade these avoided emissions much as other commodities are traded (IEA, 2001).

The starting point for an assessment of the relative cost of emission reductions by different countries is to look at their emission reduction objectives. For energy related CO_2 emissions, national circumstances —such as patterns of energy production and use, energy intensity, energy prices and also economic growth and population—are important determinants of the cost of abatement. A rough estimate of the effort required of different countries can be made by comparing their emission reduction objectives with the evolution of their emissions to date. A country which has a large reduction target and whose emissions have historically increased will have to make a great effort.

In a carbon market, tradable units are either allocated by a regulatory body, or produced by GHG emission reduction projects (Bayon, *et al*, 2007). The former system often involves a "cap-and-trade" scheme, where a regulator caps participants' allowable emission levels, then issues tradable units equal to the cap. Participants who emit less GHG than allowed can sell their left-over allowances at market price. The emission cap produces the environmental benefit and might be lowered over time. Project-based transactions involve buyers purchasing emission credits from a project that can "demonstrate GHG emission reductions compared with what would have happened otherwise" (Robin Clarke, 1996). There is also an expanding retail segment that sells emission reductions to individuals and companies wanting to voluntarily offset their "carbon emission footprints." According to the World Bank's 2007 report on carbon markets: "Reports of increased interest of banks, credit card issuers, private equity funds and others in this segment suggest that it could grow exponentially if only there were a credible, voluntary standard for such assets"(Capoor, 2007).

With tradable permit systems for greenhouse gas (GHG) emissions in place in some parts of the world and actively being considered in others, increasing attention has been given to the opportunity to link these systems. Linking occurs when the government that maintains one system allows regulated entities to use allowances or credits from another system to meet domestic compliance obligations. Linking creates opportunities for inter-system trading that can reduce the aggregate cost of meeting emission targets by broadening allowance and credit markets, and by permitting emission reduction efforts to be redistributed across linked systems. In principle, this flexibility can be offered to regulated entities without compromising the effectiveness of climate change mitigation efforts because emission reductions have the same effect on atmospheric GHG concentrations no matter where in the world they occur.

While linking can reduce the long-run cost of mitigating climate change, it also has other implications. In particular, linking can have distributional and emissions implications and can reduce a government's control over the impacts of its tradable permit system. Thus, in considering linkages, governments may have to weigh linking's implications for potentially competing policy objectives, much as will be required in developing other elements of climate policy. Because linking's implications depend on the type of link that is established and the specific characteristics and design of the linked systems, in the near-term, some links will be more attractive and easier to establish than others. Importantly, those links that may be the easiest to establish—links between cap-and-trade systems and emission reduction credit systems such as the Clean Development Mechanism—likely can provide much of the near-term cost-saving and risk-diversifying advantages that linking can offer..

Carbon Credits—The Fundamental Driver of Emissions Trading

Carbon credits correspond to a determined tradable quantity of greenhouse gas (GHG) emissions. They are essentially a permit to emit one metric tonne of CO_2 into the atmosphere. The goal is to achieve a zero net increase in GHG emissions, because each tonne of increased emissions is 'offset' by project-based GHG reductions (http://en.wikipedia.org/wiki/Carbon_credit). They are a key component of national and international emissions-trading schemes. Credits can be exchanged between businesses or individuals, or bought and sold in international markets at the prevailing market price. Carbon credits create a potential market for reducing greenhouse emissions by giving a monetary value to emissions, in effect making emissions a commodity. Credits can be used to finance carbon reduction schemes between trading partners and around the world. They are normally quoted in metric tonnes of carbon dioxide equivalent, and are used to offset emissions from any process that emits GHGs, whether in industry, transportation or the household. Carbon credits are used in signatory countries to the Kyoto Protocol to meet emission reduction targets announced annually through international conventions. Carbon credits are created when GHG emissions are reduced below some business-as-usual baseline. This baseline allowance is set by governing bodies such as the United Nations (through the UN Framework Convention on Climate Change) or, in North America, the Chicago Climate Exchange. The basic requirements are that emission reductions are measurable and last at least five years. The potential agency problem is addressed through the mandatory use of independent third parties to verify that emission reductions are real (UNFCC, 2005).

Carbon Currency

A way of assigning value to emissions reduction that is central to the economics of carbon trading. Under the Kyoto Protocol, credits may be emissions allowances originally allocated or auctioned as part of a national cap-and-trade program or emissions offset (http://opencarbonworld.com/wiki/carbon-credit.html). In a cap-and-trade scheme, policymakers set a cap on total emissions and the market sets a price for emissions through trading. Carbon credits are permits that represent emissions reductions. These credits are awarded to countries or firms that emit at or have reduced their emissions to a level below the quota or cap assigned to them by an emissions regulatory regime. Emitters above the cap purchase credits in the carbon market. Credits are bought and sold in international markets and can be exchanged between businesses. The marketplace finds the monetary value of greenhouse gas emissions. It is also possible for individuals and companies who would like to reduce their carbon footprint to purchase carbon credits in voluntary markets. Sellers in the voluntary market aggregate credits from individual carbon projects. These units tend to face a less rigorous validation process than credits sold through the Clean Development Mechanism and therefore have less value because of their lower quality.

Carbon Credits under Kyoto Protocol

The Kyoto Protocol established caps on the maximum quantity of greenhouse gas emissions permitted for Annex I developed and developing countries. These countries set internal quotas on emissions and oversee this responsibility through their own national 'registries', which are required to be validated and monitored for compliance by the UNFCCC. Each operator of a polluting installation is allocated an allowance of credits. Generally, each unit gives the owner the right to emit one metric ton of CO_2e. Operators that have not met their quotas can sell their unused allowances as carbon credits, while businesses that are about to exceed their quotas can buy the extra allowances as credits, privately or on the open market. Businesses alter their decision-making to find the most cost-effective way of operating under these regulations, either by investing in cleaner business practices or by purchasing credits from another operator with excess capacity.

The European Union Emission Trading Scheme and its member states follow the Kyoto template. The "Linking Directive" permits the use of Kyoto flexible mechanisms within the EU ETS. Similar schemes are under consideration in the United States, which has not ratified Kyoto, and in Australia, whose ratification came into force in March 2008.

The Kyoto Protocol includes three flexible mechanisms which offer countries or operators in developed countries means of acquiring carbon credits.

- Under Joint Implementation (JI), a developed country with relatively high costs of domestic emission reduction can fund carbon projects in another developed country.
- Under the Clean Development Mechanism (CDM), a developed country can sponsor a greenhouse gas reduction project in a developing country where the cost of greenhouse gas reduction project activities is usually much lower, but the atmospheric effect is globally equivalent. The developed country is given credits towards meeting its emission reduction targets, while the developing country receives the benefits of sustainable development and investment.
- Under International Emissions Trading (IET), countries can trade in the international carbon credit market to cover their shortfall in allowances. Countries with surplus credits can sell them to countries with capped emission commitments.

These carbon projects can be created by a national government or by an operator within the country. In practice, most transactions of this type are not performed by national governments directly, but by operators who have been set quotas by their country.

Emissions Market

For trading purposes, one allowance or CER is considered equivalent to one metric tonne of CO_2 emissions. These allowances can be sold privately or in the international market at the prevailing market price. These trade and settle internationally and hence allow allowances to be transferred between countries. Each international transfer is validated by the UNFCCC. Each transfer of ownership within the European Union is additionally validated by the European Commission.

Climate exchanges have been established to provide a spot market in allowances, as well as futures and options market to help discover a market price and maintain liquidity. Carbon prices are normally quoted in Euros per tonne of carbon dioxide or its equivalent (CO_2e). Other greenhouse gases can also be traded, but are quoted as standard multiples of carbon dioxide with respect to their global warming potential (GWP). These features reduce the quota's financial impact on business, while ensuring that the quotas are met at a national and international level.

Currently, there are at least six exchanges trading in carbon allowances: the Chicago Climate Exchange, European Climate Exchange, Nord Pool, PowerNext, Multi Commodity Exchange and National Commodity and Derivatives Exchange. Recently, NordPool listed a contract to trade offsets generated by CDM carbon projects called Certified Emission Reductions

(CERs). Many companies now engage in emissions abatement, offsetting, and sequestration programs to generate credits that can be sold on one of the exchanges.

Unchecked, energy use and hence emission levels are predicted to keep rising over time. Thus the number of companies needing to buy credits will increase, and the rules of supply and demand will push up the market price, encouraging more groups to undertake environmentally friendly activities that create carbon credits to sell.

An individual allowance, such as a Kyoto Assigned Amount Unit (AAU) or its near-equivalent European Union Allowance (EUA), may have a different market value to an offset such as a CER. This is due to the lack of a developed secondary market for CERs, a lack of homogeneity between projects which causes difficulty in pricing, as well as questions due to the principle of supplementarity and its lifetime. Additionally, offsets generated by a carbon project under the Clean Development Mechanism are potentially limited in value because operators in the EU ETS are restricted as to what percentage of their allowance can be met through these flexible mechanisms.

Emissions Trading Modalities: Cap and Trade System

Emissions-trading is an administrative approach used to control environmental pollution by providing economic incentives for achieving reductions in the emissions of pollutants. It is sometimes called *cap and trade.* The overall goal of an emissions trading plan is to reduce emissions. The cap is usually lowered over time—aiming towards a national emissions reduction target. In other systems a portion of all traded credits must be retired, causing a net reduction in emissions each time a trade occurs. In many cap and trade systems, organizations which do not pollute may also participate, thus environmental groups can purchase and retire allowances or credits and hence drive up the price of the remainder according to the law of demand. Corporations can also prematurely retire allowances by donating them to a nonprofit entity and then be eligible for a tax deduction.

Kyoto is a 'cap and trade' system that imposes national caps on the emissions of Annex I countries. On average, this cap requires countries to reduce their emissions 5.2% below their 1990 baseline over the 2008 to 2012 period. Although these caps are national-level commitments, in practice most countries will devolve their emissions targets to individual industrial entities, such as a power plant or paper factory. National governments, some of whom may not have devolved responsibility for meeting Kyoto obligations to industry, and that have a net deficit of Allowances, will buy credits for their own account, mainly from JI/CDM developers. These deals are occasionally done directly through a national fund or agency, as in the case of the Dutch governments ERUPT programmes, or via collective funds such as the *World Bank's Prototype Carbon Fund* (PCF).

Kyoto enables a group of several Annex I countries to join together to create a so-called 'bubble', or a cluster of countries that is given an overall emissions cap and is treated as a single entity for compliance purposes. Several non-Kyoto carbon markets are already in existence as well, and these are likely to grow in importance and numbers in the coming years. These include: the New South Wales Greenhouse Gas Abatement Scheme, the Regional Greenhouse Gas Initiative (RGGI) in the United States, the Chicago Climate Exchange, the State of California's recent initiative to reduce emissions, the commitment of hundreds of US mayors to adopt Kyoto targets for their cities, and the State of Oregon's emissions abatement program. Taken together—these initiatives point towards a series of linked markets rather than a single carbon market. Important decisions that are to be made about what a cap-and-trade program will look like, including:

- What emissions will be capped by the program (*scope of program*), and what entities will be required to hold allowances equivalent to emissions (*point of regulation*)
- The level of the emissions cap (*stringency*)
- Whether provisions will be included to help ensure the costs of the program do not get too high or volatile (*cost containment mechanisms*)
- Whether the program should be linked with similar trading programs (*linkage*) how allowances are to be distributed (allowance *distribution*).

The first step in setting up a cap-and-trade program is deciding which greenhouse gases and emissions sources are covered and who is responsible for holding allowances. Some sectors that might be included under the cap are electric power, manufacturing, transportation, or fossil fuel use. One of the recommendations of the Kyoto protocol is that a *cap-and-trade* scheme is established in a signatory country. It would work like this: companies are told how much CO_2 they can emit (the cap). If they produce less than the cap, they have surplus credits for sale. If they emit more than their cap they can buy credits from other businesses that come in under their cap (the trade). Trade takes place in an over the counter market, or via a Carbon Credit Exchange trading market (http://www.cmawebline.org/documents/JAMAR-v6.1-%20Carbon%20Strat%20Mang%20Accounting.pdf).

For a proper understanding on the modalities of emission trading a good understanding of *carbon-economics* encompassing a number of elements of business accounting and finance are required, such as the economic modeling of demand and supply of carbon credits and allowances, forward and spot pricing, risk management business value, capital allocation and the possible International Financial Reporting Standards (IFRS, 2002) directives for accounting for carbon and related transactions. In addition, taxation issues of direct carbon taxes, value-added (VAT) and goods and services (GST) taxes, and transfer pricing implications of carbon trading need to be considered.

Policymakers have many options as they consider how to achieve greenhouse gas (GHG) reductions, but two approaches are most prominent: traditional command-and-control regulation, in which regulatory authorities direct how emissions limits will be achieved, and market-based approaches, which harness the forces of supply and demand to change behavior and achieve environmental goals. One proven market-based approach is cap and trade.

In a cap-and-trade program, the government determines which facilities or emissions are covered by the program and set an overall emission target, or "cap," for covered entities. This cap is the sum of all allowed emissions from all included facilities. Once the cap has been set and covered entities specified, tradable emissions allowances (rights to emit) are distributed (either auctioned, or freely allocated, or some combination of these). Each allowance authorizes the release of a specified amount of greenhouse gas emissions, generally one ton of carbon dioxide equivalent (CO_2e). The total number of allowances is equivalent to the overall emissions cap (e.g., if a cap of one million tons of emissions is set, one million one-ton allowances will be issued). Covered entities must submit allowances equivalent to the level of emissions for which they are responsible at the end of each of the program's compliance periods.

Allowance trading occurs because firms face different costs for reducing emissions. For some emitters, implementing low-emitting technologies may be relatively inexpensive. Those firms will either buy fewer allowances or sell their surplus allowances to firms that face higher emission control costs. Since a ton of carbon dioxide (CO_2) emitted from one source has the same warming effect as a ton emitted from any other, the location of a given emissions reduction does not matter. By giving firms a financial incentive to control emissions and the flexibility to determine how and when emissions will be reduced, the capped level of emissions is achieved in a manner that minimizes overall program costs.

Although a critical and effective component of any comprehensive solution to climate change, cap-and-trade programs alone cannot achieve the GHG emission reductions required to stabilize the climate. Addressing climate change requires a combination of market mechanisms with other policy measures, including incentives and standards. For example, in order to begin rapidly cutting emissions, certain technologies may require additional supportive policies to push them to their full potential. In addition, some emission sources of GHGs cannot easily be covered by a cap-and-trade program and will need to be addressed using other policies.

Figure 1(A&B) illustrates the economic benefits of trading by means of a simplified example by means reducing the emissions (CO_2) of a Company A by 1000 MT @ $ 2/MT and selling the same to a Company B which ca reduce emissions @ $ 6/MT; resulting in Profit of Company A by $ 2000 and savings of Company B by $ 2000 correspondingly. It also gives an overview of Emission Trading under a regulated cap and trade system.

A key advantage of cap and trade (and market mechanisms in general) is that it provides an incentive for continuous innovation in emissions reduction.

Fig. 1(A&B): Cost-Effective Options of Selling and Buying Carbon Credits

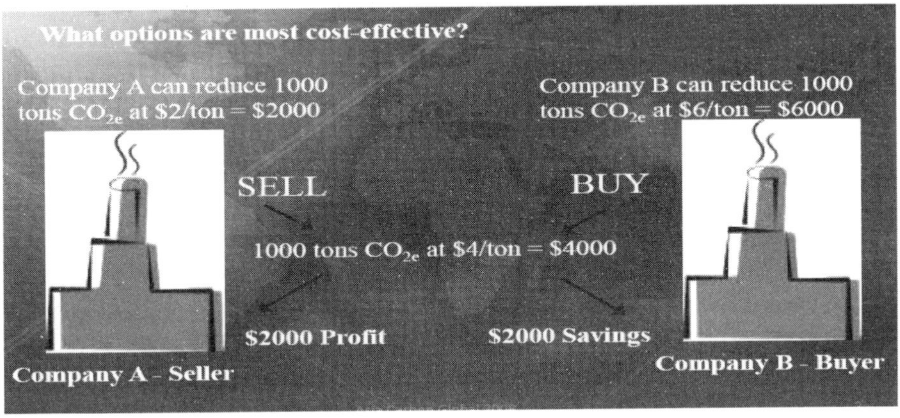

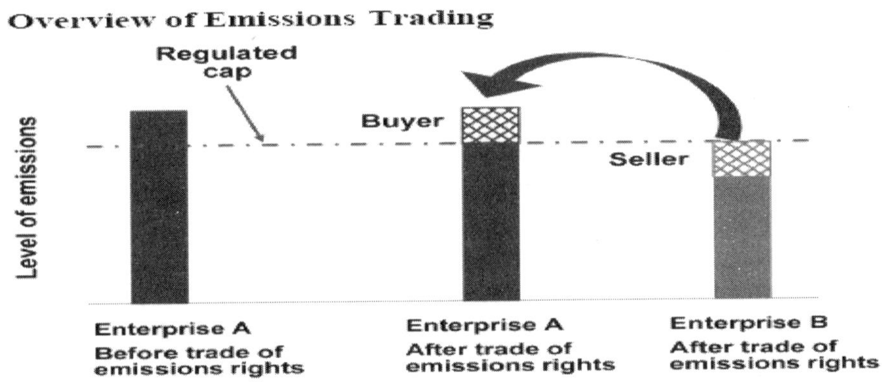

Under traditional command-and-control regulation, there is no incentive to go beyond the regulatory standard. In fact, there may actually be a disincentive to do so because demonstrating the feasibility of additional effort may result in more stringent future regulation. In a cap-and trade-program, a firm that can reduce emissions at a cost lower than the allowance price either reduces its compliance cost (because fewer allowances need to be purchased) or frees up allowances that can be sold to others. This financial incentive drives the private sector to continually innovate and seek new emission-reducing technologies that regulators might not anticipate under more prescriptive command-and-control regulations. In a market system, such emission-reducing innovation can set the stage for deeper emission cuts over time. This is particularly important because meeting the challenge of climate change will require new technology to achieve the very deep emission cuts that are necessary.

Cap and Trade Market Design

Important decisions have to be made about what a cap-and-trade program will look like, including what emissions will be capped by the program (• *scope of program*), and what entities will be required to hold allowances equivalent to emissions (*point of regulation*); the level of the emissions cap (• *stringency*); whether provisions will be included to help ensure the costs of the program do not get too high or volatile (*cost containment mechanisms*); whether the program should be linked with similar trading programs (*linkage*); and how allowances are to be distributed (• *allowance distribution*).

Scope and Point of Regulation: The first step in setting up a cap-and-trade program is deciding which greenhouse gases and emissions sources are covered and who is responsible for holding allowances. Some sectors that might be included under the cap are electric power, manufacturing, transportation, or fossil fuel use. In theory, market-based programs are most cost-effective if they cover all GHGs in all major emitting sectors because including more sources and greenhouse gases offers a broader range of opportunities for low-cost reductions. However, including sources that are small or difficult to monitor can make the program too administratively complex; these sources may be addressed more efficiently through other regulatory mechanisms.

After deciding which emissions are covered by the program, policymakers must decide who is responsible for surrendering enough allowances to match their emissions every compliance period. This is known as the "point of regulation," where compliance is demonstrated by submitting allowances. Which entities are required to submit allowances to cover emissions determines whether a cap-and-trade system is defined as an upstream, downstream, or a product- or load-based program.

Upstream: A pure upstream, economy-wide system for CO_2 would place a cap on the total amount of carbon contained in fossil fuels and other products used in the economy. It would require importers or suppliers of fossil fuels to submit allowances to cover the carbon in the products they sell. The key argument made in favor of an upstream approach is that one can achieve greater coverage of emissions at a smaller number of sources.

Downstream: Under a downstream, source-based system, the covered entities are direct emitters of greenhouse gases (generally large emitters such as power plants or manufacturing facilities). The key arguments in favor of a downstream approach are that (1) to-date, most experience with cap and trade has been based on downstream regulation, and this institutional familiarity makes such an approach less risky; and (2) downstream sources generally have more emission reduction options available, and are thus in a better position to respond to the requirements.

Product or Load-Based Cap and Trade System

In a product or load-based cap-and-trade system, the covered entities are responsible for all the emissions associated with the production of electricity, natural gas, or other product that they provide to customers. Many argue that, regardless of which entities are covered, an effective cap-and trade-program should follow some basic design criteria, such as:

- Simple, consistent, and transparent rules;
- Accurate emissions measurement, monitoring, and reporting, preferably done electronically and including Public access to emissions data;
- Sound auditing practices to ensure that emissions are being accurately reported;
- Consistent enforcement with real penalties for non-compliance and inaccurate reporting; and
- Unrestricted trading of allowances and minimal transaction costs.

An offset represents an emissions reduction credit generated by an entity not included under a cap that can be sold to cap entities and used in the same manner as an allowance. In Figure 2 above, Emitter A is included under a cap-and-trade system with an overall cap of 300 tons. If offsets are permitted under the program, an entity outside of the cap (Emitter B) can make an emissions reduction of 100 tons, creating 100 reduction credits that can be purchased by Emitter A to offset a 100 ton increase in its own emissions. Although emissions from the capped entity total 400 tons, Emitter B offsets 100 of those tons, so that, on net, the same emission reductions are achieved.

Existing and Developing Emissions Market Instruments

The existing as well as Developing Market Instruments for Emission Trading may be summed-up as follows:

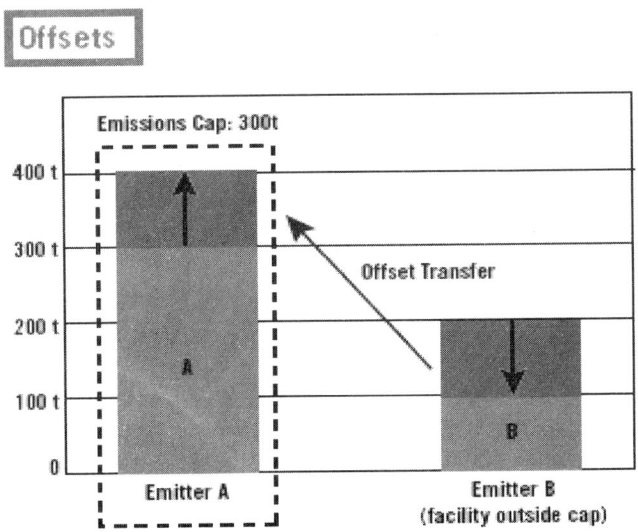

Fig. 2: Carbon Offsets Transfers under Cap and Trade System.

- Verified Emission Reductions (VERs) – Voluntary markets (compliance driven plus retail)
- Certified Emission Reductions (CERs)—primary market
- CERs (secondary market)
- Assigned Amount Units (AAUs) – International Emissions trading
- European Union Allowances (EUAs)—EUETS
- Emission Reduction Units (ERUs)—Joint Implementation
- Domestic Emissions Trading Units—still questionable about how this will develop in many countries such as those in:
  - Canada (permits and offsets) and provincial programs (ex. Alberta)
  - United States (RGGI, AB32, Western States Initiative

Allocating Carbon Allowances

National regulations specify which individual firms are covered by the cap—whether private companies, public sector installations, or both. Each firm is given its own cap consistent with an aggregate reduction in CO_2 emissions. The firms are then allocated the appropriate number of carbon allowances. At the end of the period covered by the regulations, each firm must either reduce its CO_2 emissions to a level at or below its individual cap, or deliver allowances equal to the excess amount of emissions. Firms that fail to meet their targets are fined.

Trading Facilitates Cost-effective Reductions

Once allowances are allocated, the stage is set for trading to begin. Each firm must determine the most cost-effective means to reduce (or abate) its carbon emissions. The firm may invest to improve its production processes. Or it may judge that it is cheaper to buy allowances in the carbon market sufficient to cover its excess emissions. Figure 2 provides an example of how this trading process works. Enterprise A has a level of CO_2 emissions that are above its regulated cap, and has limited or uneconomic abatement opportunities. Enterprise B has a low-cost opportunity to reduce CO_2 emissions and can therefore sell its excess allowances to Enterprise A. At the end of this transaction, both parties are within their caps. Physical trading in carbon instruments takes place either on an organized exchange or over the counter, similar to other commodities. A number of carbon exchanges have emerged in Europe and North America that trade spot, forward, and option contracts in carbon. In Canada, for example, the Montreal Exchange launched the Montreal Climate Exchange in 2006 to provide a location for Canadian firms to trade in CO_2 allowances. The Toronto Stock Exchange and the Winnipeg Commodity Exchange have also expressed interest in this new market.

Monitoring and Reporting to Ensure Accountability

As with any market, a cap-and-trade market may not develop efficiently without the necessary infrastructure. Experience suggests that a successful cap-and-trade system relies on the following elements:

- Registration of the ultimate owner of allowances or credits
- Independent monitoring of emissions from a facility
- Reporting of emissions to a central authority over a given period
- Verification of the level of emissions and confirmation of reductions

These components are critical for promoting both public and business confidence in the system, allowing the market to function properly, and minimizing uncertainty. While some of these functions may be provided by financial intermediaries, a central authority may have responsibility for some or all of the functions.

Reconciliation to Ensure Compliance

At the end of the period covered by the scheme, regulators must reconcile a firm's emissions against its holdings of allowances to ensure compliance. The firm must either deliver the equivalent number of allowances for its excess emissions, or pay some predetermined fine if they are short. This fine effectively sets an upper limit on prices for emissions. Under the EU-ETS, for example, the penalty is EUR 40 per missing allowance under Phase I (2005–07), and EUR 100 under Phase II (2008–12). Without effective compliance and enforcement, the incentives will not be in place to ensure an aggregate reduction in emissions. In the case of the U.S. Acid Rain Program, compliance has been very high, with over 99 per cent of firms meeting their targets, thanks to rigorous monitoring, reporting requirements, and stringent penalties. By 2007, the Environmental Protection Agency reported that sulphur dioxide emissions were 41 per cent below their 1980 levels, while emissions of nitrogen oxides were 57 per cent below their 2000 levels.

Emissions Trading and Potential Barriers

Emissions-trading strategy should utilize financial market best practice and spread market exposure over the year for trading a net long or net short position. This exposes the firm to an average price traded value over the year and not a single one off price. The larger the short or long exposure the closer the position should be actively managed by financially experienced personnel. Options, floating price forwards and swaps can also be used to ensure an average annual price exposure and minimize transaction costs. An effective and active trading strategy will help ensure market trends are tracked and market timing of transactions are effective. There are few barriers to cap and trade emissions trading so long as markets evolved especially across Europe. Liquidity has improved and one common market across Europe has evolved for emissions which is more akin to financial markets like foreign exchange than flow based commodity markets where there are still massive barriers to the creation of an effective pan European traded market (PowerGen Europe, 2006).

Emissions Caps in a Pluri-lateral Trading Regime

Countries participating in a pluri-lateral trading regime must establish enforceable domestic caps on emissions of gases from sources and sectors participating in the regime. These caps can be established either through an allowance system, or through a command regulatory system under which reductions by sources of emissions below the levels required by regulation are recognized as credits that can be traded to other sources to help them meet their regulatory requirements. The Kyoto Protocol to the United Nations Framework Convention on Climate Change authorizes four cooperative implementation mechanisms—bubbles, emissions trading, joint implementation and the clean development mechanism (CDM). The provision on emissions trading, the focus of this report, allows trading of 'assigned amounts' among the so-called Annex B Parties, while leaving the drafting of implementation details to subsequent conferences of the Parties. Three distinct trading possibilities emerge from this authorization: trading among countries with domestic emissions trading systems, trading among countries without domestic trading systems, and trading among countries with and without domestic emission trading systems. The case for a tradable entitlements system is based on the advantages that it would offer over other politically feasible alternatives. In the short term, it offers the possibility of reaching environmental goals at a lower cost than would be possible if each country were limited to reduction options within its own borders. Making it easier to reach these goals may encourage more countries to sign the Protocol, and should increase compliance with those goals. Finally, *Tradable entitlements* facilitate the development and implementation of novel approaches to climate change control (http://www.unctad.org/en/docs/poditctedm9.en.pdf).

A plurilateral GHG trading system should include the core essential requisites of a successful international emissions trading market, compatible with the different choices made by the different participating countries regarding the design of their domestic emissions trading systems, as discussed above. Fortunately, the core requisites are relatively few in number, and can accommodate substantial diversity in domestic trading system design. The key requisites of an international GHG emissions trading system are the same as for a domestic system: to establish enforceable limitations on net emissions by domestic sources covered by the system; to establish a homogenous and fungible TEU (tradeable emissions unit) commodity defined and protected by well-established property rights and by ownership rules that allow anyone to acquire or hold TEUs; to provide for free transferability of TEUs without prior regulatory review or approval; to accord recognition in each participating country of TEUs from the other participating countries for purposes of determining compliance by domestic sources with domestic limitations on their emissions; to establish an integrated system for tracking TEU trades and holdings; and to provide adequate assurances that net emissions are accurately determined and reported and that sources comply with their emissions limitation obligations (http://www1.law.nyu.edu/kingsburyb/spring03/globalization/SandsStewartPlurilateral_March24.pdf). Securing these requisites will, of course, be somewhat more difficult in an international context involving several countries than in a purely domestic situation, but the difficulties are far from insurmountable.

Carbon Market Statistics

The Carbon Market statistics in terms of volume and value of carbon credits traded during 2005 as well as in the first quarter of 2006 is summed up below in Table 1.

Table 1: Carbon Market at a Glance, Volumes and Values for 2005 and 2006 (until September 30)

	2005		Q1-3-2006	
	Volume ($MtCO_2$)	Value (MUS$)	Volume ($MtCO_2$)	Value (MUS$)
		Allowances		
FUETS	324.31	8,204.48	763.90	18,839.79
NSW	6.11	59.13	16.19	184.07
CCX	1.45	2.83	8.25	27.15
UK-ETS	0.30	1.31	2.26	9.27
Sub-Total	332.17	1.31	2.26	9.27
		Project-based Transactions		
CDM	359.08	2,651.44	214.26	2,260.96
JI	20.85	100.89	11.86	93.88
Other compliance	4.51	36.72	7.92	60.02
Sub-Total	384.44	2,789.05	234.05	60.02
Total	716.61	11,056.79	1,022.39	21,465.87

Source: The World Bank, State and Trends of Carbon Market, Washington D.C., 2006.

Increase in GHG Emissions since 1990

Given below (Table 2) is a list of the change in GHG emissions from 1990 to 2004 for some countries that are part of the Climate Change Convention as reported by the United Nations: http://unfccc.int/files/essentialbackground/background_publications_htmlpdf/application/pdf/ghg_table_06.pdf):

Comparing total greenhouse gas emissions in 2004 to 1990 levels, the US emissions were up by 16%, with irregular fluctuations from one year to another but a general trend to increase. At the same time, the EU group of 23 (EU-23) Nations had reduced their emissions by 5%. In addition, the EU-15 group of

Table 2: Change in GHG Emissions from 1990 to 2004 for some selected Countries

Country	Change in GHG Emissions (1990-2004)	EU Assigned Objective for 2012	Treaty Obligation 2008-2012
Germany	–17%	–21%	–8%
Canada	+27%	N/A	–6%
Australia	+25%	N/A	N/A
Spain	+49%	+15%	–8%
United States	+16%	N/A	N/A
Norway	+10%	N/A	+1%
New Zealand	+21%	N/A	0%
France	–0.8%	0%	–8%
Greece	+27%	+25%	–8%
Ireland	+23%	+13%	–8%
Japan	+6.5%	N/A	–6%
United Kingdom	–14%	–12.5%	–8%
Portugal	+41%	+27%	–8%
EU-15	-0.8%	N/A	–8%

nations (a large subset of EU-23) reduced their emissions by 0.8% between 1990 and 2004, while emission rose 2.5% from 1999 to 2004. Further complicating the debate over the Kyoto Protocol is the fact that CO_2 emissions growth in the US was far ahead of that of the EU-15 from 1990-2000, but from 2000-2004, America's rate of growth in CO_2 emissions was eight percentage points lower than from 1995-2000, while the EU-15 saw an increase of 2.3 points. From 2000-2004, the United States' CO_2 emissions growth rate was 2.1%, compared to the EU-15's 4.5%. That happened while the US economy was expanding 38% faster than the economies of the EU-15 while experiencing population growth at twice the rate of the EU-15. As of year-end 2006, the United Kingdom and Sweden were the only EU countries on pace to meet their Kyoto emissions commitments by 2010.

While the UN statistics indicate that, as a group, the 36 Kyoto signatory countries can meet the 5% reduction target by 2012, most of the progress in greenhouse gas reduction has come from the stark decline in Eastern European countries' emissions after the fall of communism in the 1990s. The desirability to cut on emissions by the developed economies vis-à-vis developing countries will be evident from the fact that both the quantum of emissions as well as per capita emissions are a lot higher in respect of USA, Canada, Russia, Japan and to some extent Germany and the UK as compared to China and India as may be seen from the following Statistics (Table 3) of UNFCC (2004).

Table 3: Total CO_2 Emissions vs. per Capita Emissions of Developed and Developing Economy

Countries	Total CO_2 Emissions (Billion MT in '04)	per Capita Emissions (MT in '04)
USA	6.049	20.04
Canada	0.639	20.00
Russia	1.524	10.50
Japan	1.254	9.80
Germany	0.860	9.70
U.K.	0.587	9.70
China	5.010	3.80
India	1.342	1.20

The World's Biggest Carbon Market

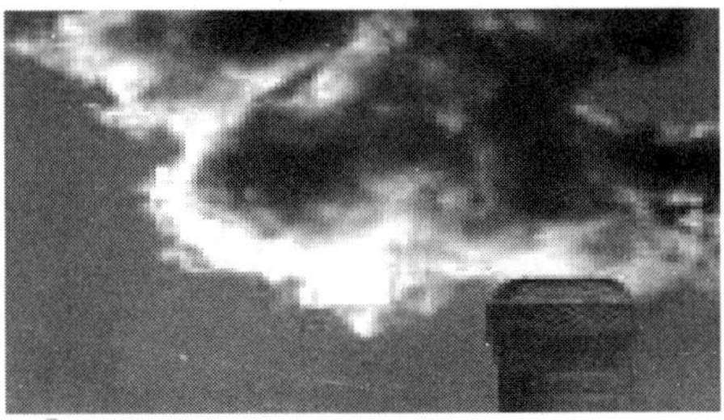

The market for CDM credits is growing fast. At present it is worth nearly $20bn a year, but this is expected to grow to over $100bn within four years. More than 1,000 projects have so far been approved, and 2,000 more are making their way through the process (http://www.buzzle.com/articles/197644.html).

The CDM certificates facilitate the funding of clean technology investments by Third World companies that are expanding their operations. Western companies can buy the certificates to offset their own pollution. But it turns out that in reality most of the funds go to coal and oil companies, builders of destructive dams and other enterprises that are not green in the slightest. The research that revealed the practices is of major importance not least because policymakers are set to review the CDM in the near future as the Kyoto Protocol expires in 2012. CDM credits are the world's largest offset market, with annual trading last year totaling around EUR40 billion. Most credits are currently traded on the European Trading System (ETS) by European countries and companies but when the US starts to participate, something that's more or less a given, trading will rise to over EUR 100 billion within two years easily.

Carbon has the potential, as stated in one of the NY Times article, to become the world's biggest market. Here's why. Consider this: Every year humans generate about 38 billion tons of carbon dioxide. Now take a look at a chart of the price of carbon on the Chicago Climate Exchange (Fig.3).

At its current price of about $3.50 per ton, the potential carbon market stands at roughly $133 billion (38 billion x $3.50). But only a fraction of those emissions are regulated. For starters, one can invest in companies that reduce emissions simply by the nature of their business. Companies that produce clean energy will soon be profiting on two fronts—they'll be selling their power *and*

Fig. 3: Price Movement of Carbon Credits on the Chicago Climate Exchange

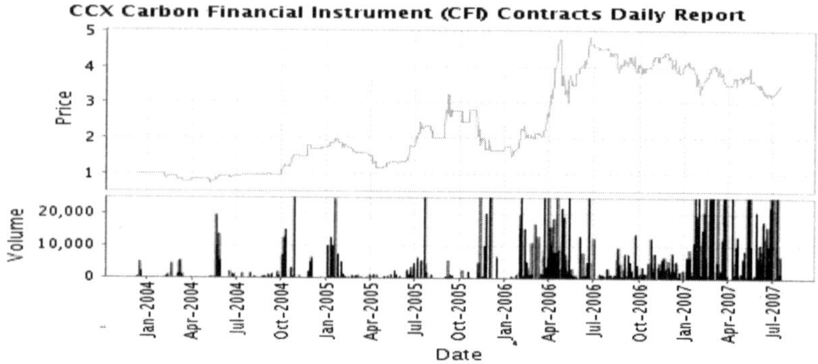

Source: http://www.google.com/finance?q=NASDAQ:EEI

the carbon credits they acquired while making it. And now, some companies are trying to make their fortunes by finding ways to reduce the daily emissions of dirty operations and selling the credits they earn in the process. These companies are not only getting consulting fees, but a portion of the carbon credit proceeds as well. Now, let us take a look at Ecology and Environment Inc. (AMEX: EEI), which offers a range of environmental consulting services; including environmental planning, management, and regulatory compliance support (Fig. 4). Ecology and Environment, Inc. (EEI) is an environmental consulting and testing firm. The Company offers a range of environmental consulting services, including environmental audits; environmental impact assessments; terrestrial, aquatic and marine surveys; air quality management and air toxics pollution control; environmental engineering; noise pollution evaluations; wastewater analyses; water pollution control; industrial hygiene and occupational health studies; archaeological and cultural resource studies; environmental infrastructure planning, and air, water and groundwater monitoring (http://www.google.com/finance?q=NASDAQ:EEI).

Fig. 4: EEI's Environmental Consultancy Support over the Years (2006-07)

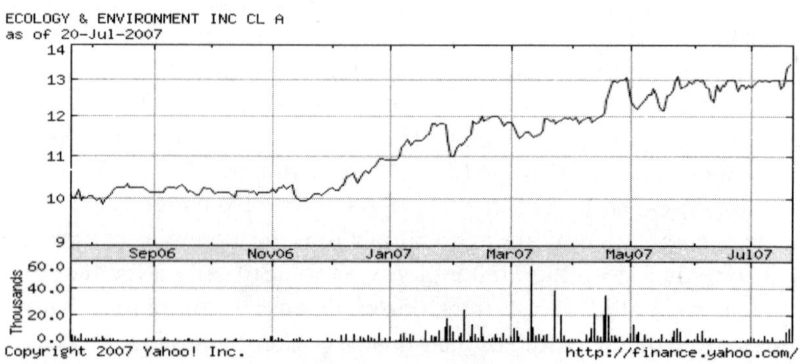

Source: http://www.google.com/finance?q=NASDAQ:EEI

In one year, this company is up 34%—six percentage points higher than the record-breaking Dow—and has increased its annual net income by over 62% (Fig. 4). As more governments begin to cap carbon emissions and initiate trading schemes, there will need to be regulatory bodies that measure and confirm reduced emissions. And those agencies will need new instruments and technologies to measure and record.

Carbon Market Overview

Carbon Transactions are defined as purchase contracts whereby one party pays another party in return for GHG emissions reductions or for the right to release a given amount of GHG emissions that the buyer can use to meet its compliance—or corporate citizenship—objectives vis-à-vis climate change mitigation. Payment is made using one or more of the following forms: cash, equity, debt, convertible debt or warrant, or in-kind contributions such as providing technologies to abate GHG emissions. Carbon transactions can be grouped into two main categories:

- *Allowance-based transactions,* in which the buyer purchases emission allowances created and allocated (or auctioned) by regulators under cap-and-trade regimes, such as Assigned Amount Units (AAUs), under the Kyoto Protocol or EUAs under the EU-ETS. Such schemes combine environmental performance (defined by the actual level of caps set) and flexibility, through trading, in order for mandated participants to meet compliance requirements at the lowest possible cost;
- *Project-based transactions,* in which the buyer purchases emission credits from a project that can verifiably demonstrate GHG emission reductions compared with what would have happened otherwise. The most notable examples of such activities are under the CDM and the JI mechanisms of the Kyoto Protocol, generating CERs and ERUs respectively.

Carbon cap-and-trade regimes currently in place allow, for the most part, for the import of credits from project-based transactions for compliance purposes. This helps to achieve the environmental target cost effectively through access to mitigation potentials from additional sectors and additional countries. Once project-based credits are issued and are finally delivered where and when desired for compliance, then they are at that time fundamentally the same as allowances. Unlike allowances however, project-based credits are compliance assets that need to be "created" through a process that has certain risks inherent with it (regulation, project development and performance, for instance) and can involve significantly higher transaction costs. Such risks are addressed through contractual provisions that define how they are allocated between parties, and, along with other factors, are reflected in the value of the transaction. Through the second half of 2006, a secondary market for CERs has grown in activity, bringing to buyers standardized compliance-grade assets coming with guaranteed deliveries for firm volume deliveries.

Under this definition, transactions are excluded whereby one party acquires rights on future potential carbon assets among other components of a transaction: for instance, when a company acquires a stake in another company whose assets might eventually be developed into carbon assets or when a company acquires a stake in a carbon project developer's portfolio. For instance, the EU ETS is focused on CO_2 emissions from major energy-intensive installations. The so-called Linking Directive allows for the import of CDM emission reduction credits from activities located in developing countries and tapping mitigation potential in the industry, in the waste management sector, in the mining sector, etc. "Residual" difference between those units pertains to the regulatory framework under which they are used for compliance purposes. Some of the rules indeed may limit the substitutability of these units at a given point in time. In order to understand the carbon markets, it is important to recognize the differences between two fundamentally different types of carbon commodities, allowances and offsets, and the systems that create them. The first, allowances, are created by cap-and-trade systems. The second offsets or carbon credits are created by baseline-and-credit systems.

Under a cap-and-trade system, an overall cap is set to achieve emissions reductions. Each of the participants within a cap-and-trade system (usually countries, regions or industries) is allocated a certain number of allowances based on an emissions reduction target. In a cap-and-trade system the cap constitutes a finite supply of allowances, set by regulation and political negotiation. These allowances are then neither created nor removed, but merely traded among participants. This finite supply creates a scarcity and drives the demand and price for allowances.

A cap-and-trade system aims to internalize (some of) the costs of emissions, and thus drives actors to seek cost-effective means to reduce their emissions. The challenge in a cap-and-trade programme is to determine the appropriate level at which to set the cap, which should be stringent enough to induce the desired level and rate of change, while minimizing overall economic costs. A baseline-and-credit system in contrast, does not entail a finite supply of allowances. It does not involve projects that are implemented under the umbrella of a cap-and-trade system. Rather, more credits are generated with each new project implemented. These credits can then be used by buyers to comply with a regulatory emission target, to "offset" an emitting activity (such as an airline flight), or to be a "carbon neutral" organization with zero "net" emissions. In a baseline-and-credit system a carbon offset buyer can only legitimately claim to offset his emissions if the emissions reductions come from a project that would not have happened anyway.

This concept is called 'additionality' in the carbon markets, and refers to the requirement that "reductions in emissions are additional to any that would occur in the absence of the certified project activity" (Kyoto Protocol in Article

12.5). Under a cap-and-trade system it is the cap and the allocations rules that drives demand, and determines the level of emissions reduction. Activities that are undertaken in response to the pressure of the cap therefore do not need to prove that they are additional

Cap-and-trade systems often allow for a certain number of offsets to come from emissions reductions that are generated by projects that are not covered under the cap (i.e. from baseline-Market Overview and-credit systems). Under a cap-and-trade system the covered sources (for example power producers) have an obligation to reduce their emissions. If these covered sources cannot buy offsets, they will have to reduce their emissions in some other way (e.g. by buying allowances or by increasing efficiency in their plants). If they can buy offsets and these come from projects that are fully additional, then the offsets replace reductions that the cap-and-trade participant would have had to otherwise achieve himself. In other words, under a cap-and-trade system, offsets do not lead to emissions reductions beyond the target set by the cap but only cause a geographical shift in where the emissions reduction occurs. Therefore, non-additional offsets sold into a cap-and-trade system will actually lead to an *increase* in emissions since the buyer will not have reduced his emissions and the seller will not have offset this increase in emissions.

In a voluntary system, on the other hand, individuals and companies are not required to reduce their emissions. One can therefore assume that they would only do so to a limited extent. The availability of offsets enables them to go beyond what they would have done anyway to reduce their own emissions. The availability of offsets in the voluntary market may therefore lead to additional emissions reduction that would not have happened without the availability of offsets. Buyers in the voluntary market can only claim a unique, incremental "offset"reduction if the reduction is additional. Yet even without additionality tests, the offset market might induce reductions that would not have happened otherwise, because the market will bring investment to some projects at the margin. But without clearly established additionality, there is no one-to-one correspondence between each credit sold and an additional tonne of reductions.

Trend of Global Carbon Market

Carbon emissions-trading has been steadily increasing in recent years. According to the *World Bank's Carbon Finance Unit*, 374 million metric tonnes of carbon dioxide equivalent (tCO_2e) were exchanged through projects in 2005, a 240% increase relative to 2004 (110 $mtCO_2e$) which was itself a 41% increase relative to 2003 (78 $mtCO_2e$). In terms of dollars, the World Bank has estimated that the size of the carbon market was 11 billion USD in 2005, 30 billion USD in 2006, and 64 billion in 2007 (http://www.wbcsd.org/DocRoot/wrHqIUtkoNq4sC2wodX8/linkages.pdf) (Table 4).

Table 4: Trend of Global Carbon Market: Transaction Volumes and Values

Markets	Volume ($MtCO_2e$)		Value (US$million)	
	2003	**2007**	**2006**	**2007**
Voluntary OTC Market	14.3	42.1	56.5	250.4
COX	10.3	22.9	38.3	72.4
Total Voluntary Markets	**24.6**	**65.0**	**96.7**	**330.8**
EU ETS	1,044	2,061	24,436	50,097
Primary CDM	537	551	5,804	7,426
Secondary CDM	25	240	445	5,451
Joint Implementation	16	41	141	499
New South Wales	20	25	225	224
Total Regulated Markets	1,642	2,918	31,051	63,697
Total Global Market	1,667	2,983	31,148	64,028

Source: Ecosystem Marketplace, New Carbon Finance, World Bank.

The Marrakesh Accords of the Kyoto protocol defined the international trading mechanisms and registries needed to support trading between countries, with allowance trading now occurring between European countries and Asian countries. However, while the USA as a nation did not ratify the Protocol, many of its states are now developing cap-and-trade systems and are looking at ways to link their emissions trading systems together, nationally and internationally, to seek out the lowest costs and improve liquidity of the market. However, these states also wish to preserve their individual integrity and unique features. For example, in contrast to the other Kyoto-compliant systems, some states propose other types of greenhouse gas sources, different measurement methods, setting a maximum on the price of allowances, or restricting access to CDM projects. Various proposals are being investigated to see how these systems might be linked across markets, with the *International Carbon Action Partnership (ICAP)* as an international body to help co-ordinate this.

From a market value of US$10.9 billion in 2005, Table 4 shows that the annual value of global trading in carbon markets nearly tripled to US$30.1 billion in 2006. This market consists of two types of assets: project-based emissions credits (or offsets), and emissions allowances from national cap-and-trade schemes. Project-based emissions credits (or offsets) are created through two mechanisms set up under the Kyoto Protocol, namely Joint Implementation (JI) projects and Clean Development Mechanism (CDM) projects. In essence; firms receive credits for investing in projects that reduce CO_2 emissions in foreign countries. Table 5 shows that trading in project-based emissions credits represented 18 per cent of the carbon market by value traded in 2006.

Table 5: An Overview of the Global Carbon Market-2006

Category	Volume ($MtCO_2e$)	Value (US$m)	% of value
Project-Based			
Clean Development Mechanism	475	5,257	17%
Joint Implementation	16	141	1%
Other	17	79	0%
Emissions Trading			
European Union	1,101	24,357	81%
New South Wales + ACT	20	225	1%
Chicago Climate Exchange	10	38	0%
Total	1,639	30,097	100%
World emissions in 2003	27,500		

Note: $MtCO_2e$=millions of tonnes of carbon dioxide equivalent.
Source: World Bank,United Nations Statistics Division.

An emissions allowance represents a permit to emit one tonne of CO_2 or its metric tonne equivalent in other greenhouse gases. Emissions allowances are created under national Cap-and trade systems, where the supply of carbon allowances for a given period is set by a national regulator to achieve a targeted reduction in overall emissions over a given time horizon. These allowances are allocated to a group of firms covered by a mandatory or voluntary firm-level cap. Table 4 shows that trading of emissions allowances represented 82 per cent of the carbon market by value traded in 2006, with almost all the volume in the European Union. Cap-and-trade schemes are seen as a cost-effective and flexible way to reduce carbon emissions. By creating an allowance that can be traded, a cap-and-trade scheme transforms a constraint (the obligation to cut emissions) into an asset (a new commodity to be traded alongside other commodities such as oil, natural gas, and electricity). This approach provides flexibility for firms that can choose the most cost-effective strategy to meet their individual emissions targets.

An important benefit of the cap-and-trade schemes is that emissions trading become a source of information for firms on the costs involved in reducing CO_2 emissions, and for policy-makers on the appropriate level and the pace to achieve national targets. A high price for allowances suggests that firms are finding it difficult and expensive to reduce emissions, while a low price suggests that there are relatively inexpensive opportunities to reduce emissions.

Monitoring and Reporting for Accountability of Carbon Credits

As with any market, a cap-and-trade market may not develop efficiently without the necessary infrastructure. Experience suggests that a successful cap-and-trade system relies on the following elements:

- registration of the ultimate owner of allowances or credits
- independent monitoring of emissions from a facility
- reporting of emissions to a central authority over a given period
- verification of the level of emissions and confirmation of reductions

These components are critical for promoting both public and business confidence in the system, allowing the market to function properly, and minimizing uncertainty. While some of these functions may be provided by financial intermediaries, a central authority may have responsibility for some or all of the functions.

Flexibility Instruments of Kyoto Protocol for Emission Trading

A. International Emission Trading (IET) Program

The International Emission Trading (IET) Program provides the basic trading structure on which the other Kyoto Flexibility Mechanisms—the Clean Development Mechanism (CDM) and Joint Implementation (JI)—are built. This section therefore begins with a discussion of IET before moving on to the other programs.

1. Program Design Choices (NOx)

Much as the (NOx) Budget Trading Program did for state governments, the Kyoto Protocol allocates a GHG emissions "budget" to certain industrialized nations (the Annex I Parties) and allows them to achieve it through an international emissions trading program. Under the IET program, a transfer increases the acquiring country's holdings of Assigned Amount Units (AAUs) and decreases that of the transferring country, thereby keeping the total allowed emissions constant. International Emission Trading is thus a closed cap-and-trade program for Annex I Parties. Features such as the Clean Development Mechanism and the ability of Annex I entities to engage in transfers of their own add some open-market qualities. These elements will be explained further below.

As in any other cap-and-trade system, each of the Annex I Parties must ultimately, at the relevant reconciliation date, hold sufficient allowances to cover all of its emissions. This means that each country must keep its actual emissions below its assigned amount, as adjusted through the IET transfers. In this respect, the IET program may be said to apply to national economies as a whole. As implemented by Annex I Parties to date, the program has a downstream focus in the sense that it requires actual reductions in emissions (or enhanced removals through sinks) and not simply changes in upstream behavior, such as fossil-fuel production. Thus, IET is an economy-wide, closed, cap-and-trade program with a downstream focus.

2. Program Implementation Rules

A pure cap-and-trade program, IET does not need many of the

implementation rules common to baseline-credit systems. It contains no certification requirement, no baseline calculation methods, and no additionality mandate. The key program rules focus instead on the transactions themselves—what can be transferred, when can it be transferred and, most importantly, who can transfer and trade? The following discussion will similarly focus on these points. As to the "what" question, IET treats all Kyoto Units as fungible and allows them to be transferred among Annex I Parties. Assigned amount units (AAUs), emission reduction units (ERUs), certified emission reductions (CERs) and removal units (RMUs) can all be exchanged for one another on a one-to-one basis. Annex I Parties can only formally transfer and/or acquire Kyoto Units after the first commitment period begins on January 1, 2008, and they meet other eligibility criteria. However, trading of the ownership rights to these emissions assets can commence at any time. Such trades will necessarily be in the form of forward trades, i.e., commitments to transfer an asset at some time in the future.

The more interesting questions relate to who can transfer and trade Kyoto Units. The International Emissions Trading system is set up to allow Annex I Parties to transfer Kyoto Units to one another. Thus, the initial answer to the question of who can transfer and acquire Kyoto Units is that the Annex I Parties can do so. Upon such a transfer, the transferring nation will experience a decrease in the Kyoto Units in its national account and the acquiring nation will experience an increase. Upon further examination, the situation is not quite that simple. There are a number of complicating factors that lawyers practicing in this area should be familiar with. To begin with, only those Annex I Parties that meet certain eligibility requirements are authorized to transfer and acquire Kyoto Units under the IET system.

Specifically, a Party must:

Be a Party to the Kyoto Protocol. Annex I nations that have not ratified the Protocol are not eligible to engage in IET. This includes, among others, the United States and Australia.

1. Any permits or allowances created by national GHG trading schemes in such countries will not be credited in the Kyoto system;
2. Have had its assigned amount calculated and recorded in accordance with the rules for doing so set out in Decision 19/CP.7;
3. Have in place a national system for estimating anthropogenic GHG emissions and removals;
4. Have established a national registry for tracking their assigned amounts;
5. Have submitted accurate annual inventories of GHG emissions and sinks;
6. Have submitted the supplementary information on assigned amounts and made any necessary adjustments to its assigned amount.

Annex I Parties that do not meet these eligibility requirements cannot transfer or acquire Kyoto Units through the IET system. A private entity wishing to transfer Kyoto Units under the authority of an Annex I Party can only do so if the Party itself is eligible to make such a transfer.

Another complicating factor is that even those Annex I Parties that are eligible to engage in the transfer of their Kyoto Units can only do so to a limited extent. The drafters of the Marrakesh Accords were concerned that some Annex I Parties would be too eager to transfer their Kyoto Units, especially if these emission rights rose in value. They worried that some Annex I Parties might transfer too many Units and be left with an insufficient number to live up to their Kyoto commitments. The Accords therefore provide that each Annex I Party hold in its national registry at least an amount of Kyoto Units equivalent to its "commitment period reserve." The commitment period reserve is calculated as 90 percent of a Party's assigned amount "or 100 percent of five times its most recently reviewed inventory, whichever is the lowest." Only those parties with a fully stocked commitment period reserve (and the authorized entities trading out of their national registry systems) can transfer Kyoto Units.

In addition to limiting Parties' ability to sell Kyoto Units, the Marrakesh Accords also restrict their capacity to acquire Units. The origin of this limitation stems from the concern that some Parties would rely too heavily on reductions made in other countries and would not undertake the hard work necessary to achieve emission decreases on their own. To address this concern, the Accords provide that the acquisition of Kyoto Units through the flexibility mechanisms should be "supplemental to domestic actions" and that domestic measures and policies should "constitute a significant element of the effort" that each Annex I Party makes to meet its Kyoto commitment. This has come to be known as the "supplementarity" principle, since it declares that the use of the flexibility mechanisms for compliance purposes should be supplemental to domestic actions. Annex I Parties may be hesitant to acquire Kyoto Units where doing so could be seen to go against the principle of supplementarity. Another important issue is whether Annex I Parties can authorize private entities to transfer and/or acquire Kyoto Units on their behalf. Most Annex I Parties do not retain exclusive control over the Kyoto Units in their national accounts. Instead, they essentially create their own national cap-and-trade system (or a regional system in the case of the EU) in which a portion of their national allocation serves as a cap for covered industries (with other portions being reserved for sectors that are not part of the trading system, such as transportation). Emitters covered by the trading scheme hold their allowances, in the form of Kyoto Units, in their own accounts listed on the national registry of their home nation. They can use these allowances to meet their own obligations under the national cap-and-trade system. In addition, an Annex I Party can authorize legal entities to transfer Units through the IET system to authorized legal entities in another Annex I Party, who must maintain a current list of those legal entities that it has authorized to engage in such activities and must make the list available to the public. When legal entities engage in such a transaction, this will affect both the holdings of the legal entities themselves and the holdings of their home nations. On one level, the transaction is a transfer between the legal entity that is selling the Units and the one that is purchasing them. The Units leave one legal entity's national registry account and enter that of the other entity.

On another level, it is a transfer between nations. The Units leave the national account of the Annex I Party in which the transferring legal entity holds its registry account, and enter the national account of the Annex I Party in which the acquiring entity holds its registry account. Thus the transaction is simultaneously a transfer between legal entities and a transfer between Annex I Party nations. Given this dual nature, it can only occur if the two Annex I Parties are legally able to engage in IET trading and if these Parties both agree to and authorize the transfer and acquisition. This circumstance makes it imperative for any private entity that wants to transfer Kyoto Units to make sure that the Annex I Party that is authorizing the transfer and/or acquisition has itself complied with the eligibility and commitment period reserve requirements set out above. The UNFCCC Secretariat's list of Annex I Parties that have met the eligibility requirements should prove to be a useful resource for making this assessment.

Another question of special importance to American companies is whether legal entities incorporated or domiciled in non-Party nations (such as the United States) can transfer and/or acquire Kyoto Units. On the national level, only those parties to the UNFCCC that have ratified the Kyoto Protocol can transfer and/or acquire Kyoto Units. Thus, it is quite clear that non-Party nations cannot engage in such transfers or acquisitions. But what of legal entities domiciled in those non-Party nations? Could an Annex I Party allow them to open an account in its national registry and employ it to acquire and/or transfer Kyoto Units? While the answer is not yet fully clear, it can be argued that Annex I Parties can open their registries to non-Party entities. The Marrakesh Accords state that Annex I Parties can authorize legal entities to trade Kyoto Units. They do not distinguish between legal entities that are domiciled in an Annex I Party and those that are not. Arguably, this silence implies that an Annex I Party could authorize all such entities to open national registry accounts and to transact in Kyoto Units.

Assuming that an Annex I Party had the authority to do this, would it actually do so? Some environmental groups argue against giving legal entities from non-Party nations the privilege of transferring or acquiring Kyoto Units. They maintain that such firms should not benefit from participation in the Kyoto Flexibility Mechanisms because they do not face any obligations under the Protocol. It remains to be seen whether Annex I Parties will adopt this position and, as a matter of national law, deny registry accounts to companies domiciled in non-Party nations, or whether they will follow a more open policy. Current indications are that at least some of them will follow the more open approach. Even if Annex I Parties adopt the more restrictive approach, non-

Party entities may still have avenues open to engage in Kyoto Unit transactions. For example, a multinational parent corporation domiciled in a non-Party nation that has a subsidiary domiciled in an Annex I Party could use that subsidiary to transfer or acquire Kyoto Units. The subsidiary could open a national registry account and use it to transfer and/or acquire Kyoto Units on behalf of its parent. Nothing in the Marrakesh Accords expressly prohibits this. Moreover, the subsidiary will presumably have emission reduction obligations of its own and so will not be as susceptible to the argument that it is "free-riding" on the Kyoto Flexibility Mechanisms. International trade agreements may also prevent Annex I Parties from treating such subsidiaries differently for emissions trading purposes than they treat their own companies. Canada has said that it will allow subsidiary companies from non-Party States that are located in Canada to hold accounts in the Canadian National Registry on the same basis as Canadian companies. Companies without a subsidiary in an Annex I Party could create one. That is, they could establish a company in an Annex I Party and use it to transfer, hold or acquire Kyoto Units.

Another option would be for a legal entity from a non-Party country to develop an arrangement with a legal entity with access to a national registry account to transfer, hold or acquire rights on its behalf. This would require contractual arrangements to be entered into among the cooperating parties to ensure that the Kyoto Units were transferred and/or acquired as directed. The party with the national registry account would likely charge a commission for such a service. Any limits on the "transfer" of Kyoto Units would be unlikely to apply to the "trade" in such Units. While the Kyoto Protocol and Marrakesh Accords expressly regulate the transferring and acquiring of Kyoto Units, they say nothing about the trading of them. In theory, the doors are wide open to trading in Kyoto Units. States, private companies, brokers, and others should all be able to engage in trading of these assets regardless of their place of domicile. Moreover, trading should be able to begin at any time. Unlike IET transfers, it need not had to wait until the beginning of the first commitment period completed in 2008.

B. The Clean Development Mechanism (CDM)

The second Kyoto flexibility instrument is the Clean Development Mechanism (CDM). This program is structured quite differently from IET, as reflected in its program design choices and program implementation rules.

1. Program Design Choices

The most salient feature of the CDM is that it is a baseline-credit program. It gives developing nations that are Parties to the Kyoto Protocol, and their authorized entities, credit for reducing GHG emissions below baseline levels (or removing GHG beyond baseline sink levels). It then allows them to sell these credits, known as Certified Emission Reductions (CERs), to developed Annex I Parties who can use them to comply with their Kyoto commitments or can transfer them within the IET system. The CDM is thus a baseline-credit add-on to the overall IET cap-and-trade regime.

Participation in the CDM is voluntary. In addition, as will be further described below, the CDM allows a wide array of legal entities to participate as producers/transferors and/or acquirers of CERs. These attributes give the CDM features of an open-market system. Yet there are some limits on participation. The projects must be located in developing nations that are Kyoto Protocol parties. They cannot take place in the United States, since it is a developed nation and is not a party to the Protocol. While the system is quite open, it is not all-inclusive either.

Like the IET, the CDM is a downstream system that focuses on actual GHG emission reductions and removals. It is also an economy-wide program that recognizes or removals from many economic sectors and does not limit itself to one or two specific industries. Finally, the CDM allows its credits to be banked from one Kyoto commitment period to the next, although it places some limits on this activity. Parties can only bank CERs up to an amount that equals 2.5% of the Party's assigned amount.

Program Implementation Rules

(a) Who can Transfer and Trade?

The CDM is set up to allow developing countries that are not listed in Annex I of the UNFCCC but are parties to the Kyoto Protocol (non-Annex I Parties) to convey credits to Annex I Parties that have binding commitments under the Protocol. Thus, as a general rule, non-Annex I Parties can carry out project activities and convey the resulting CERs, and Annex I Parties can acquire these credits. These Parties must satisfy some basic eligibility conditions before they can participate in CDM. First, they must designate a national authority to provide oversight of CDM projects. Second, as was already mentioned, they must be Parties to the Kyoto Protocol. Finally, Annex I Parties that seek to acquire and use CERs must meet the six eligibility requirements for participation in any of the Kyoto Mechanisms described above.

(b) Can legal entities transfer and acquire CERs?

The rules here largely follow those described above in reference to IET. Where authorized by their government to do so, legal entities domiciled in an Annex I Party nation can transfer and acquire CERs and can serve as "project participants" (PP), a broad term encompassing project developers, purchasers of CER rights, and project financiers. However, if the authorizing Party is not itself eligible to participate in CDM, then it cannot authorize its legal entities to do so. Legal entities should accordingly make certain that the authorizing Party meets the eligibility requirements set out above.

With respect to legal entities that are not domiciled in a Kyoto Party nation (non-Party entities), the rules again largely track those discussed above in reference to IET. Both the Kyoto Protocol and the Marrakesh Accords refer only to the participation of legal entities and do not expressly refer to the domiciles of such entities. Some read this as allowing Parties to authorize non-Party entities to acquire and transfer CERs and otherwise participate in CDM. As described above with respect to IET, Kyoto Parties may seek to limit such participation through their national policies. The experience to date suggests that this is unlikely. For example, the Netherlands has announced that it will approve entities "from the whole world" for participation in CDM projects as long as the entities meet general approval criteria. The Netherlands would thus permit non-Party entities to participate directly in CDM projects. In addition, a number of American and other nonparty entities have already begun participating in CDM projects through subsidiary companies domiciled in Annex I Parties, or in CDM host countries. If these trends continue, domicile should not prove a barrier to participation in CDM projects.

The CDM and IET situations do differ in one significant way. The CDM Executive Board (EB) registers and certifies CDM projects (see below). There is no comparable IET requirement. Potentially, the EB could object to non-Party entity participation in CDM projects and, on this basis, refuse to register or certify such projects. Yet there are reasons to believe that it will not do so. The EB will primarily be made up of representatives from developing countries that are likely to favor broader participation in CDM. Moreover, participation by a non-Party entity presupposes that an Annex I Party has authorized this participation. The EB is unlikely to override the wishes of such a Party. These factors accordingly suggest that the EB is unlikely to pose a significant obstacle to non-Party entity participation. Practitioners should watch for future EB actions and decisions on this topic.

(c) Project Eligibility

The Kyoto Protocol provides for a "prompt start" to CDM that allows some CDM projects to begin to accrue CERs beginning in 2000 and to apply them to the first commitment period. In this respect, they differ from Joint Implementation projects, which generally cannot begin to generate emission reduction units until the beginning of the first commitment period in 2008. Thus, the CDM is quite expansive when it comes to project timing. The program does place a few limits on the types of projects that can qualify for CERs. To begin with, the Preamble to the Marrakesh Accords decision on CDM provides that "Annex I Parties are to refrain from using CERs generated from nuclear facilities to meet their commitments under Article 3.1." This language will bar most such projects, but may allow those that the host country believes to be essential to sustainable development. Secondly, the CDM places some limits on the use of carbon sinks to generate CERs and meet Kyoto commitments. It grants units only for afforestation and reforestation activities and does not award them for maintenance of existing forests or other sinks. Moreover, it caps at 1% of base-year emissions the amount of sinks-based CERs that a Party can use to meet its obligations during the first commitment period. It does not place such a cap on the use of sinks-based CERs for future commitment periods, leaving this as a subject for future negotiations.

(d) Baseline-Credit Rules

Like other baseline-credit programs, CDM has rules for assessing and certifying project-based reductions and removals. CDM relies on a private/ public partnership between Designated Operational Entities (DOE) and the CDM Executive Board to carry out these functions. The CDM process can be boiled down to seven basic steps: (1) the Project Participants (PP) submit a Project Design Document (PDD) to the DOE; (2) the DOE assesses this document in accordance with existing protocols and policies and, if it passes muster, "validates" it; (3) the EB reviews the validation and, if it approves, "registers" the project; (4) the PPs carry out the project; (5) based on reports submitted by the PPs, and in accordance with the monitoring plan set out in the PDD, the DOE "verifies" the project's GHG reductions or removals; (6) the EB reviews the verification and, if it approves, "certifies" the reductions and/ or removals; and (7) the EB issues CERs corresponding to the amount of certified reductions or removals.

(e) Public Participation

The rules make provision for public input at various stages in this process. The Project Participant must show in the PDD that it gave members of the public an opportunity to comment on the document and that it took "due account" of these comments. The DOE must give the public an opportunity to comment on how the validation requirements apply to the project, make these comments publicly available, and take them "into account" in determining whether the project should be validated. It must also make its validation and verification reports publicly available.

(f) Project Design Document

In the Project Design Document, the Project Participant sets out the core information about a proposed CDM project. This document then serves as the basis for the DOE's evaluation and possible validation of the project design. The PDD must follow the specific format set out in the Marrakesh Accords in Appendix B of Decision. The Accords further specify the criteria under which the Department of Energy (DOE) must evaluate whether:

- The governments that are authorizing the project are themselves eligible to participate in CDM;
- The PPs have properly invited local stakeholders to comment on the PDD, have provided a summary of comments received, and have explained how they intend to respond to such comments;
- The PPs have documented their analysis of the project's environmental

impacts and, if these impacts are significant, have undertaken an environmental impact assessment in accordance with the host nation's procedures for such assessments;

- The project is going to result in GHG reductions or removals that are "additional" to those that would otherwise be achieved; The methodologies for calculating the baseline and for monitoring reductions and removals are in accordance with those that the EB has previously approved or have themselves been approved by the Board;
- The provisions for monitoring, verification and reporting are in accordance with the Marrakesh Accords Decision 17/CP.7, its Annex and the relevant decisions of the Conference of the Parties serving as the meeting of the Parties (COP/ MOP); and
- The project activity conforms to all other requirements contained in Decision 17/CP.7, its Annex and the relevant decisions of the COP/ MOP.

The Executive Board has published a model Project Design Document as well as guidance explaining how such a document should be prepared. This useful information is available at the CDM Web site at http://cdm.unfccc.int/Reference/Documents.

The Project Design Document, inter alia, should consider the CDM Cycle Approach of the Program Formulation as shown below diagrammatically (Fig. 5).

(g) Baseline Calculation

A Project Participant must include a baseline calculation in its Project Design Document. The Marrakesh Accords define a "bottom-up" approach to the development of baselines. They call on Project Participants, working

Fig. 5: CDM Project Cycle Pattern as shown in the Project Design Document

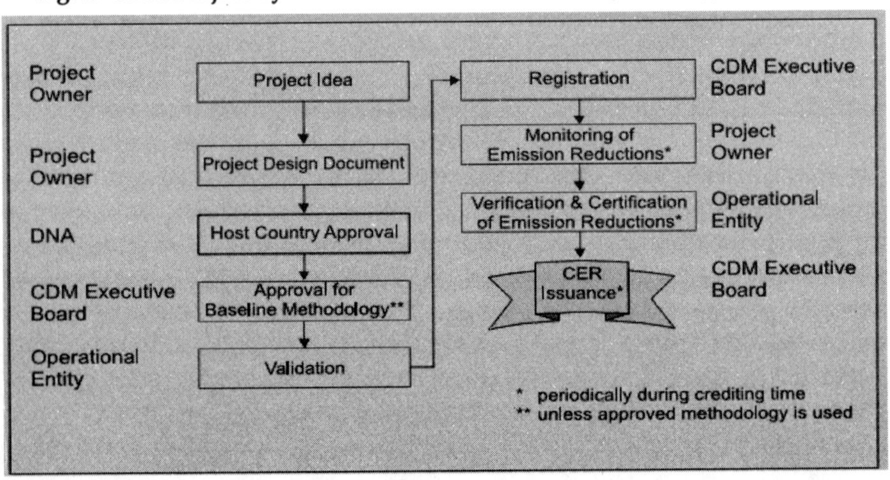

through the DOEs, to either propose a new baseline calculation methodology to the EB or to adopt a methodology that the Board has previously approved. The Accords provide some instructions to PPs, and to the Board, on the development of these methodologies. To begin with, they define the baseline scenario as that which "would occur in the absence of the proposed project activity." They further specify that a valid baseline calculation:

- Must be transparent and conservative and should take uncertainty into account;
- Must be established on a project-specific basis;
- Must take into account relevant national and/or sectoral policies and circumstances;
- May include scenarios where GHG emissions are projected to rise above current levels; and
- Must be defined in a way that does not allow CERs to be earned for decreases in activity levels outside the project activity itself, or due to force majeure.

In addition to this list of requirements, the Accords require project participants to choose from among three possible approaches to baseline calculation: (1) existing actual or historical emissions; (2) emissions given installation of economically attractive technology, taking into account barriers to investment; and (3) "the average emissions of similar project activities undertaken in the previous five years, in similar circumstances, and whose performance is among the top 20 percent of their category."

The Marrakesh Accords instruct the Executive Board to develop further guidance on the calculation of baselines. Pursuant to this mandate; the EB has established a CDM Methodologies Panel that has begun to publish accepted methodologies. Project participants can adopt these methodologies and incorporate them into their PDD. Alternatively, they can propose new methodologies to the EB for approval. The approved methodologies, recently proposed methodologies, and instructions for proposing a methodology are available at http://cdm.unfccc.int/methodologies/PA methodologies. The CDM Methodologies Panel also fields questions from DOEs regarding the application of approved methodologies and publishes its responses to these queries. These documents constitute regulatory guidance on how to utilize the methodologies (http://cdm.unfccc.int/methodologies/PAmethodologies/ Clarifications and http://cdm. unfccc.int/Reference/Guidclarif).

(h) Monitoring

The Project Design Document must also include a monitoring plan. This plan sets out the methods for collecting and archiving the data and for carrying out the calculations needed to determine whether project activities are leading to GHG reductions or removals beyond baseline conditions. As with the baseline calculation, Project Participants can either adopt an approved monitoring methodology or propose a new one to the EB. The EB publishes approved monitoring methodologies much as it does baseline methodologies and issues clarifying guidance in response to queries from DOEs. The Marrakesh Accords set forth the basic parameters for a monitoring plan.

Such plans must provide for:

- The collection and archiving of all data necessary to estimate GHG emissions within the project boundary during the crediting period;
- The collection and archiving of all data necessary to determine the baseline GHG emissions within the project boundary;
- The collection and archiving of data on the extent to which project activities will cause sources outside the project boundary to increase their GHG emissions (i.e., leakage);
- The collection and archiving of information related to the assessment of the project's environmental impacts, including trans-boundary impacts;
- Quality assurance and control procedures for the monitoring process;
- Procedures for the periodic calculation of GHG emission reductions and of leakage effects; and Documentation of all steps in all calculations used in developing the above information.

(i) 'Additionality'

The Kyoto Protocol requires CDM projects to achieve emission reductions "that are additional to any that would occur in the absence of the certified project activity." Projects that cannot demonstrate additionality will not receive credit and will not produce CERs. The CDM program thus resembles the Offset Trading Program and many other baseline-credit systems in requiring that reductions be additional to those that would otherwise have occurred. The Marrakesh Accords provide a general definition of additionality for the purposes of CDM, stating that "CDM project activity is additional if anthropogenic emissions of greenhouse gases by sources are reduced below those that would have occurred in the absence of the registered CDM project activity." This straightforward definition requires the DOE to assess what GHG emissions would have been in the absence of the project and compare them to the emissions level achieved with the project in place. The CDM Executive Board, in its sixteenth meeting, fleshed out this definition in a document titled "Tool for the Demonstration and Assessment of Additionality." In this document, the EB sets out a step-by-step approach to assessing additionality that includes the following:

- Identification of realistic and credible alternative scenarios that Project Participants might otherwise engage in and that provide similar outputs or services (e.g., electricity, cement, etc.) as those provided by the CDM project activity;
- An "investment analysis" that determines whether, absent the revenue

from the sale of CERs, the proposed project activity would actually be less economically or financially attractive than other comparable alternatives;

- A "barrier analysis" that examines whether the proposed project faces barriers that prevent the implementation of the project activity but do not prevent the implementation of at least one of the alternatives;
- A "common practice analysis" that examines whether the project activity already represents common practice in the relevant sector or region. This part of the assessment is intended as a check on the investment and barrier analyses;
- An explanation of how the benefits and incentives associated with the CDM will alleviate the economic and financial hurdles or other identified barriers that would otherwise prevent the project activity from being undertaken.

(j) Leakage

As mentioned previously, the CDM monitoring plan must measure the extent to which project activities result in emissions of GHG outside the project boundary. This requirement seeks to address the issue of "leakage" where efforts designed to decrease GHG emissions in one setting result in increased emissions elsewhere. The Accords define leakage as "the net change in anthropogenic emissions by sources of greenhouse gases which occurs outside the project boundary, and which is measurable and attributable to the CDM activity." They require that the emission reductions claimed with respect to a given project be adjusted to account for any increased emissions caused by leakage.

Emissions Trading with Flexible CDM Approach

Emissions-trading is a way of introducing flexibility into a system where participants have to meet emissions targets. These participants may be countries (as in the case of the Kyoto Protocol), or companies. Participants can buy units to cover any emissions above their targets, or sell units if they reduce their emissions below their targets. The presence of a market for these units creates a value for emissions reductions which stimulates investment in the most cost-effective areas. Emissions-trading leads to a reduction in compliance costs compared to meeting the same target through domestic/internal means only. The Clean Development Mechanism (CDM) of the Kyoto Protocol allows projects in developing countries to generate emission credits if they result in emission levels lower than would otherwise be the case; these credits can be marketed and eventually counted against a developed country's emission obligation. The IEA provides analysis on the effectiveness of the different emissions trading scheme options, both at international and domestic level.

Categories of Tradable Permit Systems

Tradable permit systems that can be linked fall into two categories: cap-and-trade systems and emission reduction credit systems (credit systems). Under a cap-and-trade system, the total emissions of a group of regulated sources are capped by creating a limited number of tradable emission allowances and requiring those sources to secure and surrender a quantity of allowances equal to their emissions. The European Union's Emissions Trading Scheme is an example of such a system. Under a credit system, entities that voluntarily undertake particular emission reduction projects are awarded credits that can be sold to participants in cap-and-trade systems. The Clean Development Mechanism (CDM) is an example of such a system.

Types of Linkages

Several types of links can be established between systems depending on whether only one or both linked systems recognize the other's allowances. This however can be broadly categorized as Direct and Bilateral systems.

(i) Direct links between systems can be either "one-way" (unilateral), in which allowances can flow in only one direction, or two Throughout this report, we use "allowances" to refer collectively to allowances and credits, except where a distinction between allowances and credits is necessary.

(ii) Two-way links can be bilateral, when established between just two systems, or multilateral, when established among multiple systems.

If a system establishes a one-way link by recognizing another system's allowances, and the former system's allowance price is the higher of the two, inter-system trading will occur until the systems' prices converge at an intermediate level. If the former system's price is the lower of the two, there will be no incentive for inter-system trading. If systems establish a two-way link, allowances will be sold from the system with the lower allowance price to that with the higher price until prices converge. Along with affecting each system's allowance price, this trading alters the distribution of emission reductions across the systems. Even if two systems are not directly linked, they can be indirectly linked through separate links with a common third system. Through trading between each system and the common third system, the supply and demand for allowances in one system can affect that in the other system even though the two systems are not directly linked.

Implications of Linkages

By broadening markets for allowances and credits, linking increases the liquidity and improves the functioning of those markets. Linking also can reduce the costs of the linked systems by making it possible to shift where emission reductions occur across the systems. Just as allowance trading *within* a system allows higher-cost emission reductions to be replaced by lower-cost reductions within that system, trading *across* systems allows higher-cost reductions in one system to be replaced by lower-cost reductions *in another system*. While the cost savings from linking are valuable in their own right, in the context of global climate change mitigation efforts, a particularly important benefit of linking is the opportunity it offers to establish "common but differentiated responsibilities" across tradable permit systems without increasing the cost of achieving global emission targets.

Linkage between Cap-and-Trade and Credit System

A one-way link between a cap-and-trade system and a credit system can offer potentially significant cost savings by giving the cap-and-trade system's participants access to low-cost emission reductions, such as emission reduction opportunities in developing countries and opportunities to reduce net emissions that are challenging to incorporate directly in a cap-and trade system (e.g., biological sequestration). In addition, such a link's distributional impacts ought to elicit broad support because the link can only serve to reduce the cap-and-trade system's allowance price, and increase the price that entities in the credit system receive for their credits. While it offers potentially significant cost savings, the main concern regarding such a one-way link is that some of the traded credits will overstate the amount of emission reductions associated with their generation in the credit system. If this occurs, the use of credits in a cap and-trade system will lead to increased emissions that are not fully offset by reductions in the credit system. The primary reason for this concern is the challenge of verifying that projects awarded credits in a credit system represent additional emission reductions beyond what would occur if they were not granted credits—the so-called additionality problem.

The key tradeoff presented by a link with a credit system is between the cost savings it can offer and its emissions implications. In addressing this tradeoff, a balance can be struck through the standards that are established to award credits in the credit system or to allow their use in the cap-and-trade system. Also, this tradeoff is affected by the cap-and-trade system's characteristics and design. If the system employs a "safety valve," any emissions impacts of linking with a credit system would be mitigated if the link limits the use of that safety valve, and the associated increase in emissions. Likewise, the emissions impacts of such a link would be mitigated if it reduces emissions leakage from the cap-and-trade system.

While some credit systems may be linked with only one cap-and-trade system, others, such as the CDM, may be linked with multiple systems. Pre-existing links between a credit system and other cap-and-trade systems significantly affect the implications of establishing a new link with that credit system. Because participants in the newly-linked cap-and-trade system must compete with other cap-and-trade systems for credits, pre-existing links can reduce the cost savings from the new link if the newly-linked system has a relatively low allowance price. Preexisting links also can mitigate the possible emissions implications of linking with a credit system. This is because some of the offsetting emission reductions associated with the use of credits in the newly-linked system will not occur in the credit system itself, but rather in other cap-and-trade systems from which the credits are bid away. The fact that some of the credits used in the newly-linked system otherwise would be used in another cap-and-trade system renders moot any additionality concerns associated with the use of those credits. The new link would only influence which cap-and-trade system uses those credits, not whether they are used. Through the indirect links that they create, one-way links between multiple cap-and-trade systems and a common credit system can achieve some and perhaps much of the cost savings that would be realized from direct links among those cap-and-trade systems.

C. *Joint Implementation*

Joint Implementation allows one Annex I Party to invest in an emission reduction or sequestration project located in another Annex I Party, and then to use the resulting credits (known as Emission Reduction Units or ERUs) to meet its own Kyoto Protocol commitments. In this way, it allows two Annex I Parties to "jointly" pursue meeting their GHG reduction obligations.

1. Program Design Choices

While IET is a pure cap-and-trade program and CDM a base line credit initiative, JI blends aspects of both. Under JI, Annex I Parties that wish to transfer ERUs must reduce their emissions below a baseline amount (or sequester GHG beyond baseline conditions) and must certify these reductions to produce credits. In these respects, the program resembles a project-based, base line credit system much like CDM. Yet when it comes time for a Party that has reduced project-based emissions to transfer the resulting ERU, the process resembles a cap-and trade system. The Kyoto Protocol states that the JI program allows "any Party" that has undertaken an emission reduction "commitment" under the Kyoto Protocol (i.e., Annex I Parties) to transfer or acquire ERUs. This language limits the program to Annex I Parties and expressly excludes those Parties to the UNFCCC, such as the United States, that did not ratify the Protocol and do not have a binding commitment. In this regard, JI is a "closed" system that limits participation to a specified group of participants. Yet the Protocol goes on to state that a Party "may authorize legal entities to participate, under its responsibility, in actions leading to the generation, transfer or acquisition" of ERUs. In practice, this opens the door to participation by private and public legal entities as long as they do so in cooperation with, and with authorization from, the Annex I Parties involved in the ultimate transfer of ERUs.

Other program design choices are more clear-cut. Joint Implementation is an economy-wide program that allows ERUs to be generated from many sectors and does not narrow its focus to one or two industries. The program has a downstream focus in that it requires the projects to reduce or sequester GHG and does not generally grant ERUs for reductions in upstream activity (such as reductions in fossil fuel production). Finally, the JI program places important limits on the banking of ERUs. Under the Marrakesh Accords, Parties may "bank" ERUs acquired in one commitment period and carry them into the next. However, they may only do so in an amount equal to 2.5% of their initial assigned amount.

Program Implementation Rules

The JI program's implementation rules reflect its dual nature as both a cap-and-trade and baseline-credit initiative. Due to JI's cap-and-trade element, rules on who can transfer, hold and acquire units are relevant. Given its baseline-credit dimension, rules on the determination of baselines, additionality and certification also play an important role, much as they do in the Clean Development Mechanism.

(a) Who Can Transfer and Trade?

As stated above, Annex I Parties are generally authorized to transfer, hold and acquire ERUs. However, the JI program does limit these activities. First, as is true in both the IET and CDM programs, Annex I Parties that wish to transfer and/or acquire ERUs must meet the eligibility requirements set forth in the Marrakesh Accords and described above. Parties can initiate JI projects prior to fulfilling all of these eligibility requirements, but they must meet them in order to ultimately transfer any ERUs that are generated. Second, as with transfers of AAUs, a Party's use of ERUs for compliance purposes should be supplemental to its domestic actions to reduce GHG emissions. Other than these limits, Annex I Parties are generally free to transfer and/or acquire ERUs. Legal entities can hold and transfer ERUs only as long as they do so with the approval of the Annex I Party in whose nation they hold a registry account. Moreover, the JI Project will only come to fruition—in terms of an actual transfer of ERUs—if that Annex I Party is eligible to participate in JI.

(b) Project Eligibility

One important difference between JI and CDM is that, whereas CDM allows CERs to be generated beginning in 2000, JI allows Annex I Parties to issue ERUs only for reductions or removals achieved after the first commitment period begins in 2008. As with CDM, participants in the JI program are supposed to refrain from using nuclear projects to generate ERUs. The JI program expressly recognizes GHG removals achieved through sinks projects. The Marrakesh Accords confirmed this and specified that such sinks projects must conform to the rules set out in Article 3 of the Kyoto Protocol.

(c) Baseline-Credit Rules

JI has rules governing baselines, additionality, certification and other such elements of the baseline-credit approach. However, JI differs from other baseline-credit systems (such as CDM) in that it creates two separate "tracks" for evaluating whether a given project meets these requirements and establishes different sets of rules for each. Track 1 is for JI projects in which the host (transferring) Party meets all six eligibility criteria for participating in the Kyoto Flexibility Mechanisms. In such situations, the host Party is thought to be in a position to evaluate its own transactions. It is accordingly allowed to establish its own policies for setting baselines, monitoring reductions and assessing additionality. It is also authorized to verify that these policies have been complied with and, once it has done so, to issue the ERUs all without international oversight. As mentioned above, upon transfer of the ERUs, the corresponding number of AAUs will be subtracted from the host Party's national account and will be added to that of the acquiring Party. This fact, combined with the host Party's need to have sufficient AAUs to cover its emissions at the end of the commitment period, are thought to provide a sufficient check on the system.

The Track 2 procedure applies where the host Party does not meet all six eligibility criteria. In these situations host Parties are not authorized to verify their own reductions. A third-party verifier—known as an Independent Entity (IE)—reviews the Project Design Document (PDD) and "determines" whether it is consistent with the JI program policies and procedures. After the project has been implemented, and upon receiving a monitoring report from the Project Participant, the IE determines whether the project has actually resulted in the GHG reductions and/or removals. Each of these determinations is to be made publicly available. An international body—the JI Supervisory Committee—establishes protocols for determining baselines, monitoring, etc., serves as the accrediting body for the IEs, and reviews the IE's determinations.

(d) Project Design Document

With respect to the first determination—the evaluation of the PDD—the Marrakesh Accords instructs the IE to assess whether:

- The parties involved have approved the project.
- The project would result in GHG reductions or removals that are "additional to any that would otherwise occur."
- The project has an appropriate baseline and monitoring plan that is in accordance with JI policies.
- The Project Participants have analyzed the project's environmental impacts and, if they are significant, have conducted an environmental impact assessment.

(e) Baseline Calculation

The Marrakesh Accords specifies that a JI project baseline is to be established:

- On a project-specific basis and/or using a multi-project emission factor.
- In a transparent manner.
- In a way that takes into account relevant national and/or sectoral policies and circumstances.
- In such a way that decreases in activity levels outside the project itself or due to force major cannot lead to the generation of ERUs. Using conservative assumptions and taking uncertainties into account.

(f) Monitoring Plan

The Marrakesh Accords also spells out the basic contours of the Monitoring Plan that each Project Participant will have to build into the PDD. Such a plan must provide for:

- The collection and archiving of the data necessary to estimate project-based GHG reductions and removals during the crediting period.
- The collection and archiving of the data necessary to determine emission and sink baseline levels.
- Identification of the ways in which the project could lead to increased emissions or reduced removals outside the project boundaries (i.e., leakage).
- The collection of information on the project's environmental impacts.
- Quality assurance and control procedures for the monitoring process.
- Procedures for the periodic calculation of emission reductions and sink removals attributable to the project and of any leakage effects.
- Documentation of all steps involved in these calculations.

(g) Additionality

Like CDM, the JI program requires that GHG reductions and removals be "additional to any that would otherwise occur" in the PDD Project. Participants must demonstrate that their reductions and/or removals are in fact additional. The IE will assess this as part of determining that the PDD meets program requirements.

A Framework for Carbon Accounting and Emissions reductions: an Australian Model

Australia's greenhouse gas accounts follow the international format. They can be re-arranged to provide a clearer basis for designing emission reduction policies and make them easier to understand and more transparent, especially in relation to non-fossil fuel emissions. This is essential to raise awareness, improve understanding and engage public support. The re-arrangement also highlights gaps in the data needed for full carbon accounting (http://www.climatechange.gov.au/ncas/publications/strategic/index.html).

Annual accounts (flows)

1. Separate fossil carbon from biocarbon

Fossil carbon is the carbon locked in fossil fuels, such as coal, oil and gas. Biocarbon is the carbon associated with living systems. The two are functionally different, especially because biocarbon is stored in the living landscape and stocks and flows are relevant, as well as emissions and uptake. For fossil carbon, for all practical purposes, only emissions need be accounted for.

2. Separate 'green' carbon from 'production' carbon

Biodiverse green carbon is the carbon held in natural ecosystems (including biomass, dead wood litter, and soil). It is the jewel of terrestrial carbon stores, ecologically permanent because it is resilient, self-reproducing and adaptive. Production carbon is the carbon held in agricultural and other production systems (including plantations and waste) where the primary purpose of management is not protection but there may be opportunities to improve carbon storage without compromising production.

3. Separate Emissions from uptake (removal of CO_2 from the atmosphere)

Fossil carbon only generates emissions. Biocarbon generates both emissions and uptake – for clarity, these must be disaggregated, not reported as net figures.

4. Account fully for all Emissions and uptake (full carbon accounting)

The UNFCCC accounts are moving towards full carbon accounting and should be used to underpin policy and action. Major gaps that remain include CO_2 emissions and uptake by soils in production systems, and emissions and uptake by non-forest vegetation and by forests in protected areas (parks and reserves).

5. Develop Accounts for Biocarbon Stocks

Deforestation and degradation have caused up to 35% of the increase in atmospheric CO_2 concentrations since the industrial revolution. This carbon debt should be reflected in a biocarbon stocks account, together with current stocks and potential for recapture. Issues such as the treatment of transient events (climatic events, fire etc) need to be worked out, as has been done for corporate financial accounting systems.

Emission Reduction Targets

Set separate targets and trajectories for fossil carbon emissions, biocarbon emissions and biocarbon uptake. This gives clarity, allows different policies to be applied for each target appropriate to its characteristics, reduces uncertainties (e.g. the impact of biocarbon offsets on a fossil carbon market), and reduces the potential for perverse outcomes. It focuses on reducing emissions as the primary goal and narrows the space for argument (e.g. whether regrowth of protected forests should be defined as 'anthropogenic).

Emissions Reduction Policies

Emissions trading should be applied to fossil carbon emissions only, with no 'offsets' from uncovered sectors. This confines the scheme to sectors with relatively uniform characteristics: only CO_2 emissions (not uptake) are caught; measurement is reasonably reliable; a small number of players are involved; and the CO_2 can be regarded as a commodity (unlike biocarbon where the CO_2 ought to be valued according to its source – emissions from clearing a rainforest are not equivalent to emissions from logging a plantation).

REDD (reducing emissions from deforestation and degradation) Plus

The primary aim for green carbon is to protect the stores in perpetuity, first by eliminating avoidable emissions (native forest clearing and logging are both largely avoidable) and then by enhancing uptake.

REDD—reducing emissions from deforestation and degradation—should be applied to green carbon, both forest and non-forest, in line with international moves. Regulation, together with transitional funding for affected industries and workers, is the simplest and most effective mechanism.

REDD Plus—managing biocarbon stores in perpetuity—should apply to green carbon stores and to opportunities for improving carbon storage in production systems. The sequestration potential as native vegetation recovers to full carbon carrying capacity and from soil conservation is large.

REDD Plus Fund—expenditure on reducing biocarbon emissions and enhancing uptake should be proportionate to the benefits (in 2006, it accounted for 32% of emissions, 22% of net emissions). Income should be accumulated from a variety of sources including restructuring existing government expenditures for these areas, a tranche from the emissions trading scheme, voluntary contributions, and industry investments. The Fund would provide a new income stream for land managers to look after native vegetation carbon stores for the long-term, give incentives to conserve soil carbon in production systems, and invest in major new R&D programs. It should be managed by an expert body (similar to the Future Fund or Building Australia Fund).

Sectoral Policies and Boundary Issues

Sectoral policies are likely to be needed to complement market based policies. Emissions trading should be augmented with energy efficiency standards and renewable energy feed-in laws. The agriculture and waste sectors are diverse, include multiple small players, and need specifically designed policies. Boundary issues for emissions trading need to address the treatment of fuelwood, bioenergy and biomass feedstock, with policies in place before the scheme begins.

Carbon Offsets

A Carbon Offset is a financial instrument aimed at a reduction in greenhouse gas emissions. Carbon offsets are measured in metric tons of carbon dioxide-equivalent (CO_2e) and may represent six primary categories of greenhouse gases. One carbon offset represents the reduction of one metric ton of carbon dioxide or its equivalent in other greenhouse gases. There are two markets for carbon offsets. In the larger compliance market, companies, governments, or other entities buy carbon offsets in order to comply with caps on the total amount of carbon dioxide they are allowed to emit. In 2006, about $5.5 billion of carbon offsets were purchased in the compliance market, representing about 1.6 billion metric tons of CO_2e reductions (The World Bank, State and Trends of the Carbon Market, 2007).

In the much smaller voluntary market, individuals, companies, or governments purchase carbon offsets to mitigate their own greenhouse gas emissions from transportation, electricity use, and other sources. For example, an individual might purchase carbon offsets to compensate for the greenhouse gas emissions caused by personal air travel. In 2006, about $91 million of carbon offsets were purchased in the voluntary market, representing about 24 million metric tons of CO_2e reductions (Ecosystem Marketplace, State of the Voluntary Carbon Markets 2007).

Offsets are typically achieved through financial support of projects that reduce the emission of greenhouse gases in the short- or long-term. The most common project type is renewable energy, such as wind farms, biomass energy, or hydroelectric dams. Others include energy efficiency projects, the destruction of industrial pollutants or agricultural byproducts, destruction of landfill methane, and forestry projects. Some of the most popular carbon offset projects from a corporate perspective are energy efficiency and wind turbine projects (EcoSecurities, ClimateBiz, Carbon Offsetting Trends Survey, 2008)

Carbon offsetting has gained some appeal and momentum mainly among consumers in western countries who have become aware and concerned about the potentially negative environmental effects of energy-intensive lifestyles and economies. The Kyoto Protocol has sanctioned offsets as a way for governments and private companies to earn carbon credits which can be traded on a marketplace. The protocol established the Clean Development Mechanism (CDM), which validates and measures projects to ensure they produce authentic benefits and are genuinely "additional" activities that would not otherwise have been undertaken. Organizations that are unable to meet their emissions quota can offset their emissions by buying CDM-approved Certified Emissions Reductions. Offsets may be cheaper or more convenient alternatives to reducing one's own fossil-fuel consumption. However, some critics object to carbon offsets, and question the benefits of certain types of offsets (Gillenwater *et al*, 2007).

Features of Carbon Offsets

Carbon offsets have several common features:

- *Vintage.* The vintage is the year in which the carbon reduction takes place.
- *Source.* The source refers to the project or technology used in offsetting the carbon emissions. Projects can include land-use, methane, biomass, renewable energy and industrial energy efficiency. Projects may also have secondary benefits (co-benefits). For example, projects that reduce agricultural greenhouse gas emissions may improve water quality by reducing fertilizer usage (United States Environmental Protection Agency, Carbon Sequestration in Agriculture and Forestry).
- *Certification regime.* The certification regime describes the systems and procedures that are used to certify and register carbon offsets. Different methodologies are used for measuring and verifying emissions reductions, depending on project type, size and location. For example, the Chicago Climate Exchange uses one set of protocols, while the CDM uses another. In the voluntary market, a variety of industry standards exist. These include the Voluntary Carbon Standard and the CDM Gold Standard that are implemented to provide third-party verification of carbon offset projects.

Sources of Carbon Offsets

The CDM identifies over 200 types of projects suitable for generating carbon offsets, which are grouped into broad categories. These project types include renewable energy, methane abatement, energy efficiency, reforestation and fuel switching.

Renewable Energy offsets commonly include wind power, solar power, hydroelectric power and biofuel. Some of these offsets are used to reduce the cost differential between renewable and conventional energy production, increasing the commercial viability of a choice to use renewable energy sources. Renewable Energy Credits (RECs) are also sometimes treated as carbon offsets, although the concepts are distinct. Whereas a carbon offset represents a reduction in greenhouse gas emissions, a REC represents a quantity of energy produced from renewable sources. To convert RECs into offsets, the clean energy must be translated into carbon reductions, typically by assuming that the clean energy is displacing an equivalent amount of conventionally produced electricity from the local grid. This is known as an indirect offset (because the reduction doesn't take place at the project site itself, but rather at an external site), and some controversy surrounds the question of whether they truly lead to "additional" emission reductions and who should get credit for any reductions that may occur (Gillenwater, Michael (a) & (b), 2008)

Carbon Credit Offsetting

Smoke and steam vapour from electric power station in New York (Source: http://science.howstuffworks.com/carbon-capture.htm/printable)

A 'Carbon credit' becomes a 'Carbon offset' when the value of the credit is invested in projects that reduce emissions. The idea is that the avoidance of one tonne of CO_2 released negates the effect of one tonne of carbon released somewhere else. Offsetting is a way for individuals and companies to compensate their own greenhouse gas emissions by investing in projects that remove carbon from the atmosphere or reduce or eliminate GHG emissions compared to the conventional methods (http://en.wikipedia.org/wiki/Carbon_offset). By offsetting emissions level for an activity (e.g. a flight), the idea is that the activity becomes "carbon neutral". Carbon offsets are generated by renewable energy, efficiency projects and other projects that avoid the release of CO_2 or other greenhouse gasses. The idea of paying for emission reductions elsewhere instead of reducing one's own emissions is known from the closely related concept of emissions trading. The commercial system has contributed to the increasing popularity of voluntary offsets among private individuals and also companies. Offsets may be cheaper or more convenient alternatives to reducing one's own fossil-fuel consumption. There is a debate therefore about the effectiveness of credits to actually reducing overall emissions, which is the goal needed to mitigate climate change. There can be three main arguments for the way Carbon offsetting helps to mitigate climate change:

- By financing renewable technologies and efficiency measures.
- By raising awareness of our lifestyle's impact on the climate.

The Kyoto Protocol has sanctioned Carbon offsetting as a way for governments and private companies to earn carbon credits which can be traded on the market. The Protocol established the CDM to validate and measure projects to ensure they produce authentic benefits. Organisations that have difficulties in meeting their emissions quota are able to offset by buying CDM-approved *Certified Emissions Reductions (CERs)*. The CDM encourages projects that involve, for example, sustainable power generation and changes in land use, although not all trading countries allow their companies to buy all types of credit.

Activities Generating Carbon Offset Credits

Many types of activities can generate carbon offset credits. In general they can be divided into two types:

1. Projects that prevent carbon from entering the atmosphere and

2. Projects that remove carbon from the atmosphere after it has been released (carbon sequestration)

Renewable energy such as installations of solar, small hydro, geothermal, and biomass energy fit into the first category because they are displacing the use of fossil fuels. Other types of preventative offsets available for sale on the market include those resulting from energy efficiency projects, methane capture from landfills or livestock, and destruction of potent greenhouse gases such as halocarbons.

Biological carbon sequestration works after the fact. Through the process of photosynthesis, all plants absorb carbon dioxide from the atmosphere; release oxygen molecules; and store carbon in plant tissues, especially roots. As plants die, carbon molecules remain underground unless disturbed by tillage or any operation allowing carbon atoms to combine with oxygen and escape into the atmosphere as carbon dioxide. Many soils in our region were native grasslands or forests prior to cultivation, containing high amounts of carbon. After years of tillage, a lowered equilibrium level of carbon in the soil has been attained, where the amount of carbon sequestered by plants annually is about equal to the amount lost in the atmosphere. Converting to no-till crop production, long-term grass seeding practices, implementing prescribed grazing, planting trees, and sustainably managing forests result in higher levels of carbon being stored in the soil. Landowners now can earn income in the carbon credit market for storing carbon, thereby reducing greenhouse gas emissions.

Marketing of Carbon Offsets

Natural Capital—a leading US based firm, develop projects for both the Chicago Climate Exchange and the Voluntary Carbon Standard protocols. Natural Capital is an aggregator member of the CCX and a registry member for the VCS. Each protocol has distinctive eligibility and trading requirements that make them suitable for particular projects.

- The Chicago Climate Exchange (CCX) is North America's only, and the world's first, greenhouse gas (GHG) emission registry, reduction and trading system for all six greenhouse gases (GHGs)—carbon dioxide, nitrous oxide, methane, hydrofluorocarbons, perfluorocarbons and sulfur hexaflouride. CCX is a self-regulatory, rules-based exchange designed and governed by CCX members. Members make a voluntary but legally binding commitment to reduce GHG emissions.
- Project carbon credits are certified to CCX specifications and bundled into 100 metric ton units called Carbon Financial Instruments CFI's that are exclusively traded on the exchange platform or through over the counter transactions. The Chicago Climate Exchange offers a classic commodity trading platform to efficiently buy and sell carbon offset credits.
- The Voluntary Carbon Standard (VCS) program provides a robust, new

global standard and program for approval of credible voluntary offsets. VCS offsets must be real (have happened), additional (beyond business-as-usual activities), measurable, permanent (not temporarily displace emissions), independently verified and unique (not used more than once to offset emissions). Projects developed to VCS standards generate Voluntary Carbon Units (VCU's) which are defined as 1 metric ton carbon dioxide. These types of credits are certified to VCS standards and entered onto a registry where they are tracked through the life of the credit.

Who's Who in a Carbon Offset Project

Designing, implementing and operating a carbon offset project requires the involvement of a large number of parties, stakeholders and authorities. Even though the parties involved differ from project to project some general categories and types of stakeholders can be defined as follows.

Project Owner

The operator and owner of the physical installation where the emission reduction project takes place can be any private person, company or other organization.

Project Developers

A person or organization with the intention to develop an emission reduction project could be the project owner, a consultant or specialized services provider.

Project Funders

Banks, private equity firms, private investors, non-profit organizations and other organizations may lend or invest equity to fund a project. Some of the standards have rules to what kind of funding, aside from the offset revenue, are acceptable for an offset project.

Stakeholders

Stakeholders are individuals and organizations that are directly or indirectly affected by the emission reduction project. Stakeholders include the parties interested in developing a specific project (e.g. owner, developer, funder, local population, and host community), parties affected by the project (e.g. local population, host community environmental and human rights advocates) and national and international authorities.

Standards Organization

In the absence of national and international legislation, standard organizations define a set of rules and criteria for voluntary emission reduction credits.

Brokers and Exchanges

In the wholesale market, emission offset buyers and sellers can have a transaction facilitated by brokers or exchanges. Exchanges are usually preferred for frequent trades or large volumes of products with standardized contracts or products, while brokers typically arrange transactions for non-standardized products, occasionally traded and often in small volumes.

Trader

Professional emission reduction traders purchase and sell emission reductions by taking advantage of market price distortions and arbitrage possibilities.

Offset Providers

Offset providers act as aggregators and retailers between project developers and buyers. They provide a convenient way for consumers and businesses to access a portfolio of project offsets.

Final Buyers

Individuals and organizations purchase carbon offsets for counter-balancing GHG emissions. Therefore, the final buyer has no interest in reselling the offset but will prompt the retirement of the underlying carbon offset.

The Role of the Voluntary Carbon Market

After giving a brief overview about how offset projects are developed, we now examine how the voluntary markets differ from CDM and how the standards that have been developed for the voluntary market approach carbon project management. Key differences exist between the mandatory and voluntary markets. Unlike the former, voluntary markets do not implement any particular policy mandates. The mandatory and voluntary markets occupy different but overlapping niches.

In order to better understand the voluntary market, it is helpful to ask what role it should play in protecting the climate and contributing to sustainable development. Compared to the compliance market, trading volumes are minimal in the voluntary market. The voluntary market does currently not make significant contribution to reducing GHGs. Furthermore, effective future climate policy will necessarily involve a gradual transition from voluntary to mandatory action, and eventual regulation (through allowance markets or other policies) of many of the actors currently involved in the voluntary market. While there will likely always be a voluntary offset market to serve those individuals or companies who want to push the envelope beyond what is possible through internal reductions and evolving regulation, a key role of the voluntary market is to shape the rules and procedures for offsets in future compliance markets. In other words, the voluntary market can be used as a testing ground for procedures, methodologies and technologies. The voluntary market can help achieve emissions reductions with projects that are too small for CDM, projects set in countries without a Kyoto target, or reductions that are ineligible for CDM for formal reasons other than quality (e.g. China CDM requires major Chinese ownership in project).The opinions on how the voluntary market can best do this, vary significantly. To clarify this ongoing discussion, a distinction is made below between three main points of view.

A. Voluntary Carbon Market Should Closely Build upon CDM

There are those, among them the governments of the UK and Norway, who argue that under the current market situation voluntary buyers can minimize their risk by buying compliance credits because the legal and procedural requirements for CERs are already well established. The current voluntary offset market is seen as potentially undercutting the compliance market with cheaper offsets that are not clearly additional and sending the wrong price signals. Since the public and the media often do not distinguish between the compliance and the voluntary market, there is also a risk of damaging the reputation of compliance markets. To secure quality and transparency in the voluntary market, it is argued that voluntary offset standards should closely follow CDM procedures and apply them to VERs (e.g. the CDM approach to additionality, the documentation of reductions, and the monitoring and verification processes). Standards that share this viewpoint include VER+ and the Voluntary Offset Standard (VOS).

B. Voluntary Market should be More Stringent than CDM

Some have taken this argument even further and have created standards with the explicit goal of enhancing the quality of offsets from both markets by requiring explicit social and environmental benefits as well as strict accounting standards. Standards that espouse this viewpoint include the Gold Standard and the Climate Community & Biodiversity (CCB) Standard.

C. Voluntary Market should complement and be different from CDM

On the other end of the spectrum are those who argue that voluntary offset standards should be less stringent and bureaucratic than the standards in the mandatory markets. They agree that the voluntary market can serve as a testing ground for future policy but they argue that in order to preserve the voluntary market's creativity and innovation it must be protected from too many bureaucratic requirements. They distinguish between the compliance market, where regulatory obligations must met, and the voluntary market, were no such obligations exist and where the emphasis is on creating a market for innovative projects with as little administrative burden as possible. Most carbon offset providers who do not use a third party standard but follow their own procedures fall under this category. The Voluntary Carbon Standard (VCS) also adheres more closely to this viewpoint. Although VCS incorporates many of the CDM procedures and guidelines, it is in principal a standard that looks to loosen the requirements for VER projects to allow for more flexibility and innovation.

Scale of the Compliance and Voluntary Carbon Offset Markets

While the voluntary market in carbon offsets expects very rapid growth over the next few years, recent reports suggesting that demand could reach ~400 million tonnes of CO_2-equivalent each year by 2010 from a 2006 figure of ~20 million tonnes, it is still a smaller market than the compliance market which currently totals, in terms of the CDM, ~155 million tonnes each year, and which is expected to have delivered close on 2 billion tonnes of CO_2-equivalent in offsets by 2012. Any tougher carbon/GHG targets accepted internationally following the expiry of the first Kyoto phase in 2012 would probably serve to increase the scale of the CDM or its successor still further. One of the strengths of the voluntary market highlighted to us in evidence is its ability to be a source of 'innovation' for the credits and projects operating outside the compliance market.

The rules and regulations of the compliance market are strict and often problematic for many smaller projects which do not have the funds, time or expertise to cope with the burden of administration. Methodologies for new project types are slow to develop and evolve. Away from these constraints, the voluntary market is able to develop projects in different countries from where most of the compliance market projects are based, in particular in Africa where currently only 1% of CERs originate. Often, voluntary market projects have more 'value-added' characteristics than those found in the compliance market, such as additional environmental or sustainability benefits. The voluntary market also supports a diversity of projects and can act as a testing and learning ground for new projects wanting to enter the compliance market. This is particularly the case for forestry and land-use change projects. There are very few such projects in the compliance market at the moment, but developers are working on methodologies which they hope will allow them to be part of the CDM.

Key Elements of Carbon Offset Standards

All sides have contributed to the discussion on the role the voluntary carbon market can play to further climate protection. Numerous new standards and registries have been introduced over the last couple of years and the competition among carbon offset standards has increased dramatically since large financial institutions, businesses, and industries have gotten involved in the carbon trade. In the next section we will discuss the elements that are necessary to create an effective carbon offset standard.

> *"Carbon offsets are an intangible good, and as such their value and integrity depend entirely on how they are defined, represented, and guaranteed. What the market lacks are common standards for how such representations and guarantees are made and enforced"* (Broekhoff, 2007)

Clearly, no standard can ever be perfect, and as pointed out in the discussion above each of the currently available standards is based on a particular view of the voluntary offset market. Yet it is safe to say that notwithstanding these differences, the best and most successful standards will be those that are simple yet rigorous and have very wide support from carbon project developers, offset traders and buyers, environmental NGOs and the financial industry. A complete and full-fledged carbon offset standard must include the following three components:

- Accounting Standards
- Monitoring, Verification and Certification Standards
- Registration and Enforcement Systems

Accounting standards ensure that offsets are "real, additional, and permanent." They include definitions and rules for the elements that are essential during the design and early implementation phase of a project. These include additionality and baseline methodologies, definitions about accepted project types and methodologies, validation of project activity etc Monitoring, Verification and Certification Standards ensure that offset projects perform as was predicted during the project design. Certification rules are used to quantify the actual carbon savings that can enter the market once the project is up and running. There is sometimes a lag time between the start of a project and when it starts producing carbon offsets. This is especially true for forestry projects – the trees have to grow for a few years before they have absorbed enough carbon that can be quantified and sold. Monitoring, verification and certification happen after validation and implementation of the project. Yet procedures and protocols for monitoring and verification have to be included very early on in the project design phase Verification and certification are *ex-post* assessments of what has actually been produced, as opposed to validation which is the *ex-ante* assessment of whether a project qualifies against a standard, provided it is going to do what it promises in the project design documentation.

Registration and Enforcement Systems ensure that carbon offsets are only sold once and clarify ownership and enable trading of offsets. They must include a registry with publicly available information to uniquely identify offset projects and a system to transparently track ownership of offsets. In the following sections we discuss each of these elements in more detail and compare the Additionality and Baseline Methodologies

"Offsets are an imaginary commodity created by deducting what you hope happens from what you guess would have happened." (Dan Welch quoted in *The Guardian*, June 16 2007)

Additionality is an essential element needed to ensure the integrity of any baseline-and-credit scheme. Yet additionality is very difficult to determine in practice. Many different tools have been developed to maximize the accuracy of additionality testing and to minimize the administrative burden for the project developer. There are two distinct approaches to additionality testing: Project based additionality testing and performance standards.

Project-Based 'Additionality' Testing

Project based additionality testing evaluates each individual project on a case by case basis. The following is a short selection of additionality tests that are commonly used:

Legal and Regulatory Additionality Test (Regulatory Surplus)

If the project is implemented to fulfil official policies, regulations, or industry standards, it cannot be considered additional. If the project goes beyond compliance ("regulatory surplus"), it may be additional, but more tests are required to confirm this. For example, an energy efficiency project might be implemented because of its cost savings and would in this case not be additional.

Investment Test

This test assumes that an offset project is additional if it would have a lower than acceptable rate of return without revenue from the sale of carbon offsets. In other words, the revenue from the carbon offsets must be a decisive reason for implementing a project. The investment test is consistent with a microeconomic view of behaviour and in theory would be a perfect additionality test. But in reality there may be projects whose finances make them look non-additional that are still "additional" because of existing non-monetary barriers.

Barriers Test

This test looks at implementation barriers, such as local resistance, lack of know-how, institutional barriers, etc. If the project succeeds in overcoming significant non-financial barriers that the business-as-usual alternative would not have had to face, the project is considered additional.

Common Practice Test

If the project employs technologies that are very commonly used, it might not be additional because it is likely that the carbon offset benefits do not play a decisive role in making the project viable. Which test is best suited to validate additionality depends on the type of project. An additionality test appropriate for one type of project (e.g., a simple regulatory test for methane flaring, where there is no reason to do the project if not required by law) might not be sufficient for other kinds of projects (e.g., energy efficiency, where there could be plenty of reasons for doing a project besides complying with regulations).

The main issue with project-based additionality testing is that the determination of whether a project is additional can be quite subjective. A developer can claim that their project's IRR was too low without a carbon revenue stream, and that the carbon revenues therefore made the project viable. But who can really determine what level of IRR is acceptable to a given company, and thus whether the additionality demonstration is valid? Such additionality claims can only be tested with access to internal company information relating to the financing of the project, yet this information is in most cases confidential.

Performance Standards of Project-based 'Additionality'

Performance Standards try to address some of the weaknesses of project-based additionality tests in that they do not rely on examining each individual project but establish a threshold for technologies or processes to determine additionality. This approach is associated with simpler procedures and lower transaction costs for project developers. Performance standards are developed and/or approved by standard organizations and therefore shift much of the project developer's administrative burden to the standard organization. Drafting performance standards requires comprehensive data collection and verification, as well as regular updates. The political process to approve such performance standards may take a long time and may only be feasible for certain industries (e.g., small renewable heat and power, biomass, or small energy efficiency). Performance Standards typically use aggregated data on project or technology characteristics to establish a threshold (e.g., a performance indicator such as an emissions rate or a market indicator such as a penetration rate) that must be met or exceeded in order for a project to be deemed additional. Performance Standards include among others positive technology lists and benchmark approaches.

Benchmark Approaches

The most widely discussed of performance standards is the emissions-based (benchmark) additionality test. This test establishes a generic baseline scenario—referred to as a benchmark—against which all projects of a given type are assessed. Employing such an assessment as an additionality test presumes that technologies with emissions lower than a given emission rate standard would not be deployed in the absence of the offset programme incentive. This method works best in sectors or applications where business-as-usual technologies and fuels do not vary widely in emissions rates. In sectors like electricity generation, where emissions rates can be as low as near zero for some hydroelectric plants or relatively high for coal-based plants—both of which are conventional technologies—benchmarking emissions rates can be problematic. For example, any threshold above zero would deem all new hydroelectric or wind development additional. Several CDM baseline methodologies include benchmark approaches for calculating baselines and emission rates, but additionality must still to be established by using project-based additionality tests.

Positive Technology Lists

Positive technology lists simply define which technologies are automatically considered additional if installed in a certain geographic region. The project developer must still use a baseline methodology to determine the numbers of offsets a project will create. Again, such lists are transparent and enable faster and simpler processing of offsets. They also shift much of the administrative burden from individual project hosts to a centralized standard-setting entity. The main problem with performance standards is that they may be too simple and broad. All activities whose emissions fall below the benchmark emissions are awarded credits, regardless of whether they would have taken place anyway. Projects that are non-additional are referred to as free-riders. One proposed solution to the problem of free-riders would be to discount offsets by the number of expected free riders. For example, if a benchmark is set at the 20th percentile, we can expect 20% of projects to be free-riders. If all offsets were then discounted by 20%, the overall environmental integrity would be preserved. Yet discounting is not a perfect solution either since it may skew the results or favor non-additional projects, which by definition rely less on offset revenue.

To summarize, any additionality test, no matter how quantitative and seemingly objective, will always create some number of false positives (projects that appear additional although they are not) and some number of false negatives (projects that appear not to be additional although they are). The design of the test determines if it will err on the side of false positives or false negative. The judgment as to which is more acceptable is determined by a political process. It is important to understand that while false positives and false negatives both impair economic efficiency; only false positives undermine the environmental integrity of offsets. In other words, it is the false positives – offsets from non-additional projects – that lead to increases in emissions and therefore hamper climate protection goals. The most practical and viable option for additionality testing may mix elements of project based and benchmark approaches.

Emissions Trading Baselines

What are emission baselines and why are they needed?

The Kyoto Protocol establishes two project-based mechanisms for greenhouse gas (GHG) mitigation: the clean development mechanism (CDM) and Joint Implementation (JI). These GHG mitigation activities are intended to result in "additional" emission reductions, and investing in these activities via a JI or CDM project can give rise to "emissions credits" for sale on the international market.

To determine the number of credits that could be generated by an individual JI or CDM project, an indication is needed of what GHG emissions would have been in the absence of that project (i.e. what would have happened otherwise). The amount of GHG emitted in the hypothetical non-project scenario is referred to as a project's *baseline*. A baseline is thus a quantification of this hypothetical emission level and may be used for comparative purposes to test for the GHG "additionality" of an individual project. CDM projects will qualify for certified emission reduction units (CERs) and JI projects will qualify for emissions reduction units (ERUs) if they are additional relative to the baseline.

Why standardize Emission baselines?

Experience with project-specific emission baselines has been gained during the pilot phase of "activities implemented jointly" (AIJ), but is rather limited. The majority of emission baselines were drawn up on a project-specific basis. Analysis of such experience (OECD, 1999; Schwarze 2000) has indicated that in the absence of detailed guidelines on how to set up an emissions baseline, the methodologies and assumptions used are often incomparable, inconsistent and not transparent.

A project developer could presumably be a private entity, a non-governmental organization (NGO) or a government. Depending on what the rules establishing the CDM contain, these entities, NGOs or governments could be based either in Annex I or in non-Annex I countries (i.e. the potential range of "project developer" could be from an energy company in an Annex I country, to a host country government). The Kyoto Protocol's text on JI specifies that legal entities, in addition to Parties, are eligible to participate; it is silent on this issue in the text on the CDM.

Standardizing baselines, methodologies and/or individual parameters would help to ensure consistency in the treatment of similar projects in similar circumstances. Standardization would provide a high degree of transparency in baseline determination and could also, if developed by independent experts, limit the level of gaming/free riders (see OECD/IEA 2000 for a more detailed discussion). Compared to a project-specific approach, standardized baselines could also reduce the potentially high transaction costs associated with setting baselines, as one baseline could be applied to several projects.

The standardization of methodologies to establish emission baselines and, ideally, the use of pre-established baseline values (i.e. multi-project baselines) could also facilitate and accelerate the required governmental acceptance and approval procedures for proposed projects in potential host countries. This may be particularly true in countries with limited administrative capacity and/or experience in establishing baselines for projects. In addition, developing a methodology (or an actual value) for a multi-project baseline could help increase data collection and/or availability at a national level, which could create positive synergies with a country's other requirements under the United Nations' Framework Convention on Climate Change and its Kyoto Protocol. Furthermore, agreement on standardized baselines, methods or assumptions would also facilitate project developers' assessment of potential JI or CDM projects. Standardized emission baselines may be especially important to encourage the initiation of small JI/CDM projects.

Since a baseline represents a hypothetical future scenario, it can change substantially with assumptions about future growth in economic activities and energy demand. While a baseline should allow for a reasonable growth of emissions in line with existing plans, emissions profiles of relevant economic activities and expected economic growth in a host country, it should also ensure that possibilities to inflate the projected baseline emissions are minimal. Standardized approaches could help by providing guidance on methods and/ or assumptions (e.g. about projected level of activities) while relating them to existing planning and activity structures in a country. Ensuring environmental effectiveness and encouraging participation in the project-based mechanisms need not be contradictory aims, yet there is some tension between the two. Baselines must seek to strike a balance to be environmentally credible while at the same time providing appropriate incentives to potential project developers to invest in emission reduction projects, and transparency, which will facilitate review and build confidence about "additional" environmental benefits to result from JI and CDM projects.

In practice, the success of the project-based mechanisms will depend on their overall contribution towards reducing GHG emissions. Elaborating the project-based mechanisms in a way that only a few JI and CDM projects are implemented is not likely to yield net GHG reductions nor will it enhance sustainable development. Two key factors related to baselines are expected to influence the success of the project-based mechanisms:

(i) The baseline development process: Minimizing the transaction costs associated with establishing a baseline should provide greater incentives to initiate projects. However, the extent of possible standardization is likely to vary according to types of projects and according to sectors. For some types of projects, it may be possible to standardize the baseline value, while for others it may be possible to only standardize methodologies or some parameters. Any type of standardization or even "good practice" guidance can enhance transparency and help to limit free riding and gaming. In this way standardization is a means to achieve both lower transaction costs and improved environmental performance. The stringency of the baseline: Using an overly stringent baseline (one at the lowest end of a possible range of emission levels) may ensure that only additional JI/CDM projects are approved, but may also disqualify some

(ii) Environmentally additional projects. A less stringent baseline may mean that some non-additional credits are generated by projects, but also increases the possibility that more additional projects are undertaken. Efforts to standardize the stringency of a baseline for a given type of project will thus need to balance the desire that only truly environmentally additional projects are adopted with the desire of ensuring that all additional projects are undertaken.

Multi-Project Baselines

Calculating greenhouse gas (GHG) emissions reductions from projects requires construction of a hypothetical baseline that approximates emissions levels without the project. Such baselines can be project-specific, multi-project, or a hybrid of the two (Ellis and Bosi, 1999). Project-specific baselines are determined on a project-by-project basis using specific measurements or assumptions. Multi-project baselines use existing or estimated emissions levels from a defined set of actual or projected projects to derive a baseline level. The hybrid approach combines project-specific and standardized parameters to derive a baseline (Ellis and Bosi, 1999).

Rationale for Use of Multi-Project Baselines

The rationale for exploring the use of multi-project baselines as an alternative to project-specific baselines is to seek a balance between ensuring environmental integrity and minimizing transaction costs while encouraging emissions reduction projects. Project-by-project baselines may have higher transaction costs than multi-project baselines, reducing the number of projects that attract investment. Experience with other project evaluations has shown that construction of project-specific baselines is time-consuming, costly, and can be highly uncertain. Thus, the concept of standardized baselines across many projects, for particular sectors or given technologies, has emerged. These multi-project baselines can be used as an alternative to project-specific baselines depending upon the preference of the developer and/or the host country government. The aim of this paper is to explore alternative options for multi-project baselines.

Fig. 6: Effect of Baseline Stringency and Complexity on Project Numbers and a Project's Environmental 'Additionality'

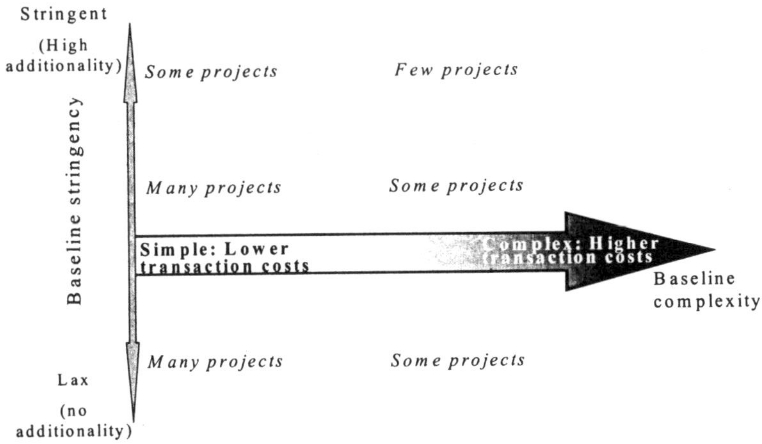

Source: Adapted from Ellis and Bosi (1999).

Project-specific baselines can be static or dynamic. Static baselines are set at the time of project approval and remain unchanged for the duration of the project, while dynamic baselines may be revised during the course of the project should new information about the baseline conditions require a re-examination of the original baseline. Multi-project baselines too could be adjusted in a similar manner if the original baseline were to undergo an unexpected change. Establishing a baseline for a particular activity, sector and/or region potentially simplifies the calculation of emissions reductions. Baselines need to be simple enough to be practical in developing countries.

What issues have to be considered in baseline standardization?

Given that baseline standardization has potentially significant advantages, the next question is what form baseline standardization should take. Baseline standardization could take a number of different forms. These include standardizing:

- Absolute baseline levels, or benchmark values (e.g. for project type X, the baseline to be used is Y kg CO_2/ton output if it is expressed in rates, or Z t CO_2/year if expressed in total CO_2 emissions saved);
- Methodologies that would apply to a group of projects (e.g. for project type P, the baseline should be equivalent to the average performance of similar recently installed equipment); and/or
- Parameters that could be used in baselines that have both project-specific and standardized components ('hybrid' baselines, e.g. for project type N, total emissions equal A + B + C. C needs to be calculated

using site-specific data, but methodologies for A and B are given).

Which form is most appropriate to use will vary according to the sector and project category. Thus, determining project categories (i.e. assessing the types of projects to which one particular baseline can be used) is in itself also an important step.

Analysis carried out by the OECD and IEA (OECD/IEA 2000) indicates a number of important parameters when assessing how to develop emission baselines for a particular project type:

- The appropriate geographic aggregation for a baseline (e.g. should a baseline be developed based on data aggregated at a local, regional, country or international level
- The length of time over which a baseline can be used to assess the emission performance of a given project (i.e. the crediting lifetime of a baseline)
- Which project boundary is appropriate (e.g. which gases and sources should be included in the baseline, whether a baseline should be for an entire process or individual process steps);
- Which data assumptions are appropriate, and the availability of this data; and
- The units in which baselines should be expressed (i.e. whether to express baselines in terms of absolute emissions, such as t CO_2, or in terms of a rate[1], such as t CO_2/GWh).

Many of these aspects are inter-linked. For example, the level of geographic aggregation of a baseline can influence project boundaries and appropriate baseline units. Variability across the different projects that are developed under a single baseline or baseline assumption will drive decisions on other standardization factors, such as aggregation, lifetime, etc. Baseline levels may vary between project types in a given sector (e.g. energy efficiency and process change projects in the cement sector). Baselines may also vary within project types (e.g. baseline levels for an electricity generation project in India may be different from a similar project in Brazil). Any baseline guidelines or reference manual will need to determine acceptable levels of variation in project types and project performances before deciding to use a particular baseline to assess the emission reductions from a particular project.

Baselines for Computing Project's Carbon Credit Benefits

In order to calculate an offset project's GHG benefits, a baseline must be established. This baseline expresses the business-as-usual scenario. In other words, it represents the counterfactual scenario of what would have happened if the project had not been implemented. The number of credits generated by the project is equal to the difference between emissions in the baseline scenario and emissions resulting from the project. The key difficulty is that the baseline scenario is a *hypothetical* scenario; by definition it describes another reality, one in which the activity is *not* implemented as an offset project. As that scenario will never occur, there is no fail-safe way to divine with certainty what the results of that scenario would have been. The baseline must be explicit and concrete enough to allow an estimation of the corresponding GHG emissions, so that the benefits of the offset project may be calculated. Baselines should be calculated conservatively so as not to overestimate the achieved emissions reductions. The baseline must be based on verifiable information sources and documented in a confirmable manner.

As with additionality, baselines can be established using project based or performance based approaches. These may either be the same as the approach used to determine additionality or different. Performance based tools may increase transparency and decrease costs; however, they must be well designed to avoid inaccuracies and to ensure environmental integrity. If the baseline is defined by a performance standard, it provides a credible estimate of reductions in aggregate. Each standard usually chooses one approach or the other, although some use a combination. Some standards prescribe upfront the methods that project developers must use to estimate baseline emissions for each type of allowable project activity (top-down). Others allow project developers to propose appropriate methods for new types of projects, following general programme guidelines. A purely bottom-up standard (like the CDM) is one in which project developers must propose, and win approval for, appropriate methods for every project category. Some programs may be a mix of top-down and bottom-up.

Baselines can be Static or Dynamic

A static baseline does not change over time, whereas a dynamic one is updated periodically based on ex-post observations, and emission reductions are calculated based on the most current baseline. Many standards have different levels of requirements for different classes of projects. For example, some might have simplified baseline methodologies for small scale projects.

The Economics of International Emissions Trading

The potential costs of limiting net GHG emissions are very large. It is therefore extraordinarily important that limitations be achieved in the most cost-effective manner basically through market mechanisms that provide incentives and business opportunities for private entities. Reducing the costs of achieving limitations can also promote the likelihood of successful international agreement on and implementation of limitations measures.

Emissions trading systems can further these objectives by capitalizing on differences in the costs of limiting emissions or enhancing sinks in different sectors and nations, and steering private sector investments to the lowest cost

GHG reducing opportunities. Emissions trading can attract consent by nations to participate in arrangements to limit GHG emissions with fewer distortionary side-effects than command regulation or emissions taxes (http://www1.law.nyu.edu/kingsburyb/spring03/globalization/SandsStewartPlurilateral_March24.pdf).

To cite specific example, let us consider two European countries, namely Germany and Sweden. Each can either reduce all the required amount of emissions by itself or it can choose to buy or sell in the market. It is possible for a country to reduce emissions using a Command-Control approach, such as regulation, direct and indirect taxes. The cost of that approach differs between countries because the *Marginal Abatement Cost* (MAC)—the cost of eliminating an additional unit of pollution—differs by country. It might cost China $2 to eliminate a ton of CO_2, but it would probably cost Sweden or the U.S. much more. International emissions-trading markets were created precisely to exploit differing MACs (http://en.wikipedia.org/w/index.php?title=Emissions_trading&action=edit&section=4).

For this example let us assume that Germany can abate its CO_2 at a much cheaper cost than Sweden, e.g. $MAC_S > MAC_G$ where the MAC curve of Sweden is steeper (higher slope) than that of Germany, and R_{Req} is the total amount of emissions that need to be reduced by a country.

On the left side of the graph is the MAC curve for Germany. R_{Req} is the amount of required reductions for Germany, but at R_{Req} the MAC_G curve has not intersected the market allowance price of CO_2 (market allowance price = P = λ). Thus, given the market price of CO_2 allowances, Germany has potential to profit if it abates more emissions than required.

On the right side is the MAC curve for Sweden. R_{Req} is the amount of required reductions for Sweden, but the MAC_S curve already intersects the market price of CO_2 allowances before R_{Req} has been reached. Thus, given the market allowance price of CO_2, Sweden has potential to profit if it abates fewer emissions than required internally, and instead abates them elsewhere.

In this example, Sweden would abate emissions until its MAC_S intersects with P (at R*), but this would only reduce a fraction of Sweden's total required abatement. After that it could buy emissions credits from Germany for the price 'P' (per unit). The internal cost of Sweden's own abatement, combined with the credits it buys in the market from Germany, adds up to the total required reductions (R_{Req}) for Sweden. Thus Sweden can also profit from buying credits in the market (Δ d-e-f). This represents the 'Gains from Trade', the amount of additional expense that Sweden would otherwise have to spend if it abated all of its required emissions by itself without trading.

Germany made a profit by abating more emissions than required: it met the regulations by abating all of the emissions that was required of it (R_{Req}). Additionally, Germany sold its surplus to Sweden as credits, and was paid 'P' for every unit it abated, while spending less than 'P'. Its total revenue is the area of the graph (R_{Req} 1 2 R*), its total abatement cost is area (R_{Req} 3 2 R*), and so its net benefit from selling emission credits is the area (Δ 1-2-3) i.e. Gains from Trade.

The two R* (on both graphs) represent the efficient allocations that arise from trading.

- Germany: sold (R* - R_{Req}) emission credits to Sweden at a unit price 'P'.
- Sweden bought emission credits from Germany at a unit price 'P'.

If the total cost for reducing a particular amount of emissions in the 'Command Control' scenario is called 'X', then to reduce the same amount of combined pollution in Sweden and Germany, the total abatement cost would be less in the 'Emissions Trading' scenario i.e. (X – Δ 123 – Δ def).

The above example applies not just at the national level: it applies just as well between two companies in different countries, or between two subsidiaries within the same company.

Cost Implications of Greenhouse Gas Mitigation

International trade holds the potential of reducing costs of controlling world emissions of greenhouse gases (GHGs) because the nations of the world experience very diff e rent costs for achieving emissions reductions on their own. However, the potential gains from trade, like the costs of compliance themselves, may be very unevenly distributed across the world's participants. While all of the parties to an agreement stand to gain collectively under trade in emissions rights as compared with "independent compliance" (i.e., each country meeting its obligations alone), non-participants in the agreement may either benefit or not depending on their own particular circumstances. The detailed rules for trading affect how effective trading could be, as well as the level of gains that would be captured in practice.

Details of the trading rules will influence both the total gains from trade and distribution of such gains. Key issues include definitions of the emissions rights to be traded, the rules for crediting carbon sinks, and regulations governing participation in the trading framework. In addition, there are economic uncertainties, such as the behavior of countries that have significant market power in supplying emissions credits, and the transaction costs associated with trading and enforcement. These effects could significantly increase the costs of mitigation compared to the most favorable case and could reduce the amount and benefits of trading.

A number of global economic models have been used to estimate the effects of emissions trading. Empirical results derived from these models can be summarized as follows:

- Costs of controlling carbon emissions would be significantly lower if trade in carbon emissions allowances were permitted than if each nation had to meet its emissions reduction responsibilities alone the broader the trade possibilities, the lower the costs of control.

- All parties with GHG emissions mitigation obligations benefit from trade. Both permit buyers and permit sellers will benefit.
- Parties without obligations may be better or worse off under a trading regime relative to a regime that does not allow trading. However, given a regime that allows trading among parties with obligations, parties without obligations will be better off trading (i.e., selling emissions reductions) than not trading.
- Because the costs of fuels could be affected by emissions control and emissions trading, countries and regions may be affected whether or not they participate in emissions reduction and in emissions trading. Parties without obligations may be either better off or worse off after obligations are established for others. For example, if emissions trading are prohibited, the prices paid to fossil fuel producers are reduced, and the energy—exporting countries are worse off relative to a no-control case. Emission trading mitigates this effect. Results for other non-participating regions are more ambiguous.
- Gains from trade are sensitive to the difference between the base case and target emissions and to the difference in marginal (incremental) abatement costs among countries. For any limit to emissions, the higher the future level of emissions is expected to be without intervention, the more difficult and costly mitigation is expected to be. Although the gains from trade depend on the differences between countries' marginal abatement costs, not their absolute level, and the gains from trade are larger for more ambitious emissions targets.
- The actual cost savings from trade in emissions are likely to be less than the theoretical savings shown in most analyses performed with integrated assessment models1 because these models do not include the various measurement, verification, trading, and enforcement costs that would characterize any real trading system. Programs must be carefully designed to assure that the potential gains from trade are realized.

There are two necessary steps for carbon emissions and reductions to be certified. Certification is needed so that the resulting credits are recognized and may be legitimately traded. We have to know what a legitimate credit is and there has to be a method for accreditation by an independent party. Certification is necessary since a carbon credit is not a deliverable commodity. The value of this intangible asset exists because of trust that buyers place in the system. The success and survival of the system is thus predicated on the credit-worthiness of emissions reductions. At the present time, the profession has gone a long way towards denying the existence of intangible assets and will have a great deal of trouble recognizing companies that are creating them.

Carbon credits are created when GHG emissions are reduced below some business-as-usual baseline. This baseline allowance is set by governing bodies such as the United Nations (through the UN Framework Convention on Climate Change) or, in North America, the Chicago Climate Exchange. The basic requirements are that emission reductions are measurable and last at least five years. The potential agency problem is addressed through the mandatory use of independent third parties to verify that emission reductions are real (UNFCC, 2005). This will create a large demand for professionals that can provide this attestation service. These transaction costs are not emphasized in discussion about which regulatory model to adopt.

The verifiability problem has not been ignored by the environmental activists. An independent third party carries out the function of verifying, monitoring, and certifying emissions reductions (Karmali, Price-Jones & Ebert, 2003). The oversight body (i.e. Supervisory Committee for the Kyoto Protocol) will only recognize emissions reductions certified by one of its accredited auditors. The qualifications to become a certified emissions auditor are not onerous. The auditing party must demonstrate technical expertise of certification criteria, possess strong internal systems and controls, and have the financial resources and liability insurance in place to carry out its duties. Periodic spot checks of the auditors are routinely carried out by the accreditation body in order to ensure ongoing compliance with existing standards. The possibility of errors and malfeasance are obvious. Market shocks due to verification failures will occur.

The verification of carbon credits is therefore a difficult issue. It is likely the profession will not allow companies to record the value of carbon credits they have created until they are sold. Another likely response will be to exclude the liability associated with the need to purchase carbon credits or invest improvements to meet standards each time the output measure changes. In both cases, the professions response results in understating reality and therefore dimming corporate response and market awareness. Continuing to uphold the current strict interpretation of a verifiability standard will hinder the professions support of environmental activism.

International Emissions Trading Issues under Kyoto Protocol

The Kyoto Protocol provisions on emissions trading and subsequent UNFCCC Conference of the parties (COP) discussions leave many aspects of the system open for future decisions. The following section clarifies several issues that are important considerations in deciding principles, modalities, rules, and guidelines for international emissions trading under Article 17 of the Protocol:

(i) participation by legal entities;

(ii) national and international systems for accounting for changes to assigned amounts as a result of trading and for verifying changes in ownership; and

(iii) Encouraging compliance through the rules of the trading system.

These issues are inter-dependent. The types of participants in the trading system will affect national systems for registering transactions. Requirements for national systems will affect the functions needed at international level. Efforts to encourage compliance may influence decisions on national systems and eligibility to participate.

A conceptual model of an international emissions trading system is developed involving the above three components. The model is based on Kyoto Protocol provisions as far as possible, and assumes minimum international institutional requirements, and strong reliance on national systems for Parties that implement domestic trading systems.

A. *Participation by Legal Entities*

The participation of private sector entities in international emission trading would increase the number of trades and improve market efficiency through greater liquidity. Private sector participation would also reduce the potential for large sellers or buyers to influence market prices in their favour. Entity trading would not alter the fact that Parties, not firms, are ultimately responsible for meeting the Kyoto Protocol commitments. Under Article 17 of the Kyoto Protocol, international emissions trading may take place among Parties with quantified national emission commitments, or "(initial) assigned amounts". Participation by authorized legal entities is not included in Article 17, nor is it explicitly excluded. The conceptual model developed for this paper assumes that legal entities will participate, and that international emissions trading will take place between governments of Parties with national emission commitments, between private sector firms, and between firms and governments. However, the eligibility of authorized legal entities to trade might be affected under international rules if there were a breach of eligibility requirements by the Party in which they are located.

B. *Domestic Emissions Trading Systems*

Parties may devolve some of the responsibility for greenhouse gas mitigation to private entities through a wide range of policies and measures, including domestic emission trading systems. Domestic emission trading systems could link directly to the international emission trading system and the other Kyoto mechanisms. In practical terms, this would mean that national governments would allow authorized entities to use parts of assigned amount (PAA), or tradable units from the other mechanisms, that are purchased from other countries to demonstrate entity compliance with domestic commitments. Similarly, national governments would allow authorized entities to sell PAA to an entity that is located in another country if their emissions are below their allowed level. Countries are likely to implement different domestic emissions trading systems, depending on their institutional capacity, policies, and legislation, cultural preferences, and national circumstances. Some might choose not to implement a domestic emissions trading system but might still want to authorize one large legal entity to trade in the international system. Other countries might choose to implement a domestic emissions trading system, allowing domestic trades among entities, but might not to allow these entities to participate in the international emissions trading system.

Key Requirements of an International Emissions Trading System

National Systems

Most aspects of national systems can and should be left to the discretion of individual countries. However, decisions at the international level may affect certain features of national systems. There is a growing consensus that eligibility requirements should be established to ensure that adequate national systems are maintained by Parties that choose to participate in international emissions trading. One key function at the national level is to account for changes to the national pool of PAA that occur as a result of trading by both government and by private entities. Changes in the national holdings of PAA throughout the commitment period will, in the aggregate, affect the final level of any Party's assigned amount. A minimum international requirement for national systems is national accounting of changes to government and entity holdings of assigned amount. Another minimum international requirement for national systems is national records of all international transactions by the government or entities. Parties' could then check their records if ownership of PAA is disputed. These records should be available for international review if necessary.

International Trading Systems

The simplest international system would simply compile and check Party reports of adjustments to their assigned amounts. The international review mechanism could investigate any inconsistencies among Party reports by checking the national records that Parties hold. If minor inconsistencies that are identified, the Parties concerned could rectify them without recourse to international procedures. A more sophisticated international system, perhaps one that might evolve in the future, could record the change of title for each PAA that is traded. An international registry could receive electronic reports of the identification numbers of each PAA traded. The international registry would hold information about all transactions. A central computer check could then confirm that each PAA is held by only one Party or one entity.

Enhancing Compliance through the Trading System

Ratification of the Kyoto Protocol will be an indication that Parties genuinely intend to meet their commitments. However, because of the legally binding nature of these commitments, many analysts and Parties have stressed the importance of responses to non-compliance.

In addition to Article 18 provisions, responses could be included in the rules for international emissions trading to enhance incentives for compliance with the Kyoto Protocol targets and to ensure the integrity of the trading system:

- Eligibility requirements and insurance approaches act as preventive measures to encourage compliance by individual countries;
- Devolving PAA (and the responsibility to reduce emissions) to entities that operate under strong domestic systems could help countries to achieve their assigned amounts at lower cost and increase the likelihood that Parties will meet their national emission commitments;
- Liability provisions for transfers and suspension of trading privileges could form part of a broader set of responses for Parties that do not comply; and
- Making entity eligibility to trade contingent on compliance by the Party in which they are located would create a domestic constituency for compliance.

Conceptual Model of a Trading System

The following conceptual model is used as a basis for the discussion of the technical questions in the rest of this paper. This model is based as far as possible on the Kyoto Protocol provisions. It assumes minimum international institutional requirements and strong reliance on domestic systems for monitoring and compliance. The model is a working example and is not recommended as the best or only possible model.

System-wide emission limit: The Kyoto Protocol places an upper limit on the total quantity of emissions in industrialized countries by establishing quantified national emission commitments or "initial assigned amounts" for the commitment period 2008 to 2012 (Article 3). Assigned amounts for each Party are listed in Annex B of the Kyoto Protocol. Individual Parties' assigned amounts may increase or decrease as a result of transactions; in aggregate, parts of assigned amount will be added to the assigned amount of a Party that buys, and subtracted from the assigned amount of the Party that sells. However, Article 17 emissions trading will not alter the system-wide emission limit among all Parties with Article 3 commitments.

Tradable units: The tradable commodity for international emissions trading is derived from national assigned amounts. The units traded will be parts of assigned amounts (PAA), expressed in tons of CO_2 equivalent (Article 3.1). In the case of Article 6 Joint Implementation (JI) among industrialized country Parties, the tradable commodity is referred to in the Kyoto Protocol as emission reduction units (ERU). Certified emission reductions (CER) is the term used in the Protocol for emission reductions that industrialized countries purchase from developing countries. ERU1 and CER from these mechanisms will be inter-changeable with PAA once they have been added to an Annex I Party's assigned amount. A global warming potential weighting index will be used to convert other greenhouse gases into CO_2 equivalent units so that they are inter-changeable. Each unit will have a unique identification number that corresponds to information about the country of origin and the commitment period for which the PAA was issued.

Gases and sinks: National assigned amounts incorporate emissions of six greenhouse gases (CO_2, CH_4, N_2O, HFCs, PFCs, and SF_6) from the energy sector, industrial processes, solvents and other product use, agriculture, and waste. Adjustments to assigned amounts for changes in carbon stocks as a result of activities under Article 3.3 (and potentially under Article 3.4) in the land-use change and forestry sector, will also be part of the tradable commodity. Transfers of PAA by governments based on national assigned amounts will therefore include emissions of all six greenhouse gases and some sequestration. *Limiting trading to certain sources:* Individual Parties could choose to limit trading by domestic entities to certain types of emission sources. The rationale would be to limit trading to certain greenhouse gas sources and sinks that can be monitored or estimated accurately. Limitations such as this on domestic firms would be a domestic decision for each Party to make and would not require rules at the international level.

Timing for beginning trading: Trading could begin at any time. Trading that occurs before the first commitment period will be in the form of forward or futures trades (contracts to buy or sell PAA for the first commitment period). However, only trades formally recorded by a Party according to the defined principles, modalities, rules, and guidelines that are to be agreed under Article 17 will be recognized for the purposes of compliance with the Protocol.

Period of validity: Any part of a Party's allocation that is not used for compliance in the first commitment period will remain valid for use or sale in the future (i.e., it can be "banked"; Article 3.13).

Traders: Under this model, both governments and legal entities will trade. National governments will decide which legal entities they will obligate to reduce emissions and authorize to trade in order to meet their obligations at lower cost. In addition, any individual or an entity that does not have an emission obligation could buy PAA and hold them or sell them to others.

Domestic trading systems: Parties may devolve some of the responsibility for greenhouse gas mitigation to private entities through a wide range of policies and measures, possibly including domestic trading systems. Under a domestic emissions trading system, this devolution of responsibility is typically administered by placing limits on entity emissions, issuing PAA to entities (e.g., by giving PAA to entities that are responsible for emission mitigation or auctioning PAA to the highest bidder).

A key aspect of domestic emissions trading systems is that Governments allow entities that have an obligation to limit their emissions to use PAA purchased from other entities (for example an entity with lower mitigation costs) to justify emissions above their initial allowed level. Similarly, national governments accept that if entity emissions are below their allowed level, the entity can sell PAA to another entity (that may have higher mitigation costs). Any emissions above entities initial allowed levels must be matched by corresponding emission decreases below the initial allowed level of other entities (or increases in sequestration). Entities that are responsible for emission mitigation are required to surrender PAA equivalent to their emissions to the government at the end of a specified period. The government monitors entity emissions and ensure that entities surrender the correct number of PAA to match their emissions. If entities do not comply the government could impose penalties under domestic law.

National accounting and recording systems: Each government that has a domestic trading system will implement a system for monitoring entity emissions and ensuring that entities surrender the correct number of PAA to demonstrate compliance. National governments should keep track of the impact of entity and government transactions on the national assigned amount and periodically report this information to the international community. National governments should also be responsible for ensuring that changes in ownership of PAA can be verified.

Table 6 below gives an example of how a national system might record transactions of different types. Party A's initial assigned amount is 100 million tons of CO_2 equivalents. Party A allocates 20 million tons of its assigned amounts to two entities; Aa and Ab receive an allocation of 10 million tons of PAA each. The government requires both Aa and Ab to limit their emissions to 10 million tons. The government authorizes them to trade PAA with other domestic entities or entities in other countries to meet their emission obligations. The government of Party A is left with 80 million tons of PAA that has not been allocated to entities. Over the commitment period, the government decides to transfer 5 PAA to Party B and to acquire 2 from Party C, leaving 77 PAA in the government's account.

Entity Aa sells 3 million tons worth of PAA, giving it an adjusted PAA holding of 7 million tons of CO_2 equivalent. Entity Aa's emissions are 7 million tons of CO_2 equivalent at the end of the commitment period, so it surrenders its entire holding of PAA to the government. Entity Ab acquires PAA equivalent to 5 million tons of CO_2 equivalent, giving it an adjusted PAA holding of 15 million tons. Entity Ab's emissions are only 10 million tons of CO_2 equivalents, so it surrenders 10 million tons worth of PAA and banks 5 million tons.

Two entities—a broker and a non-governmental organization (NGO)—have no initial PAA holding, and no requirement to surrender PAA to demonstrate compliance with an emission obligation. The broker Ac buys 11 million tons worth of PAA, sells 5, and chooses to bank 6. The NGO Ad buys and retires 10 million tons worth of PAA. The government records each transaction and keeps track of the change to the national assigned amount, which changes from 100 to 94 as a result of the aggregate international transactions by the Party and entities.

The example in Table 6 does not, however, show the details of national records. For each PAA, national governments will maintain a record of the holder, identification numbers, and buyers or sellers in their national recording systems. These records will enable a government to verify that part of its assigned amount had been bought by another Party or entity.

International systems: An important international function is to ensure that national systems meet any international requirements that are agreed under Article 17 of the Kyoto Protocol. It will also be important at the international level to ensure that unique identification numbers are issued by each Party for all PAA that could potentially be transferred. National governments will be responsible for ensuring that entities comply with their domestic emission obligations and that national emissions are within the final assigned amount. National governments will also report adjustments to holdings of PAA and to the assigned amount that occur as a result of trading. The international system will compile reports from Parties on changes to assigned amounts as a result of transfers and acquisitions (Table 6). The international review mechanism could check national records if inconsistencies are found among Parties' reported changes in assigned amount or if ownership of PAA is disputed.

Table 6: National System for Accounting for Changes to Holdings of PAA and Assigned Amount

Party	Initial holdings following distribution of PAA to entities	Transfers	Acquisitions	Adjusted holding	PAA used for entity compliance and available for national compliance	PAA retired	PAA banked
Party A initial assigned amount=100	80	–5 (to Party B)	+2 (from Party C)	77	77	0	0
entity Aa (electric utility)	10	–2 (to a utility in Party B) –1 (to entity Ac)	0	7	7	0	0
entity Ab (coal mine)	10	0	+5 (from Ac)	15	10	0	5
entity Ac (broker - no obligation	0	–5 (to Ab)	+10 (from Party C) +1 (from Aa)	6	0	0	6
entity Ad (NGO - no obligation	0		+10 (from Party C)	10	0	10	0
Total	100			115	94 (final assigned amount)	10	11
amount=100							

National and International Trading Systems: Characteristics and Modalities

Compatibility among National Systems

National systems are the most effective mechanisms for ensuring the integrity of an international trading system because they are backed by domestic law. Some national accounting and recording procedures and systems can be expected to be established under domestic emissions trading systems regardless of whether international trading by private entities is authorized and the type of domestic trading system that is implemented. Under domestic emission trading systems, governments are likely to:

- devolve responsibility for emission mitigation to some entities by placing some form of limit on their emissions;
- distribute PAA to entities (for example by giving PAA to entities that are responsible for emission mitigation, or auctioning PAA to the highest bidder);
  - keep records of private entities that have emission obligations, monitor their emissions, and check that they surrender the correct number of PAA at the end of their domestic compliance period; and
  - record changes in holdings of domestic PAA among all domestic entities, both those with and those without emission obligations.

Different types of national accounting and recording systems may be implemented depending on different Parties' cultural, institutional and legal frameworks. National systems may also vary depending on the entities they authorize to trade, domestic institutions, national capacity to establish and maintain emission monitoring systems, and systems for accounting and recording changes in PAA holdings.

International guidelines for national systems should ideally be established before significant amounts of international emissions trading takes place. Guidelines will have to take into account approaches and institutional capacity for implementing domestic emission trading systems in different countries. Ideally, incompatibilities that could arise at the interface of several national systems should be identified and considered early. Most aspects of national systems can be left to the discretion of individual Parties. It is important to avoid duplicating functions at the international level that are already needed in domestic emission trading systems. However, compatibility in some aspects of national systems would facilitate trading. International requirements should ensure that there is sufficient compatibility among national systems to facilitate transactions and to guarantee the overall environmental performance of the trading system.

For example:

1. The units traded should be easily transferable in different Parties' records (some form of electronic accounting that gathers comparable information in a consistent format could be implemented in most countries);
2. All PAA that could potentially be traded should have a unique identification number according to an internationally consistent format;
3. Minimum standards for national systems for recording and verifying all PAA holdings and subsequent changes in holdings as a result of trades, and for accounting for changes to national assigned amounts as a result of trading.

National systems for recording, verifying, and accounting are discussed below.

Accounting for Changes to Assigned Amounts

The international system must be able to verify that for every transfer that enables emissions to increase above an assigned amount in one Party there is a corresponding decrease in emissions elsewhere. The international community must be confident that the overall system limit on emissions (which underpins both the environmental and economic effectiveness of the trading system) will be achieved and that it will be possible to verify this.

The information that is relevant to changes in assigned amount includes the number of PAA used by entities to demonstrate their compliance with domestic emission obligations and transfers and acquisitions by the government. Reporting requirements under domestic law are the most effective mechanism for gathering information from private entities. The precedent established for national reporting under the UNFCCC is self-reporting by Parties. National systems are therefore a logical point for recording information that is relevant to changes to assigned amount as a result of trading by entities and Parties, at least initially.

National systems must differentiate among different types of holdings of PAA and account for them appropriately. Some holdings of PAA should not automatically lead to adjustments of Parties' assigned amounts. For example, PAA that are held by an NGO that wishes to retire them from the market permanently should not be added to the assigned amount of the Party in which the NGO is located. If any entity wished to retire or bank units for use in later commitment periods, the retired or banked units should be recorded by the government before the end of the commitment period. The government would report retired and banked units (in aggregate) to the international system at the end of the commitment period along with its report of changes to assigned amounts. This information would enable the international system to check that the overall system is in balance.

Information on changes to the national assigned amount (including net amounts sold to or bought from each Party) must, at the latest, be available when compliance is assessed after the end of the commitment period. More frequent reporting would be useful to give an indication of Parties' interim progress towards meeting their national emission commitments and likely future demand and supply of PAA. The Kyoto Protocol affirms that emission inventories will continue to be submitted annually (Article 7.3), and that this information should include the necessary supplemental information for the purposes of ensuring compliance with Article 3 (Article 7.1). It may be useful for information on transactions to be part of the supplementary information reported along with annual inventories (Corfee, 1998). Annual information 3 may be of use in considering a Party's potential risk of not complying with their assigned amount, but interim information will not indicate what a Party's final assigned amount or emissions will be at the end of the period.

Verifying Ownership

National assigned amounts (adjusted for transfers and acquisitions) are both the obligation from which the tradable commodity is derived and the basis for compliance assessment. Verifying ownership of the AA that are held by the government and obligated entities is therefore linked to national compliance. The capacity to verify all holdings of PAA (those held by governments, obligated entities and non-obligated entities) is also critical for an effective, open and transparent emissions trading market. The information that is required for verifying ownership of PAA includes the identification number of the PAA (which will correspond to information on the country of origin and period in which it was issued), the buyer and seller, and the national system in which the PAA are held. This information could be reported directly by the buyer and seller to an international registry. However, it is likely that an international institution of this sort would take some time to establish. In the absence of an international registry, Parties will need to maintain complete records of transactions by the government and by entities. If an entity or Party claims ownership of PAA, the Parties concerned could check their records of each transfer and acquisition and reconcile any inconsistencies. An international review mechanism can then check Party records if inconsistencies are not reconciled by Parties. To verify ownership, the minimum international requirement for national systems is that they have the capacity to record holdings of PAA, and changes in holdings due to trading, within an agreed period of time. A short settlement period should ideally be the basis for all transactions to facilitate the efficient operation of the international trading market. However, there could be differences in national capacity for recording transfers and acquisitions.

Enhancing Compliance through Trading System

Preventive measures to encourage compliance and responses to non-compliance with trading rules are extremely important for a trading system. The discussion below considers options for enhancing the integrity of international emission trading under Article 17 of the Kyoto Protocol. Two other OECD papers discuss the broader issue of compliance with commitments under the Protocol (Corfee, 1998 and Werksman, 1998). Further analysis of the merits of each approach or option is required on the following lines.

Commitments

There are many types of commitment in the Kyoto Protocol, such as reporting, policies and measures, technology transfer, finance, education, research, and demonstrable progress by 2005 in addition to quantified limits on emissions for Annex I Parties. Ratification of the Kyoto Protocol will be an indication that Parties genuinely intend to meet these commitments. Many analysts and Parties have stressed that responses for non-compliance are needed to reinforce the legally binding nature of the Annex I Parties' quantified limits on emissions (Werksman, 1998). Parties' commitments to report the information that is needed for assessing compliance is also very important for the international emissions trading system, particularly national systems for emission inventories under Article 5 and reporting under Article 7.

Compliance is important for the integrity of the trading system and, more generally, for the success of the Protocol. Strong and effective domestic compliance systems for entities that participate in emissions trading can provide a strong basis for the international trading system. In addition, a number of possible options to complement the compliance incentives under Article 18 of the Protocol could be included in the rules for international emissions trading. Compliance with the rules for emissions trading will also be an important aspect of a comprehensive compliance system. Possible responses for non-compliance with trading rules will need to be considered both in decisions on international rules for emissions trading under Article 17 and in decisions on non-compliance responses under Article 18. There are two main compliance approaches for enhancing the integrity of the trading system and the Protocol: the first is to establish preventive measures that place minimum conditions on participation; the second is to establish responses in cases of non-compliance. These approaches are discussed below.

Preventive Approaches to Encourage Compliance

Eligibility Requirements

There is growing recognition that Annex I Parties that wish to trade should meet and maintain minimum eligibility requirements. Eligibility requirements could include standards for the quality of emissions data and for reporting.

Such requirements could prevent Parties with lower standards of emission monitoring and reporting from trading under Article 17. This would minimize the possibility that PAA transfers from a Party with a national inventory system that does not accurately reflect emissions would be used to increase emissions in another country. Eligibility requirements for Parties wishing to participate in Article 17 international emissions trading have been proposed by Annex I Parties in submissions to UNFCCC subsidiary bodies, for example:

- Parties wishing to trade must comply with Articles 5 and 7 of the Kyoto Protocol (Umbrella Group and EU);
- Parties wishing to trade must establish and maintain a national system accounting and tracking PAA held or traded by the Party and/or its legal entities (Umbrella Group and EU);
- Parties wishing to trade must have adopted and ratified a compliance regime under Article 18 (EU only);
- Parties wishing to trade must demonstrate compliance with the rules of the emission trading system (EU only).

Minimum standards for national systems to monitor and estimate national emissions under Article 5 could be a pre-requisite for eligibility to join the international emissions trading system or to authorize entities of participate in international emissions trading. Alternatively, all Annex I Parties could initially be legible to trade, but may subsequently lose this right if they fail to comply with trading rules and other Protocol provisions. Such standards could have equity implications due to differences in national capacities which would result in richer countries benefiting most from trading.

Good quality inventory data and national systems will be required under Article 5 of the Protocol even for those countries that choose not to trade. However, for Parties that wish to participate in international emissions trading it will also be important to ensure that they have the institutional capacity for accounting and registering PAA held or traded by entities or the Party as discussed in section 0 above. Eligibility requirements for trading could provide an incentive for Parties to improve the quality of their emission data and reporting, and would facilitate compliance assessment as well as establishing a good base for trading. Emission data from entities that are monitored under domestic trading systems or other policies and measures might improve the quality of information used in national inventories. Information from independent sources that are not influenced by political considerations could be beneficial to the trading market and for compliance assessment. Eligibility requirements would also enhance the effectiveness of the market since traders would have greater confidence in the national systems run by Parties, such as the systems that track changes to assigned amount and verify the title to each holding of PAA. An important consideration, however, is that the eligibility requirements should not be so stringent as to prevent too many Parties from trading under Article 17. Domestic trading systems which effectively enforce domestic entities' obligations could also reinforce national compliance with national commitments.

Compliance Reserve

Parties could be required to hold a minimum reserve of part of their assigned amount either during or at the end of each year to ensure that they manage their emissions and assigned amount responsibly. This minimum balance could be based on an agreed portion of assigned amount per year. Under this approach a Party may have an assigned amount of 500 million tons of CO_2 equivalent for the commitment period. The Party could be required to hold a fifth (i.e. 100 million tons CO_2 equivalent) of it's of assigned amount as a minimum balance either during or at the end of 2008. In 2009 (during or at the end—depending on the approach) the Party must hold 200 million tons CO_2 equivalent, and so on.

Because of the lag between emissions and data reported in national emissions inventories, it would be necessary to review the minimum compliance reserve for each country as the emissions data comes in. For example, if the minimum balance is based on one-fifth of the Party's assigned amount (equivalent to its base year emissions) and it becomes clear that the Party has made emission reductions so that it's actual annual emissions have been lower than its base year emissions, then the compliance reserve could be decreased and more PAA could become available for possible transfer. Alternatively, the minimum balance could be based on the Party's actual cumulative emissions at the end of each year. Under a cumulative emissions approach, at the end of each year, the Party would be required to set aside part of its assigned amount equal to its emissions in that year. The compliance reserve would thus reflect cumulative annual emissions throughout the commitment period.

Possible Responses to Non-compliance

Liability: Rules governing liability must be clear from the outset as part of the trading system design. Liability for the PAA that are transferred could potentially be an important tool to provide incentives for general compliance with commitments through the rules of the trading system. Liability provisions apply to international emission trading transactions. They do not remove the need for consequences for noncompliance under Article 18. The options for liability provisions that have been discussed to date are summarized below: issuer liability (often referred to as seller liability); buyer liability; and shared liability.

Issuer liability: Under issuer liability it is the selling Party that is liable for national non-compliance. The Kyoto Protocol establishes that Parties must comply with national emission commitments as defined by their assigned amounts adjusted for PAA, ERU, or CERs transferred and acquired. This implies that issuer liability will be the norm in the absence of further decisions on liability with respect to the trading system. Under issuer liability, any part of an assigned amount acquired by a Party would be added to the buyers' assigned amount (regardless of the compliance status of the issuer). If a Party is not in compliance at the end of the commitment period, the government of that Party would have the responsibility to come into compliance or face consequences under Article 18. If the main participants in the international trading system are legal entities that operate under effective domestic monitoring and enforcement systems, issuer liability could be a simple and effective approach.

Buyer liability: The principle of buyer liability is that "any transferred 'assigned amount' should only contribute towards compliance by the acquiring Party to the extent that the transferring Party also complies" (Grubb, 1998). In effect, if the issuer Party does not comply, PAA that the Party had transferred would not be valid for use by the buyer for demonstrating compliance. The possible invalidation of PAA would discourage Parties and entities from purchasing PAA from Parties that are less likely to meet their emission commitments. PAA from potential non-compliers would be more risky (since they may not be valid for compliance) and the market would devalue PAA from Parties that are less sure to comply. Buyer liability would be more complicated to administer than issuer liability.

For example, issuers will typically be both buyers and sellers of PAA, perhaps ending up a net seller and in non-compliance for a wide variety of reasons from changes in inventory methods to non-compliance by one large entity in the country with its domestic obligations. The question of which PAA should be invalidated would have to be resolved -invalidating the last PAA to be sold or all PAA or discounting all PAA sold by the issuer proportionally; invalidating PAA sold by the government first so that domestic entities that comply are less likely to be penalized for national non-compliance. One Party's non-compliance could also cause other Parties to fall into non-compliance, thus causing a chain reaction among a number of Parties that had acquired PAA.

Shared or double liability: Under shared issuer/buyer liability, transfers made by a non-complying Party would be reduced in quantity once non-compliance was discovered. For example, perhaps only 60% of the quantity transferred would be added to the buyer's assigned amount (i.e. the PAA acquired would be discounted by 40% to enable the issuer to return to compliance). This discounting could be "across-the board" for all PAA traded or it could apply to the most recent transfers (up to the quantity by which the Party has exceeded its assigned amount).

Article 6.4 opens up the possibility of a hybrid approach in the case of joint implementation projects that would include both issuer and buyer liability:

> "If a question of implementation by a Party of the requirements referred to in this article is identified... transfers and acquisitions of emission reduction units may continue to be made after the question has been

identified, provided that any such units may not be used by a Party to meet its commitments under Article 3 until any issue of compliance is resolved."

This provision implies that any units acquired remain valid unless an issue of compliance is not resolved, leaving buyer liability as the last resort option if a question of implementation is not resolved. Non-compliers could be required to pay a penalty or to deduct the excess from the Party's assigned amount for the next commitment period. If these options fail, or are deemed inadequate given a Party's lack of capacity to reduce emissions or purchase PAA, buyer liability or shared liability could be invoked. Liability responses to non-compliance, as with other approaches, will depend on why the Party did not comply. Non-compliance due to lack of capacity to track changes in holding of PAA or to enforce entity emission obligations would merit a different response than a rogue government trader that sells the Parties assigned amount unscrupulously.

Conservative approaches: Conservative approaches for estimating the amount of PAA that can be transferred, or setting aside funds from a transfer until compliance is demonstrated, could be used as a response to non-compliance. These approaches would increase the cost of meeting the national emission commitments and reduce incentives to trade. However, further consideration of these approaches as responses to certain cases of noncompliance could be useful, for example if emission inventory and reporting requirements are not met.

Conservative estimate: Discount factors could be applied to units transferred to ensure that any trades err on the side of caution (ICF 1998). These discount factors might be set to ensure that for every unit of greenhouse gas reduction; only a percentage of a unit of PAA could be transferred. This would not be necessary for all trades, but it could be required for Parties that do not comply with article 5 and 7 requirements. It might be more appropriate to address this issue in the context of conservative estimates for emissions inventories rather than of traded amounts.

Escrow accounts: For transfers by governments that have not complied with their commitments, a portion of the proceeds from a transfer of PAA could be held in an escrow account until the seller's compliance is established (Haites, 1998). If the entire assigned amount" sold is found to be surplus when compliance for the commitment period is established, the funds would be disbursed to the national government of the selling Party. If some of the "assigned amount" sold is needed for the seller to achieve compliance, the funds could be given to an international fund as a penalty for non-compliance, or the seller could be obliged to use the proceeds to purchase PAA in order to achieve compliance.

Suspension of trading privileges: Some forms of non-compliance could warrant loss of some or all trading privileges. For example Parties that do not comply with the eligibility requirements for trading, or who willfully exceed their assigned amounts, could face suspension of their eligibility to trade internationally. Loss of trading privileges could also be a "hard" response for persistent or serious non-compliance with the monitoring and reporting commitments. Suspension of trading privileges could be invoked after other approaches had been tried, such as capacity building and expert assistance (Werksman, 1998). The suspension would last until the Party came back into compliance. Different approaches to suspension of trading privileges require further consideration. An Umbrella Group paper tabled in Bonn in June 1998 stated that a breach of Article 5 and 7 would lead to suspension of the right to transfer, but that Parties would still be able to acquire. An EU paper stated that any breach should lead to suspension of privileges to both transfer and acquire. Article 6 (concerning project-based trading) states that non-compliance with Article 5 or 7 should result in ineligibility to acquire.

Acquiring PAA is a potentially legitimate way for a Party to come back into compliance with national emission commitments. Suspending eligibility to acquire could lower the efficiency of the trading system by removing part of the demand from the market. However, suspending the right to transfer, alone, may not provide a sufficient incentive or deterrent to comply with the rules for trading, particularly for those countries who expect to be net buyers. In some circumstances, consideration may need to be given to suspending both transfers and acquisitions, at least until the initial cause is remedied. Suspending both the right to transfer and acquire would provide strong incentives for Parties to ensure that they establish and maintain good national systems for estimating and reporting emissions. If suspension of a Party's eligibility to trade also affects its entities, this approach would create a domestic constituency to lobby for compliance with monitoring, reporting requirements and trading rules.

Mixed Responses

Non-compliance with different types of commitments may warrant different remedies. Different degrees of seriousness of non-compliance may also warrant various types of remedy. Some aspects of trading make non-compliance with international emission commitments less likely. An obvious example is that emissions-trading offer potential non-compliers an alternative means to help them to achieve their target through acquiring PAA from other Parties who have surplus PAA. Such acquisitions could take place during the commitment period, or at the end of the period. An initial response to non-compliance could be to give Parties an opportunity to come into compliance through purchasing PAA within a certain period. Tougher response options could be considered for Parties that remain in non-compliance once a Party has been given adequate opportunity to fulfil its commitments. A suite of different response options for on compliance of different types, cause, degree and frequency should ideally work together in a mutually reinforcing way. Responses associated with the international emissions trading system should form part of a broader system of graduated non-compliance responses.

Green Trading Markets

Green trading encompasses all forms of environmental financial trading, including carbon dioxide, sulfur dioxide (acid rain), nitrogen oxide (ozone), renewable energy credits, and energy efficiency (megawatts). All these emerging and established environmental financial markets have one thing in common which is making the environment cleaner by either reducing emissions, using clean technology or not using energy through financial markets. Green Trading is one mechanism to accelerate change to a cleaner environment by using market-based incentives whose application is global.

Green trade organizations aim to secure sustainable supply systems and to ensure benefits to local people from national, regional and international trade. These organizations promote trade practices that adhere to the principles of sustainable development in all its facets: ecological sustainability (use and conservation), economic sustainability (productivity) and social sustainability (equity).

Green trade organizations can help improve access to national and/or international markets for NWFP producers, especially those in developing countries, and at the same time provide guidance on increasing benefit for their agricultural and forest products. In addition, they can assist in the export of products, thus helping to provide foreign exchange for the home country.

When designed properly, emissions trading allow flexibility for emitters without sacrificing the environmental objective of reducing emissions. Emitters can meet their reduction commitments entirely through their own efforts, through buying permits auctioned by the government or unused by another source, or through doing both. Ultimately, emissions-trading is driven by the desire to minimize costs and maximize profits.

Trading can significantly reduce overall compliance costs for the specific pollutants covered by the trading program if the emissions reductions that take place are the most cost-effective. The trading market functions like any other commodity market, with trade brokers, outside investors interested in the future potential of innovators and established rules of acceptable activity. According to analysts from the International Energy Agency, the use of economic instruments such as domestic and international emissions trading could reduce the costs of complying with emissions targets by up to 50%. The report on Green Trading Markets looks at the emissions trading industry and analyzes the interactions between CO_2 emission trading and markets for electricity produced from renewable energy sources in a liberalized market, including description on how to treat these issues in the electricity system model developed in the Wilmar project. This report also looks at the basics of renewable trading and what is involved in the processes. While covering the carbon dioxide trading market and the sulfur dioxide trading market, it also takes a closer look at the exchanges involved in green trading markets. This is a complete report covering the ABCs of green trading

Trading of Carbon Credits and India's Perspectives

Trading of carbon credits is actually executed in the form of CERs or *Certified Emissions Reductions*. CERs are issued in the form of certificates, just like a stock. A CER is given by the CDM Executive Board to projects in developing countries to certify that they have reduced greenhouse gas emissions by one tonne of carbon dioxide per year. For example, if a project generates energy using wind power instead of burning coal, and in the process saves (say) 25 tonnes of carbon dioxide per year, it can claim 25 CERs (one CER is equivalent to one tonne of carbon dioxide reduced). In March 2006, the United Nations Framework Convention on Climate Change (UNFCCC) CDM (Clean Development Mechanism) Registry Administrator, for the first time, forwarded CERs to the holding account of a project participant representing a milestone in the implementation of the Kyoto Protocol's CDM. The forwarding of CERs to the holding accounts of project participants provides them with the end product of their efforts to reduce emissions of greenhouse gases in developing countries. These CERs have a ready market, where transactions happen on arm's-length basis and price quoted fluctuate as per the situation of demand and supply, and also according to the negotiation skills of the two parties.

India, along with other developing nations, is at an advantage as it can implement approved CDM projects for the purposes of trading CERs. One third of the total CDM projects registered with UNFCCC are from India. India's carbon credits' trading is expected to reach $100 billion by 2010. In 2007, a total of 160 new projects were registered with UNFCCC. As a consequence, Indian industry managed to generate over 27 million carbon credits. Indian projects receive further impetus by way of investments and finance from developed nations who are potential buyers of CERs (Ghosh, 2008). The Multi Commodity Exchange of India Ltd entered into an alliance with the Chicago Climate Exchange in 2005 to introduce carbon credit trading in India. This association has integrated Indian markets with their global counterparts to cover risks associated with futures trading of carbon credits and ensuring best prices. CDM projects, mostly in key sectors such as manufacturing, energy, agriculture, mining and mineral production, would thus result in providing a boost to the Indian economy. The major CDM projects are located in Rajasthan, Andhra Pradesh, Maharashtra Karnataka, Himachal Pradesh and Punjab.

Exchange-based Trading of Carbon Credits

The role of an exchange is to facilitate emissions trading but also to guarantee each trade. In order to fulfil this role, access to the exchange floor is restricted to its members/seat-holders, who can act as buyers or sellers. In order to trade Carbon credits on an exchange like the New York Mercantile Exchange (NYMEX), ParisBourse, the International Petroleum Exchange, or the Sydney Futures Exchange, one must go through a member of the exchange. Traders on the exchange must also pay a fee *per transaction*, expressed either as a fixed amount regardless of the size of the transaction, or as a percentage of the total transaction amount. Both membership fees and transaction fees are subject to negotiation depending on the volume of transactions that a particular trader conducts through the exchange. The higher the volume is, the cheaper the cost per transaction is likely to be. Exchanges generally monitor trading to detect price distortions or market manipulation that is deemed non-competitive. Most commodities, currencies, stocks, etc. traded on an exchange are also traded in bilateral arrangements, or via brokers, but also on the floors of other exchanges. Exchange-based trading works via so-called double-auction, whereby supply and demand for a stock/commodity meet on a centralized market operated electronically or physically. Transactions are concluded once both the buyer and seller agree on a common price. Members of an exchange agree, when posting bids and offers, to buy from/sell to whoever is willing to match the offered price. This offers some guarantee of market transparency and full price competition and largely prevents market manipulations and market power of Carbon credits. The following are some of the distinct benefits of Carbon Credits Trading through Exchanges: (i) Trading emission allowances are no different from trading any other commodity. Anyone who holds an account in a central registry will be able to buy and sell allowances; (ii) An electronic emissions trading system will provide industry, gouvernements and other organisations in Canada with the opportunity to buy and sell emission reductions at a reduced overall cost; (iii) Standardization facilitates market liquidity; (iv) Futures are price transparent, ensuring fair prices; (v) Anonymity. The Trading Processes and Market Design aspects involved in Exchange Trading of Carbon Credit as transacted through Montreal Exchange are illustrated on next page.

The various prerequisites for Efficient Emissions Trading through Exchanges are as follows:

Liquidity:

- Sizable Market Volume and sufficiently large number of Market players.
- One Stop Shopping for Spot and Derivative Contacts.
- Economic Efficiency/Minimized Infrastructure Cost.

Diversity of Market Players:

- Different Market Participants with different Background and different Targets (Power Generators/Utilities, Industrialists, Traders, Exporters, Transporters, Financial Institutions, etc.)
- Cross-Border Activity (viz. Energy Transmission and or Trading).

Fig. 7: Exchange Trading Process of Carbon Credit through Montreal Exchange

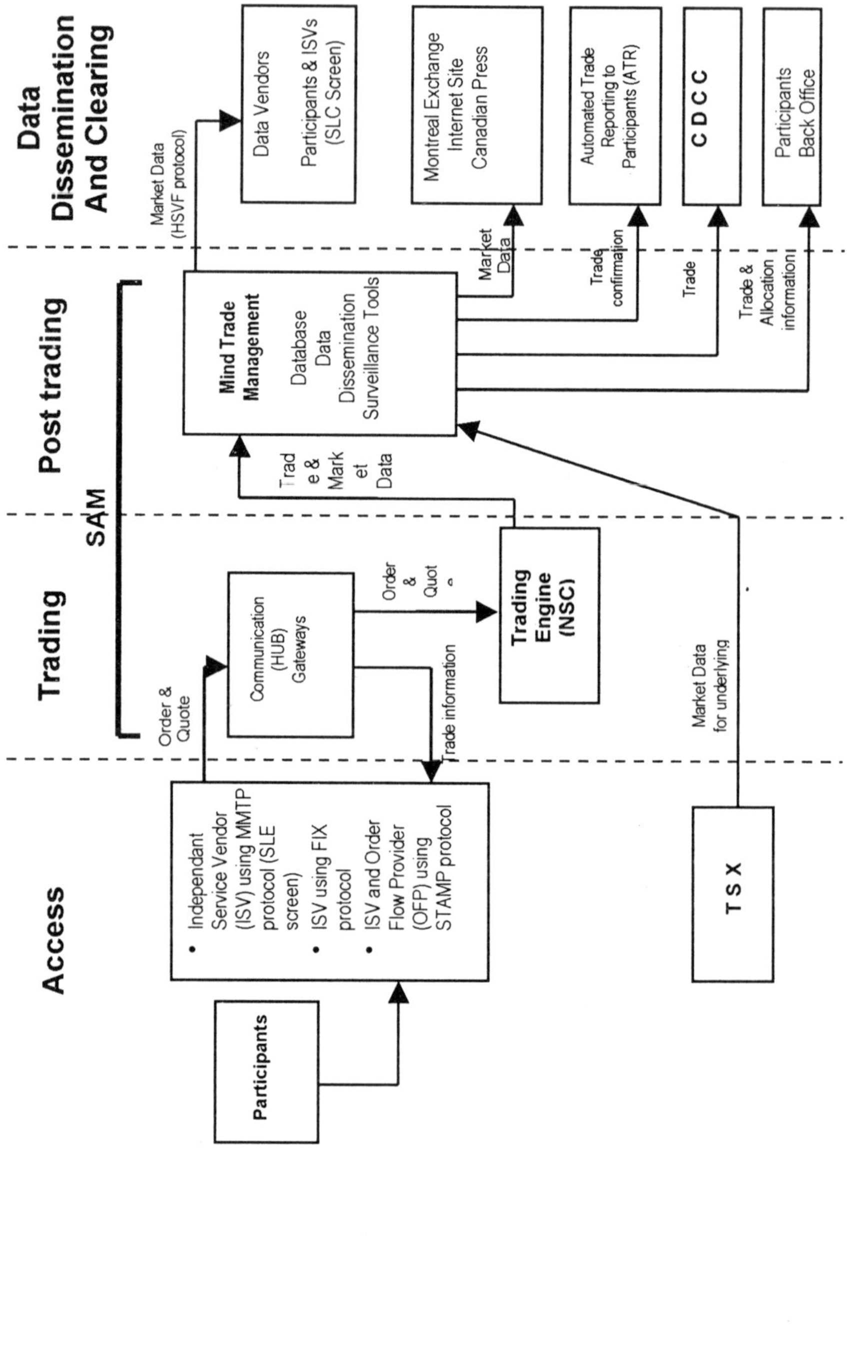

Standardization

- Standardized underlying because they reduce Market fragmentation, facilitate risk minimization and reduce Transaction costs.

MCX—Carbon Credits Contract

Headquartered in the financial capital of India, Mumbai, MCX (www.mcxindia.com) is a demutualised nationwide electronic Multi-commodity futures exchange set up by Financial Technologies with permanent recognition from Government of India for facilitating online trading, clearing and settlement operations for futures market across the country. The exchange started operations in November 2003 and presently is the top commodity exchange in the country with 70% market share for quarter ended June 30, 2007 of the total commodity exchange industry in the country. Among the top ten commodity derivatives exchanges in the world, MCX ranks among the top three bullion, energy and copper exchanges in the world in terms of contracts traded. Apart from being accredited with ISO 9001:2000 for quality management, it is the world's first and only multi-commodity exchange to have achieved ISO 27001:2005 certification, the global benchmark for information security management systems (http://www.mcxindia.com/PressRelease/MCX_PR_21jan08.pdf).

MCX offers futures trading in 56 commodities as on Sept 30, 2007, defined in terms of the type of contracts offered, from various market segments including bullion, energy, ferrous and non-ferrous metals, oils and oil seeds, cereals, pulses, plantations, spices, plastics and fibres. The exchange strives to be at the forefront of developments in the commodities futures industry and has forged eleven strategic alliances across the world, including with Tokyo Commodity Exchange, Chicago Climate Exchange, London Metal Exchange, New York Mercantile Exchange, New York Board of Trade and Bursa Malaysia Derivatives, Berhad. Financial Technologies India Ltd, a leading provider of transaction automation technologies for Equities, Derivatives, Forex and Commodity markets, is the majority shareholders of MCX. Other key stakeholders include Fidelity International, State Bank of India and its subsidiaries, National Stock Exchange (NSE) and National Bank for Agriculture & Rural Development (NABARD).

Procedural Overview of Carbon Accounting

Why conduct Accounting of Carbon Emissions?

- Provide information on which to build an effective strategy to manage GHG.
- Prerequisite for participation in GHG Trading Markets.

- Demonstrate compliance with Government Regulations, if any are in place.

Steps involved in Carbon Accounting

1. Identify boundaries
2. Identify Emission Sources to be covered
3. Select an Emission calculation approach
4. Collect activity Data and choose Emission Factors
5. Apply calculation Tools to estimate Emissions

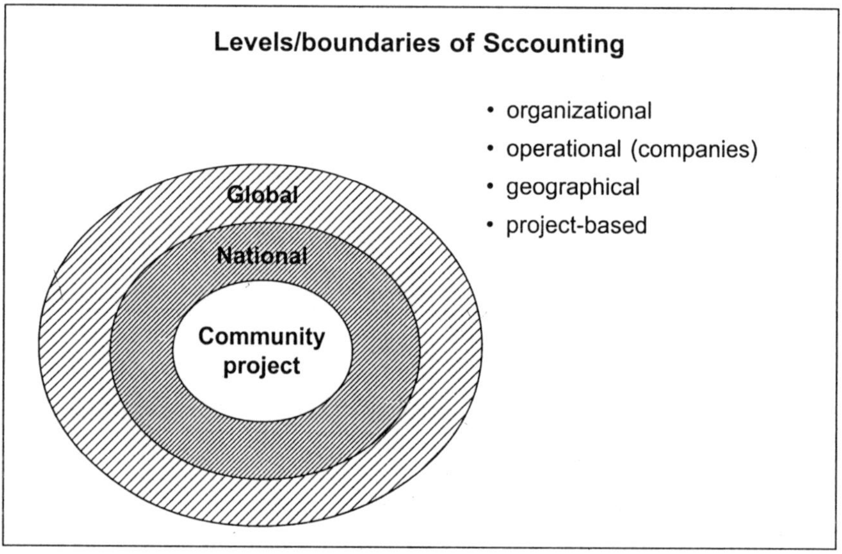

Emission Categories

Direct Emissions: Emissions with Project Control, as a direct result of Project intervention e.g. waste heat recycling and reutilization.

Indirect Emissions: Outside Project Control from changes in Secondary Energy Carriers and materials as a result of Project intervention e.g. reduced import of heat supplied by an outside heating plant/Device.

On-Site and Offsite Emissions: In and Out Project Boundary

Upstream Emissions: Materials impact of Activities that relate to the Project activity but occur before it.

Downstream Emissions: Material Impacts of activities that relate to the Project activity but occur after the Project activity.

Full-Cycle Accounting: A comprehensive Accounting of all significant emissions related to a Project.

ACCOUNTING EQUATION

$$GHG = A \times EF$$

where,

GHG = emissions (amount of CO_2 or Ch_4, etc.

A = activity data (litres of fuels, bag of cement)

EF = emission factor (kg CO_2 /litre of fuel kg CO_2 /bag cement)

$$GHG = A \times EF$$

A (activity data)—data on the **magnitude of human activity resulting in emissions or removals** taking place during a given period of time (e.g. litres of fuel consumed, etc.)

EF (emission factor)—a **coefficient that relates** the **activity data to the amount of chemical compound** which is the source of eventual emissions.

Emission factors are often based on a sample of measurement data, averaged to develop a **representative rate of emission for a given activity level** under given set of operating conditions (e.g. amount of carbon/unit activity)

Sample Emissions Calculation:

Emission from Cement Production

$$GHG = A \times EF$$

tons CO_2 = tons of cement produced × 0.4985 tons CO_2 / ton of cement produced

Emission Reduction

$$GHG = A \times EF$$

To reduce GHG emissions:

- lower A = decreasing frequency or magnitude of activity
- lower EF = shifting to more efficient less carbon intensive technology
- lower both at the same time

Emissions Reduction

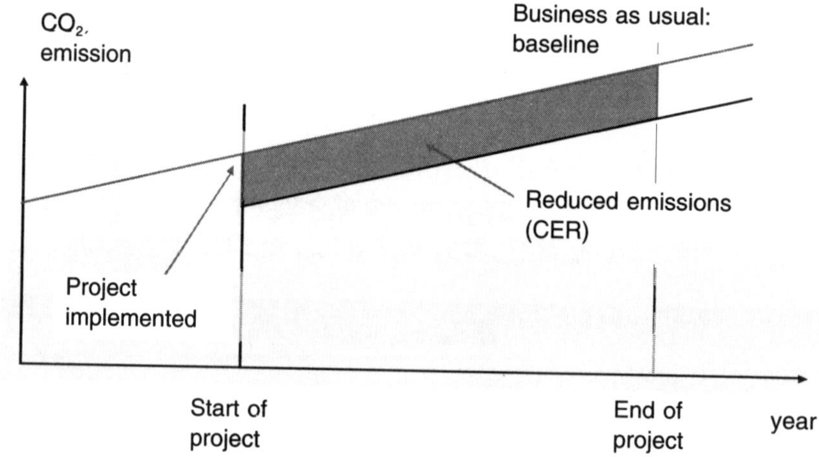

Other terms used for Scenarios

Baseline	= Business as Usual (BAU scenario)
	= Reference Scenario
	= Without Project Scenario
Project	= Alternative Project Scenario
	= With Project Scenario

Emissions Reduction Calculations

- objective is to define reference and project scenarios and determine the net difference in GHG emissions between these two scenarios

ΔGHG_{redn} = $GHG_{base} - GHG_{proj}$
= emissions reduction due to existence of the project

GHG_{base} = an estimation of emissions assuming that no alternative project was implemented

GHG_{proj} = measures the GHG emissions following project implementation

$$\Delta GHG_{redn} = GHG_{base} - GHG_{proj}$$
$$= (A^*EF)_{base} - (A^*EF)_{proj}$$

*In renewable energy projects where EFs of the project are considered zero:

$$\Delta GHG_{redn} = (A^*EF)_{base} - (A^*\overset{\text{zero}}{0})_{proj}$$
$$= (A^*EF)_{base}$$
$$\Delta GHG_{redn} = GHG_{base}$$

*hydro, solar, wind—excludes leakage and other direct and indirect emissions.

International Accounting Standards for Carbon Credits

Under the International Financial Reporting Standards (IFRS), the International Accounting Standards Board (IASB) had issued an interpretation IFRIC 3 (Emission Rights), which was withdrawn in June 2005. Thus, the IASB is still debating on an appropriate treatment for Carbon Emission Reductions (CERs). Under IFRS, most entities generating CERs treat the same as government grant covered under IAS 20 (Accounting for Government Grants and Disclosure of Government Assistance) (http://cmaindia.informe.com/forum/accounts-finance-audit-f16/accounting-for-carbon-credits-t1647.html). This is because an international agency grants the same. Accordingly, based on IAS 20 requirements, a generating entity recognizes CERs as asset once there is a reasonable assurance that it will comply with conditions attached and CERs will be received. IAS 20 gives an option to measure such grants either at fair value or nominal value. Most entities will measure the CERs at fair value to ensure appropriate matching with the costs incurred. They will recognize this in the income statement in the same period as the related cost which the grant is intended to compensate. The corresponding debit will be to intangible assets in accordance with IAS 38 (Intangible Assets).

No guidance is currently available under Indian GAAP (generally accepted accounting principles). Consequently, various practices exist (a) income from sale of CERs is recognized upon execution of a firm sale contract for the eligible credits, as prior to that there is no certainty of the amount to be realized; (b) income from CERs is recognized at estimated realizable value on their confirmation by the authorities concerned; and (c) income from CER is recognized on an entitlement basis based on reasonable certainty after making adjustments for expected deductions.

The Accounting Standards Board (ASB) of the Institute of Chartered Accountants of India (ICAI) has issued an Exposure Draft (ED) of the Guidance

Note on Accounting for Self-generated Certified Emission Reductions. The ED proposes to lay down the manner of applying accounting principles to CERs generated by an entity.

According to the ED, the generating entity should recognize CERs as asset only after receipt of communication for credit from United Nations Framework for Climate Change (UNFCCC) and provided it is probable that future benefits associated with CERs will flow to the entity and costs to generate CERs can be measured reliably. Further, such assets meet the definition of the term 'inventory' given under AS 2 (Valuation of Inventories) and, hence, are valued at lower cost and net realizable value. Only the costs incurred for the certification of CERs bring the CERs into existence and, therefore, only those costs should be included in the cost of inventory. According to the prescribed criteria, all other costs are either not directly relevant in bringing the inventory to its present location and condition or they are incurred before CERs come into existence. Thus, those costs cannot be inventorised.

The ED will result in significant cost and revenue mismatch in the financial statements. This is because entities would need to expense most of their costs as soon as incurred (with an insignificant amount being capitalized as inventory), but will recognize revenue arising from CERs only when these are actually sold. Clearly, the accounting recommended by the ICAI is very different from existing practices under Indian GAAP and, hence, every company that has significant revenue from carbon credits will have to consider the impact of the ED carefully. The treatment prescribed in the ED appears to be inconsistent with the existing Indian GAAP literature in more than one regard. The ED requirement to recognize CERs as asset only when these are credited by UNFCCC in a manner to be unconditionally available is contrary to the principles currently being followed for recognition of an asset. The ED is also inconsistent with an Expert Advisory Committee's (EAC's) opinion on export incentives. According to the EAC, DEPB credit should be recognized in the year in which the export was made, without waiting for its actual credit in the subsequent year, provided there are no insignificant uncertainties of ultimate collection. The EAC opinion is based on the application of existing accounting principles, including definition of the term 'asset' given in the Framework, which is based on the probability theory. As India is adopting IFRS and the guidance in these areas are being developed under IFRS, issuing India-specific guidance is duplicating the effort and creating more differences in how the two GAAPs are applied, which will have to be then taken care of in 2011, which is the transition date for adopting IFRS.

Accounting and Taxation of Transacted Carbon Credits

A carbon tax is an environmental tax on emissions of carbon dioxide and other greenhouse gases. It is an example of a pollution tax. The purpose of a carbon tax is to protect the environment by reducing emissions of carbon dioxide and thereby slow climate change. It can be implemented by taxing the burning of fossil fuels—coal, petroleum products such as gasoline and aviation fuel, and natural gas—in proportion to their carbon content. Unlike other approaches such as carbon cap-and-trade systems, direct taxation has the benefit of being easily understood and can be popular with the public if the revenue from the tax is returned by reducing other taxes. Alternatively, it may be used to fund environmental projects (http://en.wikipedia.org/wiki/Carbon_tax).

The emergence of the opportunity of revenue generation by taking up structured Clean Development Mechanism (CDM) projects has given a new dimension to Accounting and taxation of the transacted Carbon credits through exchanges/derivative markets. As the concept of Carbon Trading is totally new, even at the international level, there are some issues to be settled before arriving at a common opinion. The Kyoto Protocol, which came into force and became legally binding on 15th February 2005 when Russia ratified the treaty, demands a 5.2% cut in greenhouse gas emissions from the industrialized world as a whole by 2012. India, along with China and Brazil, has emerged as one of its largest beneficiaries in terms of new source of revenue. This is due to Clean Development Mechanism (CDM), which is perhaps most exciting feature of the total scheme which allows 'Annex 1 countries' (A total of 36 countries are listed in Annex 1) to meet their emission reduction targets by paying for greenhouse gas emission reduction in non-Annex 1 (developing) countries.

Most Annex 1 countries have legal binding/obligation for meeting greenhouse gas emission reduction requirements under the Kyoto Protocol. These countries, instead of reducing emissions of their own companies, can 'buy' emission reductions in non-Annex 1 countries. Article 12 of the Kyoto Protocol states: "The purpose of the Clean Development Mechanism shall be to assist Parties not included in Annex 1 in achieving sustainable development and in contributing to the ultimate objective of the convention, and to assist Parties included in Annex 1 in achieving compliance with their quantified emission limitation and reduction commitments." India, being a non-Annex 1 country, is naturally one of largest beneficiaries of the Kyoto Protocol. Studies by Crisil and CII estimate the value of the Indian CDM market at more than a billion dollars per annum (IETA, 2002).

Carbon taxes, and all environmental taxes, are "priced-based" policy instruments. Taxes increase the prices of certain goods and services, thereby decreasing the quantity demanded. This is called the "price effect." Tradable permits, or *emissions trading*, are considered a "quantity-based" environmental policy instrument. Although both policy approaches are "market-based," they operate differently—carbon taxes fix the *marginal cost* for carbon emissions and allow quantities emitted to adjust, while tradable permits fix the *total amount* of carbon emitted and allow price levels to fluctuate according to market forces.

Carbon taxes are simply direct payments to government (collection body), based on the carbon content of the fuel being consumed. Given that the primary objective of the abatement policy is to lower carbon dioxide emissions, carbon taxes make sense economically and environmentally because they tax the externality (carbon) directly. Coal generates the greatest amount of carbon emissions and is therefore taxed in greater proportion than oil and natural gas, which have lower carbon concentrations (Coal contains .03 tons of carbon per million Btu of energy, while oil and natural gas contain only .024 and .016 tons respectively).

The Case for Carbon Taxes

- A carbon tax would offer a broader scope for emissions reductions. Trading systems can only be implemented among private firms or countries—not individual consumers (transaction costs would be prohibitively high if commuters needed permits to fill up their car with gas). Carbon taxes extend to all carbon-based fuel consumption, including gasoline, home heating oil and aviation fuels. Trading systems may not be able to reach parts of the transportation and service sectors which could account for 30-50% of emissions.
- A system of tradable permits entails significant transaction costs, which include: *search costs*, such as fees paid to brokers or exchange institutions to find trading partners; *negotiating costs; approval costs*, such as delays or fees incurred during the approval process; and *insurance costs*. Conversely, taxes involve little transaction cost, over all stages of their lifetime.
- Carbon taxes have *dynamic efficiency* advantages that trading lacks because taxes offer a *permanent* incentive to reduce emissions. Technological and procedural changes, and subsequent technology diffusion, will lead to reductions in permit price (i.e. since emissions goals will be easier to meet, there will be a decrease in permit demand, and hence, a decrease in permit price). Trading systems may not be able self-adjust in response to rapid change, and thus provide the permanent incentive of a tax system to reduce emissions. In short, emissions trading must have some method of removing permits from the system or other method of ratcheting-up permit prices.
- Taxes are not susceptible to strategic behavior by firms or non-governmental organizations which may harm the contractual environment of the market. Non-governmental organizations or even private individuals that object to the concept of purchasing the "right to pollute" may purchase large numbers of permits to drive up costs of CO_2 abatement. Likewise firms may hoard permits, driving up the prices for competitors.
- Emissions trading proposals are highly complicated and technical, unlike taxes which are an extremely familiar instrument to policymakers. Many technical issues would need to be resolved before trading could begin, including treatment of sinks, different GHGs,

monitoring, enforcement, etc. Ongoing costs are also low for tax systems because of the lack of monitoring and enforcement requirements.

- Emissions trading may prevent meaningful *domestic* reductions from taking place. If the global climate system is to be stabilized, emissions reductions should take place sooner, rather than later, in the countries most responsible for the problem. This concern relates to profound equity issues among developed, developing and transitional economies.
- Carbon taxes earn revenue, which can be "recycled" back into the economy by reducing taxes on income, labour and/or capital investment. This is often referred to as a "revenue neutral" tax and may be part of a broader program of "environmental tax reform" (ETR) which attempts to shift the tax burden from "goods" like labour, to "bads" like pollution. Evidence indicates that there can be profound employment, distributional and political benefits to such an approach. Permit systems have the potential to earn revenue, but only if permits are auctioned.

Carbon Credit Trading and Taxation Norms

The need for a reduction in carbon emissions was debated at the United Nations Conference on Environment and Development (The Earth Summit) in Rio de Janeiro in 1992, resulting in the adoption of United Nations Framework Convention on Climate Change (UNFCCC), an international treaty on environment. The Kyoto Protocol, 1998, was adopted by the parties to the UNFCCC with the objective of achieving quantified emission limitations through specific policies and measures to minimizing the adverse effects of climate change. The protocol provides for various mechanisms like joint implementation, a clean development mechanism (CDM) and international emission trading to boost the cost effectiveness of climate change mitigation. India, along with other developing nations, is at an advantage as it can implement approved CDM projects for the purposes of trading CERs. One third of the total CDM projects registered with UNFCCC are from India. India's carbon credits' trading is expected to reach $100 billion by 2010. In 2007, a total of 160 new projects were registered with UNFCCC. As a consequence, Indian industry managed to generate over 27 million carbon credits. Indian projects receive further impetus by way of investments and finance from developed nations who are potential buyers of CERs.

The Multi-Commodity Exchange of India Ltd entered into an alliance with the Chicago Climate Exchange in 2005 to introduce carbon credit trading in India. This association has integrated Indian markets with their global counterparts to cover risks associated with futures trading of carbon credits and ensuring best prices. CDM projects, mostly in key sectors such as manufacturing, energy, agriculture, mining and mineral production, would thus result in providing a boost to the Indian economy. The major CDM projects are located in Rajasthan, Andhra Pradesh, Maharashtra Karnataka, Himachal Pradesh and Punjab. Trading in CERs, although at a nascent stage, has resulted in huge foreign exchange earnings for Indian suppliers. Typically, an overseas buyer requiring carbon credits to meet emission reduction targets enters into an emission reduction purchase agreement with a company engaged in an emission reducing project in a developing nation. However, on account of the ambiguity on the treatment of CERs from an indirect tax perspective, the following issues remain relevant for an Indian supplier of CERs: (i) whether CERs can be classified as goods or services for indirect tax purposes. To determine whether VAT or service tax would be applicable, it is relevant to determine whether CERs are classifiable as goods or services. Lack of clarity on the treatment of CERs for indirect tax purposes exposes companies engaged in projects generating carbon credits to the threat of significant tax demands. The Institute of Chartered Accountants of India (ICAI) is currently working on accounting norms for carbon credits.

Recognition of Income and Expenditure

Once the emissions trading scheme is up and running companies will be affected either directly, through the obligation to account for their own emissions or emissions embedded in products they sell, or indirectly. An indirect consequence arises when a company downstream from a firm with obligations faces higher prices as a result of emissions pricing. For example, businesses will face increased fuel or electricity charges because a large proportion of the cost of emissions obligations imposed on fuel companies or electricity generators is expected to be passed on to consumers in these sectors. Under current tax law, persons carrying on a business can claim a deduction against their taxable income for expenditure incurred in carrying on that business. Deductions are not available for expenditure that is of a capital nature.

When the consequences of the emissions trading scheme are indirect the costs will typically be deductible under standard income tax rules. Deductions will be for items of a revenue account nature, such as increased input costs, incurred in the course of carrying on a business. When the consequences of the scheme are direct, as they will be for oil companies in the liquid fossil fuels sector, for example, the cost is also an incident or consequence of doing business. The obligation under the scheme will arise because emissions are released in the process of production or because of the embedded emissions content of products is emitted on use by downstream consumers. Frequently, but not always, this cost will be reflected in increased income. It is an annual cost associated with the production process, from which income arises in a particular year. To conclude, one may take a view that the costs of meeting emissions obligations are an expense that should be deductible under current tax law.

Tax Planning

CER credits are indeed intangible assets, held with registry. CER credits acquired from other parties for the purposes of trading are recognized in the books at the cost of acquisition, whereas self-generated CER credits are not reflected in financial accounts. As issues for accounting of CER credits will appear for the first time in Financial Year 2006-07, it's important to disclose the accounting policy adopted for this purpose. It would be appropriate to disclose units of CER held with registry in notes bi-furcating between purchased and self-generated. As CERs are capital assets, tax liability should be admitted under the head Capital Gain, and claim for concessional rate of taxation should also be made if credit is held for more than 36 months immediately preceding the date of transfer. This gives an opportunity to take a decision about timings of sale of such credits, keeping a balance between cash flow needs, interest factor and difference in rate of tax between long-term and short-term holdings. As there would be no cost of acquisition for self-generated CER credits, section 55(2) of the Income Tax Act will come into operation, and total sale consideration will be liable for Capital Gains Tax (long term/short term) according to the period of holding. In Indian circumstances, if sale of CER credits happen to overseas buyers, of the property held overseas, such sale, though sale of 'goods', will not attract any sales tax (http://taxpolicy.ird.govt.nz/publications/files/emissions.pdf).

Cost Containment Mechanisms

There are a variety of cost containment mechanisms that can help manage the cost of compliance for covered entities in a cap-and-trade program.

Offsets

Offsets are emission reduction projects undertaken at sources outside a cap-and-trade program. An offset mechanism enables covered entities to offset their own emissions by purchasing emission reduction credits generated through projects at facilities not covered by the cap. Offsets lower the overall cost of the program by bringing in low-cost emission reduction opportunities from outside the cap. Offset projects may include landfill methane capture, afforestation, or other types of projects. Offsets should be measurable, real, and additional and have clear ownership. Regulators must also be able to verify such projects. Through the Clean Development Mechanism of the Kyoto Protocol, developed countries can use offset projects in developing countries to comply with their targets. The northeast Regional Greenhouse Gas Initiative allows certain types of offsets as well.

Allowance Distribution

Once the cap has been set and the overall design of the cap-and-trade program established, choices have to be made about the best way to distribute emissions allowances. In general, how allowances are initially distributed does not affect the emission reductions achieved by a cap-and-trade program. However, it does affect how the program's costs are distributed and can sometimes affect overall program costs. There are two basic approaches to allowance distribution: some form of free allocation, or some form of auction. A combination of auctioning and free allocation, or a shift from one to the other over time, is also possible. Regardless of which method is favored, either allowance allocation or auction revenues can be used to mitigate economic impacts (e.g., by granting allowances to emitters who are competitively disadvantaged by emission caps) or drive innovation (e.g., by using allowances or auction revenues to fund or incentivize research, development, demonstration and deployment of low-carbon technologies).

Several types of free allocation exist. Allowances can be given away for free based on participating entities' historical emissions (a method also known as "grandfathering"). Output-based methods of allowance allocation are based on the output of a product in a given sector. For example, allowances might be distributed based on megawatt-hours generated or tons of a product manufactured. Benchmarking, or setting a level of emissions (in the form of allowances) per unit, can be applied based on input or output. Allowance allocations may also be "updated" over time as input, output or emissions change. In the case of free allowance allocation, it is important to bear in mind that the point of regulation described above (where compliance is demonstrated by submitting allowances) does not necessarily need to be the same point at which allowances are initially distributed.

There are tradeoffs between simplicity and equity if allowances are distributed for free. For example, basing allocation on historical emissions is relatively simple. However, it means that some form of credit for early action would be needed to ensure that firms who took voluntary measures to reduce their emissions before the base year are not penalized for doing so. Updating has the advantage of adjusting allocation to changing circumstances. However, while fixed allocations will not affect firms' future behavior, updating encourages firms to behave in ways that will maximize their future allocation. For example, if firms believe that allowances will be distributed based on future emissions, they may try to increase their emissions in order to receive more allowances.

As an alternative to free distribution, allowances can be auctioned. Auctioning generates revenue that the government can use to provide relief for compliance or higher energy costs. The government can also use the auction revenue to reduce other taxes that may be discouraging economic growth, or to fund complementary policies. However, as with the various forms of free allocation, there are tradeoffs involved with auctioning allowances. The impact of costs on a given firm depends on the competitiveness of the industry in which the firm operates as well as that industry's regulatory environment. In some cases, auctioning may unfairly hurt participants lacking the funds to purchase enough allowances from outside the covered region. This is especially true for firms who cannot pass on some or all of the costs of their allowances to consumers. However, firms in other industries might be able to pass on their compliance costs under a cap-and-trade system. In these cases, firms would be over-compensated if most or all allowances were given away for free, which might lead to windfall profits for these firms. Auctioning some or all allowances could help avoid such windfalls. Auctioning can also help address concerns about crediting early action by firms, as it rewards those who have already reduced emissions by investing in lower-carbon technologies.

Either allowances themselves (in the case of free allocation) or auction revenues (in the case of auctioning) can be used to advance program goals under a cap-and-trade system. For example, if regulators want to promote end-use energy efficiency programs among consumers, they could either use proceeds from auctioning allowances to support efficiency projects, or distribute allowances for free to entities undertaking efficiency projects. Similarly, just as auction revenue can be used to help offset program costs, free allocation can also be used to deal with high compliance costs which might be passed on to consumers. The key difference between auction revenue and allowances is that auction revenue can more easily be used to adjust other taxes, and allowances are more easily limited to purposes more closely tied to the cap and trade program itself.

Greenhouse Gas Trading in Practice

Emissions trading systems are already proving their value as tools to address climate change by reducing emissions of greenhouse gases throughout the world, and other markets are under development. Let us have an overview of the GHG Trading practices and salient achievements under the EU Emission Trading Scheme of European Union.

The world's most ambitious and far-reaching example of greenhouse gas emissions trading is the European Union's Emissions Trading Scheme (ETS), which limits CO_2 emissions from approximately 12,000 facilities in the 25 EU member states. Launched in 2005, the ETS covers power plants and five major industrial sectors (including oil, iron and steel, cement, glass, and pulp and paper) that together produce nearly half the EU's CO_2 emissions. An initial "learning phase" (phase I) ran through 2007; a second coincides with the Kyoto Protocol compliance period (2008-12). Excess emissions incur a penalty (40 Euros/tonne in phase I, 100 Euros/tonne in phase II) and must be made up in the next phase. During the learning phase, ETS allowance prices fluctuated due to weather (affecting energy demand), shifts in energy prices, and initial over-allocation of allowances as a result of incomplete historical emissions data. Many regard these fluctuations as characteristic of a new compliance market. The EU ETS plans to adjust its allocations in the next phase and is also considering auctioning a significant portion of the allowances.

The ETS is the first program of its kind and size, and has established a functioning market in a relatively short span of time. Volume of allowance trading reached over 100 million allowances per month in early 2007, and rates of compliance with the program are high. In general, the EU ETS promotes innovation and is seen as flexible and cost-effective. European Union policymakers have said the ETS will continue beyond 2012 with or without a new international climate agreement. In January 2007, the EU commission released its proposal to commit the EU to a GHG reduction target of 20 percent below 1990 levels by 2020 and suggested that if other industrial countries follow suit—namely the United States—the EU will commit to 30 percent.

Profit from Accounting Policy Initiatives

Let us now consider how unreasonable and irresponsible it is for business leaders to accept a simple statement advocating a market solution to address environmental degradation. Society needs responsible discourse that takes into account efficient economic solutions. Accountants that have a deeper understanding of the economic and political situation can contribute to their employers and society more effectively. We all gain if the environmental agenda includes the concerned voice of accountants. As a profession, there is a great deal to be considered and studied that is missed with simplistic statements that market solutions are best. The language of business can incorporate societies need to elevate environmental issues, now that we are reaching the practical capacity of this public good. It is time for this to be treated as a professional responsibility.

The Kyoto accord left countries a great deal of flexibility with regards to the mechanisms they would use to comply with treaty obligations. Even though there is flexibility, the discussion has focused on trading carbon credits based on quota's and productivity scales. These public policy responses will be extremely difficult to incorporate into accounting policy. The accounting profession has to respond with more than a recommendation to added sustainability as a measure in the balanced scorecard. We can therefore advocate for three categories of response by professional accountants: inclusion of carbon gas emission markets in the capital budgeting process, reporting environmental impacts in financial reports, and advocating for the most efficient regulatory approach for each sector of the economy. Accountants that respond in these ways will contribute to the competitive advantage of their firms and society.

The first step is to consider future costs in capital budgeting models. A carbon credit market based on output permits is costly to implement. Many businesses can expect that the cost of this will be higher than an input tax such as the one we have on gasoline now. Since we cannot assure a change in policy direction, it is prudent to make investments with the output mechanism in mind. Retrofitting existing systems and processes to record emissions is expensive and disruptive. Firms that build in such capacity now, create a competitive advantage in the future. Costs of training are also likely to skyrocket. Firms that establish relevant training now will compound their cost advantage. The return on these investments will be a great benefit to the stockholders. This is no more than good risk assessment. Enlightened accountants will look for and emphasize these preparatory investments when they are evaluating capital budget proposals and participate in setting budgetary priorities.

The second opportunity is to report environmental assets and liabilities on the balance sheet now. There is a long learning curve required to incorporate carbon credits into the reporting process. Firms can develop experience with carbon assets by preparing procedures for determining market values and disclosing unrealized gains or losses on a pro-forma basis. They can also implement processes to collect and disclose carbon emissions much like those we have for disclosing future lease and debt payments. This disclosure will increase awareness of future costs and create stockholder confidence in firms that can give a clear accounting. Reporting on carbon offset projects, as we now report on research and development pipelines, form a third opportunity. The market will be able to value the firms' ability to create marketable credits in relation to their competition. In every case, a proactive accounting policy enhances the ability of management to see tangible benefits from environmental investments. Every time regulation is increased firms that employ enlightened accounting processes will be rewarded by a market that can assess their readiness.

A third opportunity for accounting professionals is to advocate for effective regulation. Those that are forming public policy are in a difficult position. They have a duty to take action and deal with a population of voters that do not appreciate the cost of accelerating environmental degradation. When accountants label all regulation as undesirable they miss an opportunity for considerable economic gain. Firms can gain a competitive information advantage when they involve themselves in the process. A model that regulates output efficiency will cause high implementation costs on some sectors. Firms could create a sectoral trade advantage by supporting and endorsing regulation that is less costly. For example, a US sector might gain a cost advantage over Canadian firms if it convinces US regulators to adopt a regulation, such as an input tax, that has low measurement and transaction costs.

Differential International Emissions Trading Tax Regime

Different fuels are taxed at very different rates within almost all European countries, and the same fuel is taxed at very different rates from one country to another (Newbery, 2001). According to Newbery (2001), in most cases the taxes predate environmental concerns, are not related in any systematic way to environmental damage, and do not meet minimal criteria for so doing. Coal is almost invariably the most environmentally damaging fuel, but it is usually the least heavily taxed, and in many countries its production is heavily subsidized. If road fuel taxes can to a considerable extent be justified as road user charges, there is little evidence that road taxes are set on the basis of charging the long-run marginal cost of expanding roads (Newbery, 1992). Since climate policy will be implemented under imperfections and distortions in the energy markets, one might expect the general propositions of the second best theory and the theory of trade policies to be valid for the analysis of markets for tradable emissions permits. Therefore the general proposition is that international emissions trading (IET) may be welfare decreasing when primary gains from trading are outweighed by "secondary costs" associated with preexisting distortions and market imperfections.

Certification and Verification

Since emission reductions used to generate credits under article 12 require considerably more scrutiny, a certification function is necessary to ensure that only certified emission reduction become part of the allowance system. Certified allowances should be treated in the same way as all other allowances. The certification process is one practical way to help ensure that the trading system runs smoothly. While the ultimate authority for certification should be the Conference of the Parties, the operational authority for certification can, and should, be delegated to subordinate bodies specifically designed to fulfil that function. Some certification authority could be delegated to specific governmental units within participating nations or communities of participating nations or even to private certification entities, providing that certain preconditions were met. These preconditions should include: (1) an identified organizational unit willing and able to assume the responsibility for certification; (2) the existence of sufficient enabling legislation to ensure the unit had adequate powers to carry out its mission, as well as adequate staff and resources; and (3) acceptance of, and willingness to apply, standard certification criteria. Certification and monitoring processes should be initiated promptly. Furthermore, non-Annex-B Nations that contemplates establishing private tradable entitlements systems will need guidelines on the appropriate processes of certification and monitoring. If these are provided early, the chances of achieving harmonized procedures in the future will be higher. Viable and legitimate systems created today, even if they are small, will become the de facto standard, and the early creation of such systems will highlight the changes needed in domestic institutions and procedures, allowing more time to make the necessary reforms. While certification is sufficient for transfer of a credit, use of a credit to fulfil part of an assigned amount obligation would also require verification. Whereas certification would provide assurance that a specific emission reduction or carbon absorption would be forthcoming from the project, verification would provide the assurance that these expectations had in fact materialized.

Compliance and Enforcement Norms

Multiple commitment periods offer significant opportunities for enforcing compliance. The principal tools for enforcing compliance include declaring non-compliant Parties ineligible for trading and reducing assigned amounts in subsequent commitment periods. Currently, the Protocol establishes that there will be negotiations to fix assigned amounts in subsequent commitment periods, but it has perhaps not been generally recognized how important that task is in promoting compliance within the first commitment period. Transparency is another tool for ensuring compliance, since much of domestic law presumes that enforcement will be based upon public opinion and normative pressures to comply. Transparency is an important complement to enforcement because it makes violations apparent, and the fear of detection promotes compliance. Transparency systems require the disclosure of basic information regarding obligations, actual emissions and trading activity, in order to allow judgements about compliance status. To date, transparency systems have relied heavily on self-reported data, and have worked best when data have been made available to the public. Such data need to be collected and verified at the international level, which will require the use of international public databases of self-reported and verified information, under the responsibility of the authorized body (Tietenberg *et al*, 2008).

Emissions Trading: The Tradable Entitlement

The case for a *tradable entitlements system* is based on the advantages that it would offer over other politically feasible alternatives. In the short term; it offers the possibility of reaching the environmental goals at a lower cost than would be possible if each country were limited to reduction options within its own borders. Making it easier to reach the goals may encourage more countries to sign the Protocol and would probably increase compliance with those goals. Because it separates the issue of who pays for control from who implements control, it facilitates trans-boundary cost-sharing, an item of particular importance to both the developing countries and the transition economies of Eastern Europe (Tietenberg, 1994). Tradable permits also facilitate the mobilization of private capital for controlling global warming; private capital is likely to be a critically important component of any effective global warming strategy for as long as there is insufficient public money to finance it. Tradable entitlements, also to a great measure, facilitate the development and implementation of innovative approaches to climate change control. By offering greater flexibility in how the emission reductions are achieved, tradable entitlements can significantly lower the long-term costs. Lower long-term costs may be an important element in gaining greater international acceptance for the idea of emission limits and in reducing the difficulties associated with ensuring compliance.

Trading Transaction Modalities of CERs

A transaction in Certified Emission Reductions (CERs) trading involves buying of GHG emissions credits by entities (companies, Governments, etc.) from companies or Governments involved in projects related to carbon emission reduction in developing countries. The idea is to make developed countries pay for their wild ways with emissions while at the same time rewarding countries with good behavior in this regard. For example, a company in India can prove it that it has prevented emission of x-tons of carbon, it can sell this much amount of points or carbon credits to a company in say, the US which has been emitting carbon. Consider two companies, A and B, both of which emit significant quantities of a given pollutant. Their emissions may danger air quality, and relevant authorities may decide the emissions should be reduced by a given amount, say by 10%. At first glance the task seems simple: both A and B cut their emissions by 10%. But in real world this may impose different burden on both the companies. For example, company A, may by nature of its activities, be able to reduce its emissions by 10% or more at relatively low cost. Company B, on the other hand, might find it difficult and costlý process. It is this potential difference in reduction cost between A and B that creates the market opportunity. Companies A can reduce its emissions by a required amount at relatively low cost and can then make the affordable reductions. For Company B, the reduction cost is far greater and it would welcome the way of avoiding some of the outlay. Now what if Company A agrees to reduce some additional reductions instead of Company B, provided that Company B is ready to pay a price that is above the cost but below what it would cost Company B? In these situations emissions are cut overall by the required amount Company B saves the money and Company A earns the profit for its additional reductions and the total cost is reduced (Figs. 8 and 9).

Byers of Carbon Credits

The following Pie-Chart (Fig. 10) provides country-wise break-up of carbon credits bought during January, 2005-March, 2006.

Sellers of Carbon Credits

The following Pie-Chart (Fig. 11) provides country-wise break-up of carbon credits sold during January, 2005–March, 2006:

Derivative and OTC Market of Carbon Credits

Emissions-trading is the mechanism of trading between states or firms, the right related to gas emissions such as carbon dioxide, which are considered the cause of global warming. It was adopted by the Parties to the Third Session of the United Nations Framework Convention on Climate Change (December, 1997), as a scheme to reduce greenhouse gas emissions efficiently using a

Fig. 8 & 9: Diagrammatic Presentation of Trading of CERs

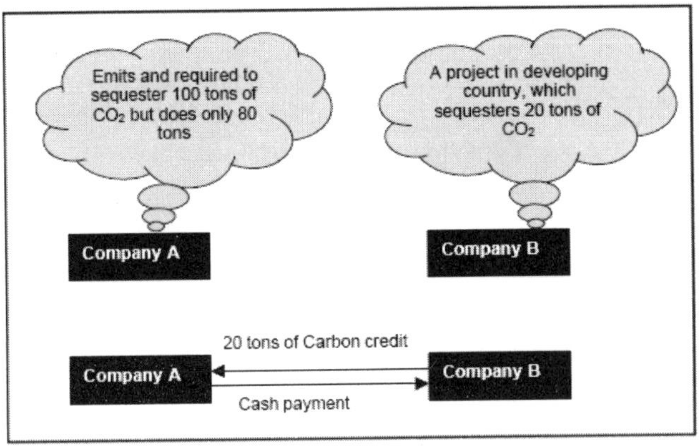

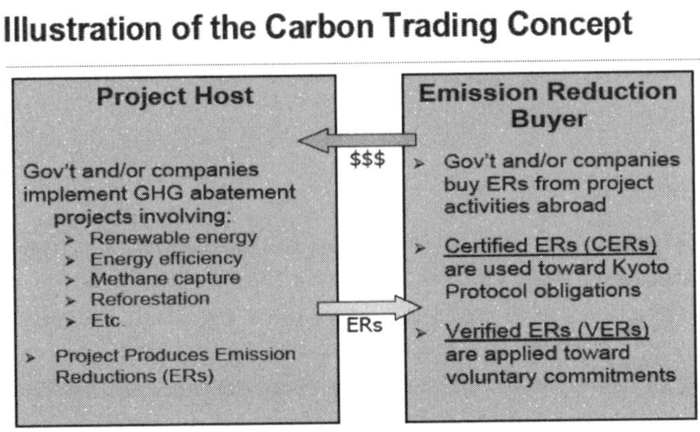

Fig. 10: Country-wise Break-up of Carbon Credits bought during January, 2005- March, 2006

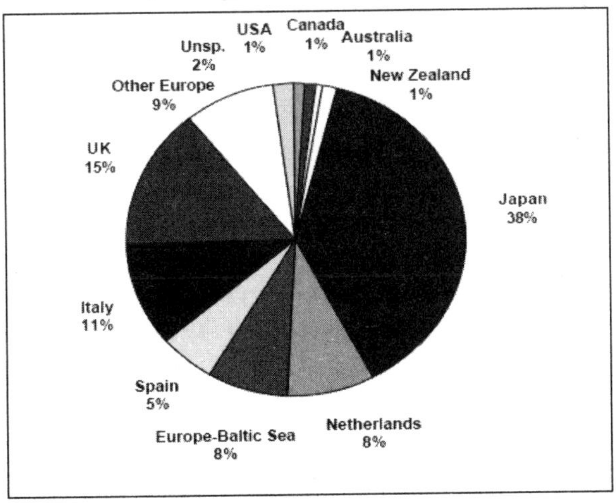

Overall volume: 453.5 million tCO_2e **January 2005 to March 2006**

Fig. 11: Country-wise Break-up of Carbon Credits Sold during January, 2005- March, 2006

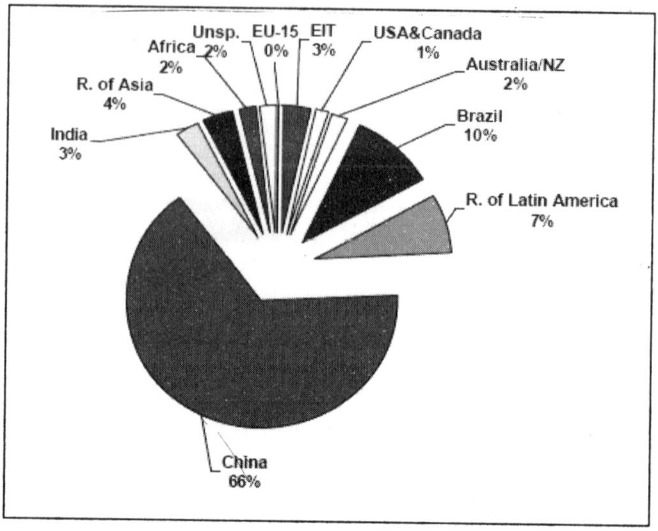

January 2005 to March 2006

market mechanism, and was incorporated into the Kyoto Protocol. The OTC market is a market where risk hedgers, risk takers, and brokers meet to deal directly in the trading of various derivative products based on various underlying assets such as interest rates, currency, equities, and commodities etc. The OTC market is also at the cutting edge of financial innovation, where new products are developed one after another based on the creativity and needs of unrestricted participants. It is believed that in the near future risk hedging techniques, such as medium-to-long term swaps, and options, will be dealt also in underlying related to emissions. OTC market in which brokers are intermediary between the demand and supply of carbon credit as it is a success factors of OTC that is standardization, simplicity, liquidity and transparency – ultimately encourage the differentiation amongst other competitors i.e. their variety, flexibility, highly structured nature and occasionally extreme complexity in terms of risk and pricing.

Futures and Forward Contracts of Carbon Credits

Most of the EUA trade is in forward and futures contracts, where the price is agreed now but allowances and payment do not change hands until a set date in the future. There is little trade in the 'spot' market. Companies that are covered under the EU ETS, so called 'compliance' buyers, buy and sell EUAs in an effort to manage their expected future needs of emissions permits based on their forward production plans of, for example, electricity, steel or cement. Half or more of the trade in EUAs has up to now been 'over the counter' (OTC) in bilateral forward deals negotiated through brokers which may have varying terms and conditions. But there is growing trade in standardized futures contracts on various electronic energy and emissions exchanges such as the ECX, EEX, Nord Pool and Powernext.

Currently, futures contracts in carbon credits are actively traded in the European exchanges. A carbon futures contract is an agreement between two parties, either to buy or sell the carbon credit at a predetermined price and at a specified period. The underlying unit (carbon credit) that is being traded would be an electronic certificate bearing a unique serial number indicating the country of origin and sequestration year. Analysts advise companies operating in the Annex I countries to take a buy position (long position) in the current market for carbon futures to mitigate the possibility of increasing prices of carbon credits. In fact, many companies actively participate in this market to manage the price risks associated with the trading in carbon credits. The most prominent risk associated with carbon trading is the failure on the part of member countries with regard to compliance with Kyoto Protocol. Keeping in view the risks associated with futures contracts in carbon credits, trading in futures contracts in carbon allowances has now become a reality in Europe with burgeoning volumes. Currently, there is active participation by public utilities, manufacturing entities, brokers, banks, and others. So, with the prospective to capture the European carbon credit market, Multi-Commodity of India Ltd. (MCX) entered into a strategic alliance with CCX in September 2005 to initiate futures carbon trading in India.

Risks Associated with Carbon Credits

The coming into being and operation of the EU-ETS, the ECX futures exchange platform, revealed that there are market- and policy-related risks for CER producers, including the supply-side risks starting from the DNA approval risk to the CER issuance risk in a complete CDM approval cycle. Apart from these risks there are a host of other risks from both the supply and demand sides that the real market players confront with. Most CDM projects by their very nature take a long time to generate the CERs and hence, face the aforesaid risks in large proportion, which if not hedged would lead to reduced realization. Under such a situation, the realization of CER generators at times may not even cover the investment put in to generate the CERs and thus, has the potential of even making a CDM project unviable in the long term. Given the long gestation period of CDM projects and the risks involved, it is rather inevitable that they pre-sell their potential credits in the futures market-preferably a domestic futures market, to avoid Forex risk attached to participation in a foreign exchange.

India as a Potential Supplier of CERs

Over 100 million tonnes of CERs are potentially available globally through

2012 The portfolio of 78 projects approved by the NCDMA to date has the potential to generate approximately 117 million tonnes of CERs (cumulative) until the end of 2012, according to the project developers' estimates. India continues to be a leader in both small and large scale projects. Two HFC-23 destruction projects and one PFC reduction project from a large aluminium smelter have been approved. In addition, projects related to renewable energy and energy efficiency and fuel switch have the potential to generate another 17and 37.4 million tonnes of CERs until 2012, respectively (http://www.cdmindia.com/gtz/publications/INDIA-%20Worlds%20Leading%20CDM%20Destination.pdf).

India, being one of the leading generators of CERs through CDM, has a large scope in emissions trading. Analysts forecast that its trading in carbon credits would touch US$ 100 billion by 2010. Currently, the total registered CDM projects are more than 300, almost 1/3rd of the total CDM projects registered with the UNFCCC. The total issued CERs with India as a host country till now stand at 34,101,315 (around 34 million), again around 1/3rd of the total CERs issued by the UNFCCC. In value terms (INR), it could be running into thousands. Various industries that have scope of generation of CERs; viz. Agriculture, Energy (renewable and non-renewable sources), Manufacturing, Fugitive emissions from fuels (solid, oil and gas), Metal production, Mining and mineral production, Chemicals, Afforestation and reforestation. Further, there has been a surge in number of registered projects in India. In 2007, a total of 160 new projects were registered with the UNFCCC indicating that more than half of all registered projects in India happened last year. It is expected that with increasing awareness this would go further up in the future. The number of expected annual CERs in India is hovering around 28 million and considering that each of these CERs is sold for around 15 euros, on an average, the expected value is going to be around Rs 2,500 crore.

Financial Accounting Issues of CDM Credits in India

India is one of the major players in the global market on the supply side of CERs. Indian companies have started getting credit of CERs and some of them have also entered into sale arrangement with buyers in the international market. As this is a new concept, it has given rise to interesting financial accounting dimensions. Issues involved are: (i) how to account for expenditure on CDM projects, (ii) whether or not to account for self-generated CERs held with registry, (iii) if credits are to be accounted, at what point of time these should be recognized in books of accounts and at what value, and (iv) how to account for sale consideration of CERs and its disclosure in accounts and notes. Answers to these questions are found within existing pronouncements of ICAI as well as Schedule VI requirements. Taxation issues will naturally follow (Agrawal, 2006).

Possible Business Models for CER transactions

This section of the paper looks at three possible models through which CERs may be transacted in India and analyses the potential tax implications of each such model. Each of these models assumes that the foreign investors' business presence in India would be in the form of a wholly-owned subsidiary ("Indian subsidiary").

A. Trader model

1. Mechanism

Under the trader model, the Indian subsidiary would purchase CERs produced from an Indian CDM project ("generator") and sell them to third party purchasers in Annex I countries. Here, there would be two related contracts: one under which the Indian subsidiary would promise to sell (export) CERs to an overseas buyer, and the other under which the Indian subsidiary would purchase the CERs from the generator. The other possibility is that the Indian subsidiary would purchase CERs produced from various Indian CDM projects and then sell them as a single package to an overseas buyer (IT Power, 2007).

2. *Legal and Tax Implications*

(a) Income tax

- The sale of CERs by the generator could either attract capital gains tax (if the right being sold is treated as a capital asset), or yield business income for the generator, which would be taxed.
- The Indian subsidiary would also earn income upon the further sale of CERs to an overseas buyer and would be required to pay income tax on the amount earned at the applicable rate.

(b) Sales tax

- The sales tax implications of a sale of CERs would vary depending upon the facts and circumstances of each case. Typically, in the case of a sale within a state of India (an intra-state sale), the generator's sales tax liability for the sale of CERs would depend upon whether the local laws of that state impose tax on the sale of intangible goods or rights. However, in the case of inter-state sales, (where the generator and the Indian subsidiary are registered in separate states), sales tax liability would have to be determined under the Central Sales Tax, 1956 (CST Act). Sales tax is typically passed through to the customer—the Indian subsidiary in this case.
- The Indian subsidiary would not be required to pay any sales tax on the sale of CERs to an overseas buyer, since sales "in the course of export" is not subject to sales tax in India.

- Even the sales by the generator to the Indian subsidiary (discussed above) could be outside the purview of sales tax if the Indian subsidiary successfully establishes a link between its contracts with the overseas buyer and the generator. This is because the CST Act views the last sale of any goods immediately preceding the actual export of those goods as a "sale in the course of export", if such last sale took place in furtherance of an agreement in relation to such export. This issue will have to be kept in mind at the time of drafting the CER sale/ purchase contracts.

B. Broker Model

1. Mechanism

Under the broker model, the Indian subsidiary would neither purchase nor sell the CERs. Instead, it would introduce prospective buyers and sellers/ generators of CERs for a commission, which could be earned either in the form of cash or CERs. This business model may be preferable in transactions involving large volumes of CERs, where the Indian subsidiary would not want to act as a trader and expose itself to the risk of downward market fluctuations between the time it buys the CERs from the generator and sells them to an overseas buyer (The British Council-IT Power, 2007).

2. Legal and Tax Implications

(a) Income tax

- As stated above, the generator would have to pay either capital gains tax or income tax depending upon whether the proceeds arising out of the sale of CERs are treated as capital gain or business income.
- The commission received by the Indian subsidiary (whether in the form of cash or CERs) would be treated as business income and chargeable to income tax at the applicable rate.

(b) Sales tax

- The sale of CERs by the generator to the overseas buyer would be in the "course of export" and would therefore not be chargeable to sales tax.
- The Indian subsidiary would not be required to pay any sales tax since it would not be selling any CERs.

C. Primary Owner Model

1. Mechanism

Under this business model, the Indian subsidiary would become a capital investor in the main CDM project by providing either funds or technology to the project and by taking a share in the CERs produced. This would make the Indian subsidiary the primary owner of its share of the CERs produced. The Indian subsidiary would then sell these CERs to overseas buyers or to other traders of CERs within India.

2. *Legal and Tax Implications*

(a) Income tax

- The transfer of CERs by the generator to the Indian subsidiary, in lieu of the investments made by the Indian subsidiary in the CDM project, may be viewed by the tax authorities as the relinquishment of rights in such CERs by the generator, which could have tax implications for the generator. If CERs are treated as a capital asset by the tax authorities, any relinquishment of such capital asset or any rights therein by the generator would be subject to capital gains tax. This model would be similar to the trader model from a tax perspective, except that the Indian subsidiary would be exposed to the additional risk associated with the operations of the generator and the Indian subsidiary's investments therein.
- The income that the Indian subsidiary would earn from such sale of CERs to an overseas buyer would be chargeable to income tax at the applicable rate.

(b) Sales Tax

- Once the Indian subsidiary holds the CERs as a primary owner, and sells them to an overseas buyer (in the course of export), such sale would be exempt from sales tax for the reasons discussed above.

D. MARKAL: A Model to Support Greenhouse Reduction Policies

MARKAL is a generic model that represents all energy demand and supply activities and technologies for a country with a horizon of up to 40 years. It is a techno-economic model which assembles in simple but economically consistent way technological information (conversion-efficiency, investment-and variable costs, etc.). As the model is formulated as a dynamic optimization model, it can produce alternative developments for energy supply and demand that achieve CO_2 emission reduction goals at least cost (http://www.belspo.be/belspo/home/publ/pub_ostc/CG2131/rappCG22syn_en.pdf).

Simultaneously, the model makes prospective energy and emission balances, tests the potential of new energy technologies and contributes to R&D policy formulation. Finally the model is well suited to approach the burden sharing issue between sectors of the same country in a transparent and scientific way. Compared to ad-hoc models which are more specific to a country or a sector and which use another modelling technique, it presents three important advantages:

- Due to its transparency it promotes the communication between experts with different sectoral or technological background (it is the place where engineers and economists understand each other),
- It is easily verifiable: its results can be related to assumptions regarding technological data and economic parameters,
- It is comparable at an international level: as many countries use the same model, its results can be immediately compared with results from other countries.

The first Belgian version has been developed by CES-KULeuven and VITO in the first GLOBAL CHANGE program of the DWTC-SSTC and has already been used intensively for policy support.

At present the model is used in 30 countries for policy analysis purposes. It is a collaborative effort coordinated by the ETSAP (IEA) network. The ETSAP-network (Energy Technology Systems Analysis Programme) is an agreement within the International Energy Agency which concentrates its work on "Energy Options for sustainable Development". The ETSAP-network is in charge of the maintenance of most of MARKAL model software (database-management system and model specification) and organizes two workshops per year where the experience with case-studies of some 20 countries is compared. Results from common case-studies are presented in international forums, organized ea. by IEA and can contribute to the negotiations within the United Nations Framework Convention on Climate Change (FCCC). This international network contributes particularly to a continuous development of the model in many directions and the network is currently involved in the development of a new Markal, called TIMES. Markal is a partial equilibrium model, which is complementary to other models. The complementarity is mainly related to the following three types of models:

Detailed sectoral models: a model for one sector can be more detailed for the technologies or the type of behaviour of the economic agents (e.g. micro-simulation models which represent the behaviour of a representative sample of approximately a hundred households); this type of model allows to evaluate more correctly other instruments with a more short term impact (information, specific norms, ...); these results can be used as an input to energy models as MARKAL.

National general equilibrium models: these are economic models which allow evaluating the macroeconomic impact of a CO_2 policy (e.g. GEM-E3 model of the EU). These models can study such questions as the use of the revenue from a CO_2 tax, the double-dividend discussion, the total impact on employment, etc. and deliver a basic forecast for the demand for energy services (an input for MARKAL).

International energy-economy models: these models evaluate the world impact of CO_2 emission reduction options in terms of burden sharing, climate damage, exhaustion of resources, etc. (e.g. DICE model of Nordhaus, MERGE model of Manne & Richels and GEM-E3 for the EU)—Markal can deliver inputs to such models and can make use of some of their results (e.g. feedback on the price of resources).

The MARKAL model can contribute to the following problems:

- propose minimum cost solutions for CO_2 reductions and in this way contribute to the burden-sharing within a country,
- compute prospective energy and emission balances,
- evaluate the role of new technologies for CO_2 reductions and contribute to the setting of R&D priorities,
- evaluate the impact on the costs and on emissions of different types of regulations, standards and taxes.

The research has focused on three components to achieve its objective:

- The maintenance of the Markal model
- The development of the model
- The development of scenarios and policy analysis with Markal and the results obtained regarding these different aspects are described hereafter.

The MARKAL Model for Environmental Accounting in Energy and Material Systems

The MARKAL dynamic linear programming model has been used to assess future energy and materials strategies for CO_2 emission reduction. In the dynamic systems approach, interactions between technologies, technology dynamics, and capacity constraints are taken into account. Storage of materials in products during the product life is considered. This model can be used to analyze the relation between current materials consumption and future waste problems. The model can also be used to analyze the impact of energy policies on materials production and materials use. MARKAL results show that future material flows will be significantly affected by CO_2 emission reduction policies. Integrated energy and materials policies can decrease CO_2 emission reduction costs compared to stand-alone energy policies. Results are discussed for two studies in which the model has been applied: a study for the Netherlands and a study for the Western European steel industry. An integrated energy and materials model for Western Europe is currently being developed (http://gasunie.eldoc.ub.rug.nl/root/1997/2039025/).

Long-term UK Carbon Reductions using the UK MARKAL and MARKAL-Macro Energy System Models

Since the publication of the 2003 Energy White Paper "Our energy future —creating a low carbon economy" (DTI, 2003), the UK Government has been assessing the key longer-term challenges for UK energy policy (http://

www.law.monash.edu.au/regstudies/synopsis.scenarios-sensitivities-on-long-term-carbon-reductions-using-the-uk-markal-and-markalmacro-energy-system-models.pdf). Notably these include reducing carbon dioxide (CO_2) emissions to mitigate the impacts of climate change, ensuring that the UK has an energy resource that is both clean and secure with the move to increased dependence on imported energy, and maintaining low cost energy service provision to aid competitiveness of UK firms. The results of the Energy Review, described in "The energy challenge" (DTI, 2006a), were published in July 2006. This document set out the types of challenges that need to be addressed to ensure that the UK could move to a low carbon and energy secure economy.

A new Energy White Paper, published in May 2007, set out the policy framework and initiatives for ensuring that these long-term energy policy objectives can be met. To support the activities under the Energy Review and proposals for the 2007 Energy White Paper, DTI and DEFRA commissioned a series of analyses with the new UK MARKAL and MARKAL-Macro models. Under the research portfolio of the UK Energy Research Centre (UKERC), these partial and general equilibrium energy systems optimization models have been substantially developed and extended by the Policy Studies Institute (PSI) together with AEA Energy and Environment. The focus of the DTI/DEFRA analysis was to investigate and quantify long-term carbon constrained scenarios, focusing on characterization of uncertainty in energy supply, technology pathways and cost implications. Since the publication of the 2003 Energy White Paper "Our energy future – creating a low carbon economy" (DTI, 2003), the UK Government has been assessing the key longer-term challenges for UK energy policy. Notably these include reducing carbon dioxide (CO_2) emissions to mitigate the impacts of climate change, ensuring that the UK has an energy resource that is both clean and secure with the move to increased dependence on imported energy, and maintaining low cost energy service provision to aid competitiveness of UK firms. The results of the Energy Review, described in "The energy challenge" (DTI, 2006), were published in July 2006. This document set out the types of challenges that need to be addressed to ensure that the UK could move to a low carbon and energy secure economy.

Results focus on a selected set of MARKAL-Macro (M-M) model scenarios, utilizing an integrated set of UKERC assumptions and data:

- Base-case, CO_2 emissions in 2050 constrained to 60% of 2000 levels (C-60), and alternate CO_2 emission trajectory (SLT) implemented linearly from 2010;
- Resource import (high and low) price scenarios, from DTI projections;
- Technology scenarios: restricted innovation (limited to either 2020 or further to 2010 levels of improvement), no-nuclear, no-CCS or no-nuclear/CCS scenarios.

In all, over 50 full scenarios sets were run for this project. Results from additional scenario runs (including standard model runs) are used to further discuss key tradeoffs between mitigation pathways. Key outputs included primary and final energy mixes, sectoral contributions to CO_2 reductions, detailed technology selection in the electricity and transport sectors, the role of demand side reductions, CO_2 prices, energy system costs and GDP impacts. Presenting a set of model results paints a complex picture of how the UK energy system could develop under deep long-term CO_2 constraints. These runs illustrate possible technology pathways, energy system interactions and resultant energy and macro cost implications. However the MARKAL and M-M runs do not constitute forecasts, rather they are cost optimal solutions based on a set of integrated assumptions in a systematic what-if analysis of the future evolution of the UK energy systems to meet long-term CO_2 reduction targets. Furthermore the scenarios do not constitute a formal or structured assessment of the breadth of uncertainties in future UK energy scenarios, but illustrate the role of key drivers that are relevant to policy assessment of the economic and technological implications of potential UK low carbon energy futures.

Principal Findings

- A 60% reduction in UK CO_2 emissions entails radical changes in resource and infrastructure use, and investment in new technology portfolios.
- This long-term transition requires a strong CO_2 price signal with a central M-M model estimate of £105/TCO_2 by 2050 (within a range of £65/TCO2 to £176/TCO_2 for the key sensitivities covered in this report);
- The resultant impacts (from a relatively smaller energy sector) on the UK economy are more modest with a range of annual GDP losses in 2050 ranging from 0.3% to 1.5% (equivalent to £B7.5 to £B42). The higher cost estimates are strongly influenced by more pessimistic assessments of future low carbon technologies
- Energy system trade-offs are pervasive under alternate assumption sets. These include the use of natural gas vs. coal under low or high resources prices, upstream technological change vs. end-use energy reductions under innovation optimism, and electricity vs. transport CO_2 reduction pathways, based on the timing of emissions reduction requirements.
- These trade-offs illustrate endemic uncertainties in future resources, infrastructures, technologies and behaviour. One example is that it is not possible to robustly project a dominant technology class within the future electricity portfolio.

System Evolution

In general the MARKAL and M-M model base-cases represent a low energy (and emissions) growth in the UK (from 6,152PJ final energy in 2000 to

6,272PJ in 2050). This is due to, even in the base case, pervasive technological change towards cost effective energy saving technologies (in the absence of an economy-wide CO_2 constraint). Natural gas and coal constitute the dominant base-case primary energy fuels.

Under CO_2 constraint scenarios, further reductions in energy consumption are substantial, falling to around 5,250PJ in 2050. This intensity improvement reflects a range of mechanisms including upstream and end-use efficiency, use of technical conservation measures and a pure behavioural demand reduction in response to rising energy prices. These mechanisms combine with energy pathway and fuel switching in all sectors to meet CO_2 reduction of around 375 $MTCO_2$ by 2050. In the CO_2 constrained scenarios, natural gas strengthens its position as the largest component (35%-40%) of primary energy (especially in end-use sectors). Coal use remains significant at around 20%, based primarily on CCS applications. Nuclear retains a share of base-load generation in many but not all constrained scenarios.

Oil and refined oil use see significant reductions due to efficiency and fuel switching (to hydrogen) in various transport modes. Finally renewables see a significant growth in all scenarios (up to a 30% primary energy share), the rate of which is largely based on the availability of other zero-carbon and efficiency options. All upstream and end-use sectors contribute to the stringent 60% reduction target. The electricity sector is a major source of carbon reductions. When electricity emissions are allocated to end-use sectors, industry and service sectors produce the deepest reductions at 24% and 27% respectively of year 2000 emissions. All hydrogen production is utilized by the transport sector, and transport is the last sector to decarbonise, retaining 55% of year 2000 CO_2 emission levels.

Technology Pathways

A 60% reduction in UK CO_2 emissions entails radical changes in resource and infrastructure use, and investment in new technology portfolios. In general, under alternate scenario assumptions there is a trade-off between emission reductions from the electricity sector (notably fuel switching and efficiency gains), the buildings and industrial end-use sectors (notably demand reductions, conservation and efficiency gains) and the transport sector (notably hybrids, hydrogen and bio-fuels). Alternate scenarios generate alternate portfolios in electricity production. Notably, uncertainties in the future costs and characteristics of both new CCS and nuclear technologies mean it is impossible to robustly project that one technology is dominant. In addition, wind generation plays an increasingly important role.

Without either nuclear or CCS the electricity portfolio transforms again to be dominated by offshore wind, supplemented by higher costs renewables (including marine) with base-load requirements met using natural gas and bio-gas CCGT plants. That the technology pathway evolution is inherently uncertain and path dependant is also illustrated in the SLT case, where the earlier CO_2 constraint results in a non-nuclear future, with emission reductions coming from more mature wind technologies and bio-fuels in transport. In the base case, transport final energy consumption is already transformed in the absence of a carbon price signal, including modal shifts towards petrol and diesel hybrid vehicles. Post 2030, buses, then HGVs, and subsequently LGV's evolve into hydrogen vehicles, based on the relative costs of the technologies and importantly the infrastructure requirements per mode. In the CO_2 constrained cases, consumption of transport fuels (petrol and diesel) is reduced further. An interesting (non-intuitive) finding is that hydrogen delivers lower levels of consumption than in the base case, as hydrogen production must be carbon neutral. Bio-fuels play an increasingly important role, with over 20% of all transport fuels by 2050.

In general there is trade-off between upstream emission reduction options and demand reductions, as seen in the cases with restricted upstream and technology options, as well as the "distance to target", as seen in the lower and higher resource price sensitivities. Conservation measures are taken up to their available limit, and combine with the purely behavioural change in energy service demands which cluster around 10-15% reductions and contribute significantly to lowering marginal CO_2 prices (e.g., comparing vs. standard MARKAL results).

Economic Impacts

Such a transformation of the UK energy system under a 60% CO_2 reduction policy necessitates a high carbon price signal. By 2050 the central constrained case generates a marginal CO_2 cost of £105/tCO_2 (or £385/tC). Without endogenous demand reductions (i.e., comparing the M-M model to the standard MARKAL model), this marginal price increases to £135/tCO_2 (or £495/tC). Scenarios with a greater or lesser distance to target (i.e. the low or high resource price cases) give a higher or lower CO_2 price. The highest marginal costs arise in those scenarios where innovation across a broad range of technologies has been restricted (up to £176/tCO_2 or £645/tC).

In terms of energy systems cost projections using the M-M model; explanations are complicated by energy service demand reductions giving a smaller energy system in the CO_2 constraint cases vs. the increasing per unit costs of the energy sector. In the core C-60 case the energy system cost in 2050 is still £0.6 billion lower than the base case, although in the (2010) restricted innovation case costs rise to an increase of £7.2 billion in 2050. Without the complexity of overall demand reduction, the standard model generates abatement costs (scenario C-60) in 2050 of £B8.8 with a low estimate of £B2.1 if all optimistic technological assumptions are employed and a high estimate of £B19.8 if innovation is restricted to 2010 levels. In terms of GDP impacts, the central CO_2 constraint case (C-60) gives an annual GDP reduction of 0.72% by 2050 (or a B£20.3 reduction from a projected UK GDP of B£2,807). The SLT case generates greater GDP losses (£B22.2) than the standard case – this is due to the cumulative emission reductions being greater (7,205 vs. 6,460 $MTCO_2$) and despite a smoother abatement path. The GDP reduction range in 2050 varies between 0.3% and 1.5% (£B7.5 to £B42.0) which is equivalent to other estimates for long-term CO_2 reductions (Stern, 2006). The higher cost estimates are strongly influenced by more pessimistic assessments of future low carbon technologies.

It is stressed that the M-M model is likely to give a lower bound to macro-economic costs, due to bottom-up optimism over future technologies and energy efficiency/conservation, together with non-consideration of trade impacts and transitional costs. Furthermore the simplicity of the M-M linkage with no government sector means that and recycling of revenues, cannot be investigated. Lastly the costs are for UK abatement with no option to purchase international emissions credits. The costs and availability of long-term emission credits is extremely uncertain and not considered in this analysis.

Future UK MARKAL modelling development will use enhanced spatial and temporal detail to further investigate the development of new infrastructures, operational details of the UK energy system and the impact of innovation. Additional modelling work will further disaggregate the role of consumer and firm behaviour in energy service demands. Finally additional UKERC scenario modelling will link analysis of UK policy objectives of low carbon and energy security.

The Emerging International Carbon Market

The overall value of the global carbon market was estimated at over $10 billion in 2005. In the first quarter of 2006 alone, the value of carbon market transactions reached $7.5 billion. This growth is driven by high prices in the EU ETS market for allowances. EUAs worth $8.2 billion were traded in 2005. This corresponded to 322 million tCO2e and represented almost a 40-fold increase over the previous years' volumes. Recent findings indicate that EU governments had been conservative and may have over-allocated allowances to installations for the EU ETS phase I (2005–07). This resulted in a volatile market when first year reports were made public in May, 2006. It is now widely anticipated that there will be much tighter compliance limits in phase II (2008–2012). This would result in strong demand for both EU allowances and CERs. In addition to the EU ETS, several EU governments have established national programs to purchase carbon credits to meet their Kyoto Protocol compliance targets. Demand from the Japanese private sector for carbon credits remains strong. The Japanese Government has announced new plans to purchase CERs, adding further to demand. Other demand drivers include the establishment of a regional greenhouse gas initiative in the northeastern United States, the continuing operation of the New South Wales carbon market in Australia and the Chicago Climate Exchange. The supply of carbon credits has improved, particularly in Asia. The PRC and India are now leading suppliers of CERs.

Political leaders from major industrialized countries and global business leaders from a range of industrial sectors have affirmed that market-based approaches and an efficient and effective "carbon market" are key elements of a successful global climate regime. These approaches create a market value for greenhouse gas (GHG) emissions and a price signal to guide action by countries, businesses and other entities (Warren Bell and John Drexhage, 2005). The G8 Gleneagles Plan of Action on climate change, clean energy and sustainable development specifically flagged "tradable certificates and trading of credits" and "project-based and voluntary offset mechanisms" as elements of a carbon market that need further policy development. The World Economic Forum Roundtable, in a statement released just prior to the G-8 Summit, urged governments to "establish a long term, market-based policy framework extending out to 2030...." and "define greenhouse gas emissions rights through a *cap and trade system* or other market-based mechanisms."

Market-based mechanisms, and emissions trading in particular, have been singled out by political and business leaders because they have the potential to significantly reduce costs and increase the feasibility of achieving the deep long-term reductions required to address the risks of climate change. These approaches can also mobilize investment in low-carbon energy technologies and promote technology transfer to less developed countries. The carbon market is a means rather than an end, and its success will be measured by how it assists countries in meeting their climate change and sustainable development commitments and priorities. For some, efficiency and cost minimization will be critical. For others it will be equally important how the international carbon market helps to mobilize financing for clean technologies worldwide.

Internationally, the primary market-based approach is the framework established under the Kyoto Protocol (the Protocol). It sets out an international emissions trading system+ and two project-based mechanisms, Joint Implementation (JI) and the Clean Development Mechanism (CDM), which together are designed to provide the countries that are Parties to the Protocol with additional means for meeting emissions targets. These "flexibility mechanisms" are widely credited with having put a market value on GHG emission reductions and creating new markets and investment opportunities, even before the Protocol's entry into force. The Protocol has driven the creation of an emerging mosaic of international and domestic trading schemes and project-based trading. It is important to recognize that while the Protocol creates a framework for action the implementation of emissions trading – and especially the involvement of private entities – depends primarily on domestic policy decisions.

The International Emissions trading market is currently dominated by the European Union Emission Trading Scheme (EU-ETS), which began on January 1, 2005. The first phase of the scheme will last until 2007. From 2008, the scheme will coincide with the Protocol's first commitment period. Partial linkages to Kyoto's project-based mechanisms (JI and CDM) have already been established and further linkages may be possible – for example with domestic schemes in non-EU countries. Domestic emissions trading schemes have been or are being introduced in a number of countries, including the UK, Norway and Canada, to help meet Kyoto targets. Linkages with the Kyoto mechanisms and other domestic schemes will likely be important elements of these systems. While Japan has yet to decide whether to implement emissions trading in its domestic policy mix, both the Japanese government and major industries have been significant early buyers in the CDM and JI credits market.

Looking Beyond 2012

If the carbon market is to play a significant role in helping to achieve the deep GHG emission reductions required over the next 20 years, decision-makers will need to consider how best to broaden and deepen the reach of that market. The broader the reach of the carbon market – through greater participation of countries and/or coverage of key sources and sectors and/or companies – the more efficient and effective it will be in minimizing costs and mobilizing investment. Broad coverage can also address concerns about leakage of emissions-intensive economic activity between countries as well as competitiveness more generally. Broadening and deepening the market will likely require a more flexible range of options to meet widely differing national circumstances and needs. The criteria for the years after 2012 is how to broaden participation in the carbon market in developing countries in a way that minimizes cost uncertainties and promotes economic growth and investments in low-carbon technology. In particular, the engagement of less advanced developing countries in the carbon market needs to be supported. The CDM will likely continue to play an important role and may expand its scope to address policy-based crediting and a broader range of projects. Efforts are underway to strengthen and streamline the CDM. A strengthened CDM can help build practical experience and confidence with market-based approaches across the developing world and position the CDM to contribute beyond 2012.

Focus on Cost-effectiveness of Developing Countries

Cost-effectiveness is a crucial requirement for meaningful agreements on international climate change policy. This is also borne out in the wording of the Framework Convention of Climate Change and, in particular, the Kyoto Protocol (KP); UNFCCC (1992) and UN (1997). However, the KP – as it stands after COP7 in Marrakesh – is not cost-effective, although it may eventually turn out to be the only politically feasible, 'most cost-effective', first step in international climate change policy. The successor to the COP7 version of the KP may be a renegotiated protocol, if the COP7 version fails to be ratified by enough countries to enter into force, or it may be the treaty to be designed for a second commitment period. Four dimensions in which cost-effectiveness may be improved in a treaty that succeeds the KP are discussed here. They all relate to international emissions trading (IET) which is likely to be the most significant instrument for attaining cost-effective reductions in aggregate greenhouse gas (GHG) emissions. It is important for a climate treaty to be able to attract as many developing countries to IET as possible and achieve this as soon as possible. This would have to occur at essentially no cost to them. Only with developing countries onboard can the world community get full access to their low-cost options for emission reductions.

Carbon Financial Markets: a Bio-based Economic Opportunity

In recent years opportunities for participation in carbon credit trading markets have been growing. The Chicago Climate Exchange (CCX) now boasts more than six million trades per month. A recent summary of the *State and Trends of the Carbon Market 2006* prepared for the World Bank's BioCarbon Fund reports a rapid increase in corporate participation in the carbon market. One opportunity lies in the Michigan landscape where forestry projects or rural conservation tillage practices can link Michigan farmers into carbon financial markets where they can sell their carbon offsets to industries wishing to reduce their carbon risk and carbon footprint. An additional exciting opportunity lies in prospects for leveraging this growing carbon financial market in the US and Europe to assist poor farmers in developing countries who could participate in tree planting or vegetation regeneration projects and earn revenues that can, in turn, catalyze economic development within their communities, increase rural incomes, enhance land tenure security, and stimulate natural resource conservation.

Forestry and Agro-forestry: A Focus Area

Most of the carbon market trade involves emission reduction credits but there is also growing interest in the use of trees and forests for absorbing carbon dioxide from the atmosphere. The Clean Development Mechanism (CDM) of the Kyoto Protocol and some voluntary carbon markets, such as the Chicago Climate Exchange, allow countries and companies to offset their carbon emissions by carrying out tree planting projects. There is also considerable evidence that forests and agro-forestry (the planting of trees on farms) in developing countries provide substantial benefits to rural dwellers, national economies, and the environment. Trees provide a range of products for home use such as food, timber, firewood, medicines, and fodder as well as products for sale, boosting farm incomes, rural economies, and national exports. Trees on farms and in forests can also provide a range of environmental services, such as conserving biodiversity, reduced soil erosion and sedimentation in rivers and lakes, and increased soil fertility. Moreover, what is often unrecognized is that while forested area is declining in developing countries, tree cover on farms is rapidly increasing, as farmers substitute for the tree products they formerly accessed from forests and seize market opportunities for selling tree products. Agricultural land now accounts for over double the area of forested land in Africa, giving justification to the slogan that, "the future of trees is on farms."

Sustainable development links natural resource management with tangible economic improvements in rural livelihoods. Approximately 3 billion people, half of the world's population, live below the ethical poverty level (EPL). The ethical poverty level is defined as the point at which life expectancy falls as rapidly as income, and above which life expectancy rises only slightly compared to income increases (Edwards, 2006). Currently, the global EPL is around $2.70/day. A vast majority of these individuals depend for their survival on the 400 million small-farms that are found throughout the developing world (Gates Foundation, 2006). If the appropriate markets can be created the cumulative human and natural resources that can be marshaled in reducing atmosphere carbon, and raising rural incomes are vast.

Traditional approaches to reforestation and agro-forestry, often fail because farmers are asked to make immediate investments of scarce land and labour to plant and protect trees with the uncertain hope that the trees will begin to produce benefits 5-15 years later. The ability to link tree planting with near-term payments through the emerging carbon markets, with additional payments from other tree products coming on-line in subsequent years, has the potential to positively impact millions of lives. Furthermore, once productive, the continued generation of high-value tree products (fruits, oil-bearing nuts) serves to protect the stored carbon from being harvested as fuel wood, burned and the re-released into the atmosphere. The fact that some oil-bearing nuts (e.g., *jatropha)* can be used in the manufacture of bio-diesel, thus reducing the use of fossil fuels and release of additional stored carbon, constitutes yet another positive gain of our approach to carbon sequestration.

Carbon Emissions-Credit Trading Stocks

A carbon credit is a tradable monetary security which is equal to one ton of a greenhouse gas, such as carbon dioxide. Due to greenhouse gas restrictions put in place by various countries based on the Kyoto Protocol quotas, businesses that are generating emissions that are less than their quotas can sell their credits to companies which are unable to stay below their quotas. The proceeds in essence reimburse the selling company for reducing their emissions. Trading also takes place between countries.

There are four major exchanges for the trading of carbon credits: the Chicago Climate Exchange in the United States, the European Climate Exchange in England, Nord Pool in Norway, and PowerNext in France. Louis Redshaw, who is in charge of environmental markets at Barclays Capital, said "Carbon will be the world's biggest commodity market and it could become the world's biggest market overall." Fortunately, there are several companies which are connected to carbon emissions trading. Many of these are low cap stocks and should be considered very speculative.

Agcert International (AGCTF.PK) is involved in the production and sale of greenhouse gas emission reductions and Certified Emission Reductions (CERs). The company generated a loss per share of 0.58 cents for 2006.

Alkane Energy PLC (ALKAF.PK) builds and operates methane harnessing systems from man-made sources for renewable generation of electricity with the potential of the trading of carbon emission reduction certificates through the EU ETS from 2008, if the European Union Emissions Trading Scheme (ETS) is expanded to cover methane. Operating profit and net profit were positive for the latest year, versus losses for the previous year.

Camco International (CAMCF.PK) identifies and develops projects for industrial companies that reduce greenhouse gas emissions and then assists in the sale and delivery of carbon credits into the international market. Losses per share increased substantially in 2006 over 2005.

Climate Exchange PLC (CXCHF.PK) owns the Chicago Climate Exchange, Inc., which trades in the emissions of six gasses including carbon dioxide, methane, nitrous oxide, sulfur hexafluoride, perfluorocarbons and hydrofluorocarbons, and the European Climate Exchange, which provides futures contracts and options contracts of emissions, known as Carbon Financial Instruments.

Goldman Sachs has taken a major position in this company. Pro-forma operating loss was reduced by 30% in the latest reported period.

Econergy International PLC (ECONF.PK) is involved in the origination, development and marketing of Clean Development Mechanism (CDM) projects. Although revenues increased for the latest reported year, net after tax loss went up.

Ecosecurities Group (ECGUF.PK) assists companies to create and purchase carbon credits from projects that reduce emissions of greenhouse gases.

Although the pre-tax loss increased for the year 2006, diluted loss per share dropped.

Trading Emissions PLC (TRDGF.PK) is a closed end fund that was established to acquire tradable environmental instruments in projects developed under the Clean Development Mechanism (CDM) and Joint Implementation (JI) of the Kyoto Protocol. The company had a pretax profit of 2.9 million British pounds for the first half of the fiscal year.

A Formula to Limit Emissions

Under the Kyoto Protocol, which Australia ratified in 2007, developed countries are required to limit their greenhouse gas emissions according to the following formula:

Actual emissions must be less than or equal to the assigned amount +/– carbon sinks and Kyoto emissions.

It implies from above that Australia can emit more than its assigned amount (which, incidentally, is 8 per cent above its emissions in 1990) if it can simultaneously sequester the equivalent amount in sinks.

Allowable Carbon Sinks

Under the Kyoto Protocol, allowable carbon sinks include afforestation and reforestation activities undertaken since 1990. Agricultural soils and some other sources and sinks may also be included. Compliance will be a challenge in some instances, for example, measuring changes in soil carbon on agricultural land remains extremely problematic. Geo-sequestration is another option being developed as a sink for carbon in Australia and other countries with suitable geology.

Carbon Finance

Carbon finance is a new branch of Environmental finance. Carbon finance explores the financial implications of living in a carbon-constrained world, a world in which emissions of carbon dioxide and other greenhouse gases (GHGs) carry a price (http://en.wikipedia.org/wiki/Carbon_finance). Financial risks and opportunities impact corporate balance sheets, and market-based instruments are capable of transferring environmental risk and achieving environmental objectives. Issues regarding climate change and GHG emissions must be addressed as part of strategic management decision-making. The general term is applied to investments in GHG emission reduction projects and the creation (origination) of financial instruments that are tradable on the carbon market.

At the International Level, the Carbon Finance Unit serves as IFC's in-house resource for all carbon finance-related issues, providing services directly to buyers and sellers. The Unit leads in the development of new products, and serves as an advisor on a variety of products and services to support private sector participation in the evolving carbon market (http://www.ifc.org/carbonfinance).

IFC is well-positioned to assist project sponsors with participation in the rapidly growing market for 'carbon credits'. Through the Carbon Finance Program, IFC contributes to the development of an important new market for environmental services. The Carbon Finance Unit brings IFC's extensive experience in project finance to manage long term and credit risks in emerging markets. Carbon Finance products and services include:

- Finance Products and Services
- Carbon Delivery Guarantee
- Monetization of future cash-flows from sales of carbon credits
- Debt and equity for carbon rich products and businesses
- Work with Financial Intermediaries and municipalities to help aggregate carbon credits from their various investment operations

Joint Implementation and Clean Development Mechanism

Clean Development Mechanism (CDM), is recognized through the Kyoto Protocol, allowing the offset of emissions in developed countries by the investment in emission reduction projects in developing countries like China, India or America. Joint (JI), is another mechanism, allowing investments in developed countries to generate emission credit for the same or another developed country.

Market value

The market for the purchase of carbon has grown exponentially since its conception in 1996.

The following is the estimated size of the worldwide carbon market according to the World Bank:

Carbon Market Volume: (millions metric tonnes, $MtCO_2$)

- 2005: 718 (330 in Main Allowances Markets and 388 in Project based transactions)
- 2006: 1,745 (1,134 in Main Allowances Markets and 611 in Project based transactions)
- 2007: 2,983 (2,109 in Main Allowances Markets and 874 in Project based transactions)

Dollars (millions of US$)

- 2005: 10,908 (7,971 in Main Allowances Markets and 2,937 in Project based transactions)
- 2006: 31,235 (24,699 in Main Allowances Markets and 6,536 in Project based transactions)
- 2007: 64,035 (50,394 in Main Allowances Markets and 13,641 in Project based transactions)

The World Bank Carbon Finance Unit (CFU) Contributions

The World Bank CFU uses money contributed by governments and companies in OECD countries to purchase project-based greenhouse gas emission reductions in developing countries and countries with economies in transition. The emission reductions are purchased through one of the CFU's carbon funds on behalf of the contributor, and within the framework of the Kyoto Protocol's Clean Development Mechanism (CDM) or Joint Implementation (JI).

International Carbon Markets

International trade in greenhouse gas reductions is a large and rapidly growing market:

- Motivated by requirements of the Kyoto Protocol and regional programs, and by voluntary commitments in some cases, governments and private companies in industrial countries have already committed billions dollars for emission reductions from projects outside their borders
- The international carbon market offers an important opportunity to help boost renewable energy activities in many parts of world.

The comparative scenario of the contribution of diverse renewable energy sources on earning revenues from Carbon credits is given in Table 9.

Table 9: Carbon Revenue Estimates for Renewable Energy Activities

Example Technology Application	System Capacity	Example System Cost	Example tCO_2 Reduced/Yr.	Carbon sales % of system cost
Solar Water Heating	2.4 kWth	$1,850	2.5	17%
Wind Power	10 MW	$12,500,000	21,000	21%
Hydro Power	1.5 MW	$2,000,000	6,000	38%

Carbon Risk Management

Carbon is an emerging risk that requires a clearly defined policy

In previous sections we have presented the accounting challenges and potential accounting treatments for the carbon assets and associated liabilities which arise from taking on a cap under the UK Scheme. Different accounting treatments can be adopted based upon the purpose for which the allowances are held. In broad terms allowances, as marketable instruments, could be held for:

(a) Hedging of current or future period emissions exposures as a result of taking on a cap; or

(b) Speculative purposes to benefit from expected changes in the market price of the allowance.

Shareholders require, and so management should provide, clarity over whether the allowances are being held for hedging or speculative purposes as the risk profile of the company, and its financial results, are directly affected by these choices. To illustrate this point an example has been set out below. In the example the company has entered into a forward sale of emission allowances for delivery in 2007. The diagram is based on a hypothetical transaction. Under the UK Scheme it would be expected that the number of allowances that a company expects to hold for each compliance period would be less than the forecast emissions profile. The question to be answered is whether the sale has been entered into for hedging or trading purposes. For a company to be able to demonstrate that it is buying allowances for hedging purposes it needs to be able to demonstrate that the allowances are required to meet a future emissions exposure or are within their planned hedging policy.

Key elements of the operating and control framework to enable a company to support the transaction and provide additional evidence are:

(a) A carbon risk management policy; and
(b) The availability and credibility of management information to support the implementation of that policy.

Future Carbon Risk Management Policy

The purpose of a risk management policy is to provide a clearly documented articulation of the risks, respective responsibilities and delegated authorities within the company for managing the risks specifically addressed within the policy document. When formulating the risk management policy there are many factors which should be considered as part of this process. For a carbon risk management policy this could include, *inter alia*, the following:

(a) The financial objectives of the company;
(b) The predictability of production (and consequently GHG production);
(c) The level of risk to the financial objectives which a Board of Directors is prepared to accept, i.e. the extent to which movements in allowance prices can be absorbed within profit margins;
(d) The extent to which the company can pass on the effects of movements in allowance prices to customers;
(e) The extent to which competitors are able to absorb price fluctuations or pass them on to customers;
(f) The volatility of allowance prices, i.e. the extent to which price changes may have a material impact on future cash flows;
(g) The ability of the company to hedge the relevant exposures, i.e. the availability of hedging instruments or other activities;

(h) Expectations of future prices and the potential volatility of future prices;
(i) Current market conditions; and
(j) Practical aspects such as the need for simplicity given current levels of awareness and understanding of carbon risk.

These factors are then translated into a specific delegation of authority within the policy document as a series of limits and guidance on activities and would typically include, *inter alia*, the following:

(a) Limits on the uncovered exposures within specified timeframes;
(b) The length of time over which exposures should be forecast;
(c) The frequency with which updates of forecasts should be undertaken;
(d) Identification of the certainty of the forecast exposures within designated time periods (i.e. contracted or expected production);
(e) The degree of certainty over the forecast exposures;
(f) Acceptable levels of risk on un-hedged exposures (through the use of statistical risk quantification measures such as Value at Risk or simpler stress and scenario testing);
(g) The instruments and investment methods that are authorized by the Board of Directors;
(h) Approved counterparties;
(i) Counterparty risk limits; and
(j) The reporting requirements on the risk management activities to the Board of Directors.

Management information requirements to enable effective implementation of policy

Effective implementation of a policy is reliant on the quality of the management information on which decisions are based. To be of use the management information needs to be timely, reliable and relevant. For the carbon risk manager within a company who is responsible for undertaking hedging and investment activities to reduce or limit the company's exposure, there are a number of questions which are worth considering:

(a) What data should be collected?
(b) How often should it be collected?
(c) How reliable are the actual and forecast emissions data?
(d) What estimates are utilized in the collection of the data?
(e) Is there any regular bias in the data that is submitted that requires stripping out?
(f) How far forward should forecast information be collected?
(g) How does the forecast and actual emissions data tie in with other measures and data that the company is collecting?
(h) In which jurisdictions are the emitting sources based?

To enable the effective management of carbon risk, there needs to be an assessment of whether the existing data collection mechanisms and reporting systems are sufficient for the current needs of the carbon risk manager. Equally, there is a need to assess whether the existing data collection mechanisms and reporting systems will continue to be sufficient as the market develops and companies' risk management strategies (and options for risk management) become increasingly sophisticated. With the introduction of the UK Scheme, British companies will be some of the first to face new information and control framework requirements as carbon develops into a tradable liquid commodity.

Towards an Effective Global Carbon Market

A Case of Institutional Investors Group on Climate Change (IIGCC)

According to the statements issued by the Institutional Investors Group on Climate Change (IIGCC) Global emissions of greenhouse gases must be cut significantly in order to avoid dangerous climate change with catastrophic economic and social consequences. This will require substantial and timely investment in low-carbon technologies with most of it coming from the private sector. As long term investors, we are concerned both with the overall economic costs of climate change and climate policy as well as their impact on individual investments. IIGCC are exposed to carbon markets directly and indirectly through investments in carbon funds, companies that have stakes in these markets and in assets impacted by carbon policy and markets including companies that will be affected by a price on carbon. It is believed that carbon markets can provide an important part of the solution to the climate crisis as they can catalyze innovation and drive investment in low carbon solutions. They also provide developing countries with access to technology and capital. A well-functioning *global* carbon market has the benefit of allowing emissions reductions to be achieved at lowest cost. To date, however, carbon markets have not provided investors with the strong, long-term price signals that are necessary to support large investments in low-carbon solutions.

Emission Trading Schemes: Ensuring Environmental Integrity and Increasing Predictability to Investors

The caps set for emissions trading schemes have to be consistent with internationally agreed emission reduction targets, after taking into account other policies to reduce emissions. Investors require a robust price signal: This means that caps have to be ambitious in order to create sufficient scarcity and demand and therefore a price on carbon at a level that incentives investment in low carbon solutions. Long-term policy clarity is essential to reduce much of price volatility that undermines investor confidence in carbon markets. Timeframes have to be consistent with the investment cycles of the sectors covered by the schemes. IIGCC encourages governments to learn from the experience of the EU ETS and the problems faced by long-term investors as a result of excessive volatility and uncertainty.

IIGCC favours sustained moves towards auctioning in line with the polluter pays principle and support the use of benchmarking for distributing the remaining free allowances. The process for defining these benchmarks should be transparent and timely with the following stipulations:

- Markets must be well-governed and transparent in order to increase investor confidence. To this end, we would welcome regular disclosure of emissions data and of government participation in emissions trading as well as robust market oversight.
- Emissions trading schemes should cover a wide range of sectors and countries and be open to the import of international credits, including offsets from developing countries, as this allows global emissions to be reduced in the most efficient way.
- Existing and planned schemes in different countries and regions should be consistent to allow linking of trading schemes in future and to support the creation of a global carbon market. This will increase market liquidity and allow emissions reductions to be achieved at lowest cost.
- Future integration would be facilitated by consistent trading and compliance regimes which requires strong co-ordination between governments.

Carbon Offsets and Moves Towards Emissions Trading in the Developing World

IIGCC supports the continued use of the Clean Development Mechanism (CDM) in the post-2012 climate regime as it provides a means to cut emissions at lowest cost and gives developing countries access to capital and clean technologies. However, questions have been raised about the efficiency of the CDM and we urge governments to consider a range of reforms listed below in order to scale up the mechanism. Over the longer term, the carbon market will be more effective if middle income countries, such as China and India, take on absolute binding emission reduction commitments, including full participation in emission trading schemes. The approach to additionality used in assessing CDM projects needs to be re-defined in order to ensure the environmental integrity of the mechanism as well as a more transparent and effective approach to assessing project acceptability.

The CDM needs to move beyond the project-by-project approach in order to scale up investment flows to developing countries. We support turning the project-based focus into a wholesale strategy, taking into account sectoral and/ or programmatic emission reduction objectives. The institutional framework supporting the CDM must be improved, allowing the Board to concentrate on executive tasks such as ensuring the transparency and efficiency of the operational rules for assessing projects. Even with these reforms, the CDM is unlikely to deliver the emission reductions in developing countries on the scale that is required. IIGCC encourages governments to pro-actively consider additional ways of scaling up investment in low carbon solutions in developing countries.

IIGCC Towards an Effective Global Carbon Market

This Statement provides an institutional investors' perspective on carbon markets. As global institutional investors with diversified portfolios across the world, we are concerned both with the overall economic costs of climate change and climate policy as well as their impact on individual investments.

IIGCC believes that carbon markets should form a key part of the post-2012 international agreement on climate change. Studies have suggested that without a global price on carbon, it is unlikely the emission reductions of the scale recommended by the IPCC in order to avoid dangerous climate change will be made. IIGCC supports emission trading schemes because they establish a cap that places an absolute limit on CO_2 emissions, thereby ensuring that the environmental objective is met, they reduce mitigation costs through trading by allowing emission reductions to be made where they are least expensive and they allow developing countries to participate in and benefit from efforts to reduce emissions. IIGCC believes that a new global agreement in Copenhagen should establish carbon markets as a central part of the package of policy instruments that deliver the emission reduction targets recommended by the Intergovernmental Panel on Climate Change (IPCC), i.e. 25-40% below 1990 levels by 2020 for developed countries and 50%-85% below 2000 levels by 2050 globally. The caps set for emissions trading schemes have to be consistent with internationally agreed environmental targets, after taking into account other policies to reduce emissions.

To date, however, carbon markets have not delivered the scale of investment needed and at the pace required to avoid dangerous climate change. This is because they do not provide the strong, long-term price signals at the levels that are necessary for companies and investors to commit to long-term investments in low-carbon solutions. A high degree of uncertainty remains about the post-2012 climate agreement including the nature and level of national commitments, mechanisms to deal with deforestation, and the likely reform of rules and structure of the CDM, all of which are undermining investor confidence.

Whilst carbon markets have achieved significant growth in the past years, reaching a value of $110 billion in 2008, *additional* annual investments of €200-350 billion (i.e. the amount over and above business-as-usual levels) will be necessary. In the clean energy sector alone, at least $515 billion will need to be invested annually over an extended period to prevent dangerous climate change. Carbon markets must provide companies and investors with a price signal that will give them the confidence to make long-term investments in low-carbon technologies.

A single global carbon market should be the long-term goal of policymakers as this would allow emission reductions to be achieved in the most cost-effective way. While doing so it is acknowledged that in the short and medium term, emissions trading schemes will continue to be developed on a national or regional basis and we therefore call on governments to ensure that existing or planned cap-and-trade schemes take into account these suggestions for an improved structure, but also that they facilitate the integration between different schemes over the longer term.

CHAPTER 3

Global Initiative in Abatement of Climate Change and Promoting Emissions Trading

Global Climate Change Agreement

In the sphere of climate change, there is an emerging need for global cooperation; lowering of costs for international emissions trading for all nations; standardizing and formalizing of emissions reduction pricing supported by development of new, eco-friendly and innovative technologies. Such thinking clearly shaped the design of the Kyoto Protocol—an International climate change treaty, negotiated by more than 140 nations that establish a global emissions trading system for greenhouse gases.

In the entire effort to progressively reducing the emissions of greenhouse gases there are instances of showing lesser cooperation in achieving particular targets, timetables, or mechanisms, while many would be prepared to embrace the overall architecture of global cooperation and international emissions trading (Pizer, 2007). The World Bank is now managing nearly $1 billion in various project funds. A similar fund in Japan recently collected $150 million in private funds. On the supply side, the Clean Development Mechanism (CDM) Executive Board had approved roughly 50 projects by the end of 2005, totaling slightly more than 100 million tons of carbon-dioxide over 2008-12.

Based on this experience, there seem to be three lessons for the design of a global agreement. First, the protocol's targets and timetables approach seems both ill-suited for the current suite of domestic policies and increasingly unmanageable in negotiations; a more flexible architecture is needed that can accommodate price, quantity, sectoral, and technology policies. Second, there is an unusual alliance of support for project-based crediting in developing countries. Environmentalists see this as maintaining environmental integrity in a 'cap-and-trade' program, business world sees this as a pathway alternative to domestic compliance, brokers and dealmakers see profit opportunities, developing countries see foreign aid, and industrialized-country governments see opportunities to complement (or substitute) domestic mitigation. Given the inevitable need to channel mitigation resources from industrialized to developing countries, and the reluctance of developing countries to embrace other domestic policies to price CO_2 emissions, more thought should be given as to how best these mechanisms can be expanded and improved over the years globally in a mission-mode.

The Need for a Pragmatic Domestic Policy Initiative

Through repeated interactions, international agreements can negate to some extent the domestic actions but may have problems influencing them dramatically. In the case of climate change, most countries will miss their Kyoto targets without substantial inflows of allowances or credits; all but the EU ETS have avoided Kyoto's cap-like architecture and none of the noted domestic, market-based schemes allow individual (versus government) trading or use of assigned amount units (AAUs), the currency of the Kyoto Protocol. The underlying economic rationale for the Kyoto architecture, equalization of marginal costs, is thereby defeated. Looking more objectively, a nation's compliance with international agreements is tied more to its domestic politics than the agreement's intended incentives (making it different than the familiar case of a firm complying with a domestic policy). However, despite this observed disconnect between domestic actions and Kyoto commitments and architecture, the impetus for domestic policies seems unquestionably tied to the protocol.

This experience matches with other global efforts—for instance, on ozone-depleting substances, trade, and human rights—where there has been a much clearer set of domestic responses before significant global institutions have been built around these responses. In these examples, international agreement moved alongside, if not behind, domestic policy developments without constraining them. Applied to climate change, this might look like the "bottom up" or "pledge and review" approach as discussed by Bodansky et al. (2004) rather than the current Kyoto Protocol. That is, the first step is for countries to pursue domestic climate policies consistent with their own domestic pressures, reinforced by an international agreement that prods without constraining.

As countries develop market-based domestic policies, a more controversial question is whether these countries really want international emissions trading. Standard economics says yes—everyone gains from trade. But this aggregate gain ignores other issues, distributional concerns, and the underlying technological problem. The potential importance of trade gains also is diminished by the fact that the underlying near-term policy goal is to spur technological change, not to achieve significant reductions (Weyant, 2004).

While fully pricing the emissions externality would create the right incentives for this technological change, it is not the only approach (nor may it be a politically practical one). Countries could choose lower emissions prices and larger technology incentives (Fischer and Newell, 2005). Seeking to equalize marginal costs across countries might then thwart domestic efforts to find the appropriate domestic balance between these elements.

Some economists also have pointed to concerns over trade flows as a potential negative side effect of international emissions trading (McKibbin *et al.*, 1999). Even to non-economists, there may be some concern if there are large net trade flows among industrialized countries, perhaps because such flows suggest the initial allocation was unfair. Finally, and perhaps most practically, the need for international trade is diminished if prices are already similar and modest across countries or if the various domestic programs are linked indirectly through common use of project-based crediting. For example, the noted domestic policies all exhibit prices that are more modest and similar than early analyses of autarkic Kyoto Protocol implementation, such as Weyant and Hill (1999), in part because current policies are not expected to achieve the autarkic targets. They are also all seeking to link, in one form or another, to the project-based credit market.

Credit-flows in Developing Countries

Even as countries have been reluctant to embrace international emissions trading, they have moved ahead much more rapidly on the idea of buying credits for projects in developing countries taking the support of innovative international programmes and mutually-agreed trade policies. Private and public funds have come up with billions of dollars to finance such projects. International funding and development agencies have come forward in a big way to spearhead and sponsor these initiatives. In some ways, this is analogous to how domestic trading started in the United States—first embracing offsets under the 1977 'Clean Air Act Amendments' before full-fledged trading began under the 1990 amendments. It also is far more important: The very cheapest emissions reductions exist in developing countries (Weyant and Hill, 1999) and these reductions likely will not occur through domestic action because developing countries are deservedly focused on development over climate change.

Economists have tended to argue in favor using generous caps, or even non-binding targets, to bring in developing countries (Bradford, 2002; Philibert, 2005). Yet the sense of many developing countries, emphasized by their position in the most recent Montreal meetings, is that any notion of an emissions cap eventually could be used to limit their development and growth (Kopp, 2005). Such concerns exist in addition to those over the institutional capacity for monitoring and enforcement in developing countries. This means that some form of project-based—or perhaps policy-based—crediting is likely to be the only vehicle for broadly incentivizing inexpensive opportunities in developing countries for some time.

The major handicap for project-based crediting is the question of baselines—what would have happened absent the crediting scheme and, therefore, what is the basis for generating credits? From a cost-effectiveness perspective, individual baselines are not so important as long as the credit price is fully applied at the margin for each project. There is a valid concern that overly generous baselines will tend to weaken the overall target and transfer (perhaps large) infra marginal payments to project developers. Against this, one must weigh the priority of motivating action in developing countries (perhaps laying the groundwork for an eventual domestic target) as well as incentivizing inexpensive opportunities. One possible alternative is government-to-government cooperation, such as the U.S. Asia-Pacific Partnership, though the capacity of such agreements is unclear (Pizer, 2007).

Domestic Climate Change Policy

Under the given scenario, the value in an international agreement such as the Kyoto Protocol may merit reprisal of the domestic actions or even the visible initiatives rather than elaborating mechanisms for global trading and its enforcement. Two areas where more elaborate mechanisms may be valuable in the near term, however, are in crediting programs for developing country activities and evaluation tools for effectiveness of the national policies. There is wide-ranging stakeholder support for credit programs in developing countries; at the same time, common sense economics argues that this is where the action ought to be. Meanwhile, the continuation and acceleration of the initial spate of mandatory policies hinge on appropriate responses in other countries. Compliance with the Kyoto Protocol is unlikely to be the sole metric—both because compliance will reflect politics more than economic incentives and because nations are pursuing a broad range of policies that do not fit easily into the Kyoto conventions. Neither project-based crediting nor a plurality of policy approaches is likely to be as tidy—or as efficient—as a standard emissions trading regime. Such a regime should still be viewed as a longer-term goal, and efficiency should be regularly held up as a criterion to be reviewed and evaluated.

The economic toll is already considerable; global climate change is now costing the world economy billions of dollars annually. The UN estimates this cost at over $60 billion for 2003 alone, a year of climate change induced disasters such as the killer heat wave in Europe and the massive flooding in China. And a group of insurers led by Munich Re has reported to the UN that global climate change could cost the world over $300 billion per year by 2050. Meanwhile, a report issued by the German Advisory Council on Global Change (http://www.wbgu.de/wbgu_sn2003_engl.html) has stated that measures to curb greenhouse gas emissions must be at least four times stronger than the Kyoto Protocol if they are to prevent melting of the polar ice caps. The Council warns that there is only room for another 1.4ºC of global warming before dangerous climactic changes become probable. The global mean temperature has already risen 0.6ºC since the beginning of the industrial era, so this places the danger threshold at an average global increase of 2ºC which the Council warns will be exceeded within this century if we do not pursue stringent climate protection policies. Although the Kyoto Protocol is not sufficient to prevent disaster, it would be the first step towards a long historic odyssey.

International Conventions to Address Climatic Change and Funding Emission Trading

The following provides a historic backdrop of important International conventions/Fora held from the early seventies till 2002 to address the concerns for adverse Global Climatic change and need for reduction of emissions to achieve cleaner environment:

- United Nations Conference on Human United Nations Conference on Human Environment (1972)
- Vienna Convention for Protection of Ozone Layer (1985)
- Montreal Protocol (1987)
- Intergovernmental Panel on Climate Change (1988)
- United Nations Conference on Environment and Development (1992)
- Rio Conference of the Parties to the UNFCCC (UN Framework Convention on Climate Change from 1995)
- Kyoto Protocol (1997)
- Marrakesh Accord (2001)
- World Summit on Sustainable Development (WSSD), 2002
- Global Environment Facility (GEF)
- Prototype Carbon Fund (PCF), The World Bank, 2002

Kyoto Protocol: Genesis and Implications

The *Kyoto Protocol to the United Nations Framework Convention on Climate Change* is an amendment to the international treaty on climate change, assigning mandatory emission limitations for the reduction of greenhouse gas emissions to the signatory nations.

The Kyoto Protocol is an agreement made under the United Nations Framework Convention on Climate Change (UNFCCC). Countries that ratify this protocol commit to reduce their emissions of carbon dioxide and five other greenhouse gases, or engage in emissions trading if they maintain or increase emissions of these gases. The Kyoto Protocol now covers more than 160 countries globally and over 55% of global greenhouse gas (GHG) emissions. The Kyoto Protocol lays down the following principles for compliance:

- Kyoto is underwritten by governments and is governed by global legislation enacted under the UN's aegis.

- Governments are separated into two general categories: developed countries, referred to as Annex I countries (who have accepted GHG emission reduction obligations and must submit an annual greenhouse gas inventory); and developing countries, referred to as Non-Annex I countries (who have no GHG emission reduction obligations but may participate in the Clean Development Mechanism).
- Any Annex I country that fails to meet its Kyoto obligation will be penalized by having to submit 1.3 emission allowances in a second commitment period for every ton of GHG emissions they exceed their cap in the first commitment period (i.e., 2008-12).
- By 2008-12, Annex I countries have to reduce their GHG emissions by an average of 5% below their 1990 levels (for many countries, such as the EU member states, this corresponds to some 15% below their expected GHG emissions in 2008). While the average emissions reduction is 5%, national limitations range from 8% reductions for the European Union to a 10% emissions increase for Iceland; but since the EU intends to meet its target by distributing different rates among its member states, much larger increases (up to 27%) are allowed for some of the less developed EU countries. Reduction limitations expire in 2013.
- Kyoto includes "flexible mechanisms" which allow Annex I economies to meet their GHG emission limitation by purchasing GHG emission reductions from elsewhere. These can be bought either from financial exchanges (such as the new unrelated-to-Kyoto EU Emissions Trading Scheme) or from projects which reduce emissions in non-Annex I economies under the Clean Development Mechanism (CDM), or in other Annex-1 countries under the "JI".
- Only CDM Executive Board-accredited Certified Emission Reductions (CER) can be bought and sold in this manner. Under the aegis of the UN, Kyoto established this Bonn-based Clean Development Mechanism Executive Board to assess and approve projects ("CDM Projects") in Non-Annex I economies prior to awarding CERs. (A similar scheme called "Joint Implementation" or "JI" applies in transitional economies mainly covering the former Soviet Union and Eastern Europe).

What this means in practice is that Non-Annex I economies have no GHG emission restrictions, but when a GHG emission reduction project (a "GHG Project") is implemented in these countries, that GHG Project will receive Carbon Credit which can be sold to Annex I buyers.

The Kyoto linking mechanisms are in place for two main reasons:

- The cost of complying with Kyoto is prohibitive for many Annex I countries (especially those countries, such as Japan or the Netherlands for example, with highly efficient, low GHG polluting industries, and high prevailing environmental standards). Kyoto therefore allows these countries to purchase Carbon Credits instead of reducing GHG emissions domestically; and,

- This is seen as a means of encouraging Non-Annex I developing economies to reduce GHG emissions since doing so is now economically viable because of the sale of Carbon Credits.

The entire Annex I economies have established Designated National Authorities (DNA) to manage their GHG portfolios under Kyoto. Countries including Japan, Canada, Italy, the Netherlands, Germany, France, Spain and many more, are actively promoting government carbon funds and supporting multilateral carbon funds intent on purchasing Carbon Credits from Non-Annex I countries. These government organizations are working closely with their major utility, energy, oil and gas and chemicals conglomerates to try to acquire as many GHG Certificates as cheaply as possible. Virtually all of the Non-Annex I countries have also set up their own Designated National Authorities to manage the Kyoto process (and specifically the "CDM process" whereby these host government entities decide which GHG Projects they do or do not wish to support for accreditation by the CDM Executive Board).

The objectives of these opposing groups are quite different. Annex I entities want Carbon Credits as cheaply as possible, whilst Non-Annex I entities want to maximize the value of Carbon Credits generated from their domestic GHG Projects.

Kyoto Protocol Objectives

Kyoto is intended to:

- Cut global emissions of greenhouse gases.
- Stabilization of greenhouse gas concentrations in the atmosphere at a level that would prevent dangerous anthropogenic interference with the climate system.

The Intergovernmental Panel on Climate Change (IPCC) has predicted an average global rise in temperature of 1.4°C (2.5°F) to 5.8°C (10.4°F) between 1990 and 2100. Proponents also note that Kyoto is a first step as requirements to meet the UNFCCC will be modified until the objective is met, as required by UNFCCC Article 4.2(d)

The Kyoto Protocol was adopted against a background of hugely disparate perspectives concerning the urgency of action, the costs of limitations, and the appropriate instruments. In the end, the view of the US administration prevailed upon that binding emission commitment for industrialized countries should be complemented by the use of a number of 'economic instruments' adopted for the first time at the international level. From a purely economic standpoint, the aim of the resulting agreement is to tackle the threat of climate change by establishing an efficient regulatory framework that sets an international 'price' on emissions of CO_2 and other greenhouse gases, initially focused upon industrialized countries with mechanisms for offsetting against projects in developing countries. The core mechanism for achieving this is quantified emission commitments (established for industrialized countries in Kyoto's first commitment period of 2008-12), which are given market-based flexibility through the use of emissions trading and other international economic instruments, and with negotiations on subsequent period commitments mandated to follow (http://www.ieta.org/ieta/www/pages/getfile.php?docID=129).

Kyoto Protocol: A Historic Backdrop

The first meeting of the Conference of the Parties to the FCCC (COP-1) took place in Berlin from 28 March-7 April 1995. In addition to addressing a number of important issues related to the future of the Convention, delegates reached agreement on what many believed to be the central issue before COP-1—adequacy of commitments, the so-called Berlin Mandate. The result was to establish an open-ended Ad Hoc Group on the Berlin Mandate (AGBM) to begin a process toward appropriate action for the period beyond 2000, including the strengthening of the commitments of Annex I Parties through the adoption of a protocol or another legal instrument (http://www.iisd.ca/vol12/enb1276e.html).

Ad-hoc Group on Berlin Mandate (AGBM): At AGBM-1, held in Geneva from 21-25 August 1995, delegates considered several issues, including an analysis and assessment to identify possible policies and measures for Annex I Parties and requests for inputs to subsequent sessions. They debated the nature, content and duration of the analysis and assessment and its relationship to other aspects of the process. Several developed and developing countries stressed that analysis and assessment should be conducted in parallel and not prior to negotiations of a legal instrument, but a few developing countries insisted that more time was needed, particularly to evaluate economic costs.

At AGBM-2, held in Geneva from 30 October-3 November 1995, debate over the extent of analysis and assessment continued, but delegates also heard new ideas for the structure and form of a possible protocol. Delegates considered: strengthening of commitments in Article 4.2(a) and (b) regarding policies and measures, as well as establishing quantified emission limitation and reduction objectives (QELROs) within specified time frames, advancing the implementation of Article 4.1, and possible features of a protocol or another legal instrument.

At AGBM-3, held in Geneva from 5-8 March 1996, delegates heard a number of specific proposals on new commitments for Annex I Parties, including a two-phase CO_2 emissions reduction target proposed by Germany. They also discussed how Annex I countries might distribute or share new commitments, and whether those should take the form of an amendment or protocol. Delegates agreed to compile proposals for new commitments for consideration at AGBM-4, and to hold informal roundtable discussions on policies and measures as well as on QELROs.

AGBM-4, held from 8-19 July 1996 during the Second Conference of the Parties (COP-2) in Geneva, completed its in-depth analyses of the likely elements of a protocol or another legal instrument, and appeared ready to move forward to the preparation of a negotiating text. Most of the discussions dealt with approaches to policies and measures, QELROs, and an assessment of the likely impact of new commitments for Annex I Parties on developing countries. Upon the conclusion of COP-2, delegates noted the "Geneva Declaration," which endorsed the Intergovernmental Panel on Climate Change (IPCC) conclusions and called for legally binding objectives and significant reductions in GHG emissions. COP-2 also saw a significant shift in position by the US, which for the first time supported a legally binding agreement to fulfill the Berlin Mandate. However, even as Parties prepared to strengthen commitments, COP-2 highlighted the sharpest differences between them.

AGBM-5, which met in Geneva from 9-18 December 1996, considered proposals from 14 Parties or groups of Parties regarding strengthening of commitments, advancing the implementation of Article 4.1, and possible elements of a protocol or another legal instrument. Delegates adopted conclusions that requested the Secretariat to produce a "framework compilation" of proposals for further consideration.

AGBM-6 met from 3-7 March 1997 in Bonn. Delegates met in "non-groups" to exchange views and "streamlined" the framework compilation text by merging or eliminating some overlapping provisions within the myriad of proposals. This brought the process one step, albeit a small one, closer to fulfilling its mandate. Much of the discussion centered on a proposal from the EU for a 15% cut in a "basket" of greenhouse gases by the year 2010 compared to 1990 levels. Other proposals emerged in the eleventh hour, signaling that AGBM-6, despite the hopes of many observers, had yet to foster much progress on several fundamental points.

AGBM-7 met from 28 July–7 August 1997 in Bonn. A total of 145 Parties and Observer States participated in the session, as well as 691 representatives from NGOs and the media. AGBM-7 further streamlined the negotiating text. In the absence of initial formal proposals for emissions reduction targets by the US and Japan, there was a widespread sense that most of the progress achieved at this session was limited to a reduction in the number of proposals.

The final session of the AGBM was held from 22–31 October 1997 in Bonn. As AGBM-8 began, US former President Bill Clinton included a call for "meaningful participation" by developing countries in the negotiating position he announced in Washington. With those words, the debates that shaped agreement back in 1995 resurfaced, with an insistence on G-77/China involvement once again linked to the level of ambition acceptable by the US. In response, the G-77/China used every opportunity to distance itself from any attempts to draw developing countries into agreeing to anything that could be interpreted as new commitments. Some observers thought the

Japanese proposal, combining an overall reduction target of 5% with scope for differentiation, would likely provide the outline of the eventual compromise. AGBM-8 was suspended until the day before the COP-3 opening in Kyoto to allow time to continue informal consultations on outstanding items, such as the number of GHGs to include, budget period or single-year targets, and sinks.

Status of the Agreement

The Greenhouse Emission Reduction Treaty was negotiated in Kyoto, Japan in December 1997, opened for signature on March 16, 1998, and closed on March 15, 1999. The agreement came into force on February 16, 2005 following ratification by Russia on November 18, 2004. As of December 2006, a total of 169 countries and other governmental entities have ratified the agreement (representing over 61.6% of emissions from Annex I countries). Notable exceptions include the United States and Australia. Other countries, like India and China, which have ratified the protocol, are not required to reduce carbon emissions under the present agreement despite their relatively large populations. According to article 25 of the protocol, it enters into force "on the ninetieth day after the date on which not less than 55 Parties to the Convention, incorporating Parties included in Annex I which accounted in total for at least 55% of the total carbon dioxide emissions for 1990 of the Parties included in Annex I, have deposited their instruments of ratification, acceptance, approval or accession." Of the two conditions, the "55 parties" clause was reached on May 23, 2002 when Iceland ratified. The ratification by Russia on 18 November 2004 satisfied the "55%" clause and brought the treaty into force, effective February 16, 2005.

Details of the Agreement

According to a press release from the United Nations Environment Programme:

> "The Kyoto Protocol is an agreement under which industrialized countries will reduce their collective emissions of greenhouse gases by 5.2% compared to the year 1990 (but note that, compared to the emissions levels that would be expected by 2010 without the Protocol, this limitation represents a 29% cut). The goal is to lower overall emissions of six greenhouse gases—carbon dioxide, methane, nitrous oxide, sulfur hexafluoride, HFCs, and PFCs—calculated as an average over the five-year period of 2008-12. National limitations range from 8% reductions for the European Union and some others to 7% for the US, 6% for Japan, 0% for Russia, and permitted increases of 8% for Australia and 10% for Iceland."

It is an Agreement negotiated as an amendment to the United Nations Framework Convention on Climate Change (UNFCCC, which was adopted at the Earth Summit in Rio de Janeiro in 1992). All parties to the UNFCCC can sign or ratify the Kyoto Protocol, while non-parties to the UNFCCC cannot. The Kyoto Protocol was adopted at the third session of the Conference of Parties to the UNFCCC (COP3) in 1997 in Kyoto, Japan. Most provisions of the Kyoto Protocol apply to developed countries, listed in Annex I to the UNFCCC. Emission figures exclude international aviation and shipping.

Common, but Differentiated Responsibility

The United Nations Framework Convention on Climate Change agreed to a set of a "common but differentiated responsibilities." The parties agreed that:

1. The largest share of historical and current global emissions of greenhouse gases has originated in developed countries;
2. Per capita emissions in developing countries are still relatively low;
3. The share of global emissions originating in developing countries will grow to meet their social and development needs.

In other words, China, India, and other developing countries were exempt from the requirements of the Kyoto Protocol because they were not the main contributors to the greenhouse gas emissions during the industrialization period that is believed to be causing today's climate change. However, critics of Kyoto argue that China, India, and other developing countries will soon be the top contributors to greenhouse gases. Also, without Kyoto restrictions on these countries, industries in developed countries would be driven towards these non-restricted countries, thus there would be no net reduction in carbon.

Financial Commitments

The Protocol also reaffirms the principle that developed countries have to pay billions of dollars, and supply technology to other countries for climate-related studies and projects. This was originally agreed in the UNFCCC.

Market-Based Mechanisms for Greenhouse Gas Abatement

The Kyoto Protocol provides Annex B countries (countries with reduction obligations under the Protocol) with three flexibility, or market-based, mechanisms. These mechanisms are intended in part to reduce the cost of meeting emissions reduction commitments. The flexibility mechanisms were an integral part of the compromise reached in Kyoto, which allowed Annex B countries to make the commitments they did. The premise of the flexibility mechanisms is that the costs of reducing greenhouse gas (GHG) emissions are expected to vary widely across countries. Two mechanisms are targeted at transactions between Annex B countries; the third focuses on transactions between Annex B and non-Annex B (developing) countries. Developing countries are expected to have significantly lower mitigation costs than Annex B countries and to receive a significant share of mitigation funds. This result is not assured, however, since the rules governing the market mechanisms are not yet in place and the distribution of emissions reduction opportunities and costs is not yet well understood. Some observers believe, for example, that countries like the United States possess extensive low-cost mitigation opportunities and will take advantage of them as reduction mandates come into force (http://www.ceeindia.org/greenhousegases/abate.htm).

Flexibility Mechanisms under the Kyoto Protocol

Signals of climate change are already visible around the world. Some of the projected impacts, which can adversely affect life, economy and ecosystem, are its impacts on the Asian monsoon (the main source of fresh water), agriculture, and other life-saving systems such as rivers, etc. Climate change will increase vector-borne diseases, extreme weather events, their intensity and frequency, and cause an unprecedented loss of life and property, etc. The Kyoto Protocol is an important step towards stabilization of the concentration of GHGs in the atmosphere to prevent dangerous anthropogenic interference with the Earth's climate system.

The Kyoto Protocol introduced the following three flexibility mechanisms to supplement the domestic activities of the developed (Annex I) countries. The three flexibility mechanisms are:

- Clean Development Mechanism (CDM)
- Joint Implementation (JI)
- International Carbon Trading (ICT)

Of these three, only the CDM is applicable between developed and developing countries that have ratified the protocol. The certified emission reductions (CERs) that result from CDM project activities have considerable economic use to both the host (developing) and investor (developed) countries. During the last few years the mechanisms have aroused worldwide interest in carbon investment and the carbon market, which is growing as companies and governments have started purchasing emissions credits through voluntary trading schemes, carbon investment funds, and government procurement tenders. It could be noted that of the twelve CDM projects registered by the CDM Executive Board (as of July 19, 2005), three are located in India, as follows:

- GHG emissions reduction by thermal oxidation of HFC-23 in Gujarat.
- Electricity generation from mustard crop residues in Rajasthan.
- 5-MW Dehar grid-connected small hydropower in Himachal Pradesh.

Under the Kyoto Protocol, the developing countries ratifying the protocol are eligible to participate in the CDM. India signed the protocol before COP 8 in 2002, which was hosted by the government of India in New Delhi. India offers a large potential for utilizing the CDM because of its inherent dependence on fossil fuels for development, its proactive government, and the enabling environment in place. Indeed, India is the most favoured destination for the CDM.

India is a competitive supplier for cost-effective GHG offset projects. Recognizing the possible benefits derived from such investment flows to India, the government has placed the CDM at the top of its climate change agenda. The designated national authority (DNA) for CDM projects from India has been set up in the Ministry of Environment and Forests (MoEF). This guidebook is aimed at providing CDM project developers and investors interested in host country information—which is vital when making decision on investment and the procurement of CDM benefits, rules, and regulations of the India's DNA—and reliable, updated information regarding CDM opportunities in India in various end-use sectors. The information in this guide is based on the UNFCCC guidelines, the Kyoto Protocol, the Marrakesh Accords, and the decisions agreed upon during the subsequent Conferences of Parties (COP 8, COP 9 and COP 10, held in India, Italy, and Argentina, respectively).

Importance of the Flexibility Mechanisms

Many observers estimate that GHG emissions trading could lead to a market amounting to US$100 billion per year by 2010; some observers estimate the figure to be even larger. A great deal of uncertainty surrounds these estimates and the issue of how the value of the global GHG market will be distributed across regions and countries. The rules implementing the different flexibility mechanisms B particularly the CDM B will considerably affect the size and distribution of the future GHG market. Nonetheless, the CDM is expected to be a substantial component of the overall market. Most macroeconomic models forecast that the CDM could have an annual worth of US$5-25 billion per year for the next 10 years. After that, tightened emissions targets in the second commitment period and more stringent trading rules could raise the value of GHG emissions reduction credits and significantly increase the amount of funding going through the CDM (http://www.ceeindia.org/greenhousegases/abate.htm).

Relative to other financial flows to developing countries, the CDM could be an important tool for sustainable development activities and technology transfer. For example, a CDM market of US$15 billion per year would be equivalent to 30 percent of the value of present development aid from Annex B to developing countries. It would be a smaller portion of foreign direct investment (FDI) in developing countries, currently estimated at US$240 billion. Even there, CDM funding could have a significant leveraging and steering impact on overall FDI flows. As the one of the largest developing countries, India has the potential to capture a sizable share of CDM revenues. Estimate of India's likely share of CDM revenues range from 10-25 percent of the CDM market or from less than $1 billion to more than $5 billion per year—over the next decade.

Clean Development Mechanism (CDM) Principles

- The Kyoto Protocol established the Clean Development Mechanism (CDM) under the auspices of the United Nations Framework Convention on Climate Change (UNFCCC) in December 1997. CDM was established to facilitate the implementation of greenhouse gas (GHG) emission reduction projects, and lower the cost for emission reductions for industrialized countries, while at the same time assisting developing countries in achieving technology transfer and sustainable development.
- CDM allows a wide range of GHG reduction projects including energy efficiency, fuel switching, forestation, transportation, and renewable energy such as wind, solar, hydro, geothermal and waste biogas. The development and operation of qualified CDM projects produces Certified Emissions Reductions (CERs) for the actual amount of GHG reduction achieved in the host developing country. CERs can provide significant financial benefits for CDM project development and investment.
- Outside of Kyoto compliant mechanisms, other actions taken to reduce greenhouse gas emissions are being verified and traded in the global over-the-counter market for greenhouse gas emissions. These are categorized as Non-Kyoto Compliant Reductions. Some companies are trading emission reductions in other markets, since they fail to meet all the requirements or obligations under Kyoto for certification, such as, non-compliant with "additionality" or "leakage." These so-called "verified emission reductions" (VERs) are not a standardized commodity. While they may eventually become CERs or ERUs, many of these reductions have no secondary market benefits outside of their embedded "green image value" or speculative value. Buyers therefore tend to pay a discounted price for VERs, which takes the inherent regulatory risks into account.

Over the last three years, the CDM (Clean Development Mechanism) has generated a lot of hype, hope, and skepticism. The ratification of the Kyoto Protocol hinges mainly on the resolution of issues related to the operationalization of the CDM. In the meantime, efforts have been made to identify possible opportunities for the implementation of CDM projects in developing countries. The United Nation Framework Convention on Climate Change (UNFCCC), adopted by most of the world's governments in Rio de Janeiro in June 1992, recognized that it could be a launching pad for stronger action in the future towards achieving the objective of the convention. The first review of the adequacy of developed country commitments was taken up by the Parties at the first session of the Conference of the Parties (COP 1) in Berlin in 1995, and the main theme was the adequacy of the UNFCCC by the Annex I Parties. Some 10,000 delegates, observers (NGOs), and media personnel participated in this high profile event hosted by Japan in December 1997. The conference resulted in a conscious decision to adopt a new protocol, the Kyoto Protocol, under which industrialized countries (ratifying the protocol) are required to reduce their combined greenhouse gas (GHG) emissions by at least 5.2 percent compared to 1990 levels by the first commitment period (2008–2012. This legally binding commitment promises to produce an historic reversal of the upward trend in emissions that started in the developed countries some 150 years ago.

The Clean Development Mechanism (CDM) is one of the innovative features of the Kyoto Protocol to the UNFCCC. It encourages investments in projects that provide sustainable development in developing countries, while at the same time limiting GHG emissions. The CDM was introduced in Article 12 of the Kyoto Protocol. Under this mechanism, legal entities in developed countries can invest in projects in developing countries that reduce GHG emissions. Once certified, these emission reductions can be used to meet the commitments made by the developed countries under the protocol. Although there are no immediate targets for the developing countries, it can be seen as an additional source of foreign direct investment into national mitigation projects, which would contribute to sustainable development and a decrease in worldwide GHG emission levels.

According to the Kyoto Protocol, project activities that qualify for the CDM are those related to specific GHG types and to the sources and sectors responsible for the majority of emissions, as established in Annex A of the Kyoto Protocol .As per the UNFCCC, the Kyoto Protocol is to enter into force on the ninetieth day after the date on which at least 55 countries ratify the treaty, provided they account for at least 55 percent of the total Annex I countries' 1990 emission levels. With the Russian Federation's instrument of ratification by the United Nations Secretary-General on November 18, 2004, the Kyoto Protocol entered into force on February 16, 2005, and is now fully operational.

Eligibility Criteria for CDM Participation:

The following represents the various eligibility criteria for participation in a CDM project:

- They must have ratified the Kyoto Protocol.
- They must have calculated their assigned amount, as referred to in Articles 3.7 and 3.8 and Annex B of the Protocol in terms of tonnes of CO_2-equivalent emissions.
- They must have in place a national system for estimating emissions and removals of greenhouse gases within their territory.
- They must have in place a national registry to record and track the creation and movement of ERUs, CERs, AAUs and RMUs and must annually report such information to the secretariat.
- They must annually report information on emissions and removals to the secretariat.

CDM Processes and Modus Operandi

The Clean Development Mechanism (CDM) is a mechanism based on the provision of Article 12 of the Kyoto Protocol (Fig. 1). It is a scheme for greenhouse gas (GHG) reduction through cooperation between developed countries (Annex I Parties to the United Nations Framework Convention on Climate Change (UNFCCC)), which are committed to certain GHG emission reduction targets in the Kyoto Protocol, and developing countries (non-Annex I Parties), which do not have any commitments to reduce GHG emissions. The purpose of the CDM is to assist to accomplish the GHG reduction targets of developed countries under the Kyoto Protocol, as well as to contribute to sustainable development of host countries. Under the CDM, Annex I Parties implement projects ie.g. projects of landfill gas (methane) recovery with power generation resulting in reduction of greenhouse gas emissions within the territories of non-Annex I Parties. Annex I Parties are able to acquire all or parts of the credits (certified emission reductions: CERs) which result from the projects.

Key Concepts of CDM Project

(1) Baseline

Baseline is defined as "the scenario that reasonably represents the anthropogenic emissions by sources of greenhouse gases that would occur in the absence of the proposed project activity."

Fig. 1: Clean Development Mechanism (CDM) Process

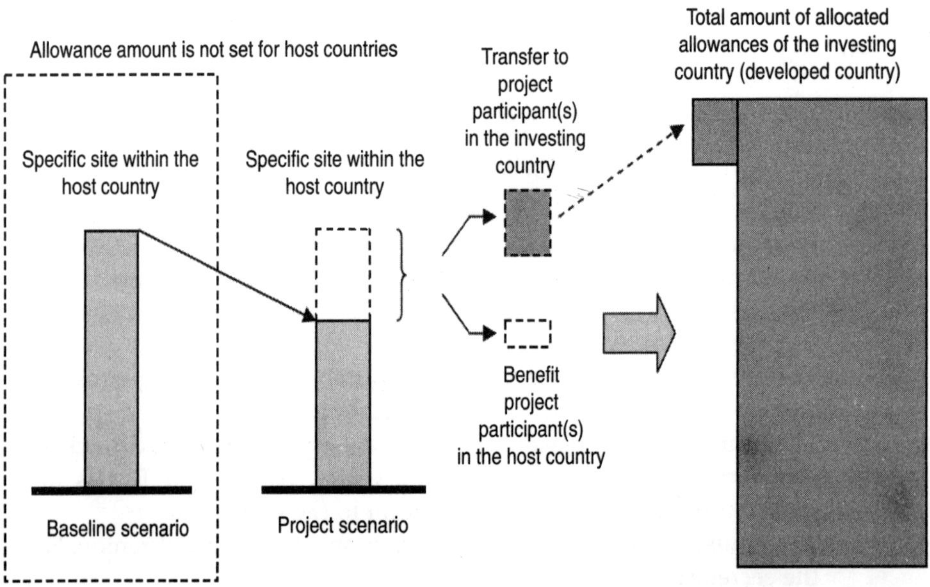

Fig. 2: Concept of Additionality

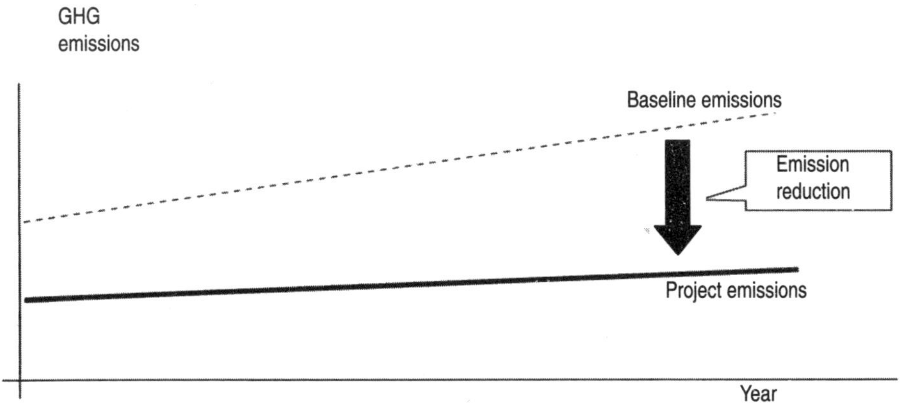

(2) Additionality

The concept of "additionality" is closely related to that of "baseline", and has to be paid particular attention to when setting up a baseline scenario and developing a methodology (Fig. 2). It is defined in the "CDM project activity is additional if anthropogenic emissions of greenhouse gases by sources are reduced below those that would have occurred in the absence of the registered CDM project activity."

Common Issues to be Resolved in CDM Projects

For effective implementation of CDM Projects, these are the common nature of issues that need to be addressed and resolved amicably:

- **Credibility?** The mechanisms are often regarded as loopholes. There is concern that, by taking advantage of the mechanisms, Parties could avoid taking real actions at home to reduce emissions. How can confidence in the three mechanisms be built up?
- **Additionality?** Participation in both JI and emissions trading must be "supplemental" to domestic actions to meet emission commitments. How is this requirement to be operationalized? There is a similar provision with the CDM—emission reductions achieved through this mechanism can only be used to meet "part" of a country's emission commitment. What is an appropriate "part"?
- **Participation of non-State entities?** The Protocol takes a different approach towards project based mechanisms and emissions trading.

CDM Project Cycle

The project cycle is common to all CDM project types, such as large scale, Small-Scale and A/R CDM activities (Fig. 3).

Fig. 3: CDM Project Cycle

1. Develop a CDM project activity

- **Project participants develop a CDM project activity.**
  - ➢ Consider various conditions associated with developing a CDM project activity from the project planning stage.
  - ➢ Draft a Project Design Document (CDM-PDD) with all required elements.

2. Approval by the Host and Investing Parties

- **Project participants obtain written approval of voluntary participation from the Designated National Authority (DNA) of each Party involved regarding the project activity that project participants wish to implement as a CDM project activity.**
  - ➢ Project participants should confirm with the host and investing parties regarding the approval process.

3. Validation and registration of the CDM project activity

- **In order to go through validation, project participants should use the methodologies approved by the EB. If no approved methodology can apply to the proposed project activity, project participants must propose a new methodology for the project and have it approved by the EB.**
- **Validation is the process of independent evaluation of a project activity by a DOE, selected by the project participants, against the requirements of the CDM on the basis of the project design document (CDM-PDD)**
- **Registration is the formal acceptance of a validated project as a CDM project activity.**

4. Monitoring of the CDM project

- **Project participants implement the CDM project activity and carry out proper monitoring necessary for calculation of GHG emission reductions, in accordance with the monitoring plan written in the PDD.**

5. Verification, Certification and Issuance of CERs

- **Project participants report to the DOE the results of monitoring of the CDM project activity and calculated emission reductions based on the results of monitoring activity.**
- **DOE verifies the results of monitoring and the resulting emission reductions.**
- **DOE certifies the emission reductions based on the results of verification.**
- **The EB issues CERs equal to the verified amount of GHG emission reductions.**
- **It is possible for a project activity, which met certain conditions, to have CERs issued for emission reductions starting as of the year 2000.**

6. Distribution of CERs

- **2% of the CERs will be deducted as the share of proceeds to assist developing country parties that are particularly vulnerable to the adverse effects of climate change.**
- **The share of proceeds to cover administrative expensed to the CDM will be deducted.**
- **Remainders of the CERs are to be distributed among project participants.**

Current positions of Governments in Implementing CDM Projects

Country-wise progress in implementation of CDM initiative aiming at reduction in the level of GHGs is summarized below on a selective basis (MarK Van Wees, 2002: http://www.capsd.nl/pages/documents/CDMcurrentdevelopments.pdf):

Australia

Despite the fact that Australia was at the time of the negotiation already one of the biggest emitters on a per capita basis (albeit the lowest on a per square kilometer basis due to low overall population density), the country was granted a limitation of an 8% increase. This is because Australia used its relative smallness, and the dependence of Australian electricity generation on coal, as a negotiation tool while other big players were negotiating. The result of the negotiation was reported in the Australian media as being to Australia's advantage.

Nonetheless, the Australian Prime Minister, John Howard, has refused to ratify the Agreement and has argued that the protocol would cost Australians jobs, due to countries with booming economies and massive populations such as China and India not having any reduction obligations. By way of example, industrial growth within China is expected to increase pollution within 9 months, and even if Australia were to shut down all of its coal fired power stations it would not negate this increase. Further, the Government takes the view that Australia is already doing enough to cut emissions; the Australian government has recently pledged $300 million over the next three years to reduce Greenhouse gas emissions. The Federal Opposition, the Australian Labor Party, is in full support of the protocol and it is currently a heavily debated issue within the political establishment. The opposition claims ratifying the protocol is a "risk free" prospect as they claim Australia would already be meeting the obligations the protocol would impose. This claim relies heavily on changes to land clearing policies that can only occur once, while ongoing emission sources have all increased substantially.

As of 2005, Australia was the world's largest emitter per capita of greenhouse gases. Analysis has projected Australia's greenhouse gas emissions at 109% of the 1990 emissions level over the period 2008–12. This is slightly above its 108% Kyoto Protocol limitation. As of 2007, the UNFCCC is reporting that Australia's 2004 greenhouse gas emissions were at 125% of 1990 levels. Australia remains committed to meeting its target despite not ratifying the Protocol.

The Australian government, along with the United States, agreed to sign the Asia Pacific Partnership on Clean Development and Climate at the ASEAN regional forum on 28 July 2005. Furthermore, the Australian state of New

South Wales (NSW) commenced The NSW Greenhouse Gas Abatement Scheme (GGAS). This mandatory greenhouse gas emissions trading scheme commenced on 1 January 2003 and is currently being trialed by the state government in NSW alone. Uniquely this scheme allows Accredited Certificate Providers (ACP) to trade emissions from householders in the state. As of 2006 the scheme is still in place despite Prime Minister John Howard's clear dismissal of emissions trading as a credible solution to climate change. Following the example of NSW, the National Emissions Trading Scheme (NETS) has been established as an initiative of State and Territory Governments of Australia, all of which have Labor Party governments. The focus of NETS is to bring into existence an intra-Australian carbon trading scheme and to coordinate policy developments to this end. According to the Constitution of Australia, environmental matters are under the jurisdiction of the States, and the NETS is intended to facilitate ratification of the Kyoto Protocol by the Labor Party if they are elected to government in the 2007 Federal Elections.

Canada

On December 17, 2002, Canada ratified the treaty that came into force in February 2005, requiring it to reduce emissions to 6% below 1990 levels during the 2008-2012 commitment periods. At that time, numerous polls showed support for the Kyoto protocol at around 70%, Despite strong public support, there was still some opposition, particularly by the Canadian Alliance, precursor to the governing Conservative Party, some business groups, and energy concerns, using arguments similar to those being used in the US. In particular, there was a fear that since US companies would not be affected by the Kyoto Protocol that Canadian companies would be at a disadvantage in terms of trade. In 2005, the result was limited to an ongoing "war of words", primarily between the government of Alberta (Canada's primary oil and gas producer) and the federal government. There were even fears that Kyoto could threaten national unity, specifically with regard to Alberta. As of 2003, the federal government claimed to have spent or committed 3.7 billion dollars on climate change programmes. By 2004, CO_2 emissions had risen to 27 per cent above 1990 levels (which compares unfavorably to the 16 percent increase in emissions by the United States during that time).

In January 2006, a Conservative minority government under Stephen Harper was elected, who previously has expressed opposition to Kyoto, and in particular to the plan to participate in international emission trading. Rona Ambrose, who replaced Stéphane Dion as the environment minister, has since endorsed some types of emission trading, and indicated interest in international trading. On April 25, 2006, Ambrose announced that Canada would have no chance of meeting its targets under Kyoto, and would look to participate in U.S. sponsored Asia-Pacific Partnership on Clean Development and Climate. "We've been looking at the Asia-Pacific Partnership for a number of months now because the key principles around it are very much in line with where our government wants to go," Ambrose told reporters. On May 2, 2006, it was reported that environmental funding designed to meet the Kyoto standards had been cut, while the Harper government develops a new plan to take its place. As the co-chair of UN Climate Change Conference in Nairobi in November 2006, Canada and its government received criticism from environmental groups and from other governments for its climate change positions. Canada's federal government has introduced legislation to set mandatory emissions targets for industry, but it will not take effect until an estimated 2050. The government has since begun working with opposition parties to improve the legislation.

A private member's bill, was put forth by Pablo Rodriguez, Liberal, aiming to force the government to "ensure that Canada meets its global climate change obligations under the Kyoto Protocol." With the support of the Liberals, the New Democratic Party and the Bloc Québécois, and with the current minority situation, the bill passed the House of Commons on 14 February 2007 with a vote of 161-113, , and is now being considered by the Senate. If passed, the bill would give the government 60 days to form a detailed plan of action. The government has flatly refused to abide by the bill, which may spark a constitutional crisis, lawsuit, or non-confidence motion once the bill becomes law, as is expected.

Peoples' Republic of China

The People's Republic of China insists that the gas emissions level of any given country is a multiplication of its per capita emission and its population. China endorses this because of the advantage it would get within the new restrictions. Because China has emplaced population control measures while maintaining low emissions per capita, it claims it should therefore in both the above aspects be considered a contributor to the world environment. China considers the criticism of its energy policy unjust. China is currently the second largest emitter of greenhouse gases, and as of first-half, 2007, is expected to become the largest by second-half 2007. In 2004 the total greenhouse gas emissions from China were about 54% of the USA emissions. China is now building on average a coal-fired power plant every week and plans to continue doing so for years. Some predictions (in 2006) are that China will emit more greenhouse gas than the USA in 2 or 3 years.

China is potentially the largest beneficiary of the Clean Development Mechanism (CDM), a market-based flexible mechanism stipulated by the Kyoto Protocol that allows industrialized nations to buy certified emission reductions produced by projects in developing countries to offset their obligations, commonly known as international "carbon trading".

The UN predicts that China will become a major carbon credits provider by 2012, because its marginal cost for emission reduction is relatively low compared to many other developing countries. According to the World Bank,

China captured $3 billion of the world's $4.8 billion in CDM subsidies last year for dozens of projects. China can not only reap remarkable revenue from participating in global carbon trading and selling certified emission reductions to organizations in developed countries, but also receive overseas fund and technology aid through CDM projects (http://www.eai.nus.edu.sg/BB333.pdf).

European Union

On May 31, 2002, all fifteen then-members of the European Union deposited the relevant ratification paperwork at the UN. The EU produces around 22% of global greenhouse gas emissions, and has agreed to a cut, on average, by 8% from 1990 emission levels. On 10 January 2007, the European Commission announced plans for a European Union energy policy that included a unilateral 20% reduction in GHG emissions by 2020. The EU has consistently been one of the major nominal supporters of the Kyoto Protocol, negotiating hard to get wavering countries on board.

In December 2002, the EU created an emissions trading system in an effort to meet these tough targets. Quotas were introduced in six key industries: energy, steel, cement, glass, brick making, and paper/cardboard. There are also fines for member nations that fail to meet their obligations, starting at 40/ ton of carbon dioxide in 2005, and rising to 100/ton in 2008. Current EU projections suggest that by 2008 the EU will be at 4.7% below 1990 levels. Transport CO_2 emissions in the EU grew by 32% between 1990 and 2004. The share of transport in CO_2 emissions was 21% in 1990, but by 2004 this had grown to 28%.

The position of the EU is not without controversy in Protocol negotiations, however. One criticism is that, rather than reducing 8%, all the EU member countries should cut 15% as the EU insisted a uniform target of 15% for other developed countries during the negotiation while allowing itself to share a big reduction in the former East Germany to meet the 15% goal for the entire EU. Also, emission levels of former Warsaw Pact countries who now are members of the EU have already been reduced as a result of their economic restructuring. This may mean that the region's 1990 baseline level is inflated compared to that of other developed countries, thus giving European economies a potential competitive advantage over the U.S. Both the EU (as the European Community) and its member states are signatories to the Kyoto treaty.

Germany

The following statement provides the Historic Backdrop of German Climate Change and Kyoto Protocol compliant initiatives:

1987 Warning of German Meteorological Society and German Physical Society on the foreseeable dangerous climate change

1990 Report of Federal Parliamentary Commission on preventing climate change, demanding a reduction of the CO_2 emissions of Germany by
30 % until 2005
50 % until 2020
80 % until 2050
compared to 1987 levels (unanimously)

1990 German government decision to reduce CO_2 emissions of Germany by 25 % until 2005

1998 German Kyoto target as part of EU burden sharing: –21% (until 2008-12, on the basis of 1990)

On June 28, 2006, the German government announced it would exempt its coal industry from requirements under the Kyoto agreement. Claudia Comfort, an energy professor at the German Institute for Economic Research in Berlin said, „For all its support for a clean environment and the Kyoto Protocol, the cabinet decision is very disappointing. The energy lobbies have played a big role in this decision." Under the Kyoto Protocol, Germany pledged to cut greenhouse gas emissions by 21 percent from 1990 levels by the period 2008-2012 (http://www.eubusiness.com/Environ/060628135510.ae1rjjju). Germany intends to reduce its absolute emissions by 40% until 2020 if the EU agrees on a 30% target. Targets for German sectors were agreed on 30 March 2004:

- Industry was granted 15 million tons of additional yearly CO_2 emissions until 2008-12 compared to its own "volontary target" ⇒ these 15 M t are an additional burden upon private households and transportation sector, questionable if this target can be reached with current policies.

United Kingdom

The energy policy of the United Kingdom fully endorses goals for carbon dioxide emissions reduction and has committed to proportionate reduction in national emissions on a phased basis. The United Kingdom is a signatory to the Kyoto Protocol. On March 13, 2007, a draft Climate Change Bill was published after cross-party pressure over several years, led by environmental groups. The Bill aims to put in place a framework to achieve a mandatory 60% cut in the UK's carbon emissions by 2050 (compared to 1990 levels), with an intermediate target of between 26% and 32% by 2020. If approved, the United Kingdom is likely to become the first country to set such a long-range and significant carbon reduction target into law.

The UK currently appears on course to meet its Kyoto limitation for the basket of greenhouse gases, assuming the Government is able to curb rising carbon dioxide emissions between now (2006) and the period 2008-12. However, annual net carbon dioxide emission levels in the UK have actually risen by around 2 per cent since Tony Blair's Labour Party came to power in 1997. Furthermore, it now seems highly unlikely that the Government will be able to honour its manifesto pledge to cut carbon dioxide emissions by 20 per cent from 1990 levels by the year 2010, unless a Climate Change Act is passed in 2006-7 and the Government takes immediate and drastic action to curb emissions over the next few years.

India

Among the major world economies, India's economy is the least energy intensive. India signed and ratified the Protocol in August, 2002. Since India is exempted from the framework of the treaty, it is expected to gain from the protocol in terms of transfer of technology and related foreign investments. At the G-8 meeting in June 2005, Indian Prime Minister Manmohan Singh pointed out that the per-capita emission rates of the developing countries are a tiny fraction of those in the developed world. Following the principle of *common but differentiated responsibility*, India maintains that the major responsibility of curbing emission rests with the developed countries, which have accumulated emissions over a long period of time.

Chicago Climate Exchange® (CCX®), the world's first global greenhouse gas emission reduction and trading system, and the only such system in North America, is launching an initiative to establish the first pilot greenhouse gas emissions trading program in India—the India Climate Exchange (ICX). The goal of this effort is to enable entities in India to become involved in greenhouse gas reductions integrated with emissions trading, for the purposes of climate change mitigation, institution building and social development. Following the example set by CCX, the India Climate Exchange will be designed through a consensus process by concerned India-based enterprises, enabling them to determine the most cost effective greenhouse gas mitigation measures, which will in turn, lead to better strategic management of energy costs, new products, and new sources of revenue, job creation and poverty alleviation (http://www.chicagoclimatex.com/content.jsf?id=1622).

Russia

Vladimir Putin approved the treaty on November 4, 2004 and Russia officially notified the United Nations of its ratification on November 18, 2004. The issue of Russian ratification was particularly closely watched in the international community, as the accord was brought into force 90 days after Russian ratification (February 16, 2005). The emission targets adopted in the Kyoto Protocol far exceed the likely level of emissions from Russia and Ukraine. These countries could sell their "bubbles" if the Protocol enters into force and industrialized countries establish an international emission trading system. Using the most recent, comprehensive scenarios for emissions of carbon dioxide from the energy system it estimated by the International

Institute for Applied Systems Analysis (IASA, 1994) that during the Protocol's 2008-2012 "budget period" the bubble will range from 9 MtC (million tons of carbon) to 900 MtC for Russia and 3 MtC to 200 MtC for Ukraine. Even scenarios with high economic growth and carbon-intensive technologies do not burst the bubble before the budget period. In the central ("middle course") scenario the total carbon bubble exceeds 1000 MtC, is worth 22 to 170 billion US Dollars (4 to 34 billion US Dollars per year), and does not burst until 2040. This flow of assets, which could exceed Russian earnings from natural gas exports ($10 billion in 19974), is comparable with projected total investments in the Russian energy system for 2008-2012. If directed towards low-carbon infrastructure investments (e.g., gas pipelines, safe nuclear power), bubble transfers could reinforce and partially lock-in decarbonization of the world energy system (http://www.iiasa.ac.at/Admin/PUB/Documents/IR-98-094.pdf).

The United States

The United States (U.S.), although a signatory to the Kyoto Protocol, has neither ratified nor withdrawn from the Protocol. The signature alone is symbolic, as the Kyoto Protocol is non-binding on the United States unless ratified. The United States was, as of 2005, the largest single emitter of carbon dioxide from the burning of fossil fuels. China is projected to take over at the top of the table by 2010. On July 25, 1997, before the Kyoto Protocol was finalized (although it had been fully negotiated, and a penultimate draft was finished), the U.S. Senate unanimously passed by a 95–0 vote the Byrd-Hagel Resolution , which stated the sense of the Senate was that the United States should not be a signatory to any protocol that did not include binding targets and timetables for developing as well as industrialized nations or "would result in serious harm to the economy of the United States". On November 12, 1998, Vice President Al Gore symbolically signed the protocol. Both Gore and Senator Joseph Lieberman indicated that the protocol would not be acted upon in the Senate until there was participation by the developing nations. The Clinton Administration never submitted the protocol to the Senate for ratification. This had been a challenge that required a 100% effort; ours, and the rest of the worlds. The world's second-largest emitter of greenhouse gases is the People's Republic of China. Yet, China was entirely exempted from the requirements of the Kyoto Protocol.

In keeping with its refusal to submit the protocol to Congress for ratification, the Bush Administration has taken no specific actions towards mitigation of climate change. In June 2002, the American Environmental Protection Agency (EPA) released the "Climate Action Report 2002". Some observers have interpreted this report as being supportive of the protocol, although the report itself does not explicitly endorse the protocol. At the G-8 meeting in June 2005 administration officials expressed a desire for "practical commitments industrialized countries can meet without damaging their economies". According to those same officials, the United States is on track to fulfill its pledge to reduce its carbon intensity 18% by 2012. The United States has signed the Asia-Pacific Partnership on Clean Development and Climate, a pact that allows those countries to set their goals for reducing greenhouse gas emissions individually, but with no enforcement mechanism.

Role of Local Governments

As of January 18, 2007, eight Northeastern US states are involved in the Regional Greenhouse Gas Initiative (RGGI) (http://www.rff.org/rff/News/Features/Regional-Greenhouse-Gas-Initiative.cfm), which is a state level emissions capping and trading program. It is believed that the state-level program will indirectly apply pressure on the federal government by demonstrating that reductions can be achieved without being a signatory of the Kyoto Protocol.

- Participating states: Maine, New Hampshire, Vermont, Connecticut, New York, New Jersey, Delaware, Massachusetts, Maryland joins June 30, 2007.
- *Observer states and regions*: Pennsylvania, District of Columbia, Eastern Canadian Provinces, Rhode Island.

On August 31, 2006, the California Legislature reached an agreement with Governor Arnold Schwarzenegger to reduce the state's greenhouse-gas emissions, which rank at 12th-largest in the world, by 25 percent by the year 2020. This resulted in the Global Warming Solutions Act which effectively puts California in line with the Kyoto limitations, but at a date later than the 2008-2012 Kyoto commitment periods. As of March 11, 2007, 418 US cities in 50 states, representing more than 60 million Americans support Kyoto after Mayor Greg Nickels of Seattle started a nationwide effort to get cities to agree to the protocol.

Cost-benefit Analysis

Economists have been trying to analyze the overall net benefit of Kyoto Protocol through cost-benefit analysis. Just as in the case of climatology, there is disagreement due to large uncertainties in economic variables. Still, the estimates so far generally indicate either that observing the Kyoto Protocol is more expensive than not observing the Kyoto Protocol or that the Kyoto Protocol has a marginal net benefit which exceeds the cost of simply adjusting to global warming. A study in Nature found that *"accounting only for local external costs, together with production costs, to identify energy strategies, compliance with the Kyoto Protocol would imply lower, not higher, overall costs."*

The recent Copenhagen consensus project found that the Kyoto Protocol would slow down the process of global warming, but have a superficial overall benefit. Defenders of the Kyoto Protocol argue, however, that while the initial greenhouse gas cuts may have little effect, they set the political precedent for bigger (and more effective) cuts in the future. They also advocate commitment to the precautionary principle. Critics point out that additional higher curbs on carbon emission are likely to cause significantly higher increase in cost, making such defence moot. Moreover, the precautionary principle could apply to any political, social, economic or environmental consequence, which might have equally devastating effect in terms of poverty and environment, making the precautionary argument irrelevant. The Stern Review (a UK government sponsored report into the economic impacts of climate change) concluded that one percent of global GDP is required to be invested in order to mitigate the effects of climate change, and that failure to do so could risk a recession worth up to twenty percent of global GDP.

Potential CDM Projects in India: Sector-wise Perspectives

Greenhouse Gas Inventory

India offers a large potential for the Clean Development Mechanism (CDM) because of India's inherent dependence on fossil fuels for development, as well as its proactive government and the enabling environment recently put in place. Indeed, India is the most favoured destination for CDM projects globally. The government of India has submitted a comprehensive national GHG inventory as part of its first national communication (NATCOM) to the United Nations Framework Convention on Climate Change (UNFCCC). This inventory has been prepared for the base year of 1994. Prior to the NATCOM, the most widely cited national GHG inventory available was that prepared for the 1990 base year in the Asia Least-Cost Greenhouse Gases Abatement Strategy (ALGAS) study of the Asian Development Bank (ADB-GEF-UNDP, 1998). The ALGAS inventory covered emissions from three sectors, namely, energy, agriculture, and forestry and land-use change. According to the ALGAS study, the energy sector is the largest contributor of carbon dioxide (CO_2) emissions in India. The CO_2 *equ*ivalent emissions from this sector are around 565.245 million tonnes, amounting to nearly 55 percent of total national emissions. These include emissions from road transport, burning of traditional biomass fuels, coal mining, and fugitive emissions from oil and natural gas.

Potential of CDM Projects in India

India is currently the world leader in development of CDM projects. Due to the effective operation of the Indian Designated National Authority for approval of CDM projects, more than 297 project proposals with an emission reduction potential of over 297 million t CO_2 have got official host country approval. So far, CDM Executive Board registers 60 Indian projects (http://cdmindia.com/India%20CDM%20Potential.pdf).

More than 200 Project Idea notes and large number of PDD's is floating in India. The wide range of possible project types and sizes in India allows international buyers to find the project of their choice or to define a broad portfolio to reduced risks.

- There is a large potential for renewable energy generation from agriculture wastes, hydro and wind.
- Thermal electricity generation and industry offer countless opportunities to improve energy efficiency, for example, regarding coal-fired power plants and of the transmission and distribution system.
- The chemical industry and aluminum production allow reductions of industrial greenhouse gases with high warming potentials.

Moreover, the availability of skilled consultants and a fierce competition of validators allow getting high-quality services for each step of the CDM project cycle at very competitive prices.

GTZ -CDM-India is receiving a number of queries every day from various sectors to develop CDM projects in India. The India market is vibrant and growing day by day. Project promoters are ready to invest making the time ripe for international buyers of CERs to invest in India.

Although there are varying estimates of the potential for CDM projects in India, it is roughly estimated to be in the range of about 300 million tonnes of CO_2 equivalent, including 90 million tonnes from renewable energy sources alone. In this section, a brief description of the important end-use sectors/areas are presented where significant potential exists in terms of improvement in both local as well as global emissions. The various end-use segments discussed in this context are (1) power; (2) industrial sector energy efficiency (including small-scale industries); (3) industrial co-generation; (4) transport; (5) renewable energy; (6) agriculture and livestock; (7) municipal solid waste; and (8) land use, land-use change, and forestry (LULUCF).

Power Sector: Technology and CDM Potential

The growth in installed power generation capacity, from 1,362 megawatts (MW) to over 100,000 MW since independence, and the electrification of more than 500,000 villages is an impressive achievement. The per capita electricity consumption has also improved significantly. It may be noted, however, that annual per capita consumption, at about 355 kilowatt-hours (kWh), is still among the lowest in the world. Still, many households in a large number of villages have no access to electricity. The Ministry of Power has set an agenda of providing power to all by 2012. To meet the projected power requirement by 2012, a capacity addition target of 46,500 MW has been tentatively fixed for central public sector undertakings under the Ministry of Power. At the state level, the state electricity boards (SEBs)/state utilities and private sector will add about 41,800 MW.

Coal is the primary source of energy for steam and thermal power generation in India. Its large-scale use also brings along with it environmental issues with respect to emissions of CO_2 contributing to the greenhouse effect, sulphur oxides (SOx) and nitrogen oxides (NOx) causing acid rain, and particulates leading to health problems. In addition, fly ash generated as a result of coal combustion poses additional disposal hazards. All these issues call for the use of steam and thermal power generation technologies, which are more efficient and lead to minimum pollution of air, water, and land. The best option to mitigate the effects of the pollutants from the use of coal is to adopt technologies and policies which help the increase in the energy efficiency of all the processes from generation to end use.

(a) Advanced Technology in the Power Sector

To improve the efficiency of power generation, integrated gasification combined cycle technology (IGCC), which has the capability of achieving up to 48 to 50 percent efficiency, will have to be adopted. This technology involves gasification of the fuel and using the gas in the combined cycle system for power generation (i.e., combination of gas turbine and steam turbine). This technology has been proven in the advanced countries for large-scale power generation applications. The capital cost of some of these plants has also been subsidized by the respective governments to prove the technology on a commercial scale. To adopt this technology for Indian coals, which have different properties compared to those in the advanced countries, a reasonable size demonstration plant of 100 to 250 MW will have to be set up. For this, the government of India may have to subsidize approximately 25 to 30 percent of the capital cost for a coal-based plant and approximately 10 to 15 percent for a refinery residue-based plant. Once the technology is demonstrated, it would be much easier to replicate this advanced technological option at Indian power plants.

Another area that holds promise is the renovation and modernization (R&M) of power plants. About 65 percent of the installed capacity in India is thermal based. The present capacity is unable to meet the demand; the country is experiencing an energy shortage of about 12 percent and a peak power shortage of about 8 percent. The characteristics of R&M that make it attractive compared to installing new capacity are its lower cost, no additional infrastructure requirement, improved efficiency, and reduction in GHG emissions. However, constraints of the nature of resources, lack of public and government determination, and absence of stringent environmental laws have always acted as a barrier towards the move.

It is important to note that, in spite of the tremendous scope and existence of technology options for mitigating the emission of GHGs, the efforts in this direction have not been significant. The reason for this is lack of capital for undertaking such measures. It is expected that the CDM will provide much needed financial resources for undertaking some of the GHG reduction measures. It should be realized, however, that uncertainties are still associated with the CDM and the eligibility of different GHG mitigation options within the power sector.

Industrial Sector Energy Efficiency

Industry in India is generally highly energy intensive, and its energy efficiency is well below that of other industrialized countries. Efforts to promote energy conservation by such industries could lead to substantial reduction of operating costs, making them more competitive globally. An analysis of the industrial energy-use pattern further reveals that around 65 to 70 percent of India's total energy consumption is accounted for by seven sectors, namely, (1) cement, (2) pulp and paper, (3) fertilizer, (4) iron and steel, (5) textiles, (6) aluminium, and (7) refineries. The scope of improving energy efficiency (and consequently CO_2 reduction) through technology transfer in India is enormous and can provide possible CDM projects.

(a) India's Small-scale Industry Sector

In addition to the above energy-intensive sectors—which are largely medium to big industries—another important segment is the small-scale industry sector, which occupies a position of prominence in India's economy. It contributes to more than 50 percent of industrial production in terms of value added, makes up one-third of total exports, and employs the largest manpower next to agriculture. Small and medium enterprises (SMEs) are the backbone of the Indian economy. With three million SMEs, this sector contributes 40 percent towards national income. It provides employment to more than 16 million people in the country and it is growing at the rate of 20 percent annually. Some of the key features of the small enterprise sector are that it is labour oriented (which provides employment to a large population of the country), it contributes to a great extent in foreign exchange earnings, and it is less capital intensive.

Growth of small enterprises has resulted in the clustering of similar types of industries in some areas and stimulating the development of localized skills. Despite these advantages, the small-scale sector is also a major concern, as it accounts for about 70 percent of industrial pollution. This is due to obsolete technologies, high waste generation factors, and less regulatory requirements. In spite of the high growth rate, unfortunately, this end-use sector is also experiencing growing industrial "sickness." The reasons for this range from technological obsolescence, information deficiency, and poor management practices to non-availability of credit. There are some highly energy-intensive sub-sectors where the cost of energy makes up a sizeable proportion of the total production cost and offers tremendous scope for energy-efficiency improvement and pollution reduction through technology upgrading.

In this context, for India's small-scale industry sectors—which are diverse, multi-sectoral entities that have issues both unique to themselves and cutting across multiple sectors—the CDM, in its present form, creates an atmosphere of opportunity. Given the problems that India's small-scale industry sectors are facing in the complex international financing and technology transfer mechanisms, and their poor capacity to modernize technology to improve product quality and reduce production costs on their own, the chances of any wide-scale technology upgrading are severely restricted. It would, therefore, be in the interest of small-scale industry to take a proactive stand and start to establish the necessary processes and procedures to initiate CDM activities and build up sufficient CDM capacity among project stakeholders (e.g., governments, enterprises, industry associations, consultants, etc.).

Energy Efficiency in the Industrial Sector

Energy-efficiency programmes across all industrial sectors hold the promise for cost-effective CO_2 abatement, as a large number of plants in India operate well below world energy-efficiency standards. The CO_2 mitigation projects in industrial sector could be broadly grouped under the following major headings:

1. Sector-specific technological options
2. Cross-cutting technologies
3. Fuel switch options
4. Recycling and use of secondary materials

Technological changes leading to reduced CO_2 emissions vary depending upon the particular industrial sub-sector, and thus cannot be generalized. As indicated earlier, more than 60 percent of total energy consumption in the industrial sector is accounted for by seven sectors. In many of industrial plants, the problem is mainly due to the vintage of the production process. A large number of plants are based on technologies developed in a period when energy efficiency was a secondary criterion of project design and engineering. Some of the progressive units have gone in for massive modernization plans, thus cutting down their energy costs and remaining competitive in the market. But in many other cases, concerns in this direction have remained confined to a conceptual level, mainly due to financial constraints. In some industrial sub-sectors, the protectionist policies of the government have proved detrimental to the implementation of energy-efficiency measures. A classic example is that of India's sugar industry. Although, in terms of size, the sugar industry is the largest in the world, a lot needs to be done in terms of its energy efficiency. There are very few sugar mills that have advanced cogeneration systems in place for efficient utilization of bagasse.

In addition to the possibilities that exist in the large industries, there are also possibilities for energy conservation in small-scale industries like foundries, brick manufacturing, and pottery, tiles, and glass manufacturing, etc. In these sectors, traditional designs are being used, and tremendous scope for energy-efficiency improvement exists through technology upgrading. Some of the possible mitigation and CDM options under various industrial sectors are given below:

Cement: The cement industry has undergone rapid technological upgrading and vibrant growth during the last two decades. Some of the plants in India can be compared in every respect with the best plants operating in the world. The industry presents a mixed picture with the existence of many old plants as well as new plants that are technologically advanced. The specific power consumption in the case of modern plants is between 68–90 kWh/ tonnes of cement, while it is between 90–120 kWh/tonne in the case of old plants. Similarly, the specific thermal energy consumption of the latest plants is as low as 665 kilocalories (kcal)/kilogram (kg) of clinker, while for the older dry process plants it can go up to 900 kcal/kg of clinker or more-depending upon the level of technology that is in place. Hence, there exists the scope for improving energy efficiency in the relatively older installations. The CO_2 intensity of India's cement industry is 0.9 tonnes of CO_2 per tonne of cement compared to the world average of 0.75 tonnes CO_2/tonne of cement, which implies that there exists the potential for CO_2 mitigation in India's cement plants. Some of the energy-efficient technological options that can be adopted in the cement plants are outlined below:

- *Fuel switching* through, for instance, use of natural gas or alternative fuels (e.g., biomass, etc.) instead of coal, petroleum coke, fuel oil, and natural gas
- *Use of mineral components to substitute clinker* (gypsum and natural pozzolanas, blast furnace slag and fly ash)
- *Raw material preparation:* Use of gyratory crushers and mobile crushers, vertical roller mills (VRM) instead of ball mills, external re-circulation systems in VRMs, and adoption of roller press technology and high-efficiency separators in the grinding circuits
- *Cement grinding:* Use of VRM with high-efficiency separators and high-pressure roller presses in various modes of operation, and use of static V separators along with dynamic separators
- *Pyro-processing section:* Installation of pre-calcinators and 5/6 stage pre-heaters with low-pressure drop cyclones, new generation coolers having better heat-recovery potential Depending upon the base-level energy consumption, the possibilities for energy saving in different plants vary—from 10 percent to as high as 30 percent when older plants are considered.

In addition, waste heat recovery from pre-heater/cooler exit gases for power generation offers enormous opportunities for reducing CO_2 emissions in the cement sector. Preliminary studies carried out by the National Council for Cement and Building Materials (NCCBM) indicate that there exists about 160 MW of cogeneration potential in India's cement industry. The corresponding reduction in GHGs is estimated at about 1.5 million tonnes CO_2/year.

Textiles: India's textile industry contributes nearly 4 percent to the country's GDP, and the export earnings are around 25 percent of the total value of exports from India. There are about 1,860 cotton and man-made fibre textile mills in the country. Energy accounts for 12–15 percent of the total cost of production at present. Some of the major energy-saving options in textile mills are as follows:

- *Adoption of new spinning processes.* It is possible to save 15–20 percent of the energy requirement in spinning by adoption of new spinning technologies like friction spinning and air-jet spinning. Modifications in ring-frame spinning machines, which account for the majority of the power consumption in a composite textile mill, can reduce energy consumption by 5–10 percent.
- *Adoption of advanced drying processes to substitute clinker* like high-speed drying machines for reducing energy consumption (20–30%). Use of radio frequency dryers can eliminate use of thermal energy in drying applications.
- *Use of solar energy for water heating.* In the textile industry, 80 percent of energy used is utilized in wet processing, with temperatures ranging from 40°C to 140°C. There is substantial scope to reduce the amount of fossil fuels used in boilers by utilizing solar thermal energy.

Based on the balance of useful energy, the potential improvement in energy efficiency in the textile sector is immense. It would be desirable, however, for India's textile industry to replace fuel oil used for process heat with renewable sources of energy (e.g., use of solar energy for water heating, etc.). Within this context, a reduction in carbon emissions would take place through the substitution of a carbon-intensive fuel by another that is more environmentally friendly.

Pulp and Paper: The Indian pulp and paper industry consists of more than 470 registered mills. While the majority of them are in the small- and medium-scale category, there are also 25 large-scale, integrated mills. 18 In general, India's pulp and paper industry is highly energy-intensive and its energy efficiency is well below that of other industrialized countries. Total CO_2 emissions from the sector have been estimated at 9.05 million tonnes, and consumption of raw materials and chemicals and generation of waste is also almost double. Energy costs represent about 20–25 percent of the total production cost, compared to 12–14 percent in the United States and Scandinavian countries. The technologies identified for improvement in India's pulp and paper industry and their energy-saving potential are given below:

- *Installation of continuous digesters.* Adoption of this technology would produce an estimated savings potential of around 690 kg of CO_2/tonne of paper (about 676,000 tonnes of CO_2 per year).
- *Other energy-saving options* in the pulp and paper industry include installation of falling film evaporators, installation of trinip presses and high-capacity chippers, oxygen delignification, disc refiners in place of conical refiners, etc.

The CDM opportunities available to India's pulp and paper sector are in the area of replacement of energy sources, as a significant portion of the energy consumption in this sector is based on non-renewable sources. Additionally, there is large potential for improving the sector's energy efficiency and for the more widespread use of cogeneration, particularly through the re-use of waste materials generated in the production process.

Iron and steel: Of the total CO_2 emissions from the industrial sector, 42 percent comes from iron and steel production (GoI, MoEF 2004). Most emissions from the steel industry occur in the production process due to the use of coking coal in the case of blast furnace-based in-line strip plants (ISPs) and non-coking coal in the case of direct reduction (Midrex) and iron smelting (COREX) based plants. In the gas-based direct reduction plants the emissions are also due to the use of natural gas. The CO_2 intensity in India's steel industry in 1995 was about 3.6 tonnes CO_2/tonne of steel, which translates to about 75 million tonnes of CO_2. Various technologies are available internationally to achieve potential reductions in CO_2 emissions, including the following:

- Integrated steel sector
  - Coke dry quenching
  - Pulverized coal injection in blast furnaces
  - Basic oxygen furnace (BOF) gas recovery
  - Continuous casting/thin section continuous casting
  - Hot rolling
- Secondary steel sector
  - Use of UHF furnaces
  - Oxy-fuel burners
  - Scrap-preheating

According to a detailed study of the CO_2 reduction potential in India's steel industry conducted by researchers from Lawrence Berkley National Laboratory, the potential CO_2 savings are 1.65 tonnes of CO_2 per tonne of steel, or 34.3 million tonnes of CO_2 per annum for the year 1995. The study estimated the potential by comparing the energy consumption of India's steel industry with benchmarks, which were an integrated steel plant operating in the Netherlands and a secondary steel plant operating in Germany (Price *et al*, 1999). During the past ten years (1992/93 to 2002/03), however, India's steel industry has invested resources into the adoption of some of the above-mentioned technologies, and as a result its specific energy consumption has been steadily declining. The CO_2 intensity of the integrated steel plants has already been reduced to around three tonnes of CO_2 per tonne of steel during this period.

Aluminium: Primary aluminium is produced by mining bauxite, refining the ore to alumina, and combining the alumina and carbon in an electrolytic cell to produce aluminium metal. India has a total installed capacity of 8, 80,000 tonnes per annum with five primary aluminium producing units (Center for Monitoring of Indian Economy 2002). The aluminium manufacturing process is electrical energy-intensive. The break-up of energy indicates that a large quantum (more than 80 percent) of energy is electrical energy and is consumed in smelting of alumina to aluminium. The major energy-saving opportunities in the sector are as follows:

- *Alumina plants:* The energy-saving potential in alumina plants is as high as 25 percent and the areas for improvement include calcinations (switch over to gas suspension calciner as opposed to rotary kilns) and waste heat utilization;
- *Aluminium smelters:* Energy conservation potential exists with regard to conversion to pre-baked systems from Soderberg systems. Other operational improvements include the following:

  Current efficiency improvements:

  - Auto operation of the cell, smaller and more frequent additions of alumina (less anode effects) and clean cathode, lower bath ratio, better quality of anode, material purity, and baking, constant current
  - Boost voltage, check on molten metal movement
  - Magnetic field: reduction in anode current density (1.1–0.8 amps/ cm^2), bigger size of anodes
  - Less frequency of anode change, quality of alumina (soda content of 0.30–0.32%, silica content of 0.009–0.01%), and physical parameters like sandy, alpha content, etc., (improved cathode life)
- *Reduction in operating voltage.* Since the scope for improvement in current efficiency is limited (Indian plants operate between 90–95% current efficiency), lowering the voltage requirements of cells represents both the largest challenge and better opportunities for improving overall energy efficiency.

Fertilizer: The fertilizer industry produces four different types of fertilizers: nitrogenous, phosphate, potash, and complex fertilizers. The nitrogenous fertilizer production is highly energy-intensive and is one of the largest consumers of petroleum-based fuels. Many of the older ammonia/urea manufacturing plants use liquid fuels as the feedstock (naphtha, furnace oil), but in newer plants, natural gas is the preferred feedstock. During the last few decades, India's fertilizer industry has witnessed many changes in feedstock and technology, which have resulted in a substantial reduction in overall energy use and improvement in energy efficiency. Most of the new plants built during the last decade incorporate state-of-the-art technologies. However, options for CO_2 mitigation still exist in many of the older plants, which include switchover from fuel oil/naphtha to natural gas and other energy conservation schemes in both ammonia and urea plants. Recent advances in process technology and catalysts have opened up opportunities for developing specific energy-saving projects in fertilizer plants that can result in lower energy intensity in fertilizer production.

Iron Foundry: In India today, there are about 7,000 foundries in the country, with an installed capacity of 3.5 Mt (million tonnes) and production of around 2.5 Mt. These units are mostly located in clusters, with the cluster size varying from less than 100 to around 400 units. The energy intensity of these units is quite high and presents an ideal situation where both energy savings and pollution reduction can be achieved through technological upgrades. The iron foundry sub-sector represents a case where technological development is not happening on its own. The constraints are essentially related to the limited capacity of the units to identify, research, develop, and absorb appropriate technologies as well as to access credit for their implementation. The overall purpose of CDM projects in the cluster should, therefore, be to decrease transaction costs to users, and thus increase the financial attractiveness of such technologies for their wider adoption. The CDM intervention in this cluster could involve establishing delivery chains for credit and technology to the small-scale units.

Brick making*:* In India, traditional technologies are typically used for brick production, and brick firing is still an energy-intensive process. The annual estimated coal consumption in the brick industry is 24 million tonnes, which forms about 8 percent of total coal consumption in India. The share of fuel in the total production cost of bricks is in the range of 35 to 50 percent. Significant scope exists for improving the energy efficiency of the brick production sector. It is possible to reduce energy consumption by 5–15 percent with incremental improvements in existing technologies. In order to ensure a sustainable supply of bricks in the future, it is imperative to improve the operations of these traditional industries besides introducing new products and processes. Without technological, financial, and institutional support, diffusion of energy- and resource-efficient technologies is expected to be very slow and may not be uniform. Similarly, the introduction and diffusion of new kiln technologies and production of alternate building materials will not happen on their own. Another big challenge associated with technology upgrades and modernization is related to development of the technological capacity of the industry. Given the fact that the brick industry has limited financial and technological capacities, the change process will be governed largely by the rate and direction of support provided by different national and bilateral programmes to upgrade and modernize it. The construction sector in India has been growing at a rate of almost 6 percent per annum since 2000 and—fuelled by the rapid growth of India's economy—is expected to grow even faster in the next few years. Burnt bricks are the most popular construction material in India and production is estimated at 120–140 billion bricks per year. India is the second largest producer of bricks after China. The brick industry in India is unorganized, with small production units of more than 100,000 clustered in rural and peri-urban areas in the country. Brick making consumes about 24 million tonnes of coal every year. Coal consumption by the brick industry is approximately 8 percent of the country's total coal consumption. The share of energy cost in the total cost of brick production is 35–50 percent. The choice of technology for firing bricks depends generally on factors such as scale of production, soil and fuel availability, market conditions, and skills available.

Co-generation: Co-generation offers an attractive solution to meet the industrial energy requirements in an efficient manner, while conserving national resources. Co-generation is the simultaneous production of power, either electrical or mechanical, and useful thermal energy from a single fuel source with an overall efficiency normally exceeding 70 percent. Co-generation offers numerous direct benefits to industrial and institutional applications but also positive carry-over benefits to utilities and society-at-large. The estimated potential of co-generation in India is between 17,000 to 20,000 MW, based on conservative estimates. It is difficult to give an accurate figure, as many small-scale industrial applications and industrial estates may also have co-generation plants, but no estimates are available for these areas. Traditionally, refineries, fertilizer plants, and sugar plants have cogenerated power and heat requirements for various industrial processes, but without exporting to the grid. There are several chemical and petrochemical processes that generate large quantities of waste heat that could be utilized for electricity generation. Given the suitability of co-generation to industry in India, the candidates for co-generation include industries with substantial combined heat and power requirements, such as sugar, textile, paper, fertilizer, food processing, chemicals, and petrochemicals. Co-generation is already being practiced widely in various industries such as pulp and paper, rayon, sugar, chemicals, and fertilizers.

Transport sector: The second most important energy-consuming sector in India is transport. Almost half of the total petroleum product consumption is accounted for by this sector in the form of high-speed diesel (HSD) and gasoline. A review of India's transport sector reveals two major structural shifts that have had a pronounced impact on energy use efficiency. First, the rail-dominant economy in the 1950s gradually gave way to a road-dominant economy in the 1990s. Second, an inadequate public transport system has led to a rapid increase of personal modes of transport and intermediate public transport, some of which are extremely energy inefficient. A reference can be made to the large number of two-stroke powered two-wheelers used as personal vehicles (the total number of registered vehicles was 23.1 million in 1996). In terms of the share of commercial energy use, road transport is most important, accounting for 81.34 percent of the total commercial energy use by this sector. Railways come next with a share of 9.35 percent, followed by water (7.29 percent), and air (2.02 percent). So in terms of an energy-efficiency intervention strategy and potential benefits, road and rail transport are considered in this section. The possibilities of efficiency improvement in road transport can be broadly categorized under two headings: (1) improvements due to design changes and (2) improvements due to inter-fuel substitution. A significant design-induced efficiency improvement is possible for the two-stroke engine categories (two- and three-wheelers). The stringent future proposed emission standards for two- and three-wheelers would force manufacturers to change the power pack to four-stroke engines. The benefits in the car and bus segment are estimated to be primarily due to fuel switching from gasoline/diesel to compressed natural gas (CNG).

Transportation is a major sector contributing to GHG emissions, which includes road, rail, airways, and waterways. The total CO_2 emissions from this sector in 1994 were 79.88 million tonnes (GoI MoEF 2004). Among the various sub-sectors, road transport is the main source of CO2 emissions and accounted for nearly 90 percent of the total transport sector emissions in 1994. Road transport is constituted of gasoline- and diesel-fuelled vehicles. According to a survey by the Indian Market Research Bureau on behalf of the Ministry of Petroleum and Natural Gas (MoPNG), the transport sector consumed nearly all the gasoline in the country (98.3 percent) (GoI: Ministry of Petroleum and Natural Gas 1998). The share of vehicle categories in gasoline consumption was two-wheelers (50.8 percent), cars/taxis (31.5 percent), and three-wheelers (13.4 percent).

Diesel is consumed in both private and public modes of transport (trucks, buses, jeeps, car/taxis, etc.) as well as in agriculture (tractors, irrigation pumps, etc.). The all-India survey by MoPNG indicated that 61.8 percent of the diesel sold through the network of retail outlets was consumed by road transport. Automotive exhaust emissions are amongst the major sources of toxic pollutants, besides producing GHG emissions like CO_2, methane (CH_4), and nitrous oxide (N_2O). The vehicular emissions standards, first introduced in India in 1991/92, were focused on reducing toxic pollutants. The standards were subsequently upgraded in 1996 and 2000.

Presently, emission norms equivalent to the Euro-I standard prevail in the entire country, while Euro-II is the standard in metropolitan cities. The government of India has made significant policy interventions, including continuous improvements in the emissions standards to improve the air quality in the urban centres in the wake of a rapidly growing vehicular population; recent years have witnessed a phenomenal growth in road transport vehicles. This trend of growing vehicle population is expected to continue with rising incomes and enhanced vehicular choices offered to consumers. The emissions of local pollutants in urban centres have, therefore, continued to grow along with rising GHG emissions.

The deterioration in urban air quality led to several responses, such as the introduction of CNG vehicles, improvement in auto fuel quality, and enhancement of road infrastructure. The government of India announced the Auto Fuel Policy in 2003, which comprehensively addresses the issues of vehicular emissions, vehicular technologies, and the provision of cleaner auto fuels in a cost-efficient manner, while ensuring security of the fuel supply. The policy includes a road map for reduction in emission standards for new vehicles. Besides proposing the enhanced quality of liquid fuels, the policy encourages the use of CNG and liquid natural gas (LNG) in cities affected by high vehicular pollution to enable vehicle owners a wider choice of fuel and technology. The policy envisages the accelerated development of alternate technologies, like battery and fuel cell-powered vehicles, and a comprehensive programme for research and development support and other measures for zero-emission vehicles. Implementation of the Auto Fuel Policy would accrue significant improvement in local air quality and also contribute to reduction of GHG emissions. However, the transport sector remains a difficult arena for CDM projects. Transport-related emissions are usually very dispersed and difficult to monitor, while baselines are hard to calculate. This makes CDM projects in the transport sector risky until standardized guidelines and processes are introduced.

Renewable Energy

The development and utilization of renewable energy has been accorded a high priority by the government of India. It is the only country to have a separate ministry looking after the promotion of renewable energy sources, and it runs one of the most diverse and biggest renewable energy programmes in the world. The government's policy has been successful in creating a fairly large and diversified manufacturing base and infrastructure (technology-support groups and facilities, as well as nodal agencies) to support renewable energy technology design, development, testing, and deployment. So far, an aggregate capacity of about 5,000 MW has been installed, based on renewable-energy technologies. In spite of the proactive stand taken by the government and the achievements made so far, however, renewable energy still remains far from the energy mainstream. For CDM options, the ones discussed in some detail are: (1) bio-fuels, (2) biomass energy (gasification), (3) wind energy, (4) hydro power, and (5) solar energy.

(a) Bio-fuels and Rural Electrification

The role of bio-fuels in meeting the future energy needs of India is gaining lot of prominence. In July 2002, the Planning Commission constituted the Committee on Development of Bio-fuels in order to accelerate the developmental pace in this emerging field of renewable fuels. The committee recently submitted its report to the commission, giving recommendations on various aspects of bio-fuel use, such as commercialization, prospects, quality standards, development of large-scale projects, marketing strategies, research and development, etc. The emphasis so far, however, is on utilizing bio-fuels for vehicular applications as a fuel extender in internal combustion engines. Keeping in mind the high demand for petroleum products in the transport sector and a steeply escalating fuel import bill to bridge the demand–supply gap, the focus on environment-friendly alternate transport fuel based on renewable energy sources seems quite justified. It is strongly felt, however, that while encouraging the use of bio-fuels for the transport sector, one should not lose sight of its possible use in other potential applications. One such area of application that holds enormous promise is in the use of bio-fuels to energize rural India.

India has embarked on a major rural electrification programme that aims to electrify more than 80,000 villages that still have no access to electricity. Out of these 80,000 villages, more than 18,000 are classified as remote, where the extension of the grid is considered unviable, either because of difficult terrain or the high cost of grid extension. It may be noted that a remote village in this context is defined as one which has less than 60 households and which is located at or more than seven kilometers from the grid. In hilly and forest areas, any village located beyond three km is considered as remote. In view of the resource constraint, the focus is on providing lighting to this target group on a priority basis. But in the longer run, the electrification programme has to go much beyond just meeting the lighting needs of rural communities. One of the major challenges in mainstreaming renewables and in renewable-based rural electrification is to link the availability of energy with productive work. This is extremely important for the long-term sustainability of such initiatives.

The Ministry of Non-Conventional Energy Sources (MNES) is responsible for the electrification of these remote villages by 2012 through the use of renewable energy-based distributed generation. The various options being considered are wind power, mini/micro hydropower (hydel turbine), solar photovoltaic, and biomass gasifiers. Along with these options, the option of biofuel-powered diesel-generating sets (either in 100% biofuel mode or in blends with high-speed diesel) also merits attention for electrification of such villages. The biofuel route holds the promise of not just improving the quality of life of people living in such remote villages by providing access to clean energy, but also stimulating regional development by generating rural employment. The opportunities for small- and micro-enterprises in the areas of cultivation of oil-bearing species of plants and downstream processing of oil seeds could be complementary to the national agenda of promoting such fuels for vehicular application. Since biofuels can be used in compression ignition engines with little or no modifications, it is a relatively simpler form of distributed generation to be managed by the local community and does not require additional servicing/repair infrastructure, which has been one of the major bottlenecks in mains treaming renewables.

With the main source of raw material for biofuels being the oil-bearing seeds of the Jatropha and *Pongamia* tree species, the consumption of biofuel as an energy source, either in stationary combustion or mobile combustion, leads to a "no net addition" of CO_2 into the atmosphere. As per the estimate, each tonne of biofuel consumed leads to a reduction of GHGs by about three times, i.e., avoids the equivalent of three tonnes of CO_2 (CO_2e). These reductions in GHGs can then be accumulated and traded as carbon credits. According to the National Mission on Bio-Fuel, around 3.35 million metric tonnes (MMT) of biofuel is required for a 5 percent blending of biofuel in petrodiesel, and around 13.38 MMT of biofuel is required for a 20 percent blending by the year 2011/12. This translates into a reduction of GHG emissions to an extent of 40 million tonnes of CO_2e in 2011/12 with a 20 percent biofuel blend, which offers great potential for CDM projects in this sub-sector. Although this data relates to fuel substitution in the transport sector, a similar advantage can be availed of by the use of biofuel for rural electrification.

(b) Biomass Energy (gasification)

Biomass energy holds enormous potential for a country like India. Being an agricultural country and the world's largest producer of sugar cane, India has abundant quantities of agricultural residues and sugar cane bagasse. It is estimated that India produces about 350 million tonnes of biomass every year, equivalent to 220 million tonnes of coal in terms of heat value. This quantum of available biomass can generate 19,500 MW of power annually. The manner in which biomass is currently utilized for energy, however, is far from ideal because it results in polluting the environment, particularly indoor air. Biomass fuel could provide a much more extensive use than at present in the field of energy generation. There could be many variants of this resource and, ideally, it could be used to meet both the electricity as well as the cooking/thermal energy needs of the rural population in the most cost-effective manner. Biomass energy includes wood, agricultural residue, leaves, etc. Power generation can be based on (1) gasification of solid biomass to generate producer gas that can be used to replace diesel in a diesel generator set, either partially or fully, and (2) direct combustion of biomass to generate steam for power generation. Both of these options would still require sustainable production and availability of biomass so as not to exacerbate deforestation.

National programmes to promote biomass energy use and disseminate biomass energy technologies have existed in India for almost two decades. Despite several programmes on biomass energy, their penetration rate has been slow. Numerous technical, economic, marketing, institutional, and social barriers have prevented alterations in biomass consumption patterns. Hence, it continues to be used in a conventional, inefficient, and unsustainable manner.

A study of these programmes reveals the complexity of issues that need to be addressed in formulating intervention strategies and designing effective measures for providing better biomass energy services. It remains a challenge as to how effective instruments could be designed to address the different barriers. Intervention measures will also need to address biomass supply issues to ensure the sustainability of providing energy services. While designing mechanisms and instruments to alter biomass consumption patterns, there is a need to address some of the characteristics of bio-energy projects, such as high transaction costs (due to their small and dispersed nature); lack of appropriate cost buy-down mechanisms of demonstrated technologies along the learning curve (by increasing their cumulative production and by market aggregation); and lack of market infrastructure to set up financing mechanisms, distribution companies, and maintenance support.

Biomass energy projects under the CDM that are oriented towards altering energy consumption patterns are mostly categorized as *small projects*. This categorization offers advantages, such as streamlined implementation of projects, and allows them to choose a standardized baseline based on a regional average or a particular technological package. Studies have proposed methods to enhance the competitiveness of small projects under the CDM. These methods include automatic additionality, formulation of a standardized baseline, project bundling, easier credit calculation methods, and expedited project registration. The potential for reducing carbon emissions would depend on the average percentage of biomass energy used, as well as the size of the biomass energy market.

Wind Energy

India has a large estimated potential of 45,000 MW for wind energy utilization. Of this, only 5 percent has actually been realized to date. India is the fifth largest wind-power producer in the world after Germany, the United States, Denmark, and Spain, with a wind-power generation capacity of 2,483 MW, of which 2,418 MW has come through commercial projects. In addition to the potential in India, there is also a fairly large potential to be tapped in South Asia and the Asia-Pacific region, where a mature Indian wind industry can play a prominent role. This is, therefore, a huge opportunity yet to be fully exploited. The wind energy potential assessment is based on a host of parameters in a dynamic environment that is continuously changing. The realizable potential is related to the power evacuation infrastructure in the country, the knowledge base in areas with sufficient wind resources, policies in place, technologies in the market, the cost of setting up wind energy projects, and financial engineering aspects. The actual potential, it is felt, could be larger. In the case of off-grid applications, that is, isolated grids, hybrid systems, small power, and water pumping, there is believed to be a huge potential yet to be assessed.

"Off-grid" applications have remained largely neglected despite a diverse range of applications that can be explored in the vast coastal, inland, and mountainous regions of India. These are lighting and small power for fishing communities, desalination, ice-making, refrigeration, battery-charging for lighthouses, remote communication, pumping of water, pumping of brine, and many other applications. These will benefit marginal and remote communities such as fishermen, salt farm workers, and island and mountain dwellers. Wider use of wind or hybrid systems in these communities will improve their lifestyles and perhaps even enhance incomes.

India has had mixed experience with wind farms set up so far; very few are performing in accordance with initial projections. A comprehensive database for wind resources now exists, but its use and extrapolation in a scientifically correct manner to plan wind energy projects continues to be an area of concern. In fact, several wind farms that were set up in the past are non-performing projects due to lack of detailing of many of these aspects. Further development of wind energy projects requires innovative financing techniques, fine-tuning of policies at the state and central government levels, technology optimization, and up-scaling of the planned projects to benefit from scales of economy.

Wind energy's potential for reducing GHG emissions makes it one of the most promising technologies for solving the long-term problem of climate change. In regards to reducing GHG emissions and contributing to sustainable development, wind energy has been identified as one of the energy sources that can achieve these dual objectives of the CDM. Additionally, small-scale wind energy projects also have a great potential to achieve sustainable development goals; however, it is unlikely that the CDM will change the prospects for them. In fact, small-scale CDM wind energy projects will face numerous institutional and financial barriers (Figs. 4 and 5). Therefore, developing countries like India will have to take several domestic actions, such as small-scale CDM projects and bundling, to bring down transaction costs in order to build up their own wind energy markets and facilitate their implementation. Then, the CDM can play a pivotal role in supporting the growth of the domestic market through the mobilization of capital flows. Thus, wind power is an interesting alternative for CDM projects in India, as it would avoid or at least postpone the construction of thermal power plants.

(d) Hydropower

The potential for small-scale hydropower plants in India is 15,000 MW. By March 2004, the installed capacity was reached to the tune of 1,603 MW, and in the tenth five-year plan it is targeted to have an additional 600 MW of small-scale power plants from 2003–07 (GoI , Planning Commission 2002). Since a large potential for small-scale hydropower plants exists in remote hilly

Fig. 4: The Middele-ground Wind Farm Copenhagen – Denmark

Fig. 5: An Integrated Windmill - Chain

areas, development of small hydropower for decentralized power generation can help in rural electrification and development of these remote areas. The government is offering a number of incentives for the development of this sector, with special emphasis on mini/micro hydel-turbine hydropower projects in remote hilly regions. However, a high investment cost for small hydropower development has impeded its penetration. Investment costs are substantially high due to terrain inaccessibility and lack of suitable transportation linkages in locations where the potential exists. Places with high potential have low demand, which implies the setting up of high-cost transmission networks. Institutional issues—such as inadequate state plan allocation, lack of coordination among planning and implementing agencies, delays in clearances and allotment of private sector projects, low priority of utilities to take on the projects, and lack of clear policy for private sector participation—have slowed growth in small hydropower generation capacity. Success of small hydro development depends to a large extent on local capacity-building programmes and setting up institutional arrangements for demonstration, training, and awareness programmes that help in technology adaptation and maintenance.

(e) Solar Energy

Solar energy represents the Earth's most abundant energy resource. In India the average global radiation is around five kWh per square metre per day, with about 250–300 clear sunny days in a year in most arts of the country, and the same can be judiciously exploited to meet ever-increasing energy requirements. Solar energy can be used in diverse ways and is divided into two principal technologies for its utilization, viz., solar thermal and solar photovoltaic. In India, solar photovoltaic (SPV), with an aggregate installed capacity of 151 megawatt peak (MWp), has a three percent contribution of the renewable energy-based power generation capacity. The solar thermal power generation potential in India is about 20 MW per km^2. Estimates indicate a potential of 800 MW per year for solar thermal-based power generation in India using the period 2010 to 2015, with worldwide advancements in parabolic trough technology. The government of India has proposed a solar thermal power plant project of about 140 MW to be set up in Rajasthan as a demonstration project. The project will consist of a 35-MW solar thermal power component using anabolic trough collectors and a 105-MW combined-cycle component using gas turbines and naphtha as fuel. This will be the first project of its kind in India and the largest such project in the world. However, the growth of solar technology in India has been primarily restricted by its high investment costs. Electricity generation costs on a life cycle basis for SPV is over ten times higher compared to coal-fired thermal power.

The commercialization of SPV technology involves high transaction costs, such as expensive and time-consuming project identification; challenging project implementation in a number of small-scale installations; the high costs of credit collection and risks associated with marketing, contracting, and information collection; conducting promotion campaigns and creating after-sales service infrastructure; and the cost of co-financing, conducting feasibility studies, and developing business plans. Solar energy is particularly appropriate for remote communities in India, where it could reduce or even replace the current stand-alone power generation systems based mainly on oil products. However, government programs require fine-tuning in order to pave the way for photovoltaic power to spread to more isolated parts of the country. Otherwise, fossil fuels will prevail within the context of mandatory universal access to electricity.

(f) Agriculture and Livestock

There are mainly two types of energy being used by the agricultural sector, namely, high-speed diesel (HSD)/light diesel oil (LDO) and electricity. Although the amount of energy consumed has increased significantly over the years, the relative share of oil and electricity remained more or less constant at around 65:35 ratio. The main areas of energy conservation in the agricultural sector are in diesel and electrically operated pump sets, with 16 million of these sets being reported to be in operation in 1991/92. The estimated savings potential considers avoiding such common drawbacks as improper selection of pumps and prime mover, improper installation, poor suction lift characteristics of the pumps, high friction losses in the foot valve and piping system, air leakage in the suction pipes, improper maintenance, and servicing, etc.

Given India's extensive rice cultivation and bovine population, the agricultural sector is a major source of GHG emissions, with methane (CH_4) as the dominant GHG. The activities responsible for methane emissions in the agriculture sector include rice cultivation, enteric fermentation in animals, manure management, and burning of agricultural crop residues. While there is some scope in decreasing emissions from paddy fields through special irrigation strategies and appropriate choice of cultivars, the main target initially could be enteric fermentation. Its methane emissions of approximately eight million tonnes per year can be reduced through proper feed management. A reduction of one million metric tonnes per year in this category means an appreciable reduction of 21,000,000 metric tonnes of CO_2e per year. The first priority, therefore, could be the area of animal fermentation. The area of CH_4 emissions from paddy fields could also figure as a priority, if the effects on changing CO_2 concentrations are considered. In the agriculture sector, GHG emissions are usually exacerbated by inefficient energy consumption practices, tools, and techniques; low digestibility of animal feeds; exposed treatment plants used for anaerobic decomposition of animal manure; cultivation of rice under continuous submergence; and the low efficiency of fertilizer use in agricultural soils.

(g) Municipal Solid Waste Management

Present municipal solid waste management and utilization in India has severe limitations. The very high rate of urban growth is a major reason for the increased solid waste management problems. Approximately, 0.1 million tonnes of municipal solid waste are generated in India every day. This translates to approximately 36.5 million tonnes of waste annually. Per capita waste generation in major Indian cities ranges from 0.2 kg to 0.6 kg per day. Urban local governments spend approximately Rs. 500 to Rs. 1,500 per tonne on solid waste for collection, transportation, treatment, and disposal. About 60–70 percent of this amount is spent on collection, 20–30 percent on transportation, and less than 5 percent on final disposal.

Waste collection efficiency in Indian cities ranges from 50–90 percent. Out of the total municipal waste collected, an average of 94 percent is dumped on land and 5 percent is composted. Methane emissions from solid waste are a major cause of concern in developed countries, and these countries have started recovering methane from landfills and re-using it for energy recovery. In India, however, there are no such facilities and solid waste is simply discarded in open dump-yards, which does not provide an opportunity for methane recovery. In view of this, other options for waste disposal and recovery of resources from the waste should be tried out so that methane emissions from municipal solid waste are minimized.

In the context of poor enforcement of regulations in India and the lack of financial resources of municipalities in the country, it is expected that there will be only slow, incremental improvement in the way municipal solid waste (MSW) is managed in the foreseeable future. This would mean that unmanaged solid waste disposal sites would continue to be the dominant means of waste management, leading to significant adverse environmental impact. Due to the stacking of waste over the years, anaerobic conditions develop and hence these dump sites generate large quantities of biogas containing a sizeable proportion of methane. Based on secondary data on the type of solid waste produced, per capita waste produced, and the bio-chemical oxygen demand (BOD) content of the waste, it is estimated that in 1994, 0.582

million tonnes of methane were emitted by this source. Also, it is expected that with the rapid development that India is presently experiencing, methane emissions from this source will rise significantly in the future.

GHG emissions produced during municipal solid waste management can be tackled by using different measures that ultimately can lead to opportunities under the CDM. The various technological options available to treat MSW are as follows:

- Sanitary landfill
- Incineration
- Composting
- Gasification
- Anaerobic digestion
- Other types

Sanitary landfill is the scientific method of dealing with MSW in which maturity of the waste material is achieved faster and hence gas collection starts even during the landfill procedure. Incineration technology is the controlled combustion of waste with the recovery of heat to produce steam that in turn can be utilized for generating power. Reductions of about 75 percent in weight and 90 percent in volume are achieved through burning. Composting is a biological process in which the organic portion of wastes is stabilized (or decomposed) to produce a useful product called compost. A gasification technology involves pyrolysis under limited air in the first stage, followed by higher temperature reactions of the pyrolysis products to generate low molecular weight gases with calorific values of 1,000–1,200 kcal/Nm^3. These gases could be used in internal combustion engines for direct power generation or in boilers for steam generation to produce power. In bio-methanation, the putrescible fraction of waste is digested anaerobically (in absence of air) in specially designed digesters. Under this active bacterial activity, the digested pulp produces the combustible gas methane and inert gas carbon dioxide. The remaining digestate is a good quality soil conditioner. Other technologies available are pelletization, pyro-plasma, and flash pyrolysis. The gasification and bio-methanation technologies are relatively new and appear to be suitable for the treatment of MSW. They do not cause atmospheric pollution, work at a higher efficiency, and at the same time have higher GHG reduction potential (approximately 1.0 to 1.2 tonnes of CO_2e per tonne of MSW treated).

(h) Land use, Land-use Change and Forestry (LULUCF)

Land-use patterns are influenced by a variety of factors such as population density, expanding urbanization, industrial growth, agriculture, grazing needs, irrigation demands, and natural calamities like floods and droughts. Despite these stresses, the area covered by forests has increased steadily due to the proactive reforestation and afforestation programmes of the government of India over the years aimed at sustainable development. India's forest cover in 1994 was assessed in 1997 by the forest survey of India through interpreting satellite imagery at 633,397 km^2, increasing to 675,538 km^2 for the year 2000 (as per the assessment conducted in 2001). An estimated 2.46 billion trees outside forests contributed an additional area of 81,472 km^2, making the total tree and forest cover to be 23 percent of the country's geographical area in 2000, while 44 percent was the net planted area. The remaining one-third is almost equally distributed between fallow land, non-agriculture land, and barren land. The forests in India are a

source of fuel and fodder for rural people, an industrial output for a growing economy, habitat for thousands of plant and animal species, a sink for CO_2 emissions, and a protective cover for soil. Considering gross CO_2 emissions and removals for the LULUCF sector, the net CO_2 emissions for India for the inventory year of 1994 was estimated to be 14.292 million tonnes of CO_2, which is less than 2 percent of national CO_2 emissions.

The Role of Government Authorities

India is a party to the United Nations Framework Convention on Climate Change (UNFCCC), and the objective of the convention is to achieve stabilization of greenhouse gas (GHG) concentrations in the atmosphere at a level that would prevent dangerous anthropogenic interference with the world's climate system. To strengthen the developed country commitments under the UNFCCC, the Parties adopted the Kyoto Protocol, which commits developed country Parties to reduce their emissions of GHGs to an average of approximately 5.2 percent below 1990 levels between the years 2008-12. The government of India has been very active in the climate change negotiations since the inception of the UNFCCC. Although it does not have a GHG emissions reduction target, it has actively taken steps to address the climate change issue.

The policy on climate change in India is based on the following basic principles:

1. It recognizes that the Earth's climate and the adverse effects of climate change are a common concern of all humanity and calls for the widest cooperation from all countries, in accordance with their common but differentiated responsibilities, respective capabilities, and their social and economic conditions.
2. The primary responsibility of reducing GHG emissions is that of developed countries and they should take the lead.
3. Economic and social development and poverty eradication are the first and overriding priorities of the developing countries.
4. The developed countries should transfer resources and technologies at favourable terms to the developing world, thereby facilitating developing countries to move towards sustainable development.

The Kyoto Protocol provides for quantified emission limitations and reduction commitments for the developed countries and mechanisms to facilitate compliance with their targets, as well as reporting and reviewing. It lists six greenhouse gases: carbon dioxide (CO_2), methane (NH_4), nitrous oxide (N_2O), hydrofluorocarbons (HFCs), perfluorocarbons (PFCs), and sulfur hexafluoride (SF_6).

The Kyoto Protocol was ratified by the Government of India in August 2002 to strengthen global efforts to meet the challenges adversely affecting the ecosystem. One of India's objectives in acceding to the protocol was to fulfil the prerequisites for implementation of projects under the Clean Development Mechanism (CDM) in accordance with national sustainable priorities, where under developed countries would implement GHG reduction project activities in developing countries—where the costs of GHG reduction project activities are usually much lower—with the purpose of assisting developing country Parties in achieving sustainable development (the ultimate objective of the UNFCCC), and to assist developed country Parties in achieving compliance with their quantified emissions limitation and reduction commitments.

The Seventh Conference of the Parties (COP 7) to the UNFCCC, held in Marrakesh, Morocco, in 2001, brought out, among other things, modalities and procedures for the CDM. It was decided that Parties participating in the CDM should designate a national CDM authority, and that as per the CDM project cycle, a project proposal should include written approval of voluntary participation from the designated national authority (DNA) of each country and confirmation that the project activity assists the host country in achieving sustainable development.

The Seventh Conference of Parties (COP-7) to the UNFCCC decided that Parties participating in CDM should designate a National Authority for the CDM and as per the CDM project cycle, a project proposal should include written approval of voluntary participation from the Designated National Authority of each country and confirmation that the project activity assists the host country in achieving sustainable development. Accordingly the Central Government constituted the National Clean Development Mechanism (CDM) Authority for the purpose of protecting and improving the quality of environment in terms of the Kyoto Protocol (http://cdmindia.nic.in/cdm_india.htm).

The composition of the "National Clean Development Mechanism (CDM) Authority" is as follows:

1.	Secretary (Environment and Forests)	Chairperson
2.	Foreign Secretary or his nominee	Member
3.	Finance Secretary or his nominee	Member
4.	Secretary, Industrial Policy and Promotion or his nominee	Member
5.	Secretary, Ministry of Non Conventional Energy Sources or his nominee	Member
6.	Secretary, Ministry of Power or his nominee	Member
7.	Secretary, Planning Commission or his nominee	Member

8. Joint Secretary (Climate Change), Ministry of Environment and Forests — Member
9. Director (Climate Change), Ministry of Environment and Forests — Member-Secretary

The National Clean Development Mechanism (CDM) Authority receives projects for evaluation and approval as per the guidelines and general criteria laid down in the relevant rules and modalities pertaining to CDM in addition to the guidelines issued by the Clean Development Mechanism Executive Board and Conference of Parties serving as Meeting of Parties to the United Nations Framework Convention on Climate Change. The evaluation process of CDM projects includes an assessment of the probability of eventual successful implementation of CDM projects and evaluation of extent to which projects meet the sustainable development objectives, as it would seek to prioritize projects in accordance with national priorities. The National Clean Development Mechanism (CDM) Authority can recommend certain additional requirements to ensure that the project proposals meet the national sustainable development priorities and comply with the legal framework so as to ensure that the projects are compatible with the local priorities and stakeholders have been duly consulted.

The Authority ensures that in the event of project proposals competing for same source of investment, projects with higher sustainable development benefits and which are likely to succeed are accorded higher priority. The Authority also carries out the financial review of project proposals to ensure that the project proposals do not involve diversion of official development assistance in accordance with modalities and procedures for Clean Development Mechanism and also ensure that the market environment of the CDM project is not conducive to under-valuation of Certified Emission Reduction (CERs) particularly for externally aided projects.

The Authority carries out activities to ensure that the project developers have reliable information relating to all aspects of Clean Development Mechanism which include creating databases on organizations designated for carrying out activities like validation of CDM project proposals and monitoring and verification of project activities, and to collect, compile and publish technical and statistical data relating to CDM initiatives in India.

The Member-Secretary of the National Clean Development Mechanism (CDM) Authority is responsible for day-to-day activities of the Authority including constituting committees or sub-groups to coordinate and examine the proposals or to get detailed examination of the project proposals.

The National Clean Development Mechanism (CDM) Authority has the powers:

(a) To invite officials and experts from Government, financial institutions, consultancy organizations, non-governmental organizations, civil society, legal profession, industry and commerce, as it may deem necessary for technical and professional inputs and may co-opt other members depending upon need;

(b) To interact with concerned authorities, institutions, individual stakeholders for matters relating to CDM;

(c) To take up any environmental issues pertaining to CDM or Sustainable Development projects as may be referred to it by the Central Government; and

(d) To recommend guidelines to the Central Government for consideration of projects and principles to be followed for according host country approval.

Structure of India's Designated National Authority

After India's ratification of the Kyoto Protocol, the central government of India constituted the National CDM Authority (NCA) for the purpose of protecting and improving the quality of the environment in terms of the protocol. The government accorded approval for setting up the DNA in December 2003, coinciding with the Ninth Conference of Parties (COP 9). News from the press indicated that the DNA would be hosted by the Ministry of Environment and Forests (MoEF) and headed by the secretary of the MoEF (The Economic Times, 2003). Thus, the MoEF, which is the nodal agency for the subject issues of climate change, was made the host of the DNA for dealing with the CDM and other climate change issues. It is the ministry that is responsible for developing an institutional framework for developing CDM projects in the country, as well as endorsing CDM projects.

The DNA was set up by the MoEF under the powers conferred on it by sub-section (1) and (3) of section 3 of the Environment Protection Act (EPA), 1986 (29 of 1986) (see appendix VIII). The composition of India's NCA is shown blow:

Composition of the National Clean Development Mechanism Authority (NCA)

1.	Secretary (Environment and Forests)	Chairperson
2.	Foreign Secretary or nominee	Member
3.	Finance Secretary or nominee	Member
4.	Secretary, Industrial Policy and Promotion or nominee	Member
5.	Secretary, Ministry of Non Conventional Energy Sources or nominee	Member
6.	Secretary, Ministry of Power or nominee	Member
7.	Secretary, Planning Commission or nominee	Member
8.	Joint Secretary (climate change), Ministry of Environment and Forests	Member
9.	Director, (climate change), Ministry of Environment and Forests	Member-secretary

The Powers and Functions of the National CDM Authority

1. The National CDM Authority (NCA) can take measures related to matters referred to in clauses (viii), (ix), (x), and (xii) of sub-section (2) of section 3 of the EPA, 1986 and may issue directions under section 5 thereof. Section 3 of the EPA outlines the power of the central government to take measures to protect and improve the environment. Section 5 under section 3 discusses the power to give directions. Under the act, the central government may issue directions to any person, officer, or any authority for non-compliance to the provisions of the act, and such person, officer, or authority shall be bound to comply with such directions. It further clarifies that issuing directions under this section includes (a) the closure, prohibition, or regulation of any industry, operation, or process; and/or (b) stoppage or regulation of the supply of electricity or water or any other service.
2. (a) The NCA accepts project proposals for evaluation and approval as per the guidelines and general criteria laid out in the relevant rules and modalities pertaining to the CDM, in addition to the guidelines issued by the CDM Executive Board, and the UNFCCC.

   (b) The CDM project evaluation process will include an assessment of the probability of successful implementation of proposed CDM projects and evaluation of the extent to which projects meet the country's national priorities and sustainable development objectives.

   India's planning processes and global climate change concerns are intricately linked. Taking care of the national planning objectives would include addressing many of the concerns about climate change. It is the prerogative of each host country to ensure that the project is in line with its national priorities, especially from the point of view of sustainable development. A planned approach to development has been a central part of India's democracy, as reflected in its national five-year plans, annual departmental plans, and prospective plans of various ministries of the central and state governments. For the last five-and-a-half decades, the guiding objectives of India's planning process have been sustained economic growth; poverty alleviation; food, health, education, and shelter for all; containing population growth; generating employment; self-reliance; and public participation in planning, programme implementation, and infrastructure development.

3. (a) The NCA can recommend certain additional requirements to ensure that CDM project proposals meet India's sustainable development priorities and comply with the legal framework, so as to ensure that projects are compatible with local priorities and that stakeholders have been duly consulted.

   (b) It must ensure that in the event of project proposals competing for the same source of investment, projects with higher sustainable development benefits and which are more likely to succeed are accorded higher priority.

4. (a) The NCA will maintain a registry of CDM projects approved and their certified emissions reduction (CER) potential, and confirm that these have been realized. NCA will also ensure the accurate accounting of the issuance, holding, and transfer of CERs by Parties.
   (b) It will carry out activities to ensure that project developers have reliable information relating to all aspects of the CDM, which include creating databases on organizations designated to carry out activities such as validation of CDM project proposals and monitoring and verification of project activities.
   (c) It will collect, compile, and publish technical and statistical data relating to CDM initiatives in India.

The member-secretary of the National CDM Authority shall be responsible for day-to-day activities of the authority, including organizing committees or sub-groups to coordinate and examine proposals, or to procure detailed examination of project proposals by experts or consulting organizations for consideration by the authority, or to engage or hire any consultant or specialist on a contract basis for such remuneration as may be approved by the central government, and submit reports to the UNFCCC and other authorities.

The National CDM Authority shall have power to do the following:

1. Invite whichever officials or experts from government, financial institutions, consultancy organizations, non-governmental organizations, civil society, the legal profession, industry, and commerce that it may deem necessary for technical and professional input and it may co-opt other members depending upon need.
2. Interact with concerned authorities, institutions, and individual stakeholders for matters relating to the CDM.
3. Address any environmental issues pertaining to the CDM or sustainable development projects referred by the central government.
4. Recommend guidelines to the central government for consideration of projects and principles to be followed for according host country approval.

The National CDM Authority must furnish a report to the central government at least once every three months to include updates on any of the following activities it is responsible for:

- Disseminate information on the NCA and CDM processes;
- Maintain records on projects under consideration for CDM benefit;
- Act as an interface between the NCA and project promoters;
- Communicate with the CDM Executive Board;
- Keep track of various baseline methodologies approved by the board

and provide further guidance on how such baselines can be adapted in the context of India;

- Other administrative roles such as collection of processing fees, etc.
- General information on the CDM;
- Status of projects received, approved, rejected, and other relevant details;
- Further develop the CDM endorsement process being followed by the DNA and periodically communicate any modifications to the public;
- Follow the PCN format for CDM endorsement;
- Links to other relevant CDM Web sites;
- Maintain relevant data in the form necessary for baselines and maintain links with relevant domestic organizations that collect authentic data, such as relevant state/central authorities;
- Collect information on domestic organizations working in the CDM field;
- Link to financial institution Web sites, insurance sector contacts, etc.

Other line ministries and government departments are invited to become involved on an "as needed" basis for their professional input (i.e., technical, sectoral, and legal experts, financial and academic institutions, and non-governmental organizations), depending on the type of project proposals received for endorsement.

The MoEF constituted the Working Group on the UNFCCC to deliberate upon measures and positions that should be taken regarding the various issues emerging out of climate change negotiations. A separate group on the Kyoto mechanisms has also been constituted. At present, the ministry bears all the costs involved in the functioning of the DNA. The NCA at present does not plan to charge project developers any fee for project assessment, approval, or other services. As well, in order to encourage India's industries to participate in the CDM, the NCA does not plan to levy a tax on CERs at the moment, but it has not ruled out the possibility of levying a tax once the CDM market develops in due course.

CDM Promotion at the State Level

Some of India's state governments have shown a keen interest in the CDM, and they have established state-level CDM promotional cells which can be divided into short term and long term frameworks. In the short term, the state-level CDM cells will focus on the following:

- Capacity building of professionals belonging to the CDM cells for developing achievable CDM projects in the respective state;
- Networking among government departments and institutions;
- Enhancing participation of the private sector in the CDM, thereby generating/developing more CDM projects;
- Enhancing its capacity and skills in examining and analyzing potential CDM projects;

- Providing project proponents with match-making services with buyers.

In the long-run, a CDM promotional cell at the state-level is expected to have capacity in the following areas:

- Understanding the CDM project concept, characteristics, project cycle, baselines, crediting period, contractual issues, banking of CERs, internal CERs registry, and associated procedures;
- Identifying achievable CDM projects and suitable promoters;
- Facilitating appropriate services for CDM project promoters in project identification and formulating potential CDM projects, baseline development, quantification of emissions reductions, identifying operational entities, and monitoring project performance;
- Continuing to explore possibilities to simplify procedures and optimize transaction costs associated with developing projects;
- Identifying and providing intensive capacity building for stakeholders for a better level of understanding of the CDM and for developing projects;
- Disseminating information on CDM criteria and procedures;
- Analyzing national/local sustainable development priorities for identifying projects in consonance with such priorities;
- Maintaining a database of technologies, consultants, operational entities, funding agencies, CER buyers, etc.
- Developing state-specific appendices to supplement the DNA's CDM manual for guidance on eligibility, baselines, estimates of transaction costs, legal issues (contracts, insurance, taxation of CERs, etc.), CERs market, project development, etc.
- Analyzing the state sectoral potential under the CDM;
- Developing a portfolio of CDM projects to offer potential buyers;
- Updating stakeholders with ongoing changes in the CDM modalities and procedures of the CDM Executive Board, Conference of Parties, and subsidiary bodies.

A few states in India have taken the lead and established state nodal cells for promoting the CDM:

- Andhra Pradesh
- Madhya Pradesh
- West Bengal

Procedures of Registration of Proposed CDM Project Activity

(UNFCCC/CCNUCC-CDM – Executive Board, Version 02/November 2003)

1. In accordance with paragraph 40(f) of the CDM modalities and procedures (CDM M&P), the request for registration of a proposed CDM project activity shall be in the form of a validation report which includes the project design document, the written approval of the host Party and an explanation of how the DOE has taken due account of

public comments received on the CDM-PDD.

2. A designated operational entity shall submit its validation report using the "CDM project activity registration and validation report form" (F-CDM-REG) (attached to these procedures) to request for registration of a proposed project activity.
3. In order to ensure transparency and efficiency of the registration process:
   (a) A request for registration will only be processed after the secretariat has determined that all information and documentation requested in the registration form has been provided by the DOE;
   (b) The date of receipt of a request for registration is the date when the deposit of the registration fee indicated in the registration form has been received by the secretariat;
   (c) A request for registration (as defined in paragraph 40(f) of the CDM modalities and procedures) shall be made publicly available through the UNFCCC CDM web site (either by a link to the DOE web site or by being directly posted) for a period of eight (8) weeks. The secretariat shall announce a request for registration of a proposed CDM project activity on the UNFCCC CDM web site and in the CDM news facility. The announcement shall specify where the request for registration can be found, the name of the proposed CDM project activity and the first and last day of the eight-week period. The secretariat shall notify the DOE requesting a registration when and where the request for registration is posted.
   (d) Unless there is a request for review, a request for registration shall, after eight weeks, be marked in the UNFCC CDM web site as "registration completed" and the corresponding proposed CDM project activity and related public documents recorded/displayed as registered.

CDM Approval Procedures and Requirements

Article 12 of the Kyoto Protocol states that the purpose of the CDM is "to assist Parties not included in Annex I in achieving sustainable development and in contributing to the ultimate objective of the Convention, and to assist Parties included in Annex I in achieving compliance with their quantified emissions limitation and reduction commitments (http://www.cdm-egypt.org/ProjectsAP.htm)." Under the CDM, Annex I countries can buy the credits for the emissions reduced through projects in developing countries and use them to meet their Kyoto Protocol targets. All CDM projects must result in a net reduction of GHGs, as in the case of energy-efficiency improvements, renewable energy generation, or carbon sequestration through afforestation and reforestation, and it is essential that GHG mitigation projects be in consonance with the host country's sustainable development priorities.

Kyoto Protocol Requirements of the CDM

CDM projects may be categorized under the energy, transportation, building, and renewable energy sectors (solar, wind, biomass). But not all projects that result in emission reductions are eligible to qualify as a CDM project. In order to qualify, projects need to have certain characteristics, which are explicit or implicit in the Kyoto Protocol. At the project level, the government of India needs to do the following:

1. Ensure that CDM proposals conform to Article 12 of the Kyoto Protocol.
2. Facilitate single-window clearance of CDM projects.

CDM projects need to be designed in such a way that they achieve the following:

- Are approved by the governments of the participating Parties and are consistent with and supportive of national environment and development priorities and strategies.
- Assist host countries (developing) to achieve sustainable development
- Are host country driven?
- Result in real, measurable, and long-term benefits.
- Reduce emissions that are additional to any that would occur in the absence of the certified project activity.
- Involve private and/or public entities.
- Have voluntary participation, approved by each party involved.
- Contribute to the ultimate objective of the United Nations Convention on Climate Change (UNFCCC).
- Benefit developing countries from project activities resulting in certified emission reductions.

Assessment of whether or not CDM projects fulfill these criteria requires comparing projects against a baseline. In many cases, it is counter-factual to construct a baseline that may never actually happen. Assessing environmental benefits also requires establishing system boundaries appropriate to the scale and complexity of the project to assess "leakage." In addition, the CDM project needs to be environmentally sound, economically efficient, and equitable.

Endorsement Criteria

As per the rules of the UNFCCC, it is the host country's prerogative to decide whether or not a proposed project activity assists in achieving sustainable development, and accordingly approve projects. Hence, the two aspects that are examined when an India's designated national authority (DNA) evaluates a CDM proposal are as follows:

- Does the project meet national priorities, set by the India's five year planning process?
- Does it contribute to sustainable development of the country?

National Priorities

Since the Stockholm Conference on the Human Environment and Development in 1972, the government of India has instituted various regulatory and promotional measures for protecting the environment and promoting sustainable development in the country. A planned approach to development has been the central process of India's democracy, as reflected in its national five-year plans, annual departmental plans, and the prospective plans of various ministries at the central and state government levels. For the last five-and-a-half decades, the guiding objectives of India's planning process have been sustained economic growth; poverty alleviation; food, health, education, and shelter for all; containing population growth; employment generation; self-reliance; and public participation in planning, programme implementation, and infrastructure development.

Sustainable Development and its Priorities

Since the adoption of the UNFCCC and Agenda 21 during the Earth Summit held in June 1992 in Rio de Janeiro, many nations around the world have initiated a range of processes aimed at achieving more sustainable development. The concept of sustainable development was first introduced by the World Conservation Union (IUCN) in 1980 (IUCN 1997). The World Commission on Environment and Development, popularly known as the Brundtland Commission, defined sustainable development in 1987 as a development process that "meets the needs of the present without compromising the ability of the future generations to meet their own needs." (Bruntland, 1997) Sustainable development can also be defined as an evolving process that improves the economy, the environment, and society for the benefit of current and future generations.

The four pillars of sustainable development are social, economic, environmental, and technological well being (World Resources Institute 1992). Poverty alleviation, economic and social development, and environmental protection are overriding sustainable development goals and priorities of India. It is the prerogative of the host party to assess whether or not proposed CDM project activities will assist sustainable development.

Projects Involving ODA

Overseas development assistance (ODA) is a grant/loan provided by developed countries to developing countries to address issues of critical importance, such as poverty alleviation programmes, etc. Using ODA by developed countries for acquiring credits for certified emission reductions (CERs) would mean that (1) high-priority issues like poverty alleviation could be neglected and diverted to sectors like transport or energy, (2) may lead to trade disputes on grounds of unfair subsidization, and (3) may force developing countries to tend to lose the ODA benefit. The Kyoto Protocol already stipulates that no funds under an ODA scheme can be diverted to CDM projects. The flow of funds to CDM projects should be additional to ODA support. India's National CDM Authority (NCA) ensures that ODA is not used either in part or in full to procure CERs from projects in India. Applicants for CDM benefits therefore need to demonstrate this aspect.

Laws and Regulatory Aspects

Areas of law implicated in the regulation and reduction of greenhouse gas emissions include tax, capital markets, trade, corporate, property, administrative, tort, criminal, finance and environmental. A multidisciplinary approach is essential to anticipating and successfully managing the risks inherent in this field. Governments at all levels increasingly are seeking to regulate the emission of greenhouse gases. Many U.S. states already have enacted legislation to address climate change, such as California's Global Warming Solutions Act of 2006 that aims to reduce statewide emissions to 1990 levels by 2020. Several cap-and-trade bills are working their way through the U.S. Congress, and the U.S. Environmental Protection Agency is assessing the implications of regulating carbon dioxide under existing environmental laws and regulations. In addition, new or expanded regulatory programs to promote low-carbon energy technologies, carbon capture and storage, and other climate-related technologies are being introduced at federal and state levels.

Worldwide, the European Union introduced a mandatory emissions trading system in 2005 and is working to extend its limits on carbon dioxide emissions to 2020. At the international level, negotiations are underway on a new climate change treaty to take effect after the Kyoto Protocol's first commitment period ends in 2012. Although the outcome is still uncertain, countries are attempting to reach a new agreement by the end of 2009 (http://www.hunton.com/practices/practice_detail.aspx?gr_H4ID=1087).

There is no legislation which specifically incorporates CDM concepts into Indian law. Accordingly, this chapter includes a general discussion on laws and regulations (as applicable in India) relating to the establishment of a business presence, foreign investment, exchange control, energy, and environmental issues. These laws and regulations apply uniformly to all inbound investment, regardless of its intended purpose or the country of its origin. This section also discusses certain contractual issues relevant in drafting and negotiating agreements to purchase certified emission reductions (CER) generated from CDM projects in India. Additionally, it addresses the issues of ownership and title to such CERs.

Establishing a Business Presence in India: Entry strategies

An overseas entity planning to invest in India may do so either as a foreign company (i.e., incorporated in any country outside India) or as an Indian company.

As a Foreign Company: A foreign company could do business in India by establishing a project office, a branch office, or a liaison office in India. In accordance with prevailing exchange control regulations, the establishment of such offices is regulated by the central bank, i.e., the Reserve Bank of India (RBI). If foreign companies secure a contract from an Indian company to execute a CDM project in India, they may set up a temporary project or site office in India, subject to specific approval from the RBI. Such approval is generally accorded in respect to projects approved by appropriate Indian authorities, where the projects are financed by inward remittances from overseas or by a multilateral or bilateral international financial institution. To qualify as a CDM project, a project would need to be approved by the Ministry of Environment and Forests (MoEF), which is the designated national authority (DNA) for India.

Liaison Office: The scope of activities permitted to be carried out by a liaison office is very limited, since they are not permitted to undertake any commercial activities or earn any income in India.

Branch Office: A branch office is allowed to perform a wider range of activities, but may not perform manufacturing or processing activities, either directly or indirectly.

Therefore, a liaison office and a branch office would not be suitable in the context of foreign investors seeking to invest in CDM projects in India.

Acquiring a stake in an existing company: Until very recently, the acquisition of shares by an overseas investor from an Indian entity would have required the approval from the Foreign Investment Promotion Board (FIPB) of the Ministry of Commerce and Industry (MoCI). This regulation has been liberalized; now, the transfer of shares from a resident entity to a non-resident entity does not require FIPB approval and is permitted under the automatic route, subject to compliance with the foreign direct investment (FDI) policy of the government of India.

Foreign Investment Policy

India's economic policies are designed to attract significant capital inflows into India on a sustained basis and to encourage technical collaboration between Indian and foreign entities. Policy initiatives taken over the last few years have resulted in inflows of foreign investment in diverse sectors of the economy. Incorporating a company in India is a two-step process. First, the registrar of companies has to approve the proposed name of the company. Then, the necessary incorporation documents such as the memorandum and articles of association of the company and other forms prescribed under the Companies Act are to be prepared. The company must be incorporated within six months of the RoC approving its name. Registration fees payable to the RoC for incorporation depend upon the authorized or nominal share capital of the company. It usually takes 10–40 days to incorporate a company from the date of submitting the name approval form to the RoC. Companies can either be incorporated as private or public limited companies. Private limited companies require a minimum of two shareholders; this number is seven for public limited companies. It is possible to incorporate a company as a private company with both the initial shareholders of such a company being foreign corporate bodies (whether public or private). As the incorporation process requires signing several documents, each foreign corporate body must, through a board resolution, authorize an individual as its representative to undertake the incorporation process, including subscribing to the memorandum and articles of association of the proposed company. Such an authorized signatory could either be a non-resident individual or a resident Indian.

(a) Economic Policies and Incentives for Foreign Investment

Following are the select notable features of India's Foreign Investment policies and incentives:

- No government approval required for FDI in virtually all the sectors/ activities, except a small negative list notified by the government.
- The government has notified "Sector Specific Guidelines for FDI" wherein investments up to specified sectoral caps are covered under the automatic route, with a few exceptions.
- The FIPB considers proposals for foreign participation that do not qualify for automatic approval.
- Decisions on all foreign investment proposals are usually taken within 30 days of application.
- Free repatriation of capital investment and profits thereon is permitted, provided the original investment was made in convertible foreign exchange.
- Use of foreign brand names/trademarks for sale of goods in India is permitted.
- Indian capital markets are open to foreign institutional investors.
- Indian companies are permitted to raise funds from international capital markets.
- Special investment and tax incentives are given for exports and sectors such as power, electronics, software, and food processing.
- "Single window" clearance facilities and "investor escort services" have been provided in various states to simplify the approval process for new ventures.

(b) Foreign Direct Investment (FDI)

India's policy on foreign direct investment (FDI) sets out the extent of foreign equity permitted in various sectors of the economy in India and the conditions attached to such investments. While foreign entities can freely invest in most sectors, there are still some sectors where the policy either prohibits foreign investments (e.g., in real estate, defence, railway transport, or atomic energy) or does not permit FDI be yond a prescribed investment ceiling (e.g., in the insurance sector, FDI is permitted only up to 26 percent). FDI in relation to most activities can be brought in through the automatic route, i.e., without any prior approval of the government of India or the RBI. For the rest, FDI is permitted with prior government approval, which is issued by the FIPB. It is anticipated that much of the CDM investment in India will be in the power and renewable energy-related infrastructure sectors, in which 100 percent FDI is permitted under the automatic route.

Foreign Investments: Mode of Participation

1. Automatic Route

(a) *New ventures:* 100 percent FDI is permitted under the automatic route for all activities except the following:

(i) Foreign equity investments beyond the specified sectoral caps or in sectors, in which FDI is prohibited,

(ii) Investment proposals for activities that require an industrial license, which is required in the following cases:

- For manufacturing those items requiring an industrial licence under the Industries (Development and Regulation) Act 1951, including those industries reserved for the public sector;
- Foreign investment in excess of 24 percent of the equity capital of a unit manufacturing items reserved for small-scale industries;
- All items which require an industrial licence in terms of the locational policy notified by the government under the New Industrial Policy, 1991, and

(iii) Investment proposals by a foreign investor for activities in a sector in which such an investor already has an existing joint venture, technology transfer agreement, or trademark agreement in India. This is in accordance with Press Note No. 1, dated January 12, 2005, which recently replaced an earlier and more restrictive set of regulations contained in Press Note 18. The new regulations provide that where a foreign investor has an existing business presence in India in the form of a joint venture, technology transfer agreement, or trademark agreement, and wishes to make a new investment in the same field, it must seek approval from the FIPB.

However, this restriction does not apply in the following cases where:

- Foreign investments are made by venture capital funds registered with the Securities and Exchange Board of India,
- Investments by either party in an existing joint venture are less than 3 percent, and
- The existing venture is defunct or sick.

(iv) Proposals relating to acquisition of existing shares in an Indian company by a foreign investor.

(b) *Existing Ventures:* Foreign investment in existing Indian companies is also permitted under the automatic route.

The following additional requirements are that:

(i) the funds for the foreign investment should be remitted in foreign currency,

(ii) The proposed foreign investment should be in a sector in which FDI is not prohibited and should not result in breach of any sectoral cap. If the investment fails to meet one or more of these requirements, the proposal will require approval from the FIPB.

(c) *Reporting requirements:* For foreign investments made under the automatic route, irrespective of whether they are in new ventures or existing companies, the company receiving the investments is required to notify and file certain documents with the relevant regional office of the RBI remittances within 30 days.

2. Non-automatic Route

The government of India considers (on merit) and gives composite clearance to proposals for foreign investment that are not covered under the automatic approval route. These include proposals seeking other industrial approvals such as industrial license, technical collaboration etc., along with approval for foreign investment.

Foreign Technology Agreements

The government has set up a Foreign Investment Implementation Authority in the Ministry of Commerce and Industry for providing a single window to foreign investors to facilitate quick translation of foreign investment approvals. Foreign entities may also participate in the Indian CDM market through the provision of technology. Transfer of CDM-related technology would take place through a foreign technology collaboration agreement between the foreign entity and the Indian company engaged in the proposed CDM project. The contribution of technology could also be by a company in which the foreign entity is a shareholder.

1. Automatic or Approval Route

Under the FDI policy, foreign technology agreements are permitted either through the automatic route or with prior approval from the government through the Project Approval Board (PAB). Government approval is required for foreign technology agreements in the following:

- Areas which attract compulsory licensing
- Areas of manufacture reserved for the small-scale sector
- By foreign entities in a field in which they already have a previous/

existing joint venture, technology transfer, or trademark agreement in India

- Where (1) lump sum payments under the foreign technology agreement exceed US$2 million or (2) royalty payable exceeds 5 percent of domestic sales and 8 percent of exports, subject to a total royalty payment of 8 percent on sales over a ten-year period.

Procedurally, applications under the automatic route must be registered with the RBI and applications submitted to the PAB, Department of Industrial Policy and Promotion, Ministry of Commerce and Industry.

1. *Payments under Foreign Technology Collaboration Agreements*

Payments under foreign technology agreements may be repatriated (as a royalty or lump sum fees) through the automatic route, i.e., without RBI approval under the following conditions:

- Lump sum payments under such foreign technology agreements do not exceed $2 million.
- Royalties payable do not exceed 5 percent of the domestic sales and 8 percent of the exports, subject to a total payment of 8 percent of sales over a ten-year period. Where these remittance limits are exceeded, permission from the MOCI is required.

2. *Industrial License*

The requirement of obtaining an industrial license for manufacturing activity is limited to the following:

- Industries reserved for public sector (two in number)
- Six industries of strategic, social, or environmental concern
- Manufacture of items reserved for the small-scale sector by non-small-scale industry units or units in which foreign equity is more than 24 percent

3. *All other Industries are Exempt from Licensing*

These exemptions are subject to certain location restrictions in metropolitan areas. In the event that location restrictions are not adhered to, the unit is required to obtain an industrial license. Industrial licensing is compulsory for investing in the following six industries of strategic, social, and environmental importance:

- Distillation and brewing of alcoholic drinks
- Cigars and cigarettes of tobacco and manufactured tobacco substitutes
- Electronic aerospace and defence equipment
- Industrial explosives including detonating fuses, safety fuses, gun powder, nitrocellulose, and matches
- Hazardous chemicals

- Drugs and pharmaceuticals

Relevant Environmental Laws and Regulations

India has a comprehensive set of environmental statutes and a well-developed jurisprudence, applied by, at times, an activist judiciary. The specific environmental laws that would impact a CDM project would depend on the details of a project. Provided below is an overview of some of the important national environmental statutes that would be of relevance to CDM projects in India.

The Water (Prevention and Control of Pollution) Act, 1974

The Water (Prevention and Control of Pollution) Act, 1974 (Water Act), vests regulatory authority in state pollution control boards and empowers these boards to establish and enforce effluent standards for units discharging pollutants into bodies of water. The boards control sewage and industrial effluent discharges by approving, rejecting, or conditioning applications for consent to discharge. The state boards also advise state governments on appropriate sites for new industries. Enforcement under the Water Act is achieved through criminal prosecution initiated by the state pollution control boards, which may also close a defaulting industrial plant or withdraw its supply of power or water by an administrative order.

The Air (Prevention and Control of Pollution) Act, 1981

The framework of the Air (Prevention and Control of Pollution) Act, 1981 (Air Act), is similar to the one created by the Water Act. Under the Air Act, all industries operating within designated air pollution control areas must obtain "consents" from the relevant state pollution control board. The states prescribe emission standards for industry after consulting the central pollution control board. State pollution control boards may close down a defaulting industrial plant or stop its supply of electricity or water. The relevant state pollution control board may also apply to a court to restrain emissions that exceed prescribed standards.

Consequences of breach by a company under the Air Act and the Water Act

Where an offence under the Air Act or the Water Act is committed by a company, every person who, at the time the offence was committed, was directly in charge of and was responsible to the company for the conduct of the business of the company, is deemed to be guilty of the offence and liable to be proceeded against. Further, if it is proved that the offence has been committed with the consent or connivance of, or is attributable to neglect on the part of a director or officer of the company, such director or officer is also deemed to be guilty of that offence and liable to be proceeded against.

The Environment (Protection) Act, 1986

The Environmental Protection Act, 1986 (EPA), is intended to provide a framework for the central government to coordinate the activities of various central and state authorities established under previous laws, such as the Water Act and the Air Act. It also delegates wide powers to bureaucrats to frame necessary rules and regulations. The EPA empowers the central government to take the necessary measures to protect and improve the quality of the environment, and to prevent, control, and abate environmental pollution. Any person who fails to comply with or contravenes any provisions of the EPA or the rules, orders, or directions issued under it, is liable to be punished, for each failure or contravention, with a prison term of up to five years or a fine of up to Rs. 100,000 or both. The EPA imposes an additional fine of up to Rs. 5,000 for every day of continuing violation. If a failure or contravention occurs for more than one year after the date of conviction, an offender may be punished with a prison term of up to seven years.

The Energy Conservation Act, 2001

The Energy Conservation Act, 2001, was enacted in September 2001 to address all matters related to the efficient use of energy and its conservation. A Bureau of Energy Efficiency (BEE) has been set up to discharge the activities entrusted under the act. The bureau is expected to investigate the energy consumption norms for each energy-intensive industry and encourage the proper labelling of energy consumption indicators on every electrical appliance. The responsibilities of the bureau also include providing guidelines for energy conservation building codes, and taking measures to create awareness and disseminate information for the efficient use of energy and its conservation. The BEE is also to develop testing and certification procedures and promote testing facilities for certification and for energy consumption of equipment and appliances.

Reforms in the Coal Sector

The government has constituted the Expenditure Reforms Commission (ERC) towards reforming the coal sector. Major recommendations of the commission are as follows:

- All restrictions on the entry of the private sector in exploration and production of coal to be removed by amending the Coal Mines Nationalization Act, 1973.
- The Coal Bearing Areas (Acquisition and Development) Act, 1957, to be amended and an independent regulatory body to be set up.
- The industry to be restructured by doing away with the holding company (Coal India Limited) and coal controller, among other things.
- The Coal Mines (Conservation and Development) Act, 1974, to be amended to empower responsibility to both public and private sectors for scientific mining, conservation, safety and health, protection of environment, etc.

- States to be permitted for developing lignite resources outside the command areas of the Neyveli Lignite Corporation.
- Design strategy to consider the role of coal in energy security.

Before January 2000, the central government was empowered under the Colliery Control Order, 1945, to fix the grade-wise and colliery-wise prices of coal. However, following the Colliery Control Order, 2000, the prices for all grades of coking and non-coking coal have been deregulated. The current basic price of coal varies from Rs. 1,450 per tonne to Rs. 250 per tone for different grades (GoI—MoEF, 2004).

The Electricity Act, 2003

The Electricity Act, 2003, enacted by the government of India, has de-licensed the generation of electricity in India. Clause (7) of the act states that "any generating company may establish, operate, and maintain a station without obtaining a license under this Act if it complies with the technical standards relating to connectivity with the grid." The salient features of the act are given below:

- The central government to prepare a national electricity policy in consultation with state governments.
- Thrust to complete rural electrification and provide for management of rural distribution by panchayats, cooperative societies, non-government organizations, franchises, etc.
- Provision for licence-free generation and distribution in rural areas
- Generation being de-licensed and captive generation being freely permitted. Hydro projects would, however, need clearance from the Central Electricity Authority.
- Transmission utility at the central as well as state level to be a government company with responsibility for planned and coordinated development of the transmission network.
- Provision for private licensees to transmit and enter into distribution through an independent network.
- Open access for transmission from the outset.
- Open access to distribution to be introduced in phases with surcharges for the current level of cross subsidy to be gradually phased out, along with cross subsidies and obligation to supply. State electricity regulatory commissions (SERC) to frame regulations within one year regarding phasing of open access.
- Distribution licensees would be free to undertake generation, and generating companies would be free to take up distribution businesses.
- The State Electricity Regulatory Commission is a mandatory requirement.
- Provision for payment of subsidy through budget.
- Trading, a distinct activity, is recognized through the safeguard of the

regulatory commissions being authorized to fix ceilings on trading margins, if necessary.
- Provision for reorganization or continuance of SEBs.
- Metering of all electricity supplied made mandatory.
- An appellate tribunal to hear appeals against the decision of the CERC and SERCs.
- Provisions relating to theft of electricity made more stringent.
- Provisions safeguarding consumer interests Ombudsman scheme for redressal of consumer grievances.

Intellectual Property Regime (IPR)

India is a signatory to the Trade Related Intellectual Property Rights Convention (TRIPS) and has a well-developed legal framework that affords world-class protection to intellectual property. Thus, intellectual property owned by a foreign investor that may be used in India in connection with its CDM project would be safeguarded under the following key legislation:

- **Patents Act, 2002.** On December 27, 2004, the central government issued an ordinance for introducing a product patent regime for pharmaceuticals and agro-chemicals from January 1, 2005, to fulfil India's obligation to the World Trade Organization.
- **Trademarks Act, 1999.** Effective since September, 2003, which affords protection to trademarks.
- **Designs Act, 2000.** Governs industrial design protection in India.
- **Copyright Act, 1957.** Protects copyrights.

Labour Law Regime

In India, both the central and state governments may enact laws relating to employment and labour. A wide range of labour statutes in India regulate the conditions of employment and impose several obligations on the employer in relation to the employees and the attitude of the government. The court is typically labour-friendly. The contravention of Indian labour laws invites different types of punishment, ranging from monetary fines to imprisonment. Employees in India are broadly categorized into (1) workmen and (2) non-workmen. This distinction is based on the nature of work performed and on certain other criteria prescribed in the Industrial Disputes Act, 1947 (ID Act). If an entity running a CDM project in India is governed by the ID Act and its employees are workmen, the closure of such a unit and removal of the workmen engaged therein must be carried out in accordance with procedures specified under the ID Act. Further, the workmen would be entitled to severance compensation determined in accordance with the ID Act. The employment terms of employees who are not "workmen" are largely governed by their employment contracts or letters of appointment.

There are specific statutes which govern wages and hours of work, primary amongst which is the Payment of Wages Act, 1936, the Minimum Wages Act, 1948, and the shops and establishments acts of various states. While the first two are central legislation applicable to the whole of India, the third is state legislation and therefore varies from one state to the other. India also has various social-securities or benefits legislation, most notably the Employees Provident Funds and Miscellaneous Provisions Act, 1952, and the Employees State Insurance Act, 1948. An employer has the option to either formulate its own policy to provide health insurance to its employees or follow the provisions contained in the Employees State Insurance Act, 1948 (ESI Act). However, if an employer chooses to have its own policy, the terms are required to be either equal to or more beneficial than those provided under the ESI Act.

Property Issues: Nature of CERs

The precise legal nature of the emissions reductions created under the Kyoto Protocol is a matter of debate in India. Even globally, the question of what exactly is being bought and sold in a transaction involving emissions reductions is under consideration. Identifying the legal nature of emissions reductions is significant for risk, title, and taxation-related issues in any CER sale and purchase transaction. Ideally, for India to benefit from emissions trading under the Kyoto Protocol, the Indian government would eventually have to "commoditize," through appropriate legislation, the cumulative benefits arising from emissions reductions, so as to facilitate international trade in CERs. Until law on the subject is enacted or executive orders are issued, a purchaser of CERs would have to secure, by way of a contractual right, any and all benefits that may arise from the fact that the seller (or some third party from whom the seller purchases the CERs) has reduced its GHG emissions.

As far as the nature of CERs is concerned, in the current legal framework, it may be appropriate to consider them as intangible assets that are capable of being transferred, bought, or sold. In this regard, we can draw an analogy between CERs and other intangible assets such as the goodwill of a business, a trademark or brand name, a statutory or contractual right to manufacture, produce, or process any article or thing, and import licenses. In certain jurisdictions, the trading of emissions reductions is regulated under securities or financial laws. At the date of writing, however, the term *securities*, (under Indian law) has not been either expanded or interpreted to include emissions reductions. Of relevance would be a consideration of the Honourable Supreme Court of India's views on the nature of import licenses, which also (like CERs) confer a right upon their holders that may be transferred for value. However, the Honourable Supreme Court was of the view that such import licenses could not be treated as securities in the absence of a specific declaration to this effect by the central government. Accordingly, the central government would have to make a specific declaration under the relevant legislation before CERs may be treated as securities.

Title to CERs

In the absence of any law in India that deals with the legal ownership of CERs or any determination by the Indian DNA in this regard, it may not be possible at this point to reach any conclusion on this subject. There is a school of thought that believes that CERs are actually owned by the countries in which they are produced, because they are instruments of an international treaty of which states are the sole parties. But there is another school of thought that believes that it is the investors in the CDM project who are the legal owners of any CERs produced from such projects.

Accordingly, if the Indian government were to take the former view, owners of Indian CDM projects would have to obtain a cessation of ownership from the Indian government to prove ownership of the CERs that they have produced. If the latter view were to be adopted, however, owners of Indian CDM projects would have the exclusive right to deal with such CERs. A relevant aspect to this debate is the fact that the Marrakesh Accords do not clarify whether the CDM Executive Board registry will open sub-accounts for public and private entities of non-Annex I Parties (host countries like India). Therefore, the determination of this issue will also have an effect on the ability of public and private entities of non-Annex I Parties to own and/or sell CERs independently of any involvement of the host country.

Contractual Issues in Carbon Trading Contracts

Until the necessary CDM-related laws are enacted in India, foreign investors in Indian CDM projects who intend to transact in CERs generated from such projects would have to rely on contractual mechanisms to protect their rights as sellers/buyers of such CERs. All parties concerned would also need to contractually address the possible consequences of unanticipated changes in Indian policy on CDM projects. Accordingly, we discuss below key issues of relevance to foreign investors entering into joint venture agreements or CER sale and purchase agreements with Indian parties.

Indian Law as Governing Law of Contract

It will not be mandatory to choose Indian law as the governing law of the joint venture agreement or the contract for sale and purchase of CERs merely because the CDM project is located in India. Indian law recognizes that parties have the contractual freedom to choose a foreign law as the governing law of the contract. However, where the contract contains no express provision with respect to governing law and no inference regarding the same can be drawn, the intention of the parties would become irrelevant and the courts would choose the governing law as the one that has the closest and most real connection to the contract. In such a situation, the courts would take into account (a) the place where contract was made, (b) the form and object of the contract, (c) the place of residence or business of the parties, and (d) courts having jurisdiction. Therefore, as Japanese investors may not be familiar with Indian law, it would be possible for them to choose the law of another country (i.e., English law) as the governing law of the contract. They would also need to ensure that the contract expressly provides for such law as the governing law for the contract. In most cases, however, the choice of Indian law should be acceptable.

Key Contractual Terms of CER Sale/Purchase Contract

In the event that the concerned parties choose Indian law as the governing law of the contract for sale/purchase of CERs, the provisions of such a contract would have to be drafted in accordance with Indian contract law, which is largely based on the concepts of English law (http://www.itpi.co.in/Resources/Clean_development_mechanisms/Business%20Plan%20for%20Establishing%20Indian%20Carbon%20Fund%200407.pdf).

A typical contract for the sale/purchase of CERs would ideally include the following clauses (these will be similar in other such contracts around the world):

- *Recitals.* These should clearly summarize the purpose and subject matter of the contract and the details of the CDM project.
- *Definitions and interpretation.* This clause should describe the terms used in the contract and contain an explanation of how to interpret and read the contract.
- *Sale and purchase.* This clause should outline the unit of measurement in which emission reductions are being calculated, and the number and vintage of the CERs to be sold from the particular CDM project.
- *Conditions precedent.* At the time of signing the contract for sale and purchase of CERs, certain regulatory approvals from the host country as well as participation requirements of Article 12 of the Kyoto Protocol may be pending. Also, the CDM project itself may not yet be in existence. It is therefore important that the obligations of the parties under the contract become enforceable only upon the occurrence of such events.
- *Delivery of CERs.* This clause should set out when the CERs accrue to the buyer, the mechanism for delivery of CERs, documents evidencing the delivery of CERs, and the point at which the legal ownership in CERs would transfer to the buyer, etc.
- *Price and terms of payment.* This clause should cover the price of CERs, manner of invoicing and payment, penalties for late payment, and responsibility for payment of taxes and statutory dues in relation to sale of CERs.
- *Representations, warranties and liability.* These clauses should cover the usual commercial warranties as well as specific warranties by the seller relating to creation of rights in CERs and warranties by the buyer relating to timely payment. Also, a clause covering indemnity against

third-party action and a limitation of liability on the part of both the parties should be included.

- ***Events of termination of the contract and consequences of such termination.*** Apart from covering common grounds of termination for both buyer and seller, this clause should address the following; (i) non-payment by buyer, and (ii) shortfall or failure to deliver CERs by the seller. In fact, one of the more contentious issues in the negotiation of a CER sale agreement is the question of how to treat the shortfall of CERs expected to be generated from the Indian CDM project. Some buyers may insist on provisions for liquidated damages and penalties. The cause of shortfall may, in fact, be outside the control of the seller, and hence sellers should seek protection from the risk of incurring penalties and damages from factors beyond their control. Sellers could also look to procure insurance for such potential shortfalls.
- ***Enforceability of liquidated damages and penalties in India.*** In India, provisions relating to liquidated damages that provide for a specific sum as a reasonable and genuine pre-estimate of damages envisaged by the parties are typically enforceable under section 74 of the Indian Contract Act, 1872. Provisions stipulating penalties on the defaulting party in a contract are also enforceable under the same section. However, courts do examine the reasonableness of such penalties and award what they consider a reasonable sum.
- ***Dispute resolution.*** In the Indian context, the parties to a contract should consider arbitration as the preferred means of dispute resolution rather than court proceedings. Foreign investors may note that under Indian contract law, parties are free to choose the governing law of arbitration in respect to substantive as well as procedural issues. Where no provision as to the governing law of arbitration is made, the courts presume that the governing law of the contract will also be the governing law of arbitration. Where, however, no mention is made regarding either the law governing the contract or the law governing arbitration, in the absence of any intention to the contrary, Indian courts presume that the law of the place of arbitration is the law governing the contract and arbitration.
- ***Force majeure.*** This clause should set out what events would constitute force majeure events and what the consequences of such events would be. Force majeure is a common clause in contracts which essentially frees one or both parties from liabilities when an extraordinary event beyond the control of the parties–such as flood, war, riots, acts of God, etc. – prevents one or both parties from fulfilling their obligations under the contract.
- ***Boilerplates.*** These clauses typically cover issues such as confidentiality provisions, severability, waiver, variation, survival, notice, assignment, etc.

Evolution of the CDM in India

The CDM project development process in India started in late 2001 when the Dutch government tender Carbon Emission Reduction Unit Procurement Tender (CERUPT) was announced. Seventeen projects sought approval from the MoEF; and an ad hoc group of officials approved 12 of them, of which the Dutch accepted six, and ultimately five were provided with an ERPA. Similar clearances were also provided for projects submitted for the Swedish Climate Investment Programme (SICLIP) in mid-2002, the Finnish tender in March 2003, and some other one-time-only opportunities (GoI Planning Commission 2003b; GoI MoEF 2004). Then came the World Bank's Prototype Carbon Fund (PCF), which accepted projects prepared or assisted by the Infrastructure Development Finance Corporation (IDFC) under their MoU of October 2002. The World Bank relied on IDFC for carrying out the promoter due diligence, and an initial investment of US$10 million was fixed. The World Bank also launched the Community Development Carbon Fund (CDCF) and the Bio-Carbon fund (BCF) (Rabo India-2003). This was after India ratified the Kyoto Protocol in August 2002 and hosted COP 8 in New Delhi (October–November 2002) when small-scale projects received support from all developing nations, the CDM movement in India picked up momentum. The working group of the planning commission of India dealt with the implementation of CDM projects in India, and submitted its report to the prime minister's office. Soon after, India announced its designated national authority (DNA) in December 2003. Subsequently, with the Russian instrument of ratification in end–2004, the field for a CDM market was well and truly established. India has already provided host country approvals for about 56 CDM projects.

The year 2005 began with the coming into force of the EU Emission Trading Scheme (ETS) and the Kyoto Protocol on February 16. There are groups like the Indian Carbon Market Group that are trying to create the necessary involvement of the major players in the CDM field in India. With the selection of India to the CDM executive board there is a great responsibility towards other developing nations as well.

Several organizations in India are into the CDM playing field and have been active in the last two–three years. They range from a wide spectrum of organizations, as shown in the indicative list below:

- Apex industry bodies (Confederation of Indian Industry (CII), Federation of Indian Chambers of Commerce and Industry (FICCI), Associated Chambers of Commerce (ASSOCHAM)
- NGOs (Development Alternatives, WWF, TERI, Winrock)
- Consulting firms (TUV, Price Waterhouse Coopers, Ernst and Young)
- ESCOs (DSCL)
- Big companies in the private and public sectors (Tata Steel, National Thermal Power Corporation (NTPC), Excel)

- International development organizations (CIDA, GTZ)
- International/multilateral lending agencies (World Bank, ADB).

Awareness about the environment and CDM—though low—has grown in the financial sector. Banks such as ICICI rely on catering to each business in a unique manner, and offer tailored and innovative financial products to suit the needs of the off-taker. Examples here are credit enhancements through trusts, securitization, and the like. Banks like the State Bank of India (SBI), Bank of Baroda (BoB), and Syndicate Bank have formed separate groups to finance energy efficiency, renewable energy, and environmental projects. The Small Industries Development Bank of India (SIDBI) supports similar themes for small and medium-sized enterprises. Institutions like the IREDA, PFC, and some others, stress emission reduction for projects funded by them.

The government of India has also reviewed insurance in the context of disaster management and is considering a unified legislation. In end–2003 the high-level task force on Insurance and Climate Change was created to identify specific national needs and concerns pertaining to insurance and related issues arising from the adverse impacts of climate change. While there is no information available yet on the strategies and approaches to be developed by the task force, it illustrates the right course of action to manage such issues.

CDM Project Financing

As far as industry in India is concerned, these are usually up-front and significant. The numbers vary, and can be as much as US$200,000 (estimated by PCF). This is actually additional investment required to be made by a company. A rule of thumb often used is that a project (other than small projects) needs to achieve a reduction of 50,000 CERs per year to be viable (http://papers.ssrn.com/sol3/papers.cfm?abstract_id=452400).

The various components of transaction costs may be listed as follows:

(a) Project documentation development
(b) Negotiation and development of legal contracts
(c) Professional fees (validation, brokerage, etc.)
(d) Host nation approval
(e) CDM registration
(f) Internal monitoring
(g) Contribution to adaptation fund (2% of CERs).

The Concept of Risk

Global climate change poses major risks and investment opportunities for shareholders and companies. The risks come primarily in two forms: physical risks and policy risks. Some statistics that are interesting and worth noting are as follows (CERES 2004; UNEP FI Climate Change Working Group 2005; Point Carbon 2004):

- Munich Reinsurance estimates that weather-related catastrophes topped $55 billion in 2003. With global warming continuing, this figure could well reach $150 billion in the next ten years.
- The French insurance company AXA estimates that 20 percent of the global GDP is now influenced by climate-related events, and that climatic risk in numerous branches of industry is more important than the risk of interest rates or foreign exchange risk.
- Market capitalization of major oil companies could decline by as much as six percent as a result of CO_2 emission constraints, while companies with large reserves of natural gas stand to benefit.
- Among auto companies, costs of compliance with climate policies could vary by a factor of 25, adding a new competitive dimension to the industry that could cause some major global companies' earnings to increase by up to eight percent while others fall by as much as ten percent.

The risk exposure of investments is of high concern for equity investors in CDM projects, as well as for the lenders that give credit for CDM projects. This section deals with risks involved for conventional project development for potential CDM project types. In India the main potential areas for CDM projects are as follows:

(a) Renewable energy technologies (mainly for biomass, wind, solar, small hydro, biomass)
(b) Energy efficiency
(c) Highly efficient clean energy technologies in the power plant sector (includes combined cycle natural gas, integrated gasification combined cycle coal power plants)
(d) Switching to alternate fuels
(e) Waste management (especially waste-to-energy projects)
(f) Oil and gas
(g) Agriculture
(h) Forestry and carbon sequestration

If one were to look at the kinds of CDM projects approved by India's DNA in the past few months, they all fall under the above list. Some of these project types are subject to high risks during project development itself. Examples here would be projects that require a power purchase agreement (PPA), and projects that need a reliable supply of fuel.

Projects requiring a PPA are typically the grid-connected renewable energy technologies (RETs) and waste-to-energy projects. In 1994 the Ministry of Non-conventional Energy Sources (MNES) issued guidelines to the state electricity boards (SEBs) to pay Rs. 2.25/kilowatt-hour (kWh) (base year 1994/95) for electricity from renewable energy technologies with an annual escalation rate of five percent. Had all states followed the guidelines, electricity from renewable energy would attract a price of Rs. 3.66/kWh in 2003.

CDM Project Risks

There are a number of requirements, including legal, and agreements that would have to be fulfilled or negotiated when developing a CDM project. Any contracts and agreements will have to be integrated into the overall project structure to ensure not only successful implementation and operation of the CDM project but also to share risks (e.g., non-delivery of CERs), responsibilities (e.g., monitoring of individual projects), and benefits (e.g., sharing of CERs). The exact types of agreements or contracts will vary project to project. However, some template documents for the sale of CERs are available as a starting point for project developers and buyers (IT Power Group, 2005).

In addition to legal requirements, there are several risk factors that have to be taken into account. These may be categorized as normal project risks and CDM-related risks of the following kinds:

- *CDM licensing/regulatory:* These risks are associated with the CDM project cycle itself as at all stages—appraisal (PDD), monitoring, verification, certification, and registration—there is risk of delay (e.g., if the application of the baseline methodology needs revising), or non-approval (e.g., for registration). The largest risk will be that the project does not make it to registration with some transaction costs already paid out and an agreement for sale of CERs already in place. There are also risks associated with the crediting period itself, with no concrete system in place post–2012 and for projects with a crediting period of 21 (3x7) years there could be risks, after seven and 14 years.
- *Political risk:* These are risks associated with those factors under the control of the government in the country of implementation. These include the risk of failure to gain host country approval, particularly as many countries are still in the process of setting up a fully functioning DNA or criteria for approval, such as sustainable development criteria that are still to be fully developed. Other risk factors include regulatory change which could affect the viability of a project and in some cases the baseline (particularly in projects to be renewed after seven years), and changes in tax which could affect the CER revenue, for example if a country introduced a tax on CERs.
- *Carbon market /financial:* As the carbon market is still non-fluid and only beginning to mature, there is still great uncertainty about the prices. This means that the sale of CERs bears the risk of market prices for CERs rising above the price set in the contract. However, it is expected that trading activity will increase with the increasing number of registered CDM projects and the full strength implementation of the European Union Emissions Trading Scheme (EU ETS). The uncertainty about the price of carbon credits will decrease as the market becomes more fluid. Furthermore, some buyers are offering more flexible and dynamic emission reduction purchase agreement (ERPA) terms with prices indexed to market prices, allowing the buyer and seller to share

this risk. Other financial risk relates to the currency used for sale of CERs, however, in some cases having revenue from CERs in a hard currency (e.g., US dollars, Euros, etc.) can offset some of the main project risk related to the currency of the host country which is likely to be more open to large fluctuations in relative value.

Furthermore, the project developer runs the normal project risks, which include:

- *Cost/delay*: These risks relate to long delays or extra costs due to licensing and permitting barriers with different probabilities at different stages of the project, such as the development, construction, start-up, and operation phases.
- *Technology:* Lower than expected performance of the project as a result of, for example, poor technology performance, could result in lower GHG reductions and therefore fewer revenues from the CDM project, and in some cases, cost penalties for non-delivery. Contractual arrangements, if properly structured, can be used to minimize risk or allocate those risks to an entity that is better able to control them. For example, where a project bundle includes a large number of small installations, such as solar systems, operation and maintenance contracts to installers or standard seller warranties and indemnities to technology providers may be included to minimize project performance risk as well as ensure regular monitoring. The price of CERs agreed upon with a carbon buyer will usually reflect risk sharing in the contract. The risks discussed above are generally found in any normal CDM project, and it is important to realize that there are additional risks when bundling is involved. These are mainly due to the increased number of parties, locations, often technologies, and a whole host of uncertainties that are increased with a project bundle compared to a single CDM project. For example, when submitting a project bundle for registration, a request for review of an individual component of the project would lead to the whole project bundle being affected. Similarly, the failure of individual projects can threaten the viability of the whole bundle, and risk mitigation techniques will be important to minimize the risk of shortfalls in expected yearly CER generation.

CDM Financing and Taxation Norms

While the next section deals in detail about the issue of taxation of CERs, one may like to know about some sample policies of the (state) governments in India on the way they would like to encourage the use of non-fossil fuels to meet energy requirements. For example, the government of Maharashtra has decided to levy a "green cess" of Rs. 0.04/kWh to industrial and commercial consumers of public and private utilities in order to promote non-conventional energy projects in the state. The funds would be utilized for promoting such projects in the state by providing infrastructure and giving subsidies to the promoters of these projects. One may also like to learn from the approach of some developed countries in their outlook towards climate change regulation. They prefer to adopt market-based instruments for their climate change policy. Specifically, they have examined the potential net benefits of shifting to a permit system for emission reduction, and the preconditions necessary for this change. Price-based instruments such as taxes and permits are ideally suited to homogenous long-term-pollutants such as carbon dioxide. They allow maximum spatial and temporal flexibility at no environmental cost. However, in the case of India, such concepts may be woven into existing environmental regulations.

Taxation Treatment of Transactions Involving CERs

The lack of clarity on the precise legal nature of emissions reductions makes it difficult to conclusively comment on whether the income received from the sale of CERs would be treated as business income or a capital gain. Though normally, the revenue stream generated from the sale of CERs would be viewed as business income, under the current income tax regime in India, such income could also perhaps be treated as a capital receipt chargeable to capital gains tax.

CERs as a Business Asset

1. *Business Asset and Business Income*

The term "business asset" is not defined under the Income Tax Act but includes all assets (such as stock-in-trade) which, when sold in the ordinary course of business, generate a revenue receipt. Profits made on such revenue receipts are taxed as business income in accordance with the provisions of the Income Tax Act.

Typically, when a person disposes of a part or the whole of his assets in the ordinary course of business or as part of a trading/business (ongoing revenue generating) transaction, the profits earned by such disposal are taxable as business income. Further, in the event that a transaction of sale is intrinsically related to the normal business of the taxpayer, profits arising from such Therefore, the likelihood of the tax authorities treating CERs as a business asset and any income generated from sale of such CERs as business income may be influenced by the fact that CERs produced by a CDM project would be intrinsically related to the normal business of the investor/project owner. Such CERs would accrue as part of an ongoing revenue-generating business activity. For most CDM projects, the generation and consequent sale of CERs may be one of the primary objectives for which the CDM project has been implemented. Separately, as discussed above, CERs could be equated to "import licenses" that are granted to persons in the course of their export and import activities. Profits made on the sale of such import licenses are taxed as business income in accordance with the Income Tax Act.

2. Taxation of business income as indicated earlier, the business income of an Indian company is taxed under the heading "profits and gains of business or profession" at the rate of 35 percent plus surcharge.

CERs as a Capital Asset

1. *Definition of a Capital Asset*

The Income Tax Act defines a "capital asset" to mean property of any kind held by an assessee, whether or not connected with its business or profession, except for the following:

(a) any stock-in-trade, consumable stores, or raw materials held for the purpose of the assessor's business or profession,
(b) personal effects,
(c) agricultural land of the type specified in this section, and
(d) certain bonds specified in the section.

Therefore, unless it can be shown that CERs would fall under these exceptions, it is likely that CERs would be treated as capital assets. Exceptions b, c and d are clearly not applicable in the case of CERs. The term "stock in trade" would also not apply to CERs since Indian courts have interpreted this term to include all goods and commodities in which an entity deals in the sense of buying and selling in the course of its business activity. An argument for CERs being treated as capital assets is that intangible assets such as the goodwill of a business, its trademark, and brand name are also treated as capital assets in accordance with the Income Tax Act.

2. *Capital Gains*

Under the Income Tax Act, any profit or gain arising from the transfer of a capital asset is chargeable to income tax under the head "capital gains". Accordingly, in the event that CERs are treated as capital assets, all monies received on account of the transfer of such CERs would be taxed as capital gains.

In this regard, it may be noted that the term "transfer" has been defined under the Income Tax Act to include, *inter-alia* sale, exchange, or relinquishment of an asset. Thus, from the above definition it is clear that not only a "sale" of CERs, but any form of "relinquishment" of such CERs or "extinguishment" of any right therein, would be treated as a "transfer" for the purpose of computing capital gains under the Income Tax Act.

3. *Computation of Capital Gains*

Under the Income Tax Act, capital gains are computed by deducting from the full value of the consideration received or accruing as a result of the transfer of the capital asset as per the following:

(a) the expenditure incurred wholly and exclusively in connection with such transfer; and b. the cost of acquisition of the capital asset and the cost of improvement thereto. In 1981, the Honorable Supreme Court of India held that if it were not possible to ascertain the cost of acquisition and/or the cost of improvement of an intangible capital asset, the gains made on the transfer of such asset would not be taxable under the Income Tax Act. This decision was interpreted to mean that gains made on the transfer of any "self-generated asset" (such as CERs) would not be chargeable to capital gains tax since the cost of acquisition/improvement of such assets would not be ascertainable. However, in response to this decision of the Supreme Court, the Income Tax Act was amended to provide that to compute capital gains, the cost of acquisition/improvement of certain specified self-generated assets such as goodwill, trademark, brand name of a business, is to be taken as "nil". Therefore, the entire consideration received for the transfer of these specified self-generated assets is charged to tax as capital gains. However, for the transfer of any other self-generated assets (that are not specified under the Income Tax Act) the consideration received for such transfer would not be chargeable to capital gains tax, unless it is possible to ascertain the cost of acquisition and/or cost of improvement of such capital assets.

Accordingly, if a part of the capital invested in a CDM project is specifically capable of being attributed to the objective of generating CERs, it may be possible to allocate a portion of such capital investment towards the cost of acquisition of the CERs, which might result in the transfer of CERs being chargeable to capital gains tax. If, in the future, CERs are brought within the purview of the specific list of self-generated assets, the above logic could be used to structure the investment in a CDM project in a manner that would reduce the incidence of capital gains tax.

Government Incentives for CDM Projects

India's Ministry of Environment and Forests (MoEF) is the nodal ministry for all matters related to climate change, both national and international, including the UNFCCC, the Kyoto Protocol, and CDM activities and processes in India. The government has constituted its designated national authority (DNA), the National CDM Authority, and has established other enabling conditions to provide incentives for CDM project developers and investors to implement CDM activities in India.

Enabling Environment

At the Central Level (GOI)

The National CDM Authority (NCA), chaired by the secretary, the Ministry of Environment and Forests, along with the secretaries of various governmental departments, provides the information on facilitating CDM activities in India on request. This enabling environment and the proactive role of the government have sound infrastructure and excellent incentives for CDM activities, particularly for project developers from India and foreign investors, who are already playing a very proactive role in CDM capacity building in India. This enabling environment at the centre facilitates project activities and works toward removing any hurdles and barriers to CDM project development. The government takes on such tasks for project developers and investors on request. It has already approved a number of CDM capacity building initiatives in India by countries like Japan (IGES), Germany (GTZ), etc.

At the State level

There is large potential for implementing the CDM in various states in India; some of these, such as Andhra Pradesh, Madhya Pradesh, West Bengal, Maharashtra, Tamil Nadu, and Rajasthan, have already set up CDM cells to strengthen their capacity for CDM activities. Such arrangements at the state level provide incentives for both project developers and investors, which supplement the incentives provided by the central government, provide facilitation at the state level to project developers and investors, and also help to overcome other hurdles and barriers.

Incentives to Establish Renewable Energy/ Power Projects

Policy Incentives

The policy incentives announced by the MNES to promote RE power projects include the following:

- No clearance is required from the Central Electricity Authority for power generation projects up to 1,000 million rupees (Rs).
- A tax holiday is allowed for RE power generation projects.
- A consolidated rate of customs duty has been made applicable for all items imported for a project under the "project imports" scheme, irrespective of the nature of the goods imported and their customs classification.
- Customs duty concessions are available for RE parts and equipment, including machinery required for renovation and modernization of power plants.
- The excise duty on a number of capital goods and instruments in the RE sector has been reduced or exempted.

To encourage investment by private and public sector companies in renewable energy electricity generation, the MNES has issued guidelines to the states, which require the relevant state electricity boards, (SEB) to ensure the following:

1. **Wheeling:** Each SEB is required to undertake to transmit on their grid

power generated by an RE power project, and make such power available to the producer for captive use or to a third party within the state at a uniform wheeling (distribution) charge of 2 percent of the energy fed to the grid, irrespective of the distance from the generating station. The third party must be a "high tension" consumer of the SEB, unless this stipulation is relaxed specifically by the SEB.

2. **Banking:** The SEB is required to permit the electricity generated to be banked for a period of up to one year.
3. **Sale of power:** The SEB is required to purchase electricity offered by the producer at a minimum rate of Rs. 2.25 per unit, with no restriction on the time or quantity of electricity supplied for sale. This rate is to be reviewed every year and will be linked to standard criteria such as the wholesale price index. Producers also have the option to sell the electricity they generate to third parties within the state at a rate to be mutually agreed upon.
4. **Exemption from duty:** Consumption of electricity generated by producers is exempt from the electricity duty.
5. **Sales tax benefits** are available to producers that own their project.
6. **Income tax benefits**, such as accelerated depreciation, are available to RE power projects.
7. **Concessions** given to industrial units in underdeveloped areas are provided, such as exemption from taxes and duties, capital subsidies, etc.
8. **Infrastructure facilities** such as access roads, water supply, cranes, power during construction period, etc., are provided, as is done with industrial estates.

Specific Fiscal Incentives

The largest number of proposals for CDM projects in India relate to biomass/co-generation projects and wind-power projects. Accordingly, following is a brief list of fiscal incentives that will be available to such projects.

1. Biomass/Co-generation Projects

- An 80 per cent depreciation may be claimed in the first year for specified equipment required for co-generation systems.
- A five-year income tax holiday with a 30 percent exemption for power generation projects with power purchase agreements.
- A customs duty leviable for RE projects of less than 50-MW capacity (under the "project import" scheme) is 20 percent ad valorem,
- Sales tax exemption is available in certain states
- Banking, wheeling, and third-party sales benefits

2. *Wind-power Projects*
   - An 80 per cent depreciation may be claimed in the first year for specified equipment required for wind-power projects.
   - Income tax holiday.
   - Concessional import duty on specified wind turbine parts.
   - Sales tax, excise duty relief.
   - Nine states have introduced specific policies to encourage such projects.
   - Banking, wheeling, and third-party sales benefits.

CDM Project Opportunities

Investors will prefer CDM projects that satisfy three basic criteria:

1. High possibility of approval by the government.
2. Low transactions costs.
3. Significant reduction potential.

Energy Sector Priorities

Despite a slow start, the GoI is now actively trying to access GEF funds. This is partly because some of the activities have a high overlap with the developmental objectives of the government. The GoI has identified the following project ideas in accordance with its development objectives.

- Removal of implementation barriers of biomass-based power generation
- Strategy for removing barriers in achieving industrial energy efficiency
- Carbon reduction through grid-interactive PV power generation
- Energy conservation through technological up-gradation of steel rerolling mills
- Solar PV diesel hybrid systems for decentralized village electrification
- Co-generation in cement plants
- Carbon emission reduction in power plants
- Decentralized rural electrification through biomass and solar energy through private participation
- Switching to non-petroleum fuels, electricity-powered and hybrid engines in transport sector
- Demand-side management in basic material and construction industry
- Improving efficiency of rural agriculture pump-sets
- Energy conservation in paper industry

As the above list shows, the priority options for the GoI are efficient technologies in power generation, centralized and decentralized renewable energy options, and energy efficiency in steel and paper industries. The text submitted by India on flexible mechanisms states that no forestry sector projects will be allowed. Preference will be given to projects in the renewable sector and those with a focus on energy efficiency improvements. Projects that improve the quality of life of the very poor from the environmental standpoint and promote state-of the art, environmentally sound technology will be sought. The energy sector mitigation options identified above were ranked by policy makers on the basis of:

- Cost of mitigation
- Feasibility of the option
- Environmental and other benefits
- Consonance with the overall development priorities of the country

Priority CDM Projects

Based on the above information, the following projects could be proposed for CDM in India.

Power Generation

1. Coal power plant using IGCC (integrated gasification combined cycle)
2. Coal power plant using PFBC (pulverized fluidized bed combustion)
3. Renovation and modernization of power plants

Renewables

4. Wind-based power generations (grid connected)
5. Solar-thermal Energy for power generation (grid connected)
6. Wind pumps for agriculture

Industrial Efficiency

7. Direct reduction processes in iron and steel industry
8. Continuous pulp digesters in pulp and paper industry
9. Demand-side management: efficient motors

Power Generation

India currently has an installed power generation capacity of about 87481 MW, which is insufficient to meet the demands of the country. The demand for power has been growing at the rate of 9 per cent a year and is likely to accelerate because of the economic liberalization policies of the GoI. TERI estimates the power demand in India will be 385000 MW by the year 2020, requiring 380 billion dollars of investment in the power sector alone. Of the total generation capacity of the nation, 71% is based on coal. With proven reserves of about 70 billion tonnes and estimated reserves of about 200 billion tonnes, coal will continue to be the dominant fuel for power generation in the country. Hence introduction of cleaner technologies for coal utilization is one of the sustainable objectives of the nation. The proposed options for power generation are:

1. Coal power plant using IGCC.
2. Coal power plant using PFBC.
3. Renovation and modernization of power plants.

Coal Power Plant using PFBC

PFBC is a clean and efficient technology for coal-based power generation. In this technology, the conventional combustion chamber of the gas turbine is replaced by a pressurized fluidized bed combustor. The products of combustion pass through a hot gas cleanup system before entering the turbine, thereby reducing the amount of CO_2 emitted through combustion. The CO_2 reduction from a per unit of power supplied at bus bar using PFBC, over the conventional technology is 0.18 kg. Replacing a conventional plant of 500 MW capacity with a PFBC plant will result in CO_2 emission reduction of 0.58 million tonnes a year. The national mitigation potential for this technological option has been calculated at 8.166 million tonnes of CO_2 annually.

As PFBC technology has a high capital cost, the cost of production is marginally higher than that from a conventional power plant. But savings in coal compensate for the higher costs. The technology results in considerably lower emissions of SO_2 and NO. The technology therefore has a good potential both from a commercial and an environmental perspective.

Coal Power Plant using IGCC

IGCC is one of the technologies being explored to improve the efficiency of power generation and reduce the environmental burden. A coal gasification combined cycle technology centres around two elements:

1. A gasification plant that converts the fuel into a combustible gas and purifies the gas.
2. A combined cycle power plant which produces synthetic gas that fuels a gas turbine whose hot exhaust gases are used to generate steam to drive a steam turbine.

The efficiency of the IGCC is the product of the gasification system efficiency and the combined cycle efficiency. The auxiliary power consumption in oxygen blown gasifier is much higher than air blown gasifier and other pulverized coal plants. It is approximately 20% of the gross output. The air separation plant is very energy intensive because of the cryogenic process involved for the production of oxygen. However, the net efficiency of this technology is significantly higher because of the higher calorific value of the synthetic gas produced. For the second-generation technologies, net efficiencies are expected to be in the range of 45-50%.

The current capacity of thermal power plants (existing and sanctioned) is 12647 MW, and the GoI has cleared another 16898.5 MW capacity for implementation. These plants are expected to have a net efficiency of 36% whereas the state-of-the-art IGCC plants have net efficiencies of 46%. Estimates based on above efficiency levels indicate that an IGCC plant will result in CO_2 emission reductions of 0.25 kg/kWh. Replacing a conventional plant of 500 MW capacity with an IGCC plant will result in CO_2 emission reduction of 0.69 million tonnes. Assuming that IGCC would be adopted for 10% of the capacity sanctioned, 1690 MW, the total potential for emission reduction using IGCC will be 2.33 million tonnes of CO_2.

The proposed IGCC power plant will reduce emissions of SO_2, NOx, and SPM (suspended particulate matter). It will also reduce solid waste disposal by nearly 70% compared to direct coal-fired plants. An IGCC plant consumes is 0.46 kg of coal for one unit of power generated as opposed to conventional plant's consumption of 0.60 kg/kWh. This reduces energy use and resource spent on producing and transporting coal. The demonstration of this technology will reduce its implementation cost. The planned activities are expected to advance the financial, technological, and economic viability of the technology, creating distribution channels for manufacturing and engineering activities related to gasification technology, advanced gas turbines, and other IGCC/power-related activities.

Renovation and Modernization of Power Plants

With increasing demand for power, the state-of the art power plants can possibly meet all the requirements but the constraint is the extremely high cost resulting from the high capital cost. Many old and inefficient power plants that contribute to the grid could be upgraded to work efficiently. For a developing country like India, where there is a shortage of financing for additions to capacity and a large number of inefficient and old plants still in operation, renovation and modernization can greatly help improve efficiency, as this just entails a replacement of parts of the plant. Most of the small plants running on coal in India operate below 30% efficiency. Increasing the efficiency could lead to huge amounts of reductions in the emission of CO_2 from this sector. But renovation and modernization has been very slow in the country mainly due to paucity of funds. The proposed project looks into the renovation and modernization of existing old inefficient power plants in the country and will result in reductions in the emissions of SO_2, NOx, and SPM. The project would reduce coal consumption which in turn would greatly reduce CO_2 emissions.

Renewables

Fossil fuel use is the largest contributor to greenhouse gas emissions, the primary cause of global climate change. India, with a large endowment of coal, has an energy system that is highly carbon intensive. Besides, large quantities of traditional biomass resources consumed for energy needs of the vast rural population are exerting pressures on forests and village woodlots. Thus, the energy system is turning out to be 'doubly unsustainable'. Renewable energy technologies (RETs), despite their techno-economic potential, have found meager deployment due to myriad barriers. Recent developments in global climate change negotiations, which culminated in the Kyoto Protocol, are likely to remove some of the vital barriers to RETs which permitted fossil fuels to externalize the environmental costs (http://www.decisioncraft.com/energy/papers/ecc/re/recc.pdf).

India has had a significant renewable energy program for nearly two decades, and is the only country to have a full-fledged national ministry to deal with renewables. Launched primarily as a response to the perceived rural energy crisis in the 1970s, the Indian renewable energy program received an impetus with the economic liberalization process that began in the early 1990s, with the emphasis shifting from purely subsidy-driven dissemination programs to technology promotion through commercial route. Thus, India has gained valuable experience in promoting RETs using different approaches, and has achieved a few successes, notably biogas and wind energy. However, a synthesis of this experience shows that a number of barriers still remain to be overcome, if RETs have to become commercially viable alternatives.

In the post-Kyoto scenario, the CDM (Clean Development Mechanism) is an instrument which offers the opportunity to enhance the deployment of RETs. In the long run, the penetration of RETs shall however depend on the market for carbon offsets and the pace of development of individual RETs. Our analysis of the long term energy and environment trajectories for India suggests that in absence of stringent climate change mitigation policies, India is likely to grow along the conventional fossil fuel path. The same can be true for most other developing nations. The policies in specific countries, especially the developing countries with do binding carbon limitation commitments shall be crucial for the generating initial "technology-push", before the market shall be ready to provide the "demand pull" in the long-run.

The potential contribution of Renewables in earning Carbon Credits from various CDM project activities can be seen for some representative countries in the Table given below.

Table: Increasing the Internal Rate of Return of Renewables through Emissions Trading

Country	Project Type	IRR w/o (per cent)	IRR w/ET (per cent)	IRR Increase (percentage points	IRR Increase (per cent)
Romania	District heating	10.5	11.4	0.9	9
Costa Rica	Wind	9.7	10.6	0.9	9
Jamaica	Wind	17.0	18.0	0.9	6
Morocco	Wind	12.7	14.0	1.3	1.0
Chile	Hydro	9.2	10.4	1.2	1.3
Costa Rica	Hydro	7.1	9.7	2.6	3.7
Guyana	Bagasse	7.2	7.7	0.5	7
Nicaragua	Bagasse	14.6	18.2	3.6	2.5
Brazil	Biomass	8.3	13.5	5.2	6.3
Latvia	Methane	11.4	18.8	7.4	6.5
India	Methane	13.8	18.7	4.9	3.6

Source: *World Bank*, July 2001.

Wind-based Power Generation (grid connected)

Wind is one of the important grid-connected renewable energy resources. The installed wind capacity as of December 1999 was 1080 MW. A huge potential remains unexploited and the GoI is addressing institutional and policy issues to increase the spread of wind based capacity. The MNES (Ministry of Nonconventional Energy Sources) to raise the share of renewable energy in total installed capacity of grid electricity generation to 10% by 2012.

India has a potential of 35,000 MW at 30 m hub height and 45,000 MW at 50 m hub height. The installed capacity is 1080 MW. Over the years the performances have increased with improving CF (capacity factors), with some projects achieving 30% CF. Concessional interest rates, favorable buy-back rates, and other fiscal policies support the wind energy projects. The availability of extra finances for the project can improve the feasibility and, hence, the rate of increase in capacity. The capacity addition in the first three years of the Ninth Plan (1997-2002) was 180 MW. Given the current requirement of 12,000 MW capacity a year to keep up with the demand, the GoI is aiming at 1200 MW from renewable sources. A more realistic capacity addition rate is been 4000 MW a year, and the target for renewable capacity additions would be 400 MW. Based on the past experience, and improving policy environment, a realistic estimate of capacity addition from wind would be 100-200 MW. The additional capacity in power generation is assumed to replace the planned thermal capacity in the country, which is expected to have design efficiencies of 38%.

Solar-Thermal Energy for Power Generation (grid connected)

The utilization of solar thermal energy for power generation is likely to be higher on the list of priorities of the Ministry of Power and that of the MNES as it has a great potential for GHG reduction. It is a clean technology and does not involve use of any conventional fuel, fossil or otherwise. The Fifteenth Electric Power Survey shows that the energy requirement at power station will be the highest for the northern region. The proposed option includes the installation of parabolic trough system to generate power using solar thermal energy. The specific option considered looks into the installation of solar thermal plants of capacity 500 MW. Given that use of solar thermal energy for power generation requires intensive solar radiation, northern region is suitable location since the Thar Desert of India falls in the region. Population is concentrated in and around urban areas of this desert region. Thus provision of space, a major requirement for setting up of solar plants, would be ideal in the region given the intense solar radiation available.

Wind Pumps for Agriculture

India has a large potential for wind energy. Apart from an estimated potential of 45 000 MW of wind-based power, the country can use wind energy directly to pump out water for irrigation and drinking purposes. If harnessed effectively, this option would reduce considerably the use of diesel and grid electricity in conventional pumps, thereby conserving fossil fuel and offsetting environmental concerns related to their use. There are about 5 million diesel and 4 million electric pumps operating in the agricultural sector. The MNES has been installing geared-type deep well wind pumps under its demonstration programme and providing financial incentives for wider dissemination of the technology. A wind pump with rotor diameter over 5 m at a windy site can replace a 5 HP diesel or electric pump. Based on this observation, the incremental cost of CO_2 saved by switching from electric to wind pumps was estimated. Assuming the requirement of water to be 10 000 litres a day, 365 days a year, it was concluded that wind pumps could meet the water requirements during lean periods. Using a discount rate of 10%, and taking the average life of the pump as 20 years, an analysis showed that the option has tremendous carbon saving potential at relatively low cost.

International Conventions and Agreements

The United Nations Framework Convention on Climate Change entered into force in March 1994 and as of July 2000 has received 184 instruments of ratification or accession. The Convention includes a commitment by developed country Parties, including economies in transition (Annex I Parties), to aim to return emissions of CO_2 and other GHGs not controlled by the Montreal Protocol to their 1990 levels by 2000. The Kyoto Protocol was adopted in December 1997 and has received 84 signatures. It will enter into force after it has been ratified by at least 55 Parties to the Convention, including developed countries accounting for at least 55 per cent of the total 1990 CO_2 emissions from this group. Meanwhile countries are to continue to carry out their commitments under the Convention. Ozone-depleting greenhouse gases (such as CFCs and HCFCs) are controlled by the Vienna Convention and the Montreal Protocol (see ozone depleting substances).

International Targets and Recommended Standards

The Kyoto Protocol sets targets for each of the developed country Parties and economy in transition Parties with a view to reducing their overall emissions of the six main GHGs by at least 5 per cent below 1990 levels in the commitment period 2008 to 2012.

Linkages to Other Indicators

This indicator is linked to many other socio-economic and environmental indicators, including GDP growth rate, energy consumption, environmental protection expenditures, and expenditures on air pollution abatement.

Underlying Definitions and Concepts

Greenhouse gases (GHGs) contribute in varying degrees to global warming depending on their heat absorptive capacity and their lifetime in the atmosphere. The global warming potential (GWP) describes the cumulative effect of a gas over a time horizon (usually 100 years) compared to that of CO_2. For example, the global warming potential of CH_4 (methane) is 21, meaning that the global warming impact of one kg of CH_4 is 21 times higher than that of one kg of CO_2. The global warming potentials of ozone-depleting greenhouse gases (such as CFCs and HCFCs) are highly uncertain, since they depend on the depletion of ozone, itself a greenhouse gas. No global warming potentials are provided for indirect greenhouse gases.

Measurement Methods

In some cases, GHG emissions can be measured directly at the source. More commonly, emissions are estimated from data on emission sources, for example oil sales data or cattle numbers, using an emission factor for each source.

Limitations of the Indicator

This indicator shows the net amount of GHGs entering the atmosphere for each reporting country each year. It does not show how much the climate will be affected by the increased accumulation of GHGs or the consequent effect of climate change on countries. Data is available and reported mainly for developed countries and economies in transition.

Status of the Methodology

Developed country Parties to the Convention have been reporting GHG data beginning with 1990 data since 1994. The IPCC has published two sets of guidelines on methodologies for the estimation of GHG inventories and further elaborated this with guidance on good practice in 2000.

Alternative Definitions and Indicators

GHG emissions can alternatively be measured on a gross instead of net basis in which case no account is taken of removal by sinks. There are a number of other gases that indirectly produce GHGs and these could also be included in the scope of the definition. The GWP potential can be calculated over different time horizons, such as 20 years or 500 years.

Intensity of Energy Use in Transportation: Setting Standards

Underlying Definitions and Concepts: Energy consumption per unit of transportation activity is a key measure of how efficiently transportation systems convert energy into human mobility and goods distribution. Because it is not meaningful to add freight and passenger travel, these types of transportation must be disaggregated. Separating the two activity measures is generally not difficult, but separating the energy consumption is often complicated.

Measurement Methods: Energy Use

Energy consumption should be measured for each kind of vehicle, including two-wheelers, automobiles, busses, small trucks, heavy trucks, and miscellaneous road vehicles, as well as trains, ships and aircraft for domestic transport, and even pipelines. In general, however, national energy balances are only disaggregated by fuel and broad traffic type: road, rail, water, and air. Considerable work is required to disaggregate road fuels consumed by vehicle type. It is important to take into account the different energy content and carbon emissions in different fuels and not simply add the weights or volumes of different fuels consumed (e.g., tonnes, or cubic metres in the case of natural gas). Some of the difficulties in disaggregating road fuels consumed by vehicle type are explained in Schipper *et al.* (1993). International air or marine transportation should not be included. Electric power consumption for rail, subway and trams, as well as electric road vehicles, should be converted to primary energy consumption, although there is no standard method for such conversion (http://www.un.org/esa/sustdev/sdissues/consumption/cpp1224m4.htm).

Indicators Used:

(i) **Vehicle Intensities:** Energy consumption per vehicle-km by vehicle and fuel type is an important indicator, as many standards for air pollution (and more recently, goals for CO_2 emissions reduction) are expressed in terms of vehicle characteristics, i.e., emissions per vehicle-km.

(ii) **Modal Intensities:** Energy use per passenger-km or tonne-km should be disaggregated by vehicle type, i.e., two-wheeler, car/van, bus, air, local and long-distance rail, subway, tram, ship or ferry for passengers; and truck, rail, ship, air for freight.

National and International Data Availability and Sources

National communications from Parties to the Convention, including both developed and developing countries are available. Developing countries report on a limited basis. At the international level, the UNFCCC Secretariat database has information based on annual data inventory submissions from developed and economy in transition countries.

Methodologies and Descriptions

Underlying Definitions and Concepts

Ozone Depleting Substance (ODS) means any organic substance

containing chlorine or bromine, which destroys the stratospheric ozone layer. Controlled substance means a substance in Annex A, Annex B, Annex C or Annex E of the Montreal Protocol, whether existing alone or in a mixture. It includes the isomers of any such substance, except as specified in the relevant Annex, but excludes any controlled substance or mixture which is in a manufactured product other than a container used for the transportation or storage of that substance. Production means the amount of listed, controlled substances produced, minus the amount destroyed by technologies to be approved by the Parties to the Montreal Protocol and minus the amount entirely used as feedstock in the manufacture of other chemicals. The amount recycled and reused is not to be considered as "production". Consumption is the sum of production plus imports minus exports of controlled substances. We are addressing apparent consumption. Weighted tonnes of ODS means the amount of ODS multiplied by their ozone depleting potential. Ozone depleting potential (ODP) is a relative index of the ability of a substance to cause ozone depletion. The reference level of 1 is assigned as an index to CFC-11 and CFC-12. If a product has an ODP of 0.5, a given weight of the product in the atmosphere would, in time, deplete half the ozone that the same weight of CFC-11 or CFC-12 would deplete. ODPs are calculated from mathematical models which take into account factors such as the stability of the product, the rate of diffusion, the quantity of depleting atoms per molecule, and the effect of ultraviolet light and other radiation on the molecules.

Measurement Methods

Weighted Tonnes of ODS for production are the sum of national annual production (in tonnes) of each controlled substance (as reported to the Ozone Secretariat in accordance with Article 7 of the Montreal Protocol) multiplied by the ozone depleting potential of that substance (as listed in Annexes A, B, C and E of the Handbook for the International Treaties for the Protection of the Ozone Layer, 2000). It can be found at: http://www.unep.org/ozone or http://www.unep.ch/ozone. Weighted Tonnes of Ozone Depleting Substances for consumption are obtained through a similar calculation using national annual consumption values (in tonnes).

(a) **Limitations of the Indicator:** Availability and accuracy of data and timely reporting will determine the country's ability to use the indicator. The indicator by itself does not reveal much about current trends in the deterioration of the ozone layer because of delays in ecosystem response.
(b) **Status of the Methodology:** For more information, please consult the Reports of the Secretariat on information provided by the Parties in accordance with Article 7 of the Montreal Protocol or the Home Page at: http://www.unep.org/ozone or http://www.unep.ch/ozone.
(c) **Alternative Definitions/Indicators:** An alternative indicator could focus on the phase out of ODS.

Agencies involved in the Development of the Emissions Indicator

Lead Agency: The lead agency is the United Nations Environment Programme (UNEP, 1999)/Ozone Secretariat.

Other Contributing Organizations: Other organizations interested in the further development of this indicator would include: The Multilateral Fund Secretariat, the Global Environment Facility (GEF) Secretariat, United Nations Development Programme (UNDP), UNEP Division of Technology, Industry and Economics (UNEP DTIE), United Nations Industrial and Development Organization (UNIDO), the World Bank, the Technology and Economic Assessment Panel to the Montreal Protocol, the Parties to the Montreal Protocol, the Organization for Economic Co-operation and Development (OECD), and members associated with the Alternative Fluorocarbon Environmental Acceptability Study (AFEAS).

Greenhouse Inventory Data

The ultimate objective of the Climate Change Convention (UNFCCC) is to achieve "... stabilization of greenhouse gas concentrations in the atmosphere at a level that would prevent dangerous anthropogenic interference with the climate system."

Estimating the levels of greenhouse gas (GHG) emissions and removals is an important element of the efforts to achieve this objective. This web page provides access to most recent data on national GHG emissions and removals as follows:

- *GHG Data—UNFCCC:* this page is an entry to the UNFCCC GHG data interface which provides access to the most recent GHG data reported by countries that are Parties to the Climate Change Convention.
- *KP Data—UNFCCC*: this is a page with GHG data relating specifically to the Kyoto Protocol; the data are as reported by countries that are Parties to the Kyoto Protocol.
- *GHG Data—Non-UNFCCC:* this page provides links to the web sites of various organizations that also collect, estimate and/or disseminate data on GHG emissions/removals.
- *Online Help:* the help page contains extensive explanatory information to assist the user in finding the right GHG data.

Regional Greenhouse Gas Initiative

The Regional Greenhouse Gas Initiative (RGGI) is the first mandatory U.S. cap-and-trade program for carbon dioxide. Currently, ten northeastern and Mid-Atlantic States are participating: Connecticut, Delaware, Maine, Maryland, Massachusetts, New Hampshire, New Jersey, New York, Rhode Island, and Vermont (http://www.rff.org/rff/News/Features/Regional-Greenhouse-Gas-Initiative.cfm)

RGGI sets a cap on emissions of carbon dioxide from power plants in the region, and allows sources to trade emission allowances. The program will begin by capping emissions at current levels in 2009, and then reducing emissions 10 percent by 2019. Sources will continuously monitor and report their emissions, and penalties for non-compliance will be enforced according to each state's rules. Member states agree to each set aside at least 25 percent of their emission allowances for public benefit purposes, such as promoting renewable energy and energy efficiency or mitigating possible increases in consumer energy prices. Many of the RGGI states have committed to or are considering setting aside a greater portion of allowances for such purposes. RGGI also allows the use of offset projects for compliance, but these projects will need to meet strict standards and will be limited to ensure that significant reductions occur at electric generators.

California and the West

In September 2006 Governor Schwarzenegger signed AB 32, the Global Warming Solutions Act. The Act caps California's greenhouse gas emissions at 1990 levels by 2020 and represents the first state-wide program in the United States that caps all GHG emissions from major industries and includes penalties for non-compliance. California is currently in the process of designing a comprehensive program to meet its goals under AB 32. The state is considering using market mechanisms, including a cap-and-trade program, as part of AB 32 implementation, and may try to link its program to RGGI and the EU-ETS markets.

California is also part of a larger, regional cap-and-trade program that is likely to emerge in the West. In February 2007 the Governors of Arizona, California, New Mexico, Oregon, and Washington signed an agreement establishing the Western Climate Initiative, a joint effort to reduce greenhouse gas emissions and address climate change. The states of Utah and Montana as well as the Canadian Provinces of British Columbia and Manitoba joined the Initiative in the following months. Under the agreement, the states and provinces jointly set a regional emissions target in August 2007 of 15 percent below 2005 levels by 2020, and by August 2008 will devise a market-based system—such as a cap-and-trade program covering multiple economic sectors—to aid in meeting the target.

Midwestern Accord

The Midwestern Greenhouse Gas Accord is a regional agreement by six governors of states in the US mid-west who are members of the Midwestern Governors Association (MGA) and the Premier of one Canadian province to reduce greenhouse gas emissions to combat climate change. Signatories to the Accord are the US states of Minnesota, Wisconsin, Illinois, Iowa, Michigan, Kansas, and the Canadian Province of Manitoba.[1] Observers of the Accord are Indiana, Ohio and South Dakota. The Accord was the fourth tier of the

MGA Energy Security and Climate Stewardship Summit Platform, signed on November 15, 2007. It establishes the Midwestern Greenhouse Gas Reduction Program (http://en.wikipedia.org/wiki/Midwestern_Greenhouse_Gas_Accord), which aims to:

- Establish greenhouse gas reduction targets and time frames consistent with signing states' targets;
- Develop a market-based and multi-sector cap-and-trade mechanism to help achieve those reduction targets;
- Establish a system to enable tracking, management, and crediting for entities that reduce greenhouse gas emissions; and
- Develop and implement additional steps as needed to achieve the reduction targets, such as a low-carbon fuel standards and regional incentives and funding mechanisms.

The GHG registry will be managed by The Climate Registry, which manages the registry for other US state schemes. One of the first actions was to convene an Energy Security and Climate Stewardship Platform to guide future development of the Midwest's energy economy.

In November 2007, six states and one Canadian Province established the Midwestern Regional Greenhouse Gas Reduction Accord, under which members agree to establish regional greenhouse gas reduction targets, including a long-term target of 60 to 80 percent below current emissions levels, and develop a multi-sector cap-and-trade system to help meet the targets. Participants will also establish a greenhouse gas emissions reductions tracking system and implement other policies, such as low-carbon fuel standards, to aid in reducing emissions. Members of the Accord include Illinois, Iowa, Kansas, Michigan, Minnesota, and Wisconsin, as well as the Canadian Province of Manitoba. As of July, 2008, staff were collecting data and running computer simulations to help determine the rules, emission levels, and costs the cap-and-trade system will impose.

Key Issues around Developing CER Projects in Asia

Relative to Latin America and South Africa (rapidly becoming a major CDM player), Asia is relatively behind in the development of CDM projects and the required governmental infrastructure to support them. Whereas Latin America pioneered the concepts that eventually became the CDM during the Climate Convention negotiation process leading up the Kyoto Protocol, there has consistently been more reticence among most Asian governments in engaging the CDM as a tool and as a domestic policy process. This said, many conventional Asian energy developers remain unaware of the CDM tool and the financial impact it can have on a variety of project types that are being pursued in the region. While the reasons for the gap are speculative and not appropriate for discussion in this forum, it can be said that there is an emerging CDM interest in the region and it can be expected that the gap will close. Given the population dynamics, projected energy demand and significant pressures in the region for better waste and air quality management, it would be surprising not to see the CDM ultimately be adopted with enthusiasm—at least among certain governments.

The engagement of governments must be considered a vital prerequisite for the CDM to flourish. Without defined and transparent process of how governments will approve projects for CDM trading, developers cannot know if they are able to financially assume any value from the CDM on their *pro-* and therefore move the commodity from the speculative to the bankable. While India has already approved several projects for trading, in the rest of the region only Malaysia has moved forward with any certitude while Thailand is only slowly reversing itself from some very strongly negative comments that it made on the mechanism late in 2002. It is expected that a significant market for Asia credits will emerge from Japan, which will potentially seek co-investment equity and debt deals to accompany carbon transactions. However, it is telling that, to date, of the millions of dollars that Japanese firms have put into emissions transactions, the highest profile direct transactions have been with Brazil, Slovakia and Kazakhstan. The transaction portfolio of the World Bank's Prototype Carbon Fund—in which Japanese investors represent some 50% of the private capital—is almost entirely in Latin America with a small amount in Africa (EcoSecurities, 2003).

The Kyoto Protocol entered into force on 16 February 2005. On 11 July 2006, Singapore formally acceded to the Protocol, enabling both the government and the private sector to participate in the Kyoto Protocol market]based mechanisms. This is in line with Singapore's drive toward becoming Southeast Asia's environmental financial centre as well as a hub for R&D and eventual marketing of renewable energy technologies. International companies looking to fund projects to cut greenhouse gas emissions are already coming to Singapore to take advantage of the country's accession to the Kyoto Protocol and its prominence as a financial centre. Amongst these companies are New York—listed energy firm AES Corp and London—listed Eco-Securities which specializes in sourcing and trading carbon credits. As Singapore gears up to play a more active role in the carbon market created by the Kyoto Protocol, Singapore lawyers can expect opportunities in this field, particularly in tandem with more traditional commercial practice areas including banking law, capital markets, and project financing. An understanding of the international context of the Kyoto Protocol is therefore essential (APCEL, 2007).

India and the Global Market for CDM projects

The international climate treaty, the Kyoto Protocol, would soon become a reality. After over seven years on the limbo, the international climate policy has achieved a breakthrough as a result of Russia's decision to ratify the Kyoto

Protocol. The Protocol that sets legally binding Greenhouse Gas (GHG) emission targets for industrialized countries entered into force on February 16, 2005. Only four industrialized countries now remain outside the treaty: the US, Australia, Liechtenstein and Monaco. The US and Australia are unlikely to come on board in the near future due to persistent opposition from their political leaders. Both countries have a vast scope for energy efficiency improvement, but are saddled with powerful fossil fuel lobbies. Moreover, their emissions have risen considerably during the last decade.

GHG Reductions Generate Revenues for Developing Countries

In the Kyoto Protocol, 38 industrialized countries and countries in transition were allocated an emissions target for the period 2008-12. The targets were differentiated between countries and range from –28 percent (Luxembourg) to +10 percent (Iceland) compared to 1990 levels. They refer to a basket of six GHGs and are denominated in CO_2 equivalent. A tonne of methane has an 'exchange rate' of 21, a tonne of the industrial gas HFC-23 as high as 11,700. (CO_2 primarily originates from burning of fossil fuels, and methane from rice cultivation and enteric fermentation in livestock among others. HFC-23 is used in the refrigeration sector).

An unprecedented feature of the Protocol is the possibility of using international market mechanisms to achieve the emissions targets. There are three "Kyoto Mechanisms" that allow buying and selling of emission permits: international emissions trading, Joint Implementation and the Clean Development Mechanism (CDM). CDM allows developing countries to implement projects that reduce GHGs and thereby generate emission permits (Certified Emission Reductions or CERs). CERs can be credited by industrialized countries toward their emission targets.

Like Canada, the EU and Japan will not be able to achieve their Kyoto targets by domestic action alone—there will be a considerable demand for CERs. By end 2004, more than \$1,000 million have been committed by governments and private companies to acquire CERs. New tenders are announced almost every month. A tonne of CO_2 equivalent currently trades at \$4-5 and the price trend is rising. So the global CER demand stands at 250 million tonnes and is likely to double within the next year as there remain large emission gaps in important countries such as Italy and Spain. The overall emissions gap for the period 2008-12 will surpass 3 billion tonnes of CO_2 equivalent. However, Russia, Ukraine and the countries of Eastern Europe have a huge surplus of emissions permits, colloquially called "hot air," due to their economic breakdown in the 1990s and could supply enough to cover the whole shortfall. Therefore, the CDM market heavily depends on the behavior of the hot air countries.

To ensure that CERs are not just "paper tonnes," CDM projects have to undergo independent scrutiny by officially accredited certifiers, the so-called "Operational Entities." An international body, the CDM Executive Board, has to register each project. There is a prescribed format to be filled by each CDM project developer; the promoter has to calculate emissions reductions according to complex rulés and make sure that key parameters are monitored throughout the lifetime of the project.

Concerns and Issues in Implementation of CDM

"To understand and overcome the risks and uncertainties associated with CDM projects, it is necessary to look at issues relating to additionality and baseline."

Concerns Regarding Additionality

To ensure that host countries refrain from introducing policies that encourage GHG reductions due to the perceived need to keep a high baseline, the Executive Board has recently decided that policies that generate emissions increase (e.g. fossil fuel or electricity consumption subsidies) can be taken into account in the baseline scenario if they were introduced before December 11, 1997. In the Indian context, this covers the electricity subsidies for farmers. Policies setting incentives for GHG reductions (e.g. mandatory renewable energy quotas, subsidies for energy efficiency and renewable technologies) can be disregarded in the baseline scenario if they were introduced after November 11, 2001. This spells trouble for India as most renewable power tariffs on the state level were already introduced long before 2001.

There is still confusion about what would be considered a 'policy' of the Government of India (GOI), because there are legal differences between a recommendation, policy, ordinance, act, directives, rules and regulations of the GOI and the states. The fact that some subjects are also concurrent complicates the matter further. For instance the Center's policies and notifications may not necessarily be accepted or implemented by the states, for instance as seen in the concurrent subject of electrical power.

Concerns Regarding Baselines

To calculate an electricity baseline, up-to-date and detailed grid data on the so-called operating and build margin have to be available. While data availability in India is much better than, for example in China, still all presently quoted baselines concerning the India power sector are approximations and at best rough guesses. A necessary condition for India to remain an attractive CDM host is that the National CDM Authority publishes and regularly updates the operating and build margin of the relevant grids.

Currently in India, many national and international NGOs and consultants are attempting to develop their own databases with and without CEA input. None of them has any access to validated data about carbon content of coals fired in Indian thermal power plants. Without this key figure all referenced regional or national grid CO_2 emission factors are mere guesstimates. It may turn out that the coal consumption and related carbon emission of Indian power plants is not as high as always reported.

Concerns Regarding Diversion of Development Assistance

It is often asked whether the ODA (Official Development Assistance) funding can be used for CDM projects. If Indian decision-makers feel that this is a problem, they can direct the Indian National CDM Authority to deny host country approval for any project using ODA and no other authority can revert this decision. Therefore, it is quite simple for the GOI to exclude any projects, which are fully, or partially ODA-funded from becoming a project entitled to earn CER under CDM.

Strengthening the Processes

India has the potential to significantly benefit from the emerging CDM market, as it is now supported by a proactive National CDM Authority and the emergence of a dynamic CDM consultancy and project development community. Indian industry has also demonstrated the capacity to tap emerging opportunities in niche markets in energy and environmental technologies. The PDD screening process is the first internationally recognized approval procedure, which attempts to define "sustainability" and to develop specific criteria to identify it. PDDs must be compiled professionally to ensure that they successfully survive the international screening process. The National CDM Authority could facilitate the project developers by setting up a baseline database for the most common project types. This would considerably reduce the documentation costs for project developers. Further, India should use CDM as an opportunity to revamp its quantitative and qualitative data collection and monitoring systems in the energy and environmental sectors. India emits only 3 percent of the world's total emissions of CO_2 and our per capita emissions are only one-fifth of the global average. However, Indian electricity utilities and industries are very CO_2 emissions-intensive due to their large coal use and somewhat obsolete technologies

Supported by a pro-active National Clean Development Mechanism (CDM) Authority, and a dynamic CDM consultancy and project development community, Indian industry should richly benefit from the emerging CDM market. However, stakeholders in India should ensure that projects and their documentation are of world standards and can survive international scrutiny. India has the potential to significantly benefit from the emerging CDM market, as it is now supported by a proactive National CDM Authority. Sector-wise projects approved by the National CDM Authority in India are presented below:

Sector	No. of Projects
Biomass/Co-generation	16
Industrial Processes/Energy Efficiency	15
Municipal Solid Waste	1
Fuel Switching	2
Renewable Energy	10
Total	44

Evaluating Greenhouse Gases and Climate Change Impacts under NEPA: A Case Study

The National Environmental Policy Act (NEPA) requires federal agencies to take a hard look at the environmental consequences of their proposed actions; however, many agencies are preparing NEPA documents with little or no analysis of greenhouse gas emissions (GHG) or the impacts of global climate change. As policy makers continue to debate how agencies should address GHG emissions and climate change impacts under NEPA, lead agencies cannot afford to wait for answers. This paper briefly explains why GHG/climate change impacts must be evaluated under NEPA and includes a recommended ten-step approach for minimizing the risk of legal jeopardy (http://www.icfi.com/docs/GHG-CC-NEPA.pdf). Every day, agencies are preparing NEPA documents with little or no guidance about how to address GHG/climate change issues. Unfortunately, all too many agencies are coming up with excuses for not addressing the issue at all or including only vague references to it. Here are some of the more common excuses:

- "NEPA doesn't cover climate change impacts"
- "NEPA only applies to impacts in the United States"
- "Climate change impacts are speculative and not foreseeable"
- "We scoped it out of our document"
- "We can't solve the global problem so why discuss it"

Agencies relying on these excuses are in denial that NEPA requires GHG/climate change impacts to be evaluated and are ignoring the emerging trends in the courts and in state legislatures.

Existing Framework under NEPA and CEQ Regulations

Under the existing framework of NEPA and the Council on Environmental Quality's (CEQ's) NEPA regulations, there is little doubt that GHG emissions and climate change impacts must be evaluated. Consider the following:

- Effects on air quality fall under NEPA's definition of *"Effects on the Human Environment"* (40 CFR 1508.8: "Effects includes ecological (such as the effects on natural resources and on the components, structures of affected ecosystems), aesthetic, historic, cultural, economic, social, or health, whether direct, indirect, or cumulative." And 1508.14: "Human environment shall be interpreted comprehensively to include the natural and physical environment and the relationship of people with that environment.").

- NEPA requires consideration of "direct," "indirect," and "cumulative" effects (40 CFR 1502.16: "The GHG emissions resulting from a proposed action are either direct or indirect effects, and therefore the resulting global climate change impacts are classic examples of cumulative effects.").
- Since climate change impacts are inherently cumulative, they fall within the cumulative effects component of NEPA's definition of significant (40 CFR 1508.27(b)]: "Whether the action is related to other actions with individually insignificant but cumulatively significant impacts. Significance exists if it is reasonable to anticipate a cumulatively significant impact on the environment.").
- NEPA requires discussions of "possible conflicts between the proposed action and the objectives of Federal, State, and local land use plans, policies and controls for the area concerned." Thus, a NEPA document must evaluate the relationship between the GHG emissions from a proposed act and any relevant state GHG emission reduction laws or policies.

As these examples illustrate, there is no reasonable interpretation of the CEQ NEPA regulations that would suggest that GHG/climate change impacts should be left out of NEPA documents. On the contrary, the trend in the federal courts and at the state level suggests GHG/climate change impacts should definitely be considered under NEPA.

Last year, the US Supreme Court held that climate change impacts are reasonably foreseeable, are caused in part by human activities and should be regulated as pollutants under the Clean Air Act (Environmental Protection Agency, 549 U.S. 497 [2007]). Additionally, the 9th Circuit Court of Appeals has already issued a ruling that a federal agency violated NEPA because it failed to adequately evaluate GHG/emissions and climate change impacts (Center for Biological Diversity)

Apart from this federal judicial activity, more than a dozen states have already enacted legislation regulating establishing reduction targets for GHG emissions. Although these state laws do not apply to federal agencies, NEPA does require that environmental documents consider the relationship between proposed federal actions and state environmental protection legislation. The Pew Center for Global Climate Change has more information on its web site regarding the issue. Thus, NEPA documents should discuss the relationship of proposed projects to applicable GHG reduction targets and should recommend mitigation measures and alternatives that would help to achieve those reductions.

Although CEQ issued draft global climate change guidance in 1997, neither CEQ nor the Environmental Protection Agency (EPA) has yet issued final guidance on how to address GHG/climate change impacts under NEPA. EPA has, however, written Nov. 2008 formal comment letters on various NEPA documents questioning why the lead agencies have not taken such impacts seriously.

Ten-Step Approach to Addressing GHG and Climate Change Impacts

Although there is little doubt that lead agencies must evaluate GHG/ climate change impacts under NEPA, in the absence of specific, mandatory requirements, the key question is:

What should an agency do – right now – to avoid putting itself in jeopardy?

The following is a recommended 10-step approach that agencies may want to follow in their consideration of GHG/climate change impacts under NEPA. These recommendations take into consideration the existing provisions of the NEPA regulations, the recent court decisions, and various state programs. They are broken down to conform to the main elements of a NEPA document.

Affected Environment

Step 1—Describe the existing global context in which climate change impacts are occurring and are expected to continue to occur in the future.

Step 2—Summarize any relevant state laws that address climate change.

Step 3—Describe any relevant national, statewide, and regional GHG inventories to which the project will contribute.

Environmental Consequences

Step 4—Quantify the project's direct and indirect GHG emissions.

Step 5—Convert the GHG emissions into carbon equivalents using an established "carbon calculator."

Step 6—Discuss whether the project would enhance or impede the attainment of applicable state GHG reduction.

Step 7—Describe the cumulative global climate change impacts to which the proposed action would contribute, in other words, the impacts of the project on climate change. (This may apply the same information as in step 1.)

Step 8—Describe how the impacts of global climate change could manifest themselves in the geographic area in which the project is proposed, and therefore potentially affect the project, in other words the impacts of climate change on the project (e.g., sea level rise could affect a coastal project).

Alternatives

Step 9—Include alternatives that would meet the project objectives but would also reduce GHG emissions

Mitigation measures

Step 10—Identify mitigation measures that would reduce GHG emissions, including both project design or operational changes and potential compensatory mitigation (e.g., carbon offsets).

Keeping in mind that the standard of adequacy for a NEPA document is whether an agency has taken a hard look at the environmental effects of a proposed action, the closer an agency comes to including all of the above steps in its analysis, the better the chance of deflecting criticism and prevailing in court should the NEPA document be challenged.

Greenhouse Gas Mitigation for Energy Efficiency

Global warming arises from the release into the atmosphere of gases that absorb the infrared radiation emitted by the earth, thus preventing the escape of the radiation into space. Examples of greenhouse gases (GHGs) include CO_2, CH_4, NOx and HFCs. The fact is that humans have produced substantial quantities of these greenhouse gases, leading to continuously degrading global environment.

As a result, GHG emissions and energy demand have risen high on the global environmental agenda—given the magnitude of GHG emissions from cities, urban energy efficiency is a significant challenge that requires special consideration. The role of city planners and the construction industry is essential as they create the necessary pre-conditions for energy savings opportunities to be realized.

Implementations of measures that are resource-efficient and mitigate negative environmental impacts are needed in all areas of human activity. The built environment is a clear example of this—buildings have a significant impact on the environment, accounting for one-sixth of the world's freshwater withdrawals, one-quarter of its wood harvest, and two-fifths of its material and energy flows. Structures also impact areas beyond their immediate location, affecting the watersheds, air quality, and transportation patterns of communities.

A deeper understanding of the issues involved is leading to changes in the way the building industry and building owners approach the design, construction, and operation of structures. Incorporation of sustainability principles within such processes is becoming a key goal in not only reducing and preventing negative environmental impacts, but also to implement energy-efficiency in buildings and construction processes. Some important aspects of energy efficient in urban areas include (a) maximizing the energy efficiency of building and infrastructure operations through the use of renewable resources, decentralized co-generation and energy cascading techniques in a manner which optimizes integrated energy flows and minimizes potential global environmental impacts such as GHG emissions, and (b) linking producers and consumers of energy and materials throughout the community, city and surrounding regions to facilitate resource exchanges and recycling networks (http://www.gdrc.org/uem/green-const/greenhousegases.html)

Therefore, the environmental implications of man-made structures—buildings, infrastructure etc. and the urban activities and consumption patterns that go into their design, construction, use, maintenance and demolition, need to be comprehended in greater detail. These issues are indeed the local beginnings of global environmental problems, and have resulted in a rethinking of how we look at cities and urban areas and of the built environment within these areas.

The increasing focus being placed on incorporating the concept of sustainability within the built environment, particularly on the impact of buildings and the construction process on the environment, has resulted in the development of a series of codes, standards, regulations and decision-support systems. These tools help in working with a range of stakeholders to achieve sustainability in the built environment.

What is missing however is a mechanism to link building and construction processes to the larger global environmental problems through a framework of goals and objectives laid against available codes, standards and regulations. We need greater focus on sustainability and environmental issues in the development of building codes and standards in light of the entire life-cycle of buildings. Built into such initiatives should be a coherent approach for energy-efficiency in the built environment, with the ultimate goal of reducing GHGs.

Adapting appropriate codes, standards and regulations for sustainability in the built environment, particularly in developing countries, requires a methodology for their scaling to different levels of economic development.

Strategy Outputs/Outcomes

1. Create enabling environments for green construction in developing countries:
   - Guidelines for Initiating, designing, constructing, maintaining, operating and demolishing buildings in an environmentally sustainable manner and the use of technologies.
   - Documentation of appropriate and indigenous building technologies and best practices Identification of sustainability indicators for the built environment and the construction industry.
   - Databases of technologies for green construction.

- Courses on green construction.

2. Stimulate the development and use of appropriate building codes and standards for sustainable building and construction:
    - Recommendations for codes and standards on sustainable design, assessment, management of building and construction.
    - Inventory of tools for environmental declaration and labeling of buildings.
    - Inventory of tools for assessing energy efficiency of buildings.
    - Databases on technologies for green construction.
3. Build required capacity to apply codes and standards among defined—institutional and professional—stakeholders:
    - Sustainable urban planning manual and guidelines.
    - Sustainable Planning Code of Practice.
    - Case studies and good practices in sustainable planning and building construction.
    - Award/Prize on good practices in green construction.

CHAPTER 4

Regulatory Regime and International Carbon Finance

Review of the Relevant Provisions on Emissions Trading in the Kyoto Protocol

The Kyoto Protocol accepts the concept of emissions trading under article 17, under which one Annex B country will be allowed to purchase the rights to emit greenhouse gases from other Annex B countries that are able to cut greenhouse gas emissions below their assigned amounts. Although Annex B to the Kyoto Protocol and Annex I to the Convention are now identical in nature, this change from Annex I into Annex B potentially allows a developing country to engage in emissions trading if it voluntarily adopts an emissions target and is inscribed in Annex B. Because the emissions trading proposal was adopted at the very end of the Kyoto negotiations, designing the relevant principles, modalities, rules and guidelines governing emissions trading have been deferred to a subsequent conference. The design of a workable emissions trading scheme is essential to the success of emissions trading. The market-based emissions trading approach can only achieve significant cost reductions in cutting greenhouse gas emissions while also allowing flexibility for reaching compliance if it is structured effectively. This study aims to facilitate the design of an international emissions trading scheme that is both workable for the Parties eligible for emissions trading and acceptable to all the Parties to the Protocol. Emissions trading transfers "assigned amount units". Ensuring that the post-transfer commitments are appropriately adjusted requires that the amount transferred should be added to the buyer's assigned amounts and deducted from the seller's assigned amounts (Article 3 (10, 11).

1. General Rules on Emissions Targets

The 1992 United Nations Framework Convention on Climate Change recognizes the principle of global cost-effectiveness of emission reduction in article 3(3) and has thus opened the way for flexibility. As it did not fix a binding emission target for any country, the need to invest in emissions reduction either at home or abroad was not pressing. In December 1997, however, industrial countries and countries with economies in transition agreed to legally binding emission targets at the Kyoto Conference and negotiated a legal framework as a protocol to the Convention—the Kyoto Protocol (United Nations Framework Convention on Climate Change, 1997). The Protocol will become effective once it has been ratified by at least 55 Parties representing at least 55 per cent of the total carbon dioxide emissions of Annex I countries in 1990.

Article 3 of the Kyoto Protocol defines the five-year commitment period (2008-12) in which the emission targets that are set out in Annex B for individual countries have to be reached. Together, Annex I countries must reduce their emissions of six greenhouse gases by at least 5 per cent below 1990 levels over the commitment period. They cover a basket of six greenhouse gases listed in Annex A: carbon dioxide, methane, nitrous oxide, hydrofluorocarbons HFCs), perfluorocarbons (PFCs) and sulphur hexafluoride (SF6). 100-year global warming potentials are used to convert gases into a common unit (article 5(3)). Emission targets relate to the base year 1990, although countries with economies in transition can use a different base year if established in their first national communication (article 3(5)). This applies to Hungary (average of 1985-87) Poland and Slovakia (1988) and Romania (1989). Croatia, Lithuania, Slovenia and Ukraine can still choose their base year, as they have not yet issued their first national communication. The base year for hydrofluorocarbons, perfluorocarbons and sulphur hexafluoride can be shifted to 1995 (article 3(8)).

Besides emissions reduction, "verifiable" sequestration through afforestation and reforestation taking into account deforestation shall be used to meet the targets (article 3(3)). Article 3(4) allows the Meeting of the Parties to include further sequestration activities (e.g. soils), for the next target period after 2012. Countries are allowed to use such sequestration for the current targets if the Meeting of the Parties takes the decision before 2008. Countries may include net emissions from land-use change and forestry in the 1990 inventory (article 3(7))—currently this only applies to Australia, Estonia and the United Kingdom.

Small countries for whom single projects have a significant impact on emissions in the target period shall benefit from special rules to be decided upon at the fourth Conference of the Parties (decision 1/CP.3, para 5(d)). If emissions during the commitment period are lower than the target, the difference may be banked for the next commitment period (article 3(13)). 'Annex I countries' are the countries of the Organization for Economic Co-operation and Development (OECD) and countries with economies in transition. These countries have committed themselves to legally binding greenhouse gas emissions targets.

2. Reporting and Accounting Rules

Article 5 (1) states that each Party to the Protocol shall install a national system for estimating emissions and removals. Inventories have to be submitted annually after 2008 via national communications (article 7(1-3)). Additional guidelines for reporting and accounting shall be developed at the Meeting of the Parties (article 7(4)).

3. Monitoring and Verification

National communications are to be reviewed by expert teams (article 8 (1)). The members are to be appointed by Parties and intergovernmental organizations (article 8 (2)). If the team identifies contentious issues, they are referred to the Meeting of the Parties (article 8 (3).

4. Cooperative Implementation

Article 3.3 of the Convention states that: «policies and measures to deal with climate change should be cost-effective so as to ensure global benefits at the lowest possible cost». To implement this provision the Kyoto Protocol has incorporated a variety of provisions for cooperative implementation mechanisms (see box 1).

Box 1: Mechanisms of Cooperative Implementation in the Kyoto Protocol

Article 4.1

"Any Parties included in Annex I that have reached an agreement to fulfil their commitments under Article 3 jointly, shall be deemed to have met those commitments provided that their total combined aggregate anthropogenic carbon dioxide equivalent emissions of the greenhouse gases listed in Annex A do not exceed their assigned amounts calculated pursuant to their quantified emission limitation and reduction commitments inscribed in Annex B and in accordance with the provisions of Article 3.

The respective emission level allocated to each of the Parties to the agreement shall be set out in that agreement."

Article 6.1

"For the purpose of meeting its commitments under Article 3, any Party included in Annex I may transfer to, or acquire from, any other such Party emission reduction units resulting from projects aimed at reducing anthropogenic emissions by sources or enhancing anthropogenic removals by sinks of greenhouse gases in any sector of the economy..."

Article 12.2

"The purpose of the clean development mechanism shall be to assist Parties not included in Annex I in achieving sustainable development and in contributing to the ultimate objective of the Convention, and to assist Parties included in Annex I in achieving compliance with their quantified emission limitation and reduction commitments under Article 3."

Article 17

"The Conference of the Parties shall define the relevant principles, modalities, rules and guidelines, in particular for verification, reporting and accountability for emissions trading. The Parties included in Annex B may participate in emissions trading for the purpose of fulfilling their commitments under Article 3. Any such trading shall be supplemental to domestic actions for the purpose of meeting quantified emission limitation and reduction commitments under that Article."

Source: The Kyoto Protocol, 1997.

Emissions Reduction Strategy under Kyoto Agreement

The Conference of the Parties to the United Nations Framework Convention on Climate Change (FCCC) took place in Kyoto in December 1997. The countries present reached an agreement on targets and timetables for reducing worldwide emissions of greenhouse gases. The following points summarize the most important aspects of the agreement (http://dsp-psd.pwgsc.gc.ca/Collection-R/LoPBdP/MR/mr150-e.htm).

- Developed countries have accepted binding targets for reducing greenhouse gas (GHG) emissions;
- Collectively, GHG emissions will be reduced to 5.2% below 1990 levels by the period 2008-12;
- Individually, reduction targets vary, with several Central and Eastern European countries and the European Union agreeing to an 8% reduction, the United States to 7%, and Canada and Japan to 6%;
- To meet the 6% target, Canada will have to reduce emissions by 25% over the business-as-usual scenario;
- The Kyoto Protocol is open for signing between March 1998 and March 1999, following which signatory countries must ratify the Protocol domestically;
- The agreement becomes binding when it has been ratified by 55 countries accounting for 55% of the developed countries' emissions.

Clearly, reaching the Kyoto target was an enormous challenge especially for an environment conscious country like Canada. In the years between signing the original FCCC in Rio in 1992 and the Kyoto conference in 1997, Canada was not on track to meet even the more modest goal of stabilizing GHG emissions at 1990 levels by 2000. Population growth and increased economic activity, among other factors, left Canada headed for estimated GHG emissions some 8% above 1990 levels by 2000.

Major changes requiring new approaches, particularly in the way the transport and the energy are used, was needed. Negotiators at Kyoto recognized the potential of emissions trading for introducing the flexibility needed if countries were to reach their emission reduction targets. The following provisions made subsequently in the Kyoto Protocol highlight this recognition:

- An international emissions trading regime will allow industrialized countries to buy and sell emission reduction credits among themselves;
- A "Clean Development Mechanism" (CDM) will allow developed countries to receive emission credits for financing emission reductions in developing countries;
- Joint implementation projects will allow developing countries to combine emission reduction efforts and share the resultant emission reduction credits;
- Excess emission reduction credits from one year will be banked for use in subsequent years;
- Countries taking steps to remove GHGs from the atmosphere through improved forestry practices will receive emission reduction credits.

To appreciate the attraction of emissions trading as an instrument for environmental policy, it is important to understand the basic concept. The trading of emission credits effectively "internalizes" the environmental cost of emitting specific air pollutants; in other words, it puts a price on emitting pollutants into the atmosphere. In the most familiar trading schemes, known as "cap and trade," regulatory authorities set a cap or target for emissions of a given pollutant or group of pollutants. They may then allocate the allowable volume of emissions among the companies or regions, or industries that produce such emissions. Over time, the cap was intended to be lowered to achieve additional emission reductions.

Each group with an emissions allowance will incur a different cost through making the changes necessary to meet the allowable level. Under an emissions trading scheme, those who can reduce emissions at the lowest cost will make the reductions. They will even reduce emissions to a level below their allowable limit, thereby creating for themselves an asset in the unused emission allowance (i.e., an emission trading credit). They can then sell the credits to other participants in the scheme for whom the cost of buying these credits is lower than the cost of implementing reductions themselves.

Contrary to some reports, this system was not meant to be equivalent to issuing licenses to pollute; rather, it lets the market to decide who makes the actual reductions and how they do so. In this way, environmental objectives can be achieved at the lowest possible cost to the economy. Although such trading is a market-based approach to reducing emissions, it is, in fact, driven by the regulatory authorities who set the cap, which, over time, can be reduced to meet increasingly stringent standards. Because the value of "producing" emission credits will therefore increase, companies will have an incentive to find technological improvements. Under the traditional "command-and-control" approach, companies were usually being encouraged only to meet the emission limit and have no incentive to exceed it.

Several jurisdictions in the United States have successfully used emissions trading systems to reduce pollution. Such a system was used, for example, to speed the elimination of lead from gasoline. It is also in use, and has been particularly effective, in implementing the U.S. *Clean Air Act* requirements for the reduction of sulphur dioxide emissions. When the *Clean Air Act* first came into force, it was estimated that compliance would cost the affected industries $5 billion (U.S.) annually with a "command-and-control" approach and only $4 billion with an emissions trading regime. More recent estimates based on the first several years of experience, however, put the compliance cost of the trading regime at just $2 billion per year. Industry representatives initially felt it would cost about $1,500 per ton to meet targets; SO_2 emission credits are currently trading for about $90 per ton.

In Canada, Ontario was the first jurisdiction to begin experimenting with emissions trading. The Pilot Emissions Reduction Trading Project (PERT), begun in 1996, is an industry-led, multi-stakeholder initiative that is developing the basic principles and program elements for creating, recognizing and trading Emission Reduction Credits (ERCs).

The Ontario approach, modeled on similar programs in operation in the Northeast United States (Massachusetts, Connecticut, New Jersey and Michigan) differed somewhat from the "cap and trade" methodology mentioned above. In this case, companies were not allocated emission permits, but rather encouraged to create ERCs by undertaking emission reductions in excess of those required by either voluntary or legislated measures. The program usually had focus on reducing emissions of the smog-precursors—nitrogen oxide (NOx) and volatile organic compounds (VOCs). It was hoped that experience gained with PERT would allow Ontario to be a part of early efforts to apply the concept to GHG emissions on a much wider scale in later years.

IEA's Climate Policy with Energy Implications

For over a decade, the IEA has been providing analyses on the energy dimension of climate change and the energy implications of the United Nations Framework Convention on Climate Change (UNFCCC) and its Kyoto Protocol. The Agency's areas of expertise include emissions trading and the CDM (from international architecture to domestic implementation issues), the links between energy security and climate policy goals and, the effects of policy uncertainty on investment, and sectoral approaches to emission reductions (http://www.iea.org/Textbase/subjectqueries/keyresult.asp?KEYWORD_ID=4106).

IEA's activities on energy efficiency and future emission scenarios are also of direct relevance to climate change policy issues. In 2005, in Gleneagles, the G8 leaders mandated the IEA to provide advice on a range of energy policy issues linked to climate change.

Since 1999, the IEA has also maintained a database of its Member countries' policies and measures to reduce greenhouse gas emissions, as well as databases on energy efficiency and renewable energy policy. Together with the OECD Environment Directorate, the IEA provides a Secretariat for the Annex I Expert Group (AIXG) on the UNFCCC, providing analysis of technical issues of relevance to the development of the Convention.

Setting Targets in International 'Cap and Trade' GHG Regulation

An international GHG cap and trade regulatory system must be based, at the operational level, on aggregate emissions limitations targets. The determination of individual countries' limitations burdens is achieved through an initial negotiated allocation, adjusted through subsequent trading. Such a system is incompatible with the use of technology-based targets or command requirements based on achievement of environmental quality standards. Environmental quality objectives, cost, and technology may, of course, enter into the selection of trading cap levels. One option, cap levels may be explicitly linked to achievement of environmental quality objectives such as those set forth in Article 2 of the UNFCCC.

A sound climate regulatory strategy must include both near-term and long-term targets. The demands of regulatory practicality, accountability, and credibility require that GHG emissions reduction targets be established and their achievement monitored for successive discrete time periods, such as five to ten years. These demands also require that countries and sources flexibility in the precise timing of reductions in order to promote cost-effectiveness. But, given the long-term character of climate problems and the GHG regulatory programs needed to address those problems, these periodic targets must be established in the context of longer term targets or goals that provide an analytical, normative and public reference point for setting successive near time targets. Basing near-term targets on long-term targets or goals is necessary to build and maintain long-term political support for regulatory programs, limit the extent of gaming in the redetermination of near-term targets, and provide a degree of longer-term regulatory guidance, assurance, and incentive for governments and firms that must make long-implemented. Fifth, governments should invest in assisting societies to adapt to a changing climate. This assistance is especially needed for poorer and more vulnerable regions of the world, which lack affordable insurance or access to adaptive technologies. These elements would be included in the first-phase plural regime sketched above.

Regulatory Instruments for Flexibility under Kyoto Protocol

Clean Development Mechanism

The Kyoto Protocol includes a new way of linking emissions reduction with economic development. A "Clean Development Mechanism" (CDM) shall be set up, which has been defined somewhat vaguely in indiscrete terms (article 12). It leads to the creation of «certified emission reductions» (CERs) (article 3 (12)). Article 12 (3) states that countries that fund projects through the CDM get credit for certified emission reductions from these projects provided benefits accrue to the host country (article 12 (3a)). Crediting shall be only allowed until a certain percentage of the emission target, which remains to be defined, is reached (article 12 (3b)). It is unclear whether crediting up to this quota is in full or only partial. Besides countries, companies are allowed to invest and execute projects (article 12 (9)). In contrast to the other flexibility mechanisms, CERs accrue for the whole period 2000-2012, not just for the commitment period (article 12(10)). On the other hand, it is unclear whether sequestration is covered.

The CDM covers its administrative budget through project revenues. Moreover, a part of these revenues shall be used "to assist developing country Parties that are particularly vulnerable to the adverse effects of climate change to meet the costs of adaptation" (article 12 (8)). It remains open as to who carries out certification of emission reductions but verification shall be carried out by independent bodies (article 12 (7)). The project criteria remain the same as for activities implemented jointly implemented (article 12 (5)).

Joint Implementation

The third option currently is most relevant concerning worldwide cost minimization project-oriented emission reduction credited to the investing country. This possibility was named «joint implementation» in the negotiations leading to the Rio Conference. In 1995, the Berlin Conference of the Parties decided on a pilot phase for joint implementation without crediting called «activities implemented jointly». By 2000, it should be decided whether activities implemented jointly would be followed by joint implementation with crediting. The Kyoto Protocol allows joint implementation between Annex I countries (article 6). It does not state, though, whether projects considered as activities implemented jointly will automatically become joint implementation projects after 2000.

Joint implementation projects shall be approved by all involved Parties and shall be «supplemental» to domestic actions (article 6 (1d)). The Meeting of the Parties shall define guidelines, verification and reporting rules (article 6 (1c)). Emission reduction units (ERUs) created through joint implementation under article 6 are treated in the same way as those from emissions trading under article 17 (article 3 (10, 11)). Emission reduction units from joint implementation do not accrue if inventories are not submitted annually or do not use the agreed guidelines (articles 5 and 7). Emission reduction units questioned through expert review teams may be transferred but are «frozen» until the question is resolved (article 6 (4)).

International and Federal Legislation

International and Federal legislation has been enacted to deal with GCC issues. The Montreal Protocol was originally signed in 1987 and substantially amended in 1990 and 1992. The Montreal Protocol governs compounds that deplete ozone in the stratosphere—chlorofluorocarbons (CFCs), halons, carbon tetrachloride, and methyl chloroform. The Protocol provided that these compounds were to be phased out by 2000 (2005 for methyl chloroform). In 1988, the United Nations and the World Meteorological Organization established the Intergovernmental Panel on Climate Change (IPCC) to assess "the scientific, technical and socioeconomic information relevant to understanding the scientific basis of risk of human-induced climate change, its potential impacts, and options for adaptation and mitigation" (IPCC, 2004).

Rules for Emissions Monitoring and Reporting

Monitoring rules are highly detailed in the U.S. SO_2 and NO*x* programs. The regulations for monitoring cover almost 300 pages and provide detailed standards for installation and certification of monitors, quality assurance and testing, handling of missing data, recordkeeping, and other features.8 Most of these rules are now incorporated into software systems at both the companies and EPA so that the reporting and review of emissions reports are highly standardized. To provide certainty and ensure consistency, EPA devotes extensive resources to answering and documenting questions that arise about monitoring requirements. EPA has an online policy manual that is largely in a question-and-answer format. It has been updated more than a dozen times over the life of the program and is now nearly 500 pages long. These detailed monitoring and reporting requirements, though complex, have provided companies with considerable certainty that if they follow the procedures, their emissions reports will be accepted in a timely manner.

Regulatory Frameworks and Markets

There exist several carbon regulatory frameworks in the financial markets, based either on international treaties or on private law contracts. Each of them defines an emission allowance trade and exchange system: the organized carbon markets. There are also alternative mutual agreement transactions systems –organized outside carbon markets: OTC—Over the Counter. The broad outline of the Regulatory Frameworks pertaining to Kyoto Protocol, Regional Regulatory Frameworks, Regulations in project phase/ Ratification phase, Regulations for Voluntary Carbon Market and the Frameworks for Trading outside regulated regime are given below in a summarized form. (see next page)

The emergence and development of the new carbon markets is heavily dependent on the evolution of regulatory frameworks.

	Regulatory Frameworks			
	Participants Industries & Countries	**Regulatory texts**	**Main markets**	**Carbon Products**
Kyoto Protocol • Quotas • Goal : –5%	5 industrial sectors are submitted: 1. Power and heat generation 2. Mineral oil refineries 3. Iron and steel 4. Pulp and paper 5. Building materials (cement, ceramics, and glass) and are situated in the 25 countries EU (listed in the appendix 1 of the Kyoto protocol)	Kyoto Protocol + *European Union Emissions Trading Scheme—EU ETS*	Organized Markets : • ECX(ICE) • EEX(Eurex) • Nord Pool (Nasdaq OMX) • Bluenext (Euronext + CDC)	• Quotas (EUA) : spot, futures & options • Carbon credits convertible into quotas (CER,ERU) : spot, future, options
	5 industrial sectors are submitted and situated in countries outside EU : Japan, Australia, NZ, Russia, Ukraine, Canada, etc. (listed in the appendix 1 of the Kyoto protocol)	Kyoto Protocol		
	India, Brazil, China, etc. (listed in the appendix 1 of the Kyoto protocol)	Kyoto Protocol regulatory regime applied to countries listed in Annex 2 : not submitted to reduction constraints		
Regional regulatory frameworks (outside Kyoto) • Quotas	Energy sector of 10 states of the USA	RGGI—Regional Greenhouse Gas Initiative	Organized Market RGGI	RGGI Quotas
• Regional Targets	10 USA's states and Canada	WCI—Western Climate Initiative	Organized Market WCI	WCI Quotas

(Contd.)

	Regulatory Frameworks			
	Participants Industries & Countries	**Regulatory texts**	**Main markets**	**Carbon Products**
	Electricity sector of New South Wales (AUS)	GGAS—Greenhouse Gas Reduction Scheme	Organized Market GGAS	GGAS Quotas
	7 USA's states and Canada	Midwestern Greenhouse Gas Reduction Accord	Organized Market in project	In project
Regulations in project phase/ Ratification phase				
	Canada	Canadian federal law under consideration	MCeX—Montreal's Climatic stock exchange (along with CCX)	Canada CO_2e Units Futures
	USA	USA federal law under consideration		
Voluntary markets				
• Regulated trade and exchange system not committed to Kyoto nor to other regional agreements	Participation on voluntary basis	Chicago Climate Exchange standard	Voluntary markets CCX CCFE—Chicago Climate Exchange Futures	Quota CFI (Carbon Financial Instruments) Credit VER, CER Spot-future-swap
• Quotas and reduction targets on voluntary basis	Participation on voluntary basis	NYMEX—Green Exchange standard	Voluntary market NYMEX—Green Exchange	Quota EUA Credit CER, VER, VCU Spot, future, option
	Participation on voluntary basis	Voluntary Carbon Standard	Voluntary markets launchings : • CDC's VCS Registry • APX • TZ1 (NZX)	Carbon credits VCU, CER
Trading outside				

Carbon Market: The Regulatory Instruments

The emerging global carbon market is growing exponentially as states and countries around the world use the power of the "market" to control greenhouse gas emissions. In the US, a voluntary market has sprung up to help provide a method to track these emission reductions for concerned individuals and businesses. Carbon offset and renewable energy credits typically represent investments in alternative energy enterprises, energy efficiency technology development and forestry- or agricultural-related (carbon sequestration) projects. Evolution Sage has made an organizational decision to only support projects that cut emissions directly, such as energy efficiency and renewable energy efforts.

Carbon Market refers to the buying and selling of emissions permits that have been either distributed by a regulatory body or generated by greenhouse gas emission reductions projects. Six greenhouse gases are generally included in "carbon" markets: carbon dioxide, methane, nitrous oxide, sulfur hexafluoride, hydro fluorocarbons and perflourocarbons. Greenhouse gas emission reductions are traded in the carbon market as carbon credits, representing the reduction of greenhouse gases to one metric ton of carbon dioxide (the most common greenhouse gas). Carbon markets enable units of pollution to be converted into units of property, making it possible to exchange pollution from one place in the world with somewhere else. This leads to polluters having to decide between accepting the cost of added pollution, changing of fuel mixes or conserving of energy.

Carbon emissions have been traded, albeit at minor levels, in the over-the-counter (OTC) market since the 1990s. In recent years, significant global governmental involvement in the stabilization of the concentration of atmospheric greenhouse gases has facilitated tremendous growth in the market for carbon emissions trading. The value of the carbon market has grown from $727 million in 2004 to $118 billion in 2008. In addition to the foundational regulatory organizations that have established the framework of the market, and the pre-requisite buyers and sellers, the magnitude and evolving complexity of the global carbon markets has attracted numerous intermediaries, such as brokers, exchanges, aggregators, and financiers, as well as other peripheral participants such as validation and verification, information and analysis, legal, and consulting service providers.

Opportunities for market participants are expected to continue to increase as the value of global carbon markets are forecast to grow by 68% per year to $669 billion in 2013 (http://www.reportbuyer.com/energy_utilities/alternative_energy/carbon_emissions_trading_markets_worldwide.html).

Baker & McKenzie—the new Financial Instruments and Exchange Law of Japan (FIEL) came into effect, opening up the Japanese carbon market to financial institutions. Emissions trading is now a regulated activity for the purposes of the Japanese financial services regulatory regime and licensed securities companies (or category 1 financial instruments firms as they are officially known under the FIEL), investment advisory firms and licensed banks are permitted to conduct certain emissions trading related activities.

The financial institutions regulated in Japan are only permitted to carry on business activities which are expressly covered by their regulatory authorizations, together with ancillary operations. They are not allowed to participate in any other business, even unregulated activities. The regulatory position in Japan prior to the introduction of the FIEL was that, subject to a very limited exception applicable only to trust banks (pursuant to which carbon credits(2) can be placed into a Japanese trust), emissions trading was not regulated under the Japanese financial services regulatory regime. Accordingly, licensed financial institutions in Japan (other than trust banks) were, in principle, not permitted to participate in the Japanese carbon market (http://www.e5.org/index.php?name=News&file=article&sid=3103).

Regulatory and Governance Aspects of Carbon Derivative Market

To combat global warming, companies and countries want to reduce their carbon footprint, thus driving an emerging growth market in carbon trading. While penetrating any immature market can be risky, there are strong opportunities for both marketplaces and financial players in global carbon trading. Carbon trading enables organizations to buy and sell carbon credits, helping to regulate carbon emissions. According to industry analyst Celent, the carbon exchange market has grown from under €9 billion in 2005 to about €25 billion in 2007 and, depending on the future regulatory framework, could reach €40 billion by 2012 and over €100 billion by 2020 (http://www.nasdaqomx.com/whatwedo/markettechnology/marketview/MarketView_2008_2/Carbon_trading__A_market_mechanism_for_tackling_pollution/).

Currently, there is both a regulated compliance segment of the carbon market and a voluntary, unregulated one. Two such regulated markets are the UN-sponsored regulated Clean Development Mechanism (CDM) and the largest regulated market, the European Union Emission Trading Scheme (EU ETS) which was created in conjunction with the Kyoto Protocol. Kyoto binds many developed countries to a cap and trade system with the goal of reducing overall emissions. Because the US did not ratify Kyoto, it is not bound.

As carbon trading matures, the size of secondary markets (where individual or repackaged carbon is sold for the second, third, or twentieth time) will vastly overshadow the primary markets (where carbon is bought and sold for the first time). Financial markets have become vastly more complex and exotic since the early 1990s, when the U.S. introduced sulfur-dioxide trading (the model policy framework for trading in carbon pollution credits). Moreover, much more than previous schemes, carbon markets will be dominated by speculators. Currently, the vast majority of carbon trading funds are set up by investors simply to make a profit; less than one-third of carbon funds have been established by companies to help them comply with carbon caps.

As speculators rush into carbon markets, they can push up carbon prices and spur the development of sub-prime assets. For example, unscrupulous intermediaries may over promise on offset projects, selling future credits based on projects that do not yet exist, do not meet standards, or which simply do not deliver promised greenhouse gas reductions. A carbon bubble can also set the stage for the kinds of financial innovation (e.g. complex securitized products) that can unwittingly spread sub-prime carbon through the broader financial marketplace. When the bubble bursts, it can have destabilizing consequences for companies and the larger economy (http://www.twnside.org.sg/title2/climate/news/TWN.Bonn.update20.doc). The financial crisis has demonstrated that significant parts of the financial system, such as the derivatives markets, are under- or unregulated. Robust derivatives regulation is one of the most important elements to ensure a well-governed carbon market, but currently most derivative proposals have focused on credit default swaps, which will have limited impacts on carbon.

A key governance challenge for carbon markets centers on the fact that carbon is a politically generated (created through government fiat) and managed market, and the assets traded are not physical commodities. In addition, there is a compliance requirement (e.g. company compliance with greenhouse gas caps) to carbon trading. It is precisely these politically generated and managed facets of carbon trading, as well as its compliance aspects that can make carbon markets especially vulnerable to lobbying and regulatory capture. For carbon trading to be successful from environmental, financial, and governance perspectives, policy makers and market regulators must be particularly insulated from corruption and political influence.

Emissions Trading and Non-compliance Regulations

The enforcement of the rules is one of the most crucial elements of a trading system. This is evidenced for example by the first analyses of the system established in the US for the trading of sulfur dioxide emission rights. Under the supervision of the Environmental Protection Agency, a very complex system for verification, tracking of allowances and enforcement has been put in place. Unfortunately, under the conditions of today's international politics and law, such a foolproof regime is not a realistic option. This does not, however, dispense with the necessity to establish a rigorous regime for the international trading of carbon dioxide. Procedures and institutions to monitor, verify, assess compliance and enforce the rules in cases of non-compliance must have top priority in the establishment of the trading regime (Hermann, 1998).

The reasons for this are twofold. First, as with other provisions dealt with above, the objective of the climate regime demands that the rather modest obligations of industrialized countries to reduce or limit emissions be fulfilled. Any failure to do so would undermine the climate regime, lead to a much-prolonged reduction in the emissions of greenhouse gases and cast serious doubt on the willingness of industrialized countries to tackle the threat of climate change. This in turn would not bode well for the much-needed involvement of developing countries at a later stage. Second, the credibility of the trading regime itself is at stake. Since trust in the market and the correct behavior of participants is one of the main pillars of this regime, the failure to detect instances of misdemeanor and ultimately to sanction such behavior would prevent the establishment of a stable and thriving market.

Such a compliance system demands several components: first, monitoring and reporting procedures must be in place. The Kyoto Protocol, in Articles 5 and 7, contains the basis for such procedures that need to be adapted to the needs of a trading regime. Those amended procedures need to provide the information required for the close tracking of trades. Compliance with these requirements must furthermore be a precondition for participation in Emissions Trading. Second, the correctness of the data submitted must be assessed. One of the cornerstones of such a procedure, the verification function, has already been incorporated in the Protocol (Article 8). The expert review teams established in this article might also be enabled to check the data submitted on trades, nationally as well as internationally. Additionally, certified auditors linked to the Secretariat could perform some of those functions.

As a third component, the appropriate reaction to a presumed infraction must be determined. This function will ultimately have to lie with the Conference of the Parties to the Kyoto Protocol. Based on a recommendation of a smaller body—like the Implementation Committee of the Montreal Protocol—the supreme body of the Kyoto Protocol will have to decide not only whether to react, but furthermore on the consequences of non-compliance. The Protocol in Article 8.5 already contains a rudimentary procedure to that effect—the Conference of the Parties shall "consider the information received and take decisions on any matter required for the implementation" of the Kyoto Protocol.

Unfortunately, Article 18 on non-compliance does not lend itself well to the establishment of a sanctions regime for Emissions Trading. Although it requires the Parties to "approve procedures and mechanisms to determine and address cases of non-compliance" with the Protocol, it contains a clause that makes "binding consequences" dependent on a formal amendment. Since an amendment to the Protocol requires ratification by at least three-fourths of the parties, its entry into force appears to be very unlikely or would at least take an unacceptably long time. Even if such "consequences" would be adopted and applied provisionally in the interim period until the amendment becomes effective, its legal status would probably be diminished due to the above mentioned clause.

The appropriate sanctions—including those with "binding consequences"—should therefore be incorporated into the trading regime. This is possible because Article 18 only talks about binding consequences "under this article". Including sanctions into the Emissions Trading regime would thus have the beneficial side effect of providing the Parties to the Kyoto Protocol with a much needed enforcement regime for the reduction and limitation commitments. The political resistance to hard sanctions notably by the US, which was responsible for the mutilation of Article 18, will still be strong. However, the necessity for such instruments will clearly be evident to US negotiators and thus meet with less objections.

Non-compliance with the rules of the trading system is conceivable in a variety of ways. Participants may fail to fulfil the reporting or monitoring obligations, they may violate the trading rules or cheat and, ultimately, they may fall into non-compliance because at the end of the commitment period they have exceeded their assigned amount of emissions. In order to provide clarity, a non-exhaustive list of possible cases of non-compliance would be helpful.

A great number of possible consequences for infractions of the rules of trading have been outlined in the literature on emissions trading, but some of them—like personal fines and imprisonment for the management of companies involved—are reserved for domestic enforcement. Should a Party decide to involve private actors, the provision of adequate sanctions at the national level should be compulsory. On the international level, financial sanctions are extremely rare. However, in order to facilitate such fines with regard to non-compliant states, a certain financial deposit should be required for the permission to trade. In case of non-compliance, such deposit might be (partly) lost and go into a fund to finance mitigation and adaptation measures in developing countries.

Emission trading itself provides the best means for sanctions. First of all, the parts of assigned amounts can be devalued, either partly or completely. There are basically two approaches, depending on whether liability lies with the seller or with both, buyer and seller (shared liability). Under the traditional rules of international law, the seller only would be responsible for the correctness of his behavior. Sanctions would thus be directed towards the country that has sold parts of its assigned amount although it has breached the rules. In the framework of domestic legal systems, this approach is usually sufficient, since penalties for non-compliance can be severe and will be enforced by the judicial and executive branches of government.

Regulatory Regime of Global Voluntary Carbon Market

In the course of 2006 and 2007, interest in climate change, carbon offsets and the voluntary carbon markets accelerated dramatically. Despite this interest, and the fact that voluntary carbon markets have effectively been operating since 1989, quantitative data surrounding this market has been sorely lacking. Because of this situation Ecosystem Marketplace and New Carbon Finance teamed up to undertake the most comprehensive analysis to date of the voluntary carbon market.

The research conducted by the New Carbin Finance Group on Voluntary Carbon Market (http://www.ecosystemmarketplace.com/documents/acrobat/StateoftheVoluntaryCarbonMarket17July.pdf) has involved a wide ranging survey with responses from over 70 organizations involved all stages of the supply chain from developers, aggregators, developers and retailers, and covering five continents. The results show that, like the early stages of the regulated carbon markets of the European Union's Emissions Trading Scheme (EU ETS) and the Kyoto Protocol, the voluntary markets are evolving rapidly. They also show that 2006 was a year of significant growth with many new retailers, brokers, and other actors entering the market. Since 2002 the number of organizations supplying carbon credits into the market has grown by 200%, with online retailers being the fastest growing sector of the marketplace. Between 2005 and 2006 the Over the Counter (OTC) voluntary offset market grew 200%.

In 2006, 23.7 million tons of carbon dioxide equivalents ($MtCO_2e$) were transacted in the voluntary carbon markets. Of this, 10.3 $MtCO_2e$ were transacted on the Chicago Climate Exchange, and our survey revealed that some 13.4 $MtCO_2e$ were transacted in the OTC market. Because it is impossible to capture all OTC transactions in a survey such as this, the actual volume traded may be considerably larger than this amount. While these numbers are small relative to volumes of transacted in the regulated carbon markets like the EU ETS, the combined voluntary markets (CCX+OTC) are larger in volume than both the Kyoto Protocol's Joint Implementation mechanism and the New South Wales Greenhouse Gas Abatement Scheme. Just as importantly, the voluntary markets are significant in that they represent an active demand by businesses and individuals for some form of action on climate change in the absence of direct regulation.

Much of the demand driving the voluntary carbon markets comes from the developed and more environmentally aware markets in North America and Europe. Survey respondents reported that 68% of their customers are based in the United States and 3% in Canada. In addition, about half of the suppliers responding to our survey were based in the U.S, and roughly 43% of carbon offsets sold in the OTC market were sourced from North American-based projects. Europe was also a major source of market demand and supply in the market, with 28% of the survey respondents' customers based in the EU and a little over 30% of suppliers based in the EU. About 10% respondents were based in Australia. As could have been expected, businesses were the largest buyers (by volume) in this market, but contrary to expectations, anticipation of future regulation did not appear to be the main motivation for purchases. According to buyers surveyed, their main motivations for participation in the market were corporate social responsibility and to "walk the talk" in terms of environmental stewardship.

Voluntary carbon markets have historically served as sources of experimentation and innovation in the carbon markets, as well as the markets most likely to reach poorer and smaller communities in developing countries. This is, in part, because they lack the bureaucracy and transaction costs of their regulated counterparts. For example, compared to the Kyoto regulatory markets the voluntary OTC markets are currently the only source of carbon finance for avoided deforestation, have a higher proportion of forestry based credits out of total market transactions than the CDM (36% vs. 1% for CDM), and a slightly higher proportion of credits sourced from Africa (6% vs. 3% for CDM). Moreover, the voluntary markets seem to be a particularly hospitable climate for smaller offset project. More respondents cited selling offset credits sourced from micro projects, generating less that 5,000 tCO_2e, than any other project type. Around 36% of offset credits in the OTC market were sourced from projects less than 100,000 tCO_2e. This provides greater opportunities for voluntary markets to contribute to sustainable development in smaller communities.

The flexibility of the voluntary markets is both a source of strength and a weakness. One of the reasons the market has very low transaction costs is that it does not require proof of quality in the same way as the regulated markets. For instance, in the OTC markets there are no widely accepted standards, processes for certification and verification, or requirements to list credits on established registries. This lowers transaction costs, but it also makes it a "buyer-beware" market where getting a handle on the quality of credits being bought can be difficult for customers.

But this is changing. The quality of offsets is—and will likely continue to be—the most important issue for both buyers and sellers in this market. In our survey, buyers indicated that the quality of offsets was more important to them than price, and sellers all agreed that addressing issues of quality would ultimately determine how (and how fast) this market continues to grow. According to suppliers, the issues that determine quality of offsets in this market include: additionality (would the reductions have happened anyway with or without the offset purchases), third party certification and verification, standards, and avoidance of double-counting and double-selling (i.e. registries).

As part of the consolidation in the market that began to take shape in 2006, various groups (from non-profits and industry associations, to offset providers and government agencies) continue work aimed at creating rigorous standards and processes as a way of ensuring confidence and quality in the market. In 2006 and early 2007, the issue of quality in the voluntary market became very visible in the form of media stories and articles questioning the validity of offsets being sold. This backlash was (at least partly) the result of the increased growth and visibility of the market, but it also helped to fuel increasing efforts on the part of those interested in the industry to strengthen quality and create standards.

Overall, the survey confirmed reports that the voluntary carbon markets are a vibrant and growing sector of the carbon markets, one with direct links to consumers, and one whose future (assuming issues of quality can be addressed) looks bright indeed. In fact, based on data we are beginning to receive, it is possible to predict record volumes for 2007. For example, in June of 2007, the Chicago Climate Exchange reported that in the past six months, it had already traded 11.8 $MtCO_2e$, more than had been traded in the entire year of 2006. If this pattern continues, the CCX is well on its way to trading more than 20 $MtCO_2e$ this year. As the number of companies and individuals who have decided to go "carbon neutral" seems to grow, the voluntary OTC market is also showing similar signs of growth. Some of the respondents to our survey reported that in 2007 they had seen a doubling, a tripling (or more) of volumes transacted. Already Dell, Delta, AEP, Google, Pacific Gas and Electric, Yahoo, Nike, Sky, Origin Energy, and various other major consumer-facing organizations have announced that they will be buying offsets from the voluntary markets.

Compliance (Regulatory) Carbon Markets

There is now a number of what are referred to as "cap and trade markets" around the world, all of which are influenced by the Kyoto Protocol. The idea of a cap and trade market is that a government or organization sets a cap on the overall emissions that can occur, and if a company, industry, or nation goes over those limits, they can buy or trade from others who have emitted below their limits. Additionally, you can buy carbon credits to compensate for exceeding your limits. The Kyoto Protocol agreement has been ratified by 163 countries and is a legally binding treaty committing industrialized countries to reduce their combined greenhouse gas emissions by 5.4% below 1990 levels by 2012. Authors of the agreement established mechanisms to be used by treaty signatories to meet greenhouse gas emission reduction targets through cost-effective means. Beyond the direct Kyoto markets, countries and regions have begun to establish emissions trading schemes. The largest of these is the European Union Emission Trading Scheme (EU ETS), launched in 2005 to facilitate participating EU countries to achieve reductions delineated in the Kyoto Protocol. The EU-ETS is already ramping up its operations: By the end of year one, an estimated 362 million tons of carbon were traded with an estimated value of US$9 billion through the EU ETS.

Commercial Emissions Trading and Carbon Funds

Some Important Terminologies and Implications

Kyoto Protocol has established market based mechanisms as a means of reducing global greenhouse gas emissions. The following pages provide examples of trading mechanisms, brokerage activities and carbon funds that have been established in response to the demand for carbon credits aiming at progressive reduction of emissions and retardation of adverse effects of global climate change:

Asia Carbon Group and the Asian Carbon Exchange

(http://www.asiacarbon.com/aboutus.htm)

- Carbon Credit Capital (CCC)
- Carbon Positive
- CD4CDM
- Chicago Climate Exchange
- Emissions Marketing Association
- European Climate Exchange
- European Union Emissions Trading Scheme (EU ETS)
- Greenhouse Gas-Credit Aggregation Pool (GG-CAP)
- ICEcap Ltd.
- International Emissions Trading Association
- International Finance Corporation
- Kyoto Protocol
- Renewable Energy Certificate System (RECS)
- The Union of the Electricity Industry
- United Nations Development Programme—Millennium Development Goals (MDG) Carbon Facility
- World Bank Carbon Finance Unit

Asia Carbon Group's primary objective is to mitigate global climate change and initiate sustainable development through the application of the Kyoto Protocol financial mechanisms in particular the Clean Development Mechanism (CDM), Joint Implementation (JI) and Emissions Trading. In addition to carbon financing, the Group is now establishing a trading platform—the Asia Carbon Exchange and also the Asia Carbon Fund for the purposes of investing in sustainable development projects globally. It is further planning to launch an exchange in Japan that will enable utilities to gain better access to carbon reduction projects.

Carbon Credit Capital (CCC)

(http://www.carboncreditcapital.com/)

Carbon Credit Capital's mission is to cultivate the supply of sustainable carbon credits from energy projects in developing countries for sale to international carbon credit buyers. It provides advice and carbon finance to GHG offset project developers seeking to create carbon credits for the regulated markets or voluntary markets. It also screens and designs projects to generate sustainable credits that create lasting benefits to host communities and carry minimum risk for buyers. Carbon Credit Capital focuses on three proven and tested reduction technologies: Biomass, Hydropower and Landfill Gas. It works with project locations in the regions of Africa, Asia and South and Central America.

CCC also serves as brokers to buyers of carbon credits, matching buyers with credits from projects most suited to their risk tolerance, financial needs and GHG exposure and structuring transactions between buyers and sellers. It works with governments and companies seeking to comply with regulated GHG reduction standards under the Kyoto Protocol or other regulatory schemes, financial institutions seeking to benefit from carbon credits as an asset class and corporations who voluntarily buy carbon credits as a strategy to raise their corporate social responsibility profile.

CarbonPositive

http://www.carbonpositive.net/

Carbonpositive arranges and manages sustainable-development projects that reduce greenhouse-gas emissions in developing countries. These projects are designed to generate Certified Emissions Reductions (CERs) under the Kyoto Protocol Clean Development Mechanism. Its website provides useful information ranging from generic information on climate change to the details of the Kyoto Protocol, carbon markets and company news.

CD4CDM

http://www.cd4cdm.org/

Capacity Development for CDM (CD4CDM) is a 4-year project with funding from the government of the Netherlands. The project is intended to help establish GHG emission reduction projects that are consistent with national sustainable development goals, particularly projects in the energy sector. It will develop national capabilities so that persons in the countries are at the project's conclusion capable of analyzing the technical and financial merits of projects and negotiating possible finance agreements with Annex 1 countries or investors.

Chicago Climate Exchange

http://www.chicagoclimatex.com/

The Chicago Climate ExchangeR (CCXR) is a greenhouse gas (GHG) emission reduction and trading pilot program for emission sources and offset projects in the United States, Canada, and Mexico. Projects also include Brazil. CCXR is a self-regulatory, rules based exchange designed and governed by CCXR Members. These members have made a voluntary, legally binding commitment to reduce their emissions of greenhouse gases by four percent below the average of their 1998-2001 baselines by 2006, the last year of the pilot program.

Emissions Marketing Association

http://www.environmentalmarkets.org/index.ww

The objective of the Environmental Markets Association is to promote market-based trading solutions for environmental management and to serve its membership in order to:

- Promote the advancement and application of policy and regulation relevant to market-based emission trading systems.
- Encourage and facilitate information exchange among members, other professional and technical groups and the public.
- Provide programs in education and training to improve both the knowledge and skills of members and the understanding and acceptance by the public.

European Climate Exchange

http://www.europeanclimateexchange.com/index_flash.php

The European Climate Exchange is a wholly-owned subsidiary of the Chicago Climate Exchange (CCX). It manages the sales and marketing for ECX Carbon Financial Instruments (ECX CFIs), listed on the ICE Futures electronic platform. ECX/ICE Futures is the most liquid pan-European platform for carbon emissions trading with standard contracts and clearing guarantees provided by LCH. Clearnet. More than 50 leading businesses, including global companies such as ABN AMRO, Barclays, BP, Calyon, E.ON UK, Electrabel, Fortis, ICAP, Morgan Stanley and Shell have signed up to trade ECX products.

European Union Emissions Trading Scheme (EU ETS)

http://europa.eu.int/comm/environment/climat/emission.htm

In January 2005 the European Union Greenhouse Gas Emission Trading Scheme (EU ETS) commenced operation as the largest multi-country, multi-sector Greenhouse Gas emission trading scheme world-wide. The scheme is based on Directive 2003/87/EC, which entered into force on 25 October 2003. Allowances traded in the EU ETS will not be printed but held in accounts in electronic registries set up by Member States. All of these registries will be overseen by a Central Administrator at EU level.

Greenhouse Gas-Credit Aggregation Pool (GG-CAP)

http://www.natsourcejapan.com/english/works/greenhouse/ggcap.html

GG-CAP is initiated by Natsource Group and is a carbon pool. Natsource Japan solicits Japanese entities for this GG-CAP which is run by its affiliate company Natsource Asset Management Corp. GG-CAP enables its participants to comply with domestic, international and voluntary greenhouse gas emissions reduction requirements by purchasing low-cost compliance instruments from a diverse portfolio of project-based emissions reductions. GG-CAP will initially procure 15 to 30 contracts totaling 30 to 50 million tons of greenhouse gas emissions reductions for delivery in the 2004-2012 periods.

ICEcap Ltd.

http://www.icecapltd.com/index.asp

ICECAP is a private sector provider of emissions credits from developing world CDM and JI projects in the emissions trading markets. In March 2004, the ICECAP Carbon Portfolio was launched, one of the world's first private sector carbon hedging vehicles. This reached first closing on 17th January 2006 with aggregate commitments of 15 million tonnes. In addition, ICECAP has developed a track record in Certified Emissions Reduction transactions through its trading business.

International Emissions Trading Association

http://www.ieta.org/ieta/www/pages/index.php

The International Emissions Trading Association (ETA) is a membership organization dedicated to the establishment of effective market-based trading systems for greenhouse gas emissions by businesses that are demonstrably fair, open, efficient, accountable and consistent across national boundaries. To achieve its goals, IETA focuses on the following areas:

- Developing components of the GHG market and trading systems
- Promoting market mechanisms and participation in GHG markets
- Developing a global GHG market
- Capacity building

International Finance Corporation (IFC)

http://www.ifc.org/ifcext/enviro.nsf/AttachmentsByTitle/p_2005SustReport_15/$FILE/ClimateChange.pdf

IFC has established Carbon Finance Facilities. The Facilities are arrangements under which IFC will purchase carbon credits for the benefit of the Government of the Netherlands under the international emission reduction transfer rules of the Kyoto Protocol. IFC presently has about $100 million under management in two facilities to purchase credits: (i) the IFC-Netherlands Carbon Facility (INCaF) operating under the rules of the Clean Development Mechanism (CDM); and (ii) the Netherlands European Carbon Facility (NECaF) operating under the rules of the Joint Implementation (JI) mechanism and managed jointly with IBRD. Payments for carbon credits purchased will be made to eligible projects in emerging markets and will provide them with additional revenue. The Government of the Netherlands will use the purchased credits to meet its GHG reduction obligations under the Kyoto Protocol. The credits will be purchased and paid for regardless of whether or not the Kyoto Protocol is in force.

Renewable Energy Certificate System (RECS)

http://www.recs.org/

RECS stand for Renewable Energy Certificate Systems and are considered as an international trading system in Europe (EU25) for renewable energy certificates. The system is initiated and coordinated by the RECS International Association in cooperation with the Association of Issuing Body (AIB). Every participating country has appointed an independent national issuing body to operate the system. This national issuing body should be a member of the AIB so certificates can be exchanged based on the same standard: the RECS. A RECS-certificate can be issued for every MWh of renewable energy as long as the trader or producer opens an account for RECS at its national issuing body. To do this, one should be a member of the RECS International association.

The RECS International Association is a non-profit European organization registered in Brussels. Its members are all renewable energy producers, traders and suppliers in Europe that either wish to have a RECS account at their national issuing body and/or wish to influence policy on governmental and system level concerning the certificate trading. RECS International acts as the representative of market players towards national and European governmental authorities and facilitates all possible events and activities that will enhance the mission statement of RECS.

United Nations Development Programme—Millennium Development Goals (MDG) Carbon Facility

http://www.undp.org/mdgcarbonfacility/docs/BookletMDGCarbonFacility.pdf

To capitalize on the potentially significant benefits of carbon finance for the developing world, UNDP is establishing the MDG Carbon Facility, a carbon-finance mechanism featuring emission offsets derived from a pool of projects designed to contribute directly to achieving the MDGs. UNDP will mobilize carbon finance and direct this towards developing a portfolio of projects that yield tangible sustainable development and poverty reduction benefits across a diverse group of developing countries.

The World Bank Carbon Finance Unit

http://carbonfinance.org/Router.cfm?Page=Home&ItemID=24675

The World Bank Carbon Finance Unit (CFU) uses money contributed by governments and companies in OECD countries to purchase project-based greenhouse gas emission reductions in developing countries and countries with economies in transition. The emission reductions are purchased through one of the CFU's carbon funds on behalf of the contributor, and within the framework of the Kyoto Protocol's Clean Development Mechanism (CDM) or Joint Implementation (JI). CFU contracts to purchase emission reductions similar to a commercial transaction, paying for them annually or periodically once they have been verified by a third party auditor. The selling of emission reductions—or carbon finance—has been shown to increase the bankability of projects, by adding an additional revenue stream in hard currency, which reduces the risks of lending or grant finance. Thus, carbon finance provides a means of leveraging new private and public investment into projects that reduce greenhouse gas emissions, thereby mitigating climate change while contributing to sustainable development.

The World Bank—List of Carbon Funds

http://carbonfinance.org/Router.cfm?Page=Funds&ItemID=24670

- Prototype Carbon Fund
- BioCarbon Fund
- Community Development Carbon Fund
- Italian Carbon Fund
- The Netherlands CDM Facility
- The Netherlands European Carbon Facility
- Danish Carbon Fund
- Spanish Carbon Fund
- Umbrella Carbon Facility

Some Institutional Issues in Carbon Trading

Numerous issues concerning carbon trading regimes have yet to be worked out, but could significantly affect the various parties involved in trading. These issues include:

- Whether countries' control regimes will be compatible with trading;
- The effect of restrictions on permit availability or demand;
- Impacts of international transfer payments;
- Measurement and reporting of emissions, sinks, and costs; and
- Accountability and enforceability

Many of these issues are examined in much greater depth in an upcoming Pew Center paper on institutional issues and trading. They are of concern here because of their impacts on the effectiveness and cost of trading and on the volume of permits traded. The key implication of this discussion of institutional issues is: "The actual cost savings from trade in emissions will likely be less than the theoretical savings shown in most analyses performed with integrated assessment models because these models do not include the various measurement, verification, trading, and enforcement costs that would be characteristic of any real trading system. On the other hand, the gains from trade could be greater than the models predict since the models may not anticipate all the control options trade would encourage."

Common Issues on Carbon Trading to be Resolved

There are certain common issues, as given below, that Parties will seek to resolve as they put their minds to the design of the cooperative implementation mechanisms:

(i) Credibility

- The mechanisms are often regarded as loopholes. There is concern that, by taking advantage of the mechanisms, Parties could avoid taking real actions at home to reduce emissions. How can confidence in the three mechanisms be built up?

(ii) Additionality

- Participation in both JI and emissions trading must be "supplemental" to domestic actions to meet emission commitments. How is this requirement to be operationalized? There is a similar provision with the CDM—emission reductions achieved through this mechanism can only be used to meet "part" of a country's emission commitment. What is an appropriate "part"?

(iii) Participation of Non-state Entities

The Protocol takes the following different approach towards project based mechanisms and emissions trading:

- The Protocol explicitly permits non-state entity participation in the project based JI and CDM. Indeed, the mechanisms only make sense if they attract new private investment to EITs (for JI) and developing countries (for the CDM). Private sector involvement, however, raises questions regarding ownership of emission reduction units (JI) and certified emission reductions CERs (CDM) and also regarding government accountability for the acquisition of such "ERUs" and "CER" by private firms, including transnational.
- The situation for emissions trading is different—the Protocol simply provides for intergovernmental trading of parts of assigned amounts through a double-entry book keeping system. Non-state entities are not mentioned. However, nothing precludes governments from setting up their own domestic trading schemes (e.g. along the lines of the US scheme for SO_2)—the indications are that many Parties are considering this option. The issue of cross-border trading by private firms (including transnational) will likely emerge as Protocol provisions on emissions trading are elaborated further—this would raise similar questions of government accountability.

(iv) Building on the AIJ Pilot Phase

- Useful experience has been amassed from the pilot phase since 1995. How can this experience be put into practice in the new mechanisms? What are the implications of JI and the CDM for the AIJ pilot phase and its review? Should AIJ now be wound down? Considered completed? Continue until CDM and/or JI start up, leading to a "seamless transition"?

(v) Linkage between the three Mechanisms

- Could there be... (1) Common rules and guidelines? (2) Comparable methodologies and reporting requirements? (3) Emergence of a "common emissions unit"? Certainly, the market will tend to equalize the unit value of reductions under the three schemes.

Focus on the CDM

The CDM was one of the "surprise" Articles emerging from Kyoto, and also one of the more innovative. It has therefore attracted considerable attention post-Kyoto. Some critical issues are worth highlighting here:

Governance: Institutional configuration of the CDM remains to be clarified. viz, what will be the nature and functioning of the executive board? Will it be located in the COP/MOP framework? The CDM system also involves different entities with various roles. e.g. certification, funding, auditing and verification. What is the relationship between these entities? How do they relate to the executive board and the COP/MOP? What is the chain of accountability between them?

Certification: The CDM has a double requirement for certification—emission reductions (Para. 5) and projects (Para. 6). Issues to be unraveled are not only methodological. For example, which organizations ("operational entities") will COP/MOP designate to certify emission reductions? Will CERs be time bound or run the length of a project's lifetime? According to what criteria will projects be certified, and who by? Note that the CDM also calls for "independent auditing and verification" of project activities (Para. 7) —who could be given this task?

Funding: What is the most appropriate funding "niche" for the CDM? The mechanism could be most effective as a channel for private funds, with the public sector focusing on project development, financial mediation or clearing house functions. A key point here is that funding entities should not, at the same time, be certifying entities, because of possible conflicts of interest. Another question concerns the "share of the proceeds" to help cover administrative expenses and adaptation costs—what are the "proceeds"? What is an appropriate "share"?

Focus on Emissions Trading

Article 17 can be considered as a "door opener" to emissions trading. Together with Articles 3.10 and 3.11, it contains enabling provisions permitting inter-governmental trading between Parties with emission commitments on the basis of simple "double-entry bookkeeping". The "buyer" will add the part of an assigned amount which it acquires from the "seller" to its own assigned amount, whilst the "seller" will subtract a corresponding amount of emissions from its assigned amount. "Principles, modalities, rules and Guidelines" for emissions trading are to be developed through subsequent negotiations. There are a number of key issues which Parties will seek to resolve as they turn their minds to the design of a workable emissions trading system. These include the following:

- *Complexity:* How elaborate should the trading system be? Could the present double-entry bookkeeping provisions evolve into a more intricate market-based regime?

- *Private sector participation:* As already noted, Article 17 does not mention non-state entities; yet we know that there is expectation in the private sector that private companies will be able to engage in emissions trading. The possible modalities for such private sector participation will therefore undoubtedly emerge as a key issue.
- *Links between domestic and international trading:* Parties may, of course, establish their own domestic emissions trading. Emissions trading could, therefore, take shape along two fronts—both bottoms up, as national and regional initiatives multiply, and top down, as the details of international emissions trading are developed under the Protocol. How national, regional and international systems will eventually interconnect is a matter requiring further thought.
- *Supplementarity:* As previously states, Article 17 requires that emissions trading be used only as a supplementary means for meeting emission commitments. This prompts the question of so-called "hot air" —some Parties were allocated targets under the Protocol which are above their current emissions and, possibly, also above their expected emissions in the target period. Other Parties who face difficulty in meeting their commitments could buy this "hot air", thereby reducing the need to implement domestic policies and measures to lower emissions.
- *Rules and Procedures:* What rules should underpin emissions trading, particularly to ensure environmental effectiveness? For example, should Parties be required to comply with other Protocol provisions—such as Article 5 on methodological issues and Article 7 on reporting—before they can trade? Should rules on liability be developed in addition to the Protocol compliance provisions? That is, who is responsible if a Party transfers part of its assigned amount, but then fails to meet its commitment, so that the basis of the trade is invalid?

These are just some of the questions which will need to be answered as the Protocol's Framework Provisions on emissions trading are fleshed out.

International Funding Initiatives on Emissions Trading European Union Emission Trading Scheme (EU ETS)

The EU Emission Trading Scheme (EU ETS) is the first international trading scheme for CO_2 in the world. Its aim is to reduce the cost of compliance to existing targets under the Kyoto Protocol. Since January 1, 2005, companies in high-energy sectors covered by the scheme must limit their CO_2 emissions to allocated levels, arranged in two periods: from 2005-07 and 2008-12 (to match the first Kyoto commitment period). In practice, the scheme is likely to cover over 10,000 installations across the European Union, corresponding to approximately forty-six percent of the total EU CO_2 emissions. The European Union Emission Trading Scheme (EU ETS) is the largest multi-national, greenhouse gas emissions trading scheme in the world and is a main pillar of EU climate policy. Under the scheme, each participating country has a National Allocation Plan (NAP) specifying caps on greenhouse gas emissions for individual power plants and other large point sources. Each facility gets a maximum amount of EU emission "allowances" (EUA) for a particular period (e.g. 2005-07). To comply, facilities can either reduce their emissions or purchase allowances from facilities with an excess of allowances. Progressively tightening caps are foreseen for each new period, forcing overall reductions in emissions. On 2nd July 2003, the European Council formally adopted the Emissions Trading Directive. The Directive laid out the framework for the European Emissions Trading Scheme (the 'EU ETS'.

Key Facts of EU-ETS

- Covers power sector energy-intensive industrial sectors—of around 46% of European emissions.
- First trading phase (2005-07) is for CO_2 with potential expansion in second phase (2008-12) to cover all greenhouse gases
- JI and CDM credits will be able to be used for compliance in the EU ETS thanks to the 'Linking Directive', agreed in 2004.
- The scheme started on 1 January 2005. From this date emissions from companies in sectors covered by the scheme will be capped across 25 European countries.
- Only CO_2 will be included in the first phase (2005-07), with the potential to expand this to the other 5 greenhouse gases (GHG) from 2008.
- The European Union's (EU's) Emissions Trading System (ETS) is a cornerstone of the EU's efforts to meet its obligation under the Kyoto Protocol. It covers more than 11,500 energy intensive facilities across the 25 EU member countries, including oil refineries, power plants over 20 megawatts (MW) in capacity, coke ovens, and iron and steel plants, along with cement, glass, lime, brick, ceramics, and pulp and paper installations. Covered entities emit about 45% of the EU's carbon dioxide emissions. The trading program does not cover emissions of non-CO_2 greenhouse gases, which account for about 20% of the EU's total greenhouse gas emissions.

Emissions Trading: EU Directives

The European Union has established a trading directive to be in place during the 2005-07 time periods in a first step, followed by a 2008-12 second period. The program foresees a mandatory CO_2 cap and trade system, in which sources are allocated a certain number of emission "allowances," based on historic performance and other parameters. Participants reducing emissions below their cap can sell the resulting excess allowances. The EU program covers

12,000–15,000 sources such as: utilities; refineries; cement, glass, and ceramics production facilities; and the pulp and paper industries. Up to 45% of EU greenhouse gas emissions are covered.

An Overview of the EU-ETS

The European Union Emissions Trading Scheme (EU-ETS), launched in 2005, sets mandatory caps on CO_2 emissions for the 27 member countries of the European Union. The EU-ETS is by far the largest source of carbon allowances, representing 99 per cent of the volumes of allowances traded globally in 2006. The EU-ETS has consecutive trading phases, with the first phase running from 2005 to 2007 and a second phase from 2008 to 2012. A review of the third trading phase of the ETS starting in 2013 is in preparation. In Phase I, the scheme is mandatory for 11,500 energy-intensive installations across the EU that accounts for almost half of Europe's CO_2 emissions. The number of allowances given to installations covered by the scheme is based on national emissions caps set by EU member states in their National Allocation Plans (NAPs). While the NAPs are calculated by national governments, these national plans must be approved by the European Commission. Banking of allowances across the first two phases covered by the plan is left to the discretion of individual member states, but because of the large oversupply in the first period, member states have refrained from allowing their installations to bank these emissions. Banking will become mandatory from the second phase onwards.

Uncertainty over supply and demand in the EU-ETS has led to very volatile prices for EU allowances, which is a typical feature of any new trading system where market players have insufficient oversight of key supply and demand dynamics. Over the past two years, the average price of spot EU allowances for delivery in Phase I was EUR 14.91 per tonne of CO_2, with a high of EUR 29.60 and a low of EUR 0.09. Prices dropped sharply by one-third on 27 April 2006, after the release of the EU verified emissions report that showed emissions were around 7 per cent lower than expected. Prices fell sharply again in September 2006 following the collapse of spot natural gas prices, their earlier high level being the main cause of EU allowance prices initially rising to the level of €29.60/ton. Member states did not allow installations to bank excess allowances from Phase I for use in Phase II, so by mid-2006 there was no longer any demand for Phase I allowances, since companies had purchased all the necessary allowances. As a result, the price of spot allowances fell close to zero and has remained there as the first phase approaches expiry at the end of 2007. By contrast, the forward price of EU allowances for delivery in 2008 has averaged EUR 20.06. Forward allowances have exhibited lower volatility than spot EU allowances as market participants anticipate improvements in the allocation and design features of the EU-ETS.

Lessons Learned from the EU-ETS

There are a number of lessons to be learned from the operation of Phase I of the EU-ETS. These lessons point to design features and operational issues that other countries should consider when developing their own emissions trading systems.

Setting the Cap

A difficult issue in a cap-and-trade system is the level to set the overall cap at on CO_2 emissions. The cap effectively sets the number of allowances that will be created under the scheme, and thereby affects the price at which they trade. If the cap is set too high, then allowances will be oversupplied and prices will be low. Firms will not have an incentive to invest in technologies or introduce operational improvements that reduce emissions, since it will be cheaper to buy allowances to meet their individual caps. If the cap is set too low and no use can be made of offsetting mechanisms from the Kyoto Protocol, prices will be too high, with potentially undesirable effects. Setting the right cap requires, foremost, reliable data about historic emissions for each installation. The lack of reliable data is seen as the main cause of the collapse in the allowance price in the EU-ETS. The diagram below shows the CERs transactions on EU Market linking EURO payment to a Chinese power plant.

EU-China-India Partnership Initiative in Emissions Trading

In September 2005, EU summits with China and India resulted in an EU-China Partnership on Climate Change and an EU-India Initiative on Clean Development and Climate Change. Through these initiatives the EU aims to scale up efforts for development, deployment and transfer of low carbon energy

Fig. 4.1: CERs Transactions on EU Market Linking EURO Payment to a Chinese Power Plant.

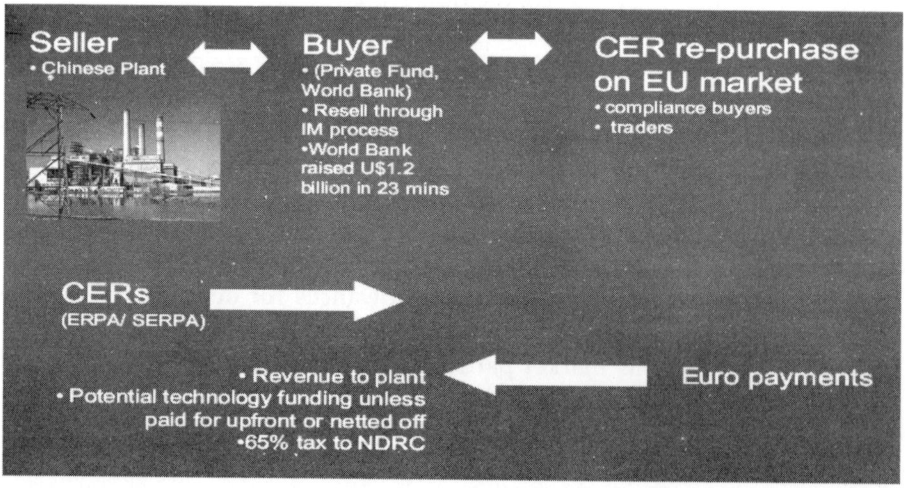

technologies as well as to increase co-operation on adaptation to climate change. On 2nd July, 2003 the European Council formally adopted the Emissions Trading Directive. The Directive laid out the framework for the European Emissions Trading Scheme (the 'EU ETS').

Linking JI/ CDM to the EU ETS

- Commission made a proposal in July 2003 for allowing use of JI and CDM credits in the EU scheme.
- Proposal adopted by European Parliament in April 2004.
- Allows companies in the EU ETS to convert JI/CDM credits into allowances to fulfil their obligations under the EU emissions trading scheme from 2005.
- Gives an incentive for companies in the EU emissions trading scheme to carry out emission-curbing JI and CDM projects—will boost global demand for JI/CDM.

Benefits of Linking JI/CDM to EU ETS

- Stimulate investments in JI/CDM projects.
- Promotion of the transfer of environmentally sound technologies to third countries.
- Contribution to host countries' Sustainable Development.
- Increased liquidity of the EU emissions trading market.
- More compliance options and lower costs for companies.

Stages of Development of EU-ETS

- 2005-07: Start-up period
  - Allowances mostly allocated for free (auctioning limited to 5%)
  - Robust emissions monitoring and verification
  - Well-performing electronic registry system
  - Sound market development
  - However, insufficiently ambitious levels for emission reductions
- 2008-12: First commitment period of Kyoto Protocol
  - Auctioning possible up to 10%
  - Commission approval given to 10 plans in November 2006, 2 more in January 2007, fair and equal treatment being given to all MSs
  - Extension of EU ETS through 'opt-in' provisions
- Subsequent five-year periods
  - Legislative review underway: *Building a global carbon market*
  - Aim: *Limiting Global Climate Change to 2℃ (3.6℉ahrenheit)*
  - EU aims for 20 % in any case and 30 % reduction in the context of sufficient action taken by others.

EU-ETS Summing-up

- The introduction of the EU ETS in 2005 is a major achievement and signifies a paradigm shift in Europe's climate policy.
- The EU ETS is followed closely across the world and a successful scheme is therefore of strategic importance for the post-2012 climate architecture.
- It is important to keep in mind that the first period of the EU ETS was established as a learning period for regulators and companies alike.
- The second trading period will see improvements in many areas.
- The review of the EU ETS is aimed at further improvements and extension of the scheme.
- The internationalization has started and will proceed over the second trading period.

Emission Trading Initiatives in China

Ever since the Kyoto Protocol has entered into force, both China's particular national circumstances and the CDM process as a whole have evolved. Up to 1st November 2008, 286 CDM projects from China completed registration at UNFCCC. During this process, China has been broadening its contribution to the CDM market gradually from non-carbon industrial processes into multiple market areas offering direct reduction of carbon emissions, primarily in the energy sector. However, the projects related to industrial processes still constituted 65 percent of the CER delivery expected from the Chinese portfolio of registered projects up to 2012. Factoring in registration risks of pipelined projects as well as project performance risks, it is projected that out of the 1.5 bn CERs expected to get issued globally up to 2012 (URC-UNEP Risoe Center 2008), some 700 to 800 million CERs are expected to be supplied by CDM projects from China (World Bank, 2009; Michaelowa, 2008). Nevertheless, on the assumption that the global CDM market will continue to operate in a manner that builds directly on the valuable experience gained under the Kyoto Protocol, China could do more to expand the scope and depth of its participation in CDM markets. Halfway through the 11th Five-Year-Plan period (2006–10), registered projects as of 1st November 2008, have generated in total only 5.5 million CERs which is expected to increase to 110 million CERs by 2010. The respective amount issued in 2010 would set off not more than 1 percent of China's CO_2 emissions in that year, emissions expected to grow more than 5 percent per annum during the 11th Five-Year-Plan period (Schwank, and Guyer 2009).

Although emissions trading are a relatively new concept and have so far only been implemented on a regional level in the US, the mechanics of the 'cap and trade' system have spread quickly to China and several credit trading initiatives have been undertaken in recent years. Since 1991, various emissions trading programs have been established in six cities in China. These programs include:

- Fluoride emissions trading in Baotou City, Inner Mongolia;
- Atmospheric pollutants' emissions trading in Taiyuan City; Shanxi Province;
- SO_2 dust, and smoke emissions trading in Kaiyuan City, Yunnan Province;
- Dust and suspended particulates emissions trading in Pingdingshan City, Henan Province;
- SO_2 and dust emissions trading in Guiyang City, Guizhou Province; and
- SO_2 emissions trading in Liuzhou City, Guangxi Province.

These programs have reduced emissions in some areas, greatly improving ambient air quality, and in some cases facilitating economic development and providing social benefits through new job opportunities.

China CDM Involvement and Status

China has already committed to Kyoto Protocol with the following perspectives:

- Although there is limited information from the government; there is a temporary guideline from the China government:
- Interim Measures for Operation and Management of Clean Development Mechanism Projects in China (available on http://cdm.ccchina.gov.cn)
- DNA office has recently been established with the Tsinghua University
- The whole CDM DNA involves several parties:
- National Development and Reform Commission (NDRC)
- Ministry of Foreign Affairs (MoFA)
- Ministry of Science and Technology (MoST)
- State Meteorological Administration (SMA)

National Development and Reform Commission (NDRC)

- Coordination on Climate Change Policies and actions adopted by various departments.
- Participating international climate change negotiation, together with Ministry of Science and Technology, and Ministry of Foreign Affairs.
- Implementing National strategy study on climate change, developing China's approaches for international negotiation.
- Administer international cooperation projects in the climate change field.
- National wide awareness rising and publicity on climate change and China's national policies.

Priority of the China government in terms of CDM projects:

1. Energy Efficiency Projects;
2. Renewable Energy Projects (Wind Energy, Solar Energy, Biomass, etc.);
3. Landfill Gas Projects:
   - All projects has to be complying with the requirement of sustainable development, which also contribute benefits to the local community
   - Projects involving HFC reduction, even though it is eligible as a CDM project according to UNFCCC, it is not as sustainable as the other projects and the China government will consider taking part of the CERs away.

General perception of the Chinese potential project is still hesitant because of:

- Lack of promotion equals to not as significant as other topics
- Limited direction from the government
- Limited guideline from the government
- Market is not developed in China
- Not a lot of project owners are willing to invest
- Not well informed (a perception of CDM equals to upfront money or an investment fund)

Australia in Meeting Kyoto Protocols

Australia's ratification of the Kyoto Protocol facilitates new opportunities for Australian businesses to participate in global carbon markets. In February 2008, there were 918 approved CDM projects around the world, with more than 1900 others waiting for official approval by the CDM's 'Executive Board'. Projects are currently underway to reduce emissions in a range of sectors, including energy industries (both renewable and non-renewable), waste handling and disposal, agriculture, manufacturing emissions, and fugitive emissions from fuels. In total, the CDM market was estimated to be worth more than US$5 billion in 2006. Australian innovation has an important role to play in global efforts to reduce emissions and the Government will be working to facilitate access by Australian companies to the emerging global market (http://www.climatechange.gov.au/international/publications/fs-kyoto.html).

The National Strategy Study Program

Another initiative that progresses the establishment of the CDM is the National Strategy Study Program. Through this capacity-building initiative, donor countries, in collaboration with the World Bank, are helping host countries explore the benefits and opportunities offered by the CDM. One of the objectives of the studies is to develop a pipeline of possible projects for different financing opportunities. Australia has contributed almost US$2.5 million to the Program for studies in developing countries of the Asia-Pacific region. Australian-funded studies are under way in Thailand, Vietnam, PNG and Indonesia. Discussions are ongoing for other studies in the Pacific.

International Greenhouse Partnerships Program

As part of Australia's commitment to ensuring a workable and cost effective CDM that would contribute to developing countries' sustainable development, the Australian Government established the International Greenhouse Partnerships (IGP) Program. The Program is working closely with partner countries from most regions of the world, including Africa, and has focused on capacity building and the establishment of commercial international collaborative projects to reduce greenhouse gas emissions. A key element of the capacity building involves the staging of training and development courses for developing countries. The courses seek to increase awareness of the technical issues associated with the establishment of CDM projects. To date, two courses have been held in Australia (in June 1999 and May 2000) which were attended by representatives from 16 newly industrializing and developing countries. To gain experience in the lead-up to the establishment of the CDM, Australia has reached agreement with seven countries from around the world—namely Mauritius, Chile, Malaysia, Fiji, Indonesia, Viet Nam and Solomon Islands—to establish thirteen projects under the Activities Implemented Jointly (AIJ) pilot phase. The projects cover a diverse range of activities including energy efficiency improvements, renewable energy in rural villages, better industrial production technology, and forest plantations using improved planting stock.

The IGP Office is currently in the process of expanding the range of project opportunities. To facilitate the establishment of new projects, funding is being provided to Australian-based organizations on a competitive basis for projects where the host country government, is willing to accord the project official status under the UNFCCC. As emission reduction credits are not available from the AIJ pilot phase, there are few incentives for business to participate at this stage. Nevertheless, the existing projects are providing some insight into the potential benefits which could flow in the form of technology transfer, capacity building and investment when the CDM becomes operational and credits become available. (Further information on the IGP Program and the existing IGP projects can be obtained from the IGP Internet site at www.isr.gov.au/resources/energy_greenhouse/igp and http://www.econ4env.co.za/archives/climate/connolly.pdf)

UK Emissions Trading Scheme (UK-ETS)

The UK Government has international climate change commitments as well as more demanding domestic targets for carbon dioxide reductions. The UK Emissions Trading Scheme (UKETS) is one of a number of initiatives employed to promote carbon reduction as part of the of the Government's Climate Change Programme. An important driver in the development of the

UKETS was the Climate Change Levy, and in particular the need to offer a flexible mechanism to those companies who had entered into Negotiated Agreements (NA) with government to reduce carbon emissions in return for an 80% rebate on their Levy payments. The UKETS came into operation on 1st April 2002. Emissions trading are a market-based mechanism designed to enable business to deliver carbon emissions reductions in the most cost-effective way. Companies those are able to reduce emissions at relatively low cost are able to sell permits to companies unable to make reductions cost-effectively.

Participation in the UK scheme is voluntary and incentivized with three separate methods of participation—cap and trade, conversion of energy efficiency gains (for those with Negotiated Agreements), and specific emission reduction projects. The government has offered financial incentives, £215 million over 5 years, to encourage wider participation in the direct entry scheme. Companies agreed to take an emissions reduction target in return for a financial incentive, thus providing valuable liquidity for the market as a whole. The scheme is also open to the 6000 companies with Climate Change Agreements. These negotiated agreements between business and Government set energy-related targets. Companies meeting their targets will receive an 80% discount from the Climate Change Levy, a tax on the business use of energy. These companies can use the scheme either to buy allowances to meet their targets, or to sell any over-achievement of these targets. Anyone can open an account on the registry to buy and sell allowances. In the first year, the Direct Participants achieved emission reductions of 4.64 million tonnes CO_2e (carbon dioxide equivalent) against their baselines and in the second year they achieved emission reductions of nearly 5.2 million tonnes CO_2e against their baselines.

Emission Trading Initiatives in Singapore

Singapore is well-equipped to facilitate the trading of carbon emissions by the extension of our oil-trading capabilities to this emerging commodity. Singapore is also uniquely positioned, both in terms of geographical location and capabilities, to support our surrounding region to meet their sustainable development and greenhouse gas mitigation needs. Asia alone represents about 70% of global CDM potential for emissions reduction. Asia Carbon is a carbon advisory, carbon finance and carbon asset management company, registered both in Singapore and the Netherlands, with operations in 9 countries including the Netherlands, Vietnam, Malaysia, India and Singapore. Asia Carbon is one of the first-movers in the CDM arena, having soft-launched the world's first CDM-focused online exchange in May 2005 at the Carbon Expo in Cologne, and secured 5.7 million CERs from regional CDM projects. They have developed an electronic trading platform and have conducted the world's first auction on CER forward trades. For the five auctions conducted offshore in the Netherlands since late 2005, global buyers were able to procure in excess of 1.3 million CERs from CDM project developers in Asia. Offshore trading volume is in excess of 20 million Euros.

CERs Potentials in Malaysia

The CDM can give a financial contribution to projects reducing GHG emissions. Projects that have the potential to reduce GHG in Malaysia include amongst others: additional to what would have otherwise occurred. This implies that it should be possible to demonstrate that the proposed project activity is not the business as usual scenario. This can be done by demonstrating that the revenues of CDM can help overcome some existing financial or other barriers (http://cdm.eib.org.my/subindex.php?menu=7&submenu=43). Secondly, several costs have to be made to register a project as a CDM project and before the tradable CERs can actually be generated. These costs are also referred to as transaction costs. These costs can vary from 40,000 US$ to as high as 150,000 US$ per project.

Table 1 below provides an overview of the expected potential of CER revenues for different types of projects in Malaysia and the corresponding amount of megawatt (MW) that can be installed from renewable energy. It should be stressed that the results are based on an assessment of the potential in the energy sector only and that the results are still preliminary. The realization of this potential will depend upon the removal of other existing barriers for these project types.

Table 1: Expected Potential of CER Revenues for Different Types of Projects in Malaysia

Project type	CERs per year in 2010	MW electricity
Biogas POME + animal manure	5,900,000	190 MW
Landfill gas	3,700,000	45 MW
Reduction of gas flaring from oil production	4,600,000	N/A
Mini hydro	70,000	25 MW
Biomass CHP	380,000	90 MW
Other projects	3,150,000	N/A
Total	17,800,000	350 MW

Host country approval can be officially registered as CDM projects and generate CERs. Without such an approval, no CERs can be generated. In case the Government does not want to support a certain type of project or technology, it can withhold national approval and thus prevent CERs to be generated and traded. Among the ASEAN countries, of late, Malaysia has emerged one of the important favoured destinations for emission trading, as could be seen from the performance statement of CERs given in Table 2.

Table 2: Impact of Renewable CDM Projects on CERs and Energy Generation in Malaysia

Project type	CERs per year in 2010	MW electricity as SREP
Biogas POME + animal manure	5,900,000	190 MW
Landfill gas	3,700,000	45 MW
Reduction of gas flaring from oil production	4,600,000	N/A
Mini hydro	70,000	25 MW
Biomass CHP	380,000	90 MW
Other projects	3,150,000	N/A
Total	17,800,000	350 MW

- CDM potential in Malaysia are associated with energy and waste management.
- CDM: tool to turn economically unviable projects into attractive investments.
- Total of 18 million tonnes CERs/yr estimated in 2010.

Dialogue on Future International Actions to Address Global Climate Change (the FAD)

The following is the key Agenda for CCCP's FAD initiatives for taking future international actions to address Global Climate Change related issues:

- Brings together senior climate negotiators from approx. 15 developed and 15 developing countries
- Meets face-to-face biannually and through workgroups via conference call to develop specific ideas in b/t meetings
- Informal, off-the-record forum to discuss mitigation and adaptation options for a possible post-2012 international framework for climate policy
- Funding for the project supported by governments of 13 countries.
- CCAP produces FAD working papers and will produce a compendium the end of 2006 on the post-2012 options discussed.
- For all presentations and finalized working papers from the process, see: www.ccap.org/international/future.htm

A Sectoral Approach to Post-2012 GHG Reductions

A sectoral approach to post-2012 GHG reductions is being taken to encourage Developing countries fulfilling sectoral e.g. steel, cement, electricity) pledges and for deriving Annex I Country targets in the following manner:

- Bottom-up method for encouraging developing country sectoral (e.g. steel, cement, electricity) pledges and for deriving Annex I country targets
- Based on analysis of what is technologically feasible and economically

cost-effective in each industrial sector both globally and in each country

- For developing countries, goal is no-lose target—pledge to reach intensity level in given sector, rewarded if achieved, no penalty if not achieved
- For A1, creates building block for next national target

Country-based Structure

- Program operated country-based in which individual countries are responsible for ensuring that the emission levels of the covered sectors meet the emissions intensity targets
- Not an industry-based or transnational programs, in which targets are established for an entire sector's operations worldwide (Watson et al., 2005).
- Program will aim to include all major developing countries, but at least 80% of sector's emissions
- Variety of approaches could be used, but key is to cover enough of the sector to address leakage concerns
- Select internationally competitive industry sectors—for many sectors, small number of countries account for a large share of emissions

Establishing Emissions Reduction Target

GHG Emission reduction Targets are computed in a country-specific manner by resorting to the following norms:

- Independent agency (e.g., IEA) assesses and defines benchmark(s) reflecting a limited number of the major processes within the sectors
  - Defined as energy intensity of commercially available technology, taking into account of cost
  - Updated every 7 years (similar to the CDM)
- Participating countries "pledge" a GHG intensity target (e.g., GHG / ton of steel) for each sector that they can meet on their own based upon the benchmark, fuel mix, impact on competitiveness of their products, and other factors (e.g., energy security, air quality, and sustainable development).
- Developed countries negotiate what support they will provide through the "technology finance and assistance package" and the impact that will have on the "no lose" target.
- Likely negotiation will produce different "no lose" targets for each country, but will begin from the same starting point—the benchmark

The following illustration (Fig. 4.2) depicts the Targets set up for Emissions reduction by selected polluting Nations in the world to comply with the requirement of Kyoto Protocol.

Fig. 4.2: Targets for Emissions Reduction by Selected Polluting Nations in the World to Comply with Kyoto Protocol Requirement

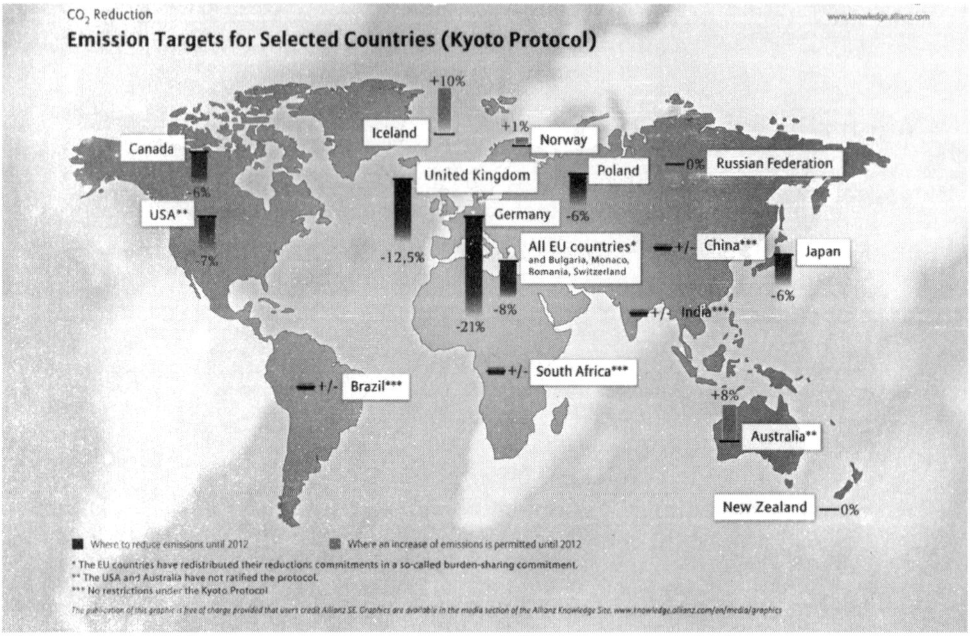

CDM Benchmarking for Developed Countries

CDM benchmarking for developed countries are formalized in the following manner:

- Retain "hard economy-wide targets"—use benchmarks as building block similar to the EU Triptych approach
- Bottom-up development of the sectors AAUs (e.g., national allocation plans) using the BAT energy-intensity benchmark
- Requiring Benchmark in A1 as a minimum threshold for new facilities
- "Levels the playing field" since the covered sector in both developing and developed countries have the same minimum requirement
- Trading allowed to provide covered sectors flexibility to improve cost-effectiveness helps to offset disadvantage that DCs have sectoral intensity targets rather than hard targets.

Growth of the Clean Development Mechanism: Summing-up

- India—459 projects in pipeline, amounting to 278 Mt CO_2eq.
- Brazil—190 projects in pipeline, amounting to 148 Mt CO_2eq.
- China—177 projects in pipeline, amounting to 519 Mt CO_2eq.
- Mexico—132 projects in pipeline, amounting to 57 Mt CO_2eq.

- Other countries –316 projects
- More than half of all CDM projects (52%) relate to renewable energy (geothermal, tidal, wind, hydro, biomass or solar)
- 15 fluorinated greenhouse gas projects (8 in China, 4 in India, 1 in Mexico) account for almost one-third of expected emission reductions
- Expected market for JI and CDM in the EU ETS of up to 1.3 billion tonnes over 2008-12.

Source: New Carbon Finance.

Emission Reduction Commitments: An Overview

According to the Kyoto Protocol countries must reduce total greenhouse gas emissions by 2012, compared to 1990 levels. This is an overview of the contribution of various countries to this total emission reduction.

Australia: rejected Kyoto but would have been allowed an 8% increase by 2012 compared to 1990 levels

Canada: 6% total reduction by 2012 compared to 1990 levels

Europe: 8% total reduction by 2012 compared to 1990 levels

***Western-European countries*:**
Austria → 13% reduction
Belgium → 7,5% reduction
Denmark → 21% reduction
Finland → 0% reduction
France → 0% reduction
Germany → 21% reduction
Greece → allowed increase of 25%
Ireland → allowed increase of 13%
Italy → 6,5% reduction
Luxembourg → 28% reduction
Netherlands → 6% reduction
Portugal → allowed increase of 27%
Spain → allowed increase of 15%
Sweden → allowed increase of 4%
Switzerland → 8% reduction
United Kingdom (UK) → 12,5% reduction

(Contd.)

***Eastern European countries*:**
Maximum 8% total reduction by 2012 compared to 1990 levels

***Japan*:** 6% total reduction by 2012 compared to 1990 levels

New Zealand: 0% total reduction by 2012 compared to 1990 levels

Ukraine: no obligated reduction

***Russia*:** no obligated reduction

CDM: Indo-German Energy Program (IGEN)

GTZ-CDM-India was established in August 2003, through an agreement between German Technical Cooperation (GTZ) and the Bureau of Energy Efficiency (Ministry of Power), Government of India, under the Indo-German Energy Programme (IGEN) as the capacity building facility that can help reduce transaction costs in the early market development process. Its objective is to foster high quality CDM projects that will successfully complete the project cycle and provide experience through 'learning by doing'. These projects should be widely replicable. Capacity building and providing support to public and private sector institutions for preparation and implementation of internationally acceptable projects under the Clean Development Mechanism is its primary aim. It actively cooperates with the National CDM Authority (NCDMA) of India for institutionalizing CDM projects in India. Till now, 60 project proposals with the potential to generate over 40 million CERs have been received by GTZ.

It supported PDD development of 10 projects; of which 6 projects have received host country approval and 3 have signed validators' contracts.

GTZ and KfW Entwicklungsbank jointly implement the Indo-German Energy Programme (IGEN). The broad focus of IGEN is to support the implementation of the Energy Conservation Act. The Energy Conservation Act implemented by the Bureau of Energy Efficiency (BEE), a statutory body under Government of India, envisages creation of cadre of professionally qualified energy managers and auditors with expertise in energy management, project management, financing and implementation of energy efficiency projects, as well as policy analysis. It is a Law to force firms to make more profit and not an Act to control and monitor energy consumption of industry (http://www.energymanagertraining.com/new_index.php). The Energy Conservation Act positively intervenes at all levels of society, such as energy-intensive large industries, manufacturers of household appliances and industrial equipment, residential households as well as engineering consultancy firms. The bilateral cooperation involves local and international professionals and supports (http://www.gtz.de/en/weltweit/asien-pazifik/indien/9521.htm):

- labelling of energy efficiency of household appliances and energy intensive industrial equipment
- certification of energy managers and energy auditors
- setting norms and standards for energy intensive industries
- transfer and promotion of cutting edge technology to reduce energy consumption
- promotion of Clean Development Mechanism and
- operation of one of the largest web portals on this subject

Results

- 1.4 Bill EUR has been invested on energy efficiency and climate protection measures.
- The energy management website is the most popular one in India with two million clicks so far.
- 4500 energy managers and 3500 energy auditors have been certified.
- Indian Certified Emission Reduction (CER) worth 160 Mill. EUR have been sold abroad.
- Rules and regulations under the EC Act have been vetted.
- Energy efficiency measures result in 150 Mill. EUR annual reduction of energy costs.
- Energy consumption labels have been introduced and accepted by the market for four major electric end-user devices.

Canada's GHG Policy Framework and Principles

Type of System/Program and Context

Canada has a goal of reducing its overall GHG emissions by 20% by 2020.

Canada's Offset System for Greenhouse Gases will serve as a compliance mechanism for industrial facilities whose GHG emissions intensity will be regulated under the Government of Canada's proposed Regulatory Framework for Air Emissions. The Regulatory Framework proposed in 2007 is expected to come into force on January 1, 2010. It will set mandatory emissions intensity reduction targets relative to 2006 levels, starting with an 18% reduction in 2010 and increasing each year by 2%—achieving a 33% emissions intensity reduction by 2020 (Environment Canada, 2008). All Canadian industrial facilities, with the exclusion of very small facilities, will be regulated and required to meet emissions intensity reduction targets. The targets will use baselines set at three different levels—facility, sector-wide or corporate—which will depend on the sector. These targets will create the primary demand for offsets generated from activities not covered under the Regulatory Framework. There is no limit on the use of offsets by a regulated facility for compliance (http://www.co2offsetresearch.org/policy/Canada.html).

Canada's GHG Policy Framework and Principles include the following developmental objectives:

- Contribute to a global effort that has to involve the major emitting countries in a cooperative effort.
- Reflect balance of near term improvements and advancing technology.
- Drive action across the economy, recognizing the close link of economic growth, energy use and emissions.
- Integrate GHG and clean air policies recognizing co-benefits and trade-offs.

Recognition of Standards/ Linkage with Other Trading Systems

Regulated facilities under the Regulatory Framework will be able to meet up to 10% of their emission reduction obligations through the purchase of Clean Development Mechanism (CDM) Certified Emission Reductions (CERs) (excluding bio-sequestration projects) (Environment Canada, 2008a). Initially, emission trading under the Regulatory Framework will be restricted to Canada. The Government of Canada has indicated its intention to move from emissions intensity targets to fixed emission caps in the 2020–2025 periods (Environment Canada, 2008a). When it does so, it will take into account developments in other countries, especially the US, with the aim of establishing a North American trading system. Such linkages, if established, are also expected to include the transfer of offset credits generated under Canada's Offset System for Greenhouse Gases (http://www.co2offsetresearch.org/policy/Canada.html).

Offset Project Eligibility and Project Types

Environment Canada has developed a preliminary "fast track" protocol eligibility list which includes external protocols from other programs that are recommended for approval under the Canadian program. A final approved list is expected in 2009 (Environment Canada, 2008c). A summary has been included below (Environment Canada, 2008a):

- Agriculture—Livestock Feeding (Alberta)
- Manure methane (Alberta, CCAR, C3)
- Soil sequestration (Alberta)
- Energy Efficiency (CDM, Alt Env/C3)
- Forestry (CCAR, Alt Env/C3)
- Fossil Fuel (CDM)
- Geological Sequestration (Alt Env/C3)
- Methane management of landfills, composting, and coalmines (Alberta, CCAR, GE AES, CDM, CCAR/C3, Alt Env/C3, GE AES)
- Renewable energy (Alberta, CDM, Alt Env/C3)
- Transportation (Alberta)
- Waste management (CDM, Alberta)

Project Approval Process

Validation and Registration

Validation is the first step of the project registration process under the Canada's Offset System for Greenhouse Gases (Environment Canada, 2008c). Project developers must submit an initial Base Protocol Plan to Environment Canada for feedback. Subsequently, project developers submit a complete project proposal, referred to as a Base Protocol Plan, to Environment Canada. With further input from technical review by experts, Environment Canada makes a final decision regarding the inclusion or exclusion of the proposed project. Once included, project developers submit an Offset System Quantification Protocol to Environment Canada and, once approved, the project is registered. Both the Base Protocol Plan and the Offset System Quantification Protocol are posted on the Offset System website by Environment Canada.

Monitoring, Verification and Certification

Environment Canada is in the process of developing detailed monitoring, verification and certification requirements, which will be published in a series of guidance documents in late 2008 (Environment Canada, 2008b). Ideas under consideration for these processes are outlined here. For monitoring and verification, project developers will have to prepare a Reduction/Removal Report at the end of the calendar year after registration or when 100,000 metric tons of GHG reductions have been achieved, whichever is earlier, and thereafter at a minimum of five-year intervals. The Reduction/Removal Report reports on the GHG reductions or removals claimed by the project developer. Additionally, the project developer must have a recognized verifier provide a Verification Report to Environment Canada. Environment Canada will review both the Reduction/Removal Report and the Verification Report. Once Environment Canada certifies that all the requirements for the issuance of credits have been met, it will authorize the deposit of offset credits (Environment Canada, 2008b).

Two types of credits have been considered for biological sequestration projects to address the risk of non-permanence. The first type—where the project proponent has an obligation to address any reversals of a sequestration project for a fixed period of time, as yet unspecified, known as a 'liability period'—is as an offset credit (Environment Canada, 2008b). At the end of the 'liability period' the project proponent would no longer have an obligation to maintain carbon storage (Environment Canada, 2008b). The second type is where a temporary credit is issued that represents one metric ton of CO_2 sequestered and stored for one year (Environment Canada, 2008b). Temporary credits would be non-fungible with offset credits and would be tracked separately (Environment Canada, 2008b).

Canada's Approach to Climate Change

Canada has produced several action plans on GHG reduction since 1992. The most recent of these is the *Climate Change Plan for Canada* (CCPC), released by the federal government in November 2002. This plan describes specific approaches and targets for a variety of sectors, such as transportation, buildings, infrastructure, agriculture, etc, some of which are existing activities from previous plans. The measures described account for a 180 Mt reduction from projected 2010 BAU levels, leaving 60 Mt of the Kyoto Gap to be dealt with in subsequent plans. The CCPC states that a Domestic Emissions Trading (DET) system will be created for Large Final Emitters (LFEs—referred to in the plan as Large Industrial Emitters). The reductions target set for this system is 55 Mt. The CCPC also outlines several principles that are relevant to the design of the DET system. Three key principles, which will be dealt with in more detail in the Issues section, are:

- Governmental jurisdictions must be respected;
- No region or jurisdiction of the country should be asked to bear an unreasonable burden in the realization of climate change goals; and
- Risk and uncertainty must be minimized.

Canadian Carbon Credit Market Prospect

Can a Canadian domestic market develop? In this regard the following three key issues are to be addressed:

- Effectiveness of Offset System in generating reductions and supply of credits;
- Size and manner of Climate Fund domestic purchases;
- Functioning of international market:
  - Effectiveness of CDM/JI in generating reductions and Kyoto credits
  - Demand from EU ETS and national purchases by EU and Japan
  - Resulting international price

Fig. 4.3: CleanAir Canada

Source: http://pm.gc.ca/eng /media_gallery.asp?media_category_id=69&media_category_typ_id=3

CleanAir Canada

'CleanAir Canada' has evolved out of the PERT (Pilot Emission Reduction Trading) Project, considered by many to be the leading body for emission reduction analysis and application in Canada. Formed in 1996, this industry-led, multi-stakeholder group set the standard for the development and implementation of a robust emission reduction trading system. Their analysis and recommendation on system design, trading rules, and review methods has helped lay a solid foundation for sound government policy.

PERT originated as a experimental program designed to reduce smog and acid rain emissions of NOx and SO_2 more quickly and cost-effectively than would otherwise occur under traditional "command and control" environmental regulation. As climate change grew as an issue of international importance, PERT adopted a multi-pollutant approach which included reductions of carbon dioxide. PERT started to review and register CO_2 emission reductions in 1998, shortly after the Third Conference of the Parties to the United Nations Framework Convention on Climate Change in Kyoto, Japan. The Ontario Ministry of Environment has been an active partner in the work of the PERT project. A negotiated Letter of Understanding has provided industry the confidence and incentive needed to take early action to reduce smog and acid rain causing emissions.

In anticipation of the fact that Ontario's new Emission Reduction Trading regulation would supersede the rules of the PERT project, the membership resolved to form a new not-for-profit, federally incorporated body called CleanAir Canada. Operational in September of 2000, CleanAir Canada was comprised of largely the same membership as the PERT project. Since then, new members have joined and interest is continuing to grow. In addition to the objectives outlined in the previous section, CleanAir Canada continues to serve as a forum for the review and registration of GHG emission reduction activities.

As no jurisdiction in Canada currently has legislation regulating either emissions of greenhouse gases (GHGs), or the trading of GHG reductions, CleanAir Canada is currently seeking a renewed Letter of Understanding with the Ontario MOE which will address the need for early reductions of greenhouse gas emissions (http://www.cleanaircanada.org/).The objective of Clean-Air Canada is to demonstrate leadership in the development and facilitation of market-based approaches to reducing, offsetting and managing emissions. Clean-Air Canada Inc. is committed to excellence in the development, operation and expanded use of local and global emission reduction market mechanisms in Canada. The multi-stakeholder membership is built on the volunteer efforts of a broad range of companies and organizations. Clean-Air Canada has developed the country's largest experienced-based knowledge centre for emission reduction market mechanisms.

Canada's Regulatory Framework for Air Emissions

On April 26, 2007, the Government of Canada released a framework for regulating air emissions that outlined a proposed emissions trading system for greenhouse gas emissions and other air pollutants. The proposal stated that the emissions trading system would be a baseline-and-credit system, and not a cap-and-trade system. The government proposes intensity-based targets that will come into force in 2010, which should lead to absolute reductions in emissions of greenhouse gases from industry by 2012, based on current growth forecasts. The government will then set national caps for 2012–15 for the emissions of various greenhouse gases relative to 2006 emissions levels for each pollutant. National caps will be broken down into sector-specific caps over the same time horizon. In a baseline-and-credit emissions trading system, a baseline is set. In this case, the baseline would be the emissions-intensity target. Facilities that reduced emissions below their target would be allocated tradable credits that they could either bank for a future compliance obligation or sell to another facility. Facilities that emitted above their target would have to buy credits from other facilities or use their own banked credits to meet their regulatory obligation.

2. The report can be downloaded at: http://www.ec.gc.ca/doc/media/m_124/report_eng.pdf.

3. A cap-and-trade system will be used for SOx and NOx only, similar to the mechanism in place for the United States.

4. For existing facilities, the emissions-intensity reduction target for each sector is based on an improvement of 6 per cent each year from 2007 to 2010, yielding an initial enforceable reduction of 18 per cent from 2006 emissions-intensity levels in 2010. Every year thereafter, a 2 per cent continuous emissions

intensity improvement will be required, resulting in an industrial emissions-intensity reduction of 26 per cent by 2015.

Targets for new facilities will be based on cleaner fuel standards. For example, suppose the emissions-intensity target is 5.0 kilotonnes (kt) CO_2e for every tonne of widgets produced. Suppose Facility A had an emissions intensity of 4.5 kt/tonne and produced 1000 tonnes of widgets during the year. It would then receive tradable credits from the government equal to the difference between the target (5.0 kt) and its actual emissions intensity (4.5 kt) times its production in that year (1000 tonnes), or (5.0-4.5) × 1000 = 500 credits received.

Suppose Facility B had an emissions intensity of 5.3 kt/tonne and produced 1200 tonnes of widgets. It would be required to remit to the government credits equal to the difference between its actual emissions intensity (5.3 kt) and the target times its production (1200 tonnes), or (5.3-5.0) × 1200 = 360 credits owed. It could buy these credits from another facility or use credits that it had banked from a previous compliance period. The government's proposal includes the possibility of importing credits from CDM projects up to 10 per cent of each firm's total target.

The emissions trading system would also include domestic offset credits. Offsets are emissions reductions that take place outside the domain of regulated activities. Offset credits, which regulated firms could use towards their regulatory obligations, would be issued for verified reductions in greenhouse gas emissions that were incremental to what would have happened without the regulatory system or other government programs.

Finally, the report suggests that linkages to North American emissions trading systems—such as the Western Regional Climate Action Initiative and the Regional Greenhouse Gas Initiative—will be actively pursued. Over time, as the international carbon market becomes more fully developed and robust, and as emissions monitoring, verification, and reporting systems evolve, the government will consider further international linkages.

To sum-up, Emissions-trading is a cost-effective and flexible mechanism to reduce carbon emissions. An effective cap-and-trade scheme has these features: a regulated absolute cap to set a clear goal; an allocation of allowances that assigns responsibility to firms; a trading facility to allow cost effective reductions; a monitoring and reporting system to ensure accountability; and reconciliation with effective enforcement to ensure compliance. Thorny issues that need to be addressed include setting the level of the cap, allocating allowances across firms, setting rules for banking allowances, and linking with other schemes or allowing the use of project-based offsets. Whatever choices are made, simplicity and transparency will ensure the most effective outcome.

Japan's Voluntary Emissions Trading Scheme (JVETS)

Japanese Ministry of the Environment has selected 34 companies and corporate groups as participants in the nation's new Voluntary Emissions Trading Scheme. They were chosen from among applicants who responded to an open invitation and screened based on their cost-effectiveness for this program. Among them are INAX Corp., Nippon Electric Glass Co., Mitsubishi Gas Chemical Co., and Yamazaki Baking Co.

Under the scheme, the ministry subsidizes the installation cost of CO_2 emissions reduction equipment to help businesses that are actively attempting to reduce greenhouse gas (GHG) emissions. In exchange for the subsidy, the participants are required to commit to a certain reduction in their CO_2 emissions. The scheme also allows them to trade CO_2 emission quotas to meet their reduction targets. Through this scheme, the ministry aims to achieve a cost-effective and substantial reduction in GHG emissions, and accumulate knowledge and experience relating to domestic CO_2 emissions trading. The total government budget for the subsidy is 2,596,340,000 yen (about U.S. $23.6 million). The total of emissions reductions promised by the individual companies for fiscal 2006 is 276,380 tons, or 21% of their average annual CO_2 emissions in the base years, fiscal 2002 to 2004. The reduction in CO_2 emissions over the officially-recognized service life of the subsidized equipment is calculated at about 3.7 million tons.

The participating companies must have their base year emissions independently verified by the end of October 2005, and make the relevant equipment operational by the end of fiscal 2005. Throughout fiscal 2006, they will implement their CO_2 reduction projects using the equipment. After the end of fiscal 2006, the actual CO_2 emissions during the year will be calculated and verified. The CO_2 emissions quota trading system will also start in April 2006. If a participant cannot achieve its reduction target despite the emissions trading, the participant may be required to refund its subsidy.

Japan's Voluntary Domestic Emissions Trading Scheme:

JVETS (Launched in May 2005)

Objectives

- To seek measures for certain and cost efficient Emissions reduction and accumulate knowledge and experiences in domestic ETS
- Subsidy will be provided to installation of new facilities for the reduction of emission in return for pledging certain amount of GHG reduction.
- Participants can utilize ETS as flexibility mechanism to achieve pledged emission reduction.

Characteristics of JVETS

I. Voluntary Scheme

- Companies/facilities participate voluntarily by legging concrete emissions reduction targets and receive the EAs from the government.
- 1/3 of cost of GHG reduction activities will be subsidized from the

government as incentive.

- But, if failed to achieve target, subsidy should be returned to the government.

II. Facility-Based (Target Facility) Scheme

- Government selects target facilities from applicants based on the *cost effectiveness of GHG reduction* activity.
- Government issues the tradable allowances (JPA) to each facility (not company)
- Target facilities face the obligation to submit the same amount of the emissions allowances as their actual GHG emissions to the government.

III. Emission Trading Scheme

- Companies/facilities can trade the JPAs in order to achieve their "pledged" targets.
- Target facilities can use the credits under the Kyoto Mechanisms (CERs/ERUs, etc.) in complying with their obligations.

JVETS: Some Highlights of Achievements

Under the JVETS as many as 32 entities/groups were selected as participants with target and 8 companies selected as trading participants with the following results: (http://www.iges.or.jp/en/cp/pdf/activity06/07.pdf):

- Total Amount of Subsidies: *JPY 2.6 bn. (FY-2005 budget is JPY 3bn.)*
- Base-year emissions: Average works out during 2002-04: *1.3 mil t-* CO_2*/ yr.*
- Estimated total amount of CO_2 emissions reduction pledged (FY2006): *276,380t-*CO_2 *(=21% of the total emissions from 32* facilities)
- Estimated (Statutory useful life of facilities) total amount of CO_2 reduction: *3.7mil t-*CO_2 *(based on the pledges)*
- Total Subsidies/Estimated (Statutory useful life of facilities) = *JPY 692/ t-*CO_2 *low cost emissions reduction!*
- Emissions allowances= Base-year emissions—Estimated/pledged reduction amount (FY2006)

JVETS: Summing up

- JVETS could be a good basis for a future Cap and trade scheme containing a lot of implications such as:
  - Policy cost
  - Incentives and Participants Behavior
  - Mandatory emissions trading are NOT an option at this moment. In future, its feasibility will be analyzed/ considered as an option, taking into account—

- Results of JVETS
- Trend in GHG emissions
- International situation (link between EU and US?)
- Co-benefit and its effect on carbon market

Lessons Learnt from the experiences in JVETS

- "In order to accumulate knowledge and experience concerning cost-efficient emissions reductions and trading, etc., the Government of Japan provides economic incentives for the corporations that make efforts to achieve reduction targets they have determined themselves and implements voluntary participation domestic emissions trading utilizing the trade of emissions quotas."
- "The domestic emissions trading system is an issue that must be comprehensively studied about the wide range of discussion points includes a comparison of the domestic emissions trading system with other methods and their effects and the impact on industrial activities and the national economy ((http://www.iges.or.jp/en/cp/pdf/activity06/07.pdf)

Proposal for Upstream Emissions Trading-Japan

Importers of fossil fuels should hold allowances to sell them in Japan. Such allowances can be traded. If we start upstream emissions trading early, for example from 2005, the total issuance of allowances will be reduced gradually from business-as-usual (BAU) emissions so that the price of allowances will go up gradually. The price of allowances will be shifted to those that demand fossil fuels. The increase in the price of fossil fuels will reduce their demand. The combustion of fossil fuels occupies 93.7% of total CO_2 emissions. HFCs, PFCs and SF_6 will also be included in this scheme because they are produced intentionally. The GHG emissions which cannot be monitored upstream are not included in this scheme (Saijo *et al*, 2003).

GHG Emissions Trading in the USA: Private Initiatives

About 40 US companies, with combined emissions almost equal to Germany, are creating a pilot market for all six GHGs to be traded at the Chicago Climate Exchange (CXX) beginning in early 2003. Eventually, the CXX wants to cooperate with the EU, Mexico and Canada to create a larger, more diverse market that will drive down prices (Silverstein, 2002; Murphy, 2002). Eight conglomerates—Alcan, BP, DuPont, Entergy, Ontario Power Generation, Pechiney, Shell International and Suncor—will establish a trading system among themselves with Environmental Defense acting as the verification body. They have mutually agreed to monitor and reduce emissions of 80 million metric tons of CO_2-equivalent (CO_2e) no later than 2010. Trading is expected to start by the end of 2002. Many major companies based or operating in the

USA have adopted voluntary emission reduction targets. Some examples include Alcoa, General Motors, Ford Motor Co., IBM, Johnson & Johnson, and Toyota. Others have developed internal emissions trading schemes (e.g. BP, DuPont, and Royal Dutch Shell) and/or have completed small trades with companies in Canada, United Kingdom and Denmark. Companies that are experimenting with offset projects in developing countries include ABB, American Electric Power, and Whirlpool. (Pew Center, 2002)

Emissions Trading to Reduce Greenhouse Gas Emissions in the United States: The McCain-Lieberman Proposal (MIT)

The Climate Stewardship Act of 2003 (S. 139) is the most detailed effort to date to design an economy-wide cap-and-trade system for U.S. greenhouse gas emissions reductions. The Act caps sectors at their 2000 emissions in Phase I of the program, running from 2010 to 2015, and then to their 1990 emissions in Phase II starting 2016. There is a strong incentive for banking of allowances, raising the costs in Phase I to achieve savings in Phase II. Use of credits from outside the capped sectors could significantly reduce the cost of the program, even though limited to 15% and 10% of Phase I and II allowances respectively. These credits may come from CO_2 sequestration in soils and forests, reductions in emissions from uncapped sectors, allowances acquired from foreign emissions trading systems, and from a special incentive program for automobile manufacturers. The 15% and 10% limits increase the incentive for banking and could prevent full use of cost-effective reductions from the uncapped sectors. Moreover, some of the potential credits might contribute little or no real climate benefit, particularly if care is not taken in defining those from forest and soil CO_2 sequestration. Analysis using the MIT Emissions Prediction and Policy Analysis model shows that costs over the two Phases of the program could vary substantially, depending on normal uncertainty in economic and emissions growth, and the details of credit system implementation (Paltsey *et al*, 2003).

Market based Mechanism of Carbon Market

Market-based mechanisms have the potential to play a large role in the global effort to address climate change under the United Nations Framework Convention on Climate Change (UNFCCC). Three such instruments—International Emissions Trading (IET), Joint Implementation (JI) and the Clean Development Mechanism (CDM)—were introduced in the Kyoto Protocol to help countries meet their targets and effectively created the carbon market. These market mechanisms are designed to help countries meet their targets in a cost-effective manner, encourage the private sector to contribute to Greenhouse Gas (GHG) emission reduction efforts, and encourage the participation of developing countries as well as stimulate sustainable development, technology transfer and investment in these countries.

Countries have entered into formal negotiations on a climate change agreement after 2012 when the Kyoto Protocol's first commitment period ends. There are expectations that a comprehensive post-2012 climate agreement will be adopted at the 15th Conference of the Parties (COP 15) in Copenhagen in December 2009. There is considerable uncertainty as to what this post-2012 regime will look like and what market instruments and mechanisms could be employed within this. The Bali Action Plan (BAP), adopted in December 2007, set out broad parameters to guide the two-year negotiating process, including mitigation, adaptation, technology and financing. In regard to mitigation, the Plan emphasized the importance of Various approaches, including opportunities for using markets, in order to enhance the cost-effectiveness of, and to promote, mitigation actions, bearing in mind different circumstances of developed and developing countries (UNFCCC, 2007).

Parties with commitments under the Kyoto Protocol (Annex B Parties) have accepted targets for limiting or reducing GHG emissions. These targets amount to a 5.2 per cent reduction in GHG emissions from 1990 levels over the five-year period 2008-2012. The targets are expressed as levels of allowed GHG emissions or assigned amounts over the commitment period. The allowed GHG emissions are divided into Assigned Amount Units (AAUs), equal to one metric tonne of carbon dioxide equivalent (CO2e). Each country receives AAUs equivalent to how many tonnes of GHGs each can emit.

Trading in carbon tentatively started as the Kyoto Protocol moved toward ratification and is now one the world's fastest growing markets. The market was worth US$64 billion in 2007 with 70 per cent within the European Union Emission Trading Scheme (EU-ETS) (Capoor and Ambrosi, 2008). The bulk of the remainder of the market was for credits generated under the CDM. Markets for voluntary offsets and trading of AAUs under the Kyoto Protocol's IET facility have recently experienced strong growth.

The May 2009 CDM pipeline is dominated by projects in the following sectors: hydro (27 per cent of total projects), biomass energy (15 per cent), wind (15 per cent) and energy efficiency own generation (9 per cent). An examination of the distribution of CERs by project type indicates that the largest sectors are hydro (17 per cent of CERs expected by 2012), hydrofluorocarbon (HFC) destruction (16 per cent), energy efficiency own generation (10 per cent), landfill gas (9 per cent) and wind (9 per cent). China is the host country for the majority of CERs issued (53.6 per cent).

Proposals to use market mechanisms to implement greenhouse gas emission reductions have revolved around three approaches: tradable permits (as "allowances" and as "credits"), carbon taxes, and joint implementation (Larry Pakerr, 2002)

Tradable Permits (Allowances): A model for a tradable permit approach is the SO_2 allowance program to reduce acid rain contained in Title IV of the 1990 Clean Air Act Amendments. The Title IV program is based on two premises. First, a set amount of SO_2 emitted by human activities can be assimilated by the ecological system without undue harm. Thus the goal of the program is to put a ceiling, or cap, on the total emissions of SO_2 rather than limit ambient concentrations. Second, a market in pollution rights between polluters is the most cost-effective means of achieving a given reduction. This market in pollution rights (or allowances, each of which is equal to one ton of SO_2) is designed so that owners of allowances can trade those allowances with other emitters who need them or retain (bank) them for future use or sale. Initially, most allowances were allocated by the federal government to utilities according to statutory formulas related to a given facility's historic fuel use and emissions; other allowances have been reserved by the government for periodic auctions to ensure the liquidity of the market.

Current SO_2 allowance trading plans between individual utilities do not shed much light on how well the existing allowance market will work over the long-term. Some individual trades between utilities and EPA-sponsored auctions have been conducted, but the current level of trading activity has not established the long-term viability of the marketplace. For a market to thrive, transactions must become sufficiently commonplace for an open, public market price to be established with limited bilateral negotiation. Based on the results of the EPA auctions conducted by the Chicago Board of Trade, allowance prices are considerably below that anticipated when the legislation was enacted. However, the six-year experience of the SO_2-allowance market may be insufficient to give much guidance on how well a CO_2-allowance market might work.

Tradable Permits (Credits): **A** tradable allowance involves future emissions. An allowance is a limited authorization to emit a ton of pollutant; allowances are allocated to an emitting facility under an applicable emission limitation at the beginning of a year. The facility decides whether to use, trade, or bank those allowances, depending on its emissions strategy. Then, at the end of the year, the agency compares an emitting facility's actual emissions with its available allowances to determine compliance.

A different approach to creating a tradable permit program is to use credits instead of allowances. A credit is created when a facility actually emits a pollutant at less than its allowable limit as defined in by the program. An example of this type of program is EPA's "Emission Reduction Credit program" (ERC) under the Clean Air Act. Under the ERC program, EPA requires that any credit created under a state program implementing emissions trading be "surplus, enforceable (by the state), permanent, and quantifiable." Thus, a state must certify the creation of the credit, unlike an allowance program, where allocation is dictated by a statutory or regulatory formula. Any CO_2 reduction credit program could build on EPA's and states' experience with the current emission reduction credit program.

The primary advantage of a credit program over an allowance program is that it does not discriminate against new sources. Allowance programs tend to allocate their allowances based on some historic baseline year. Those sources included in the baseline get their allowances free. Those future sources not included in the baseline have to pay either the older, existing sources to obtain allowances or to buy allowances at auction. With a credit program, sulfur credits can be created by any source, as the baseline is dictated by the emissions cap and yearly production, not a historical year. The disadvantage of such a system is that facility planning is very difficult as operators do not know precisely what their permissible limit will be from year to year.

Carbon/CO_2 Emissions Tax: An alternative market-based mechanism to the tradeable permit system is carbon taxes—generally conceived as a levy on natural gas, petroleum, and coal according to their carbon content, in the approximate ratio of 0.6 to 0.8 to 1, respectively. In the view of most economists, the most efficient approach to controlling CO_2 emissions would be a carbon tax. With the complexity of multiple pollutants and millions of emitters involved in controlling CO_2, the advantages of a tax are self-evident. Imposed on an input basis, administrative burdens such as stack monitoring to determine compliance would be reduced. Also, a carbon tax would have the broad effect across the economy that some feel is necessary to achieve long-term reductions in emissions.

However, in other ways, a tax system merely changes the forum rather than the substance of the policy debate. Because paying an emissions tax becomes an alternative to controlling emissions, the debate over the amount of reductions necessarily becomes a debate over the tax level imposed. Those wanting large reductions quickly would want a high tax imposed over a short period of time. Those more concerned with the potential economic burden of a carbon tax would want a low tax imposed at a later time with possible exceptions for various events. Emissions taxes would remain basically an implementation strategy; policy determinations such as tax levels would require political/regulatory decisions. In addition, a tax system would raise revenues. Indeed, one argument for—or against—such a system would be that it is a tax that would raise revenues. The disposition of these revenues would significantly affect the economic and distributional impacts of the tax.

Other tax schemes to address global climate change are also possible. For example, the European Community (EC) has discussed periodically a hybrid carbon tax/ energy tax to begin addressing CO_2 emissions. Fifty percent of the tax would be imposed on energy production (including nuclear power) except renewable; 50% of the tax would be based on carbon emissions. Some European countries have modified their energy taxation to fit the model discussed by the EC.

Currently, five European countries have carbon-based taxes. Finland imposed the first CO_2 tax in 1990 and modified it in 1994. The Finnish tax has two components: (1) a basic tax component to meet fiscal needs and (2) a combined energy/CO_2 tax component. For coal, peat, and natural gas, there is no fiscal component. The Netherlands also introduced a CO_2 tax in 1990, modified in 1992 to fit the EC model; however, the tax does not vary according to fuel type and energy use. It does include tax relief from the energy component of the tax for energy-intensive industries. Sweden introduced a CO_2 tax in 1991 on all fossil fuels, unless they are used in electricity production. In 1993, the tax scheme was modified to reduce its burden on industry. Denmark introduced a CO_2 tax in 1992 that covers fuel oil, gas, coal, and electricity (gasoline is taxed separately). Taxes paid by industry are completely reimbursed to the sector. Norway introduced a CO_2 tax in 1991 on oil and natural gas and extended it to some coal and coke use in 1992. However, there are many exemptions and the tax rate is not differentiated according to the carbon content of the fuels.

Joint Implementation: Joint Implementation (JI) is an attempt to expand the availability of cost-effective CO_2 reductions into the international sphere through a variety of different activities. Basically, a developed country (where opportunities for reducing emissions are expensive) needing CO_2 reductions to meet its obligations under any international treaty could obtain reduction credits by financing emission reductions in another country, usually a developing country (where more cost-effective reductions are available). As generally conceived, the developed country financing the reductions and the developing country hosting the reduction project would split the achieved reductions between them in some previously agreed-upon manner. Joint Implementation is a keystone of U.S. climate change policy; it was subject to considerable debate at the Conference of Parties (COP) meetings in Berlin. These discussions resulted in agreement to implement JI in a pilot phase. Projects must be compatible with and supportive of national environmental and development priorities; accepted, approved, or endorsed beforehand by the Parties' governments; and have anticipated environmental benefits and projected financing fully articulated beforehand. Credits generated cannot be used to meet the Rio Treaty year 2000 target; credit for post-2000 targets was left to the meeting in Kyoto, which included JI as one of its flexible implementation mechanisms.

International Market Mechanisms in a Post-2012 Regime

Most of the international climate discussions on expanded or new market mechanisms in a post-2012 regime are focused on the supply of credits from developing countries, taking the CDM as a starting point and seeing how it could be expanded and/or if new market mechanisms would be required. Capoor and Ambrosi (2008) have observed that the experience of the CDM has demonstrated its power as a tool to engage developing countries in mitigation activities. They noted that the CDM represents only the tip of the iceberg of the potential of market mechanisms and other approaches to mitigating and adapting to climate change in developing countries (Murphy *et al*, 2009). The Ad Hoc Working Group on Further Commitments for Annex I Parties under the Kyoto Protocol (AWG-KP) is discussing possible improvements to GHG emissions trading and the project-based mechanisms under the Kyoto Protocol for the period after 2012. The options and issues being considered are set out in an AWG-KP report (UNFCCC, 2009) and listed below.

CDM

Include other LULUCF activities (afforestation and reforestation or A/R; reducing GHG emissions from deforestation and forest degradation; restoration of wetlands; sustainable forest management and other sustainable land management activities; soil carbon management in agriculture; and revegetation, forest management, cropland management and grazing land management):

- include carbon capture and storage (CCS);
- include nuclear activities;
- introduce sectoral crediting of GHG emission reductions below a previously established (no-lose) target;
- introduce crediting on the basis of nationally appropriate mitigation actions (NAMAs);
- encourage the development of standardized, multi-project baselines;
- ensure environmental integrity and assess additionality through the development of positive or negative lists of project activity types;
- differentiate the eligibility of Parties through the use of indicators;
- improve access to CDM project activities by specified host Parties;
- promote co-benefits for CDM projects by facilitative means; and
- introduce multiplication factors to increase or decrease the CERs issued for specific project activity types.

JI

- introduce modalities for treatment of CDM project activities upon graduation of host International Carbon Market Mechanisms Parties;
- include nuclear activities; and
- promote co-benefits for JI projects under Track 2 by facilitative means.

GHG Emissions Trading

- introduce GHG emissions trading based on sectoral targets;
- introduce GHG emissions trading based on NAMAs; and
- introduce modalities and procedures for the recognition of units from voluntary GHG emissions trading systems in non-Annex I Parties for trading and compliance purposes under the Kyoto Protocol.

Cross-cutting Issues

- relax or eliminate carry-over (banking) restrictions on Kyoto units;
- introduce borrowing of assigned amounts from future commitment periods; and
- extend the share of proceeds.

Some possible elements of a framework that would enable a broad international carbon market could include:

- Binding fixed emission limits for industrialized countries; for industrialized countries unable to agree to the above, binding fixed or dynamic emission limits for some sectors in some regional groupings—or possibly economy-wide binding dynamic emission limits;
- Binding transnational sectoral emission limits (fixed or dynamic) for some key sectors represented by multinational "operators" (e.g., cement, steel, aluminium); individually customized non-binding baselines for specific sectors to generate credits, while attracting investment in key sectors of developing countries; and a project-based crediting mechanism to provide coverage of emission reduction and sink enhancement activities not already covered by other market-based mechanisms;
- While an international framework is essential, the ability of the international carbon market is to minimize costs;
- The International Carbon Bank and Exchange provide a platform that enables individual and corporate clients to keep track of greenhouse gases in a secure environment. Clients can use their account at the ICBE to establish baselines, introduce emission reductions, and acquire offsets in real-time.

International Emissions Trading Association (IETA)

The International Emissions Trading Association (IETA) is committed to establish a functional international framework for trading greenhouse gas emission reductions. IETA members seek to develop an emissions trading regime that results in real and verifiable greenhouse gas emission reductions while balancing economic efficiency with environmental integrity and social equity. IETA activities range from work achieved within working groups, IETA events and sponsored IETA events to developing IETA positions, sharing knowledge and creating networking opportunities (http://www.ieta.org/ieta/www/pages/index.php?IdSiteTree=5).

International Greenhouse Partnerships (IGP) Program

International Greenhouse Partnerships (IGP) Program is a cooperative effort by Australian industry and government to reduce greenhouse gas emissions through projects overseas. The IGP Office was set up in the Department of Industry, Tourism and Resources, Government of Australia, to implement the IGP Program and facilitate through international negotiations the establishment of the CDM and JI (http://www.isr.gov.au/content/controlfiles/display_details.cfm?objectid=6D35F6EF-3AFE-443B-979905B5A6B8E39D). The program provides assistance to help cover additional transaction costs incurred by businesses undertaking cooperative implementation projects and undertakes capacity building and training courses on the project based flexibility mechanisms.

UNIDO Knowledge Network for Industrial Technology Transfer

The UNIDO Knowledge Network for Industrial Technology Transfer (KNITT) aims to facilitate the development and application of new knowledge to 'enable' industrial activities under the project-based mechanisms (the clean development mechanism and joint implementation) of the Kyoto Protocol in developing and transition economy countries (www.unido.org/knitt/index.html). KNITT consists of inter-linked national networks each supporting the application of locally relevant knowledge to the processes of identification, screening and implementation of industrial CDM projects (assessment, transfer, adaptation, adoption and diffusion of industrial greenhouse-gas mitigation technologies) in their countries.

Emissions Reduction: Highlights of G-8 Conferences

The 33rd G8 summit took place at Kempinski Grand Hotel in Heiligendamm in the old Duchy of Mecklenburg in the Northern German state of Mecklenburg-Vorpommern on the Baltic Coast. The group of eight leaders met together from June 6 to June 8, 2007. The locations of previous G8 summits to have been hosted by Germany include: Bonn (1978, 1985); Munich (1992) and Cologne (1999). In a non-binding communiqué issued on June 7, 2007 it was announced that the G8 nations would 'aim to at least halve global CO_2 emissions by 2050'. The details which would enable this salutatory goal to be achieved were left to be negotiated. It was anticipated that the G8 Environment ministers would work together within the United Nations Framework Convention on Climate Change in a process that would also include the major emerging economies. Groups of countries would also be able to reach additional agreements on achieving the goal outside and in parallel with the United Nations process. The G8 also announced their desire to use the proceeds from the auction of emission rights and other financial tools to support climate protection projects in developing countries (http://en.wikipedia.org/wiki/33rd_G8_summit).

Leaders of the Group of Eight industrialized nations meeting in Germany have agreed to pursue "substantial" cuts in the greenhouse gases that cause global warming. Here are the details of what they agreed:

- Leaders acknowledged that "resolute and concerted international action" was urgently needed to reduce emissions and increase energy security.
- Rises in greenhouse gases must halt and be reversed, the G8 said.
- The leaders agreed to work towards a global framework for a new deal on cutting greenhouse gases by the end of next year, which would be finalized by the United Nations in 2009.

- The group did not commit directly to a target for reducing emissions but in considering a goal in the future it said it would be mindful of the EU, Canada and Japan's target of halving global emissions by 2050.
- Developed economies have to take the lead, the G8 said, but less developed countries must participate. The G8 called on emerging economies to reduce the carbon intensity of their economic development to address the increase in their emissions. The United States has insisted that big, poorer polluters like China and India, who have no obligations under the Kyoto treaty currently must start to share the burden of emissions cuts.
- The G8 welcomed a U.S. offer to host a meeting this year with major emitters.
- While offering "strong leadership" the group noted different approaches to tackling climate Change existed. But it backed the United Nations Climate Change conference in Bali, Indonesia in December and said the U.S. efforts to involve emerging economies would feed into the U.N.-led process of finding a successor to Kyoto. There were concerns that U.S.-led initiatives would undermine United Nations efforts.
- Promoting the diversification of energy supplies, increasing energy efficiency, tackling deforestation and setting up emissions-trading systems were all useful means of mitigating climate change.

The following were the key points that were agreed to be adhered to in future:

- agreeing to launch serious UN negotiations for new emissions reduction targets for the time from 2012
- energy efficiency goals for the G8 nations to reduce energy consumption by at least 20% by 2020
- affirm the need for global warming to be kept well below 2°Celsius and to cut global emissions 50% by 2050 from 1990 levels.

In the G-8 Meeting of the Major Economic Forum (MEF) held on 9 July, 2009 in L'Aquila, Italy, the leaders of the largest economies agreed to a goal of achieving at least a 50% reduction in global greenhouse gas emissions by 2050, with developed countries achieving an 80% reduction in emissions by the same date. These goals were set in the context of the scientific view that global average temperatures should not rise by more than 2ºC above preindustrial levels. The G8 leaders had already called for a 50% global emissions reduction at the 2008 G8 meeting in Hokkaido, Japan. The L'Aquila agreement expressed the G8's willingness to share a 50% reduction of global emissions with all countries, noting that global emissions would soon have to peak. The G8 said that major emerging economies would need to undertake quantifiable actions to reduce emissions below 'business as usual' levels. The declaration was unclear on what actual reduction the 80% emission reduction goal for developed countries represented, saying it would be set against 1990 emissions levels or those of more recent years. As global emissions have risen by around a fifth since 1990, a later baseline would represent a higher final emissions goal. It was recognized that the 80% emissions reduction goal would require robust mid-term reductions, although no intermediate targets were set (http://www.world-nuclear-news.org/EE-G8_endorses_emissions_cuts_and_nuclear-0907096.html).

The year 2009 is a crucial year for climate change. At a pivotal UN conference in Copenhagen at the end of 2009, governments of the world need to agree the continuation of the global climate treaty, building on and strengthening the Kyoto Protocol. The G8 leaders meeting in Italy this July have availed the opportunity to set the course for the Copenhagen conference to succeed. In doing this they realized the triple advantage of action against climate change (http://us-cdn.creamermedia.co.za/assets/articles/attachments/22098_wwfbinaryitem12911.pdf):

- Action now will start to clean up our climate act, keeping the option open for the planet to avoid catastrophic climate change.
- Action now will set the world on a sustainable energy path, which in the short run can help stabilize the economy, and in a few decades can provide ample energy for the whole planet.

Action now can help develop a low carbon economy that helps avoiding millions of climate refugees and massive cost for later adaptation to climate impacts.

Key Expectations for a Successful Copenhagen Climate Treaty

The WWF Expectations Paper for the global climate deal and an NGO proposal for the treaty, published in June 2009, outline what needs to be agreed at the Copenhagen Climate Summit in December, 2009. It lists the key asks and ideas for an adequate and fair treaty that will help keep global warming far below the danger threshold of 2°C.

The central tasks for a powerful, transformational Climate Treaty can be summarized as follows (http://us-cdn.creamermedia.co.za/assets/articles/attachments/22098_wwfbinaryitem12911.pdf):

- Global emissions must peak in the next commitment period (before 2020).
- Industrialized countries as a group should commit to binding absolute emission reduction targets at 40% below 1990 levels by 2020, the vast majority of these being achieved domestically (30-35%). They should also commit to put in place Zero Carbon Action Plans (ZCAPs) to achieve zero net emissions (at least –95%) by 2050.
- Developing countries as a group should pledge to reduce their actual emissions substantially by deviating by at least 30 % below a business as usual pathway by 2020.

- Adaptation Action Framework: Massive flows of secure and predictable funding delivered through well governed and effective funding mechanisms, plus regional climate risk insurance schemes. Together, these should be in the order of US$ 63 billion per year.
- A new institution, the *Copenhagen Finance Facility* should be set up to steer the transition to a low carbon future with binding financial support of at least US$160 billion per year by 2017 from industrialized countries.

The following diagram (Fig. 4.4) illustrates a summing-up of the outcome of various International Climate Change Negotiations held in past 17 years, starting off from Rio Earth Summit-1992.

Kyoto Greenhouse Gas Emissions Reduction Targets: Economic Issues and Prospects

The first report of scientists through the Intergovernmental Panel on Climate Change (IPCC) in 1990 indicated a likely rise in global temperatures and its potentially serious consequences. This led most participants at the 1992 Earth Summit in Rio de Janeiro to sign the United Nations Framework Convention on Climate Change (UNFCCC), by which they undertook to reduce their emissions of the main greenhouse gas, carbon dioxide, to1990 levels by 2000. However, by 1995 it became evident that few countries had managed to reduce their emissions. This failure, together with a second IPCC report warning that global warming had already begun and was on the rise prompted the need for legally binding emissions targets for greenhouse gases (GHG) and timetables for achieving the targets, leading to the Kyoto Protocol signed in December, 1997.

Under the Kyoto Protocol, 39 industrial countries (Annex I countries) committed themselves to reduce their total emissions of six greenhouse gases by 5.2 percent below 1990 levels between 2008 and 2012. However, to allow for the differences among the committing parties, the protocol sets different emissions reduction targets for different countries to achieve this overall target. Accordingly, the reduction targets for the European Union (EU), the United States, and Japan are 8 percent, 7 percent, and 6 percent, respectively. The protocol permits a number of other countries to increase their emissions over the period. Iceland's emissions will be permitted to rise by 10 percent from 1990 levels, Australia and Norway will also be allowed to increase their emissions by 8 percent and 1 percent, respectively. Importantly, Russia and Ukraine are required only to stabilize emissions at 1990 levels. In 1997, Russia's emissions were 30 percent *below* its 1990 levels, due to decline in economic activity and electricity usage since 1989. The protocol does not commit developing countries to any specific reduction targets (Hossein Farzin, 2008).

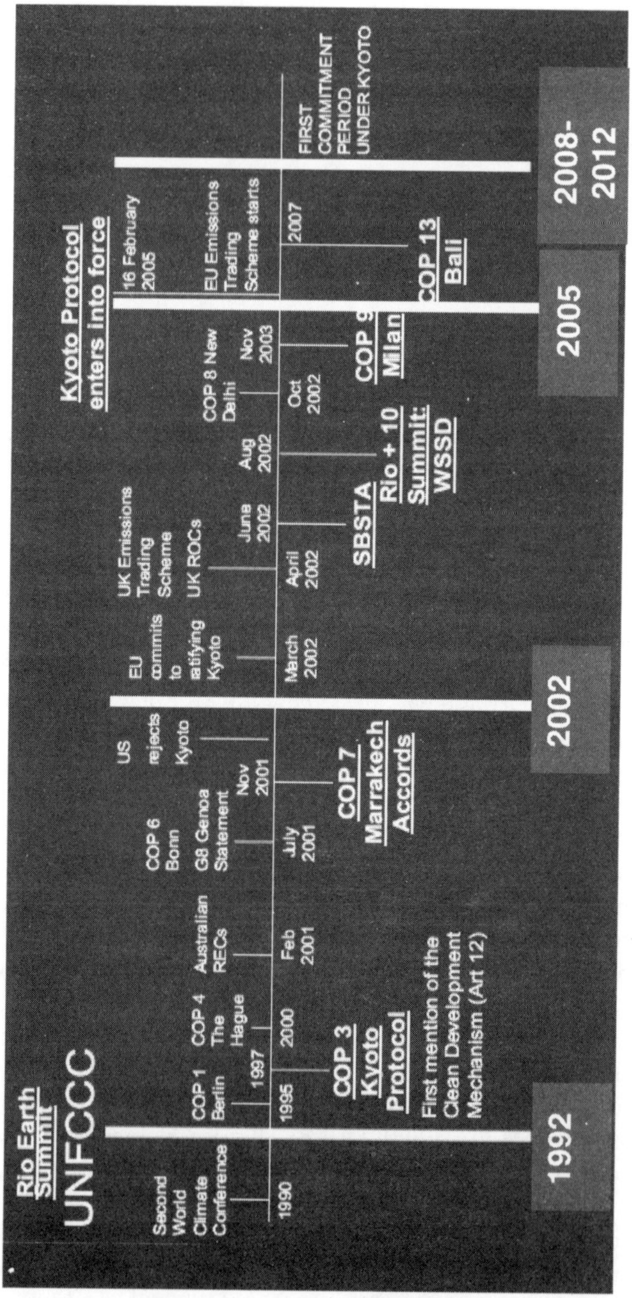

Fig. 4.4: Outcome of various International Climate Change Negotiations held in Past 17 years Starting off from Rio Earth Summit-1992

U.S. Participation Issue

The Kyoto protocol will become a legally binding agreement 90 days after it is ratified by 55 countries whose CO_2 emissions are at least 55 percent of total emissions by the Annex I parties in 1990. However, a major unresolved issue impeding the progress of the protocol toward becoming a legal agreement is U.S. participation. Since the United States is the largest producer of greenhouse gases (more than 1/3 of the emissions by Annex I countries in 1990), it is highly unlikely that without U.S. participation the protocol will come into force, or even if does, that it will succeed in stabilizing and reducing GHG. On the other hand, U.S. compliance with its Kyoto target implies a noticeably large reduction from its business-as-usual baseline projected emissions (for example, about 28 percent below its baseline forecast for 2010). A study in November 1998 by the U.S. Business Roundtable (whose members include Exxon, General Motors and Chase Manhattan Bank) estimated that compliance with the Kyoto target by reducing domestic emissions would lower U.S. GDP by 1.2 percent (or 60 billion dollars a year) from the baseline forecast for 2020. Accordingly, without further incentive to reduce compliance cost, the likelihood of the U.S. ratifying the protocol is very little, noting particularly that the compliance by European Union (EU) will require only 2.6 percent reduction from its baseline projected emissions for 2010 compared with 28 percent reduction by U.S. In turn, the asymmetry in compliance costs between U.S. on the one hand and EU and developing countries on the other, and hence the U.S. participation issue, has raised a number of other difficult issues that render the progress of the Kyoto protocol uncertain.

The following section highlights some of these issues:

Flexibility Mechanisms

To reduce compliance costs for all parties, and in particular to provide incentives for U.S. participation, the protocol allows the use of several flexible mechanisms. These mechanisms are:

- Trading CO_2 emission permits across countries;
- Joint Implementation (JI) projects among Annex I parties: through this mechanism countries or companies can invest in joint projects to reduce emissions where it is least costly to do so;
- Clean Development Mechanism (CDM), whereby Annex I countries can invest in emissions-saving projects in the non-Annex I (i.e., developing) countries and obtain certified emissions reduction to offset against their own reduction obligations. The uncertainty about the rules and procedures that should govern these mechanisms will continue to be a major obstacle in the way of bringing the Kyoto protocol into force. In particular, an important unresolved issue is the extent to which the three flexibility mechanisms should be allowed to be used to meet countries' abatement obligations as against using strictly domestic

measures. The specification of the ceiling on the use of flexibility mechanisms has profound economic implications for all parties, and hence significantly affects the likelihood of an agreement to bring the protocol into effect. Specifically, the lower the ceiling on the use of the flexibility mechanisms (i.e., the larger the percentage of abatement obligations that must be met through domestic measures), (i) the larger the overall cost of meeting the Kyoto emissions targets, (ii) the larger the differences in compliance costs across the regions/countries (specially between the high-cost countries like the U.S. and Japan on the one hand, and the EU on the other), and (iii) the larger the divergence between the domestic marginal abatement cost and the international price of permits for the countries with high abatement obligations, such as U.S. and Japan, implying smaller gains for the purchasing countries from trade in emissions permits. Thus, for example, it is estimated that, with no limits on use of the flexibility mechanisms, U.S. total cost of meeting its obligations will be reduced by as much as 85 percent from that if no emissions trading is allowed. The corresponding estimated figure for Japan is even higher, about 93 percent, while for EU as a whole it will be less than 1 percent, reflecting EU's negligible projected abatement obligation (about 3 percent of its projected 2010 emissions), because of the burden-sharing among the member states and various emission saving measures implemented since 1990.

A question of particular interest may be the impact of the Kyoto emissions targets and flexibility mechanisms on the economy of California. Clearly, an informed answer to this question entails a comprehensive benefit-cost study that, among other things, takes into account the likely share of California in the U.S. total emissions reductions burden under the Kyoto Protocol, the scope for California to use the various flexibility mechanisms to meet its share of the emissions target, and other state or federal policies (specially those pertaining to the energy sector and environmental standards) that may affect California's future GHG emissions. Also, a key consideration in such a study will be the expected long-run economic losses that California's economy could incur in a business-as-usual scenario, or, equivalently, the benefits it could reap from reducing GHG emissions. Given California's dominant share in U.S. energy consumption, its expected high growth rate of energy consumption in the next decade, and its heavy dependence on fossil fuel for transportation and electricity generation, it seems reasonable to speculate that (a) California's share of the total U.S. emissions reductions targets will be large, so that (b) the use of flexible mechanisms could considerably lower the costs of meeting its emissions reductions burden. Further, given the leading role of California in mandating and practicing high environmental standards, together with its advanced agricultural and industrial technological base, it is quite possible that compliance with Kyoto abatement targets will in the long-run encourage faster innovation and diffusion of clean technologies, thereby giving California a comparative advantage in supply of such technologies and thus indirectly boosting its economy.

In any event, the vastly unequal national gains from the flexibility mechanisms place the U.S. and Japan diametrically against the EU in defining the ceiling on the use of the mechanisms. In particular, under the pressure from industry groups and with the backing of Australia, which advocates carbon sequestration through land use practices and forestry activities (the so called "land-use change") as an additional flexibility measure, the U.S. is anxious to have maximum use of the flexibility mechanisms to allow for 85 percent of commitments to be met through them. The EU, under pressure from environmentalist groups, is worried that other industrialized countries, especially the U.S., will evade their abatement obligations by resorting to these mechanisms. It has therefore proposed that the use of flexibility mechanisms be limited by requiring that at least 50 percent of the obligations be met through individual country domestic action. However, it is uncertain that this proposal will be accepted by U.S., as it most likely raises her compliance costs by a significant factor. The EU proposal may also encounter opposition from developing countries (especially China and India), since they will be the other major beneficiaries (aside from the U.S. and Japan) from the full use of the flexibility mechanisms.

Issues Facing the Implementation of Flexibility Mechanisms

As the foregoing discussion makes clear, the prospects for the Kyoto Protocol to enter into force depend crucially on successful implementation of the flexibility mechanisms, particularly CDM. There are, however, a number of important implementation issues that ought to be resolved if the perceived gains from flexibility mechanisms are to materialize. One of the notorious issues facing CDM is that of the "additionality" test. According to the Article 12.c of the Kyoto Protocol, only projects that cause "reductions in emissions that are *additional* to any that would occur in the absence of the certified project activity" can generate emissions reduction credits under CDM. However, there is so far no agreement on the definition of standards, and hence of the baselines, for determining "additionality". Such baselines are likely to vary both across developing countries with different technological conditions and as technologies change over time. A project-by-project verification of claimed emissions reduction credits and establishment and execution of punitive rules for false claims can formidably increase the CDM transaction costs.

The distribution of benefits from CDM poses another difficulty; which opens up the following fundamental questions that need to be addressed and resolved n future:

- How the benefits of CDM projects should be equitably distributed between host developing countries and investing countries?

- Does the sovereignty of host countries entitle them to larger shares?
- Will the opportunities for CDM investments be equitably distributed across developing countries, or would the investors prefer the more developed countries over the least developed ones?
- What will be the respective roles of private sector and government in initiating and implementing CDM projects?
- Who could initiate and/or finance CDM projects?
- How should one distinguish between the projects that, as a by-product, would reduce emissions but the government would have undertaken them anyway, and the projects that generate additional reductions?
- Can a developing country's own government and/or private sector initiate, finance and execute emissions saving projects?
- Given the scarcity of public sector resources, and hence the important role for the private sector, what are the incentives for the latter to invest in such projects?
- What role should international financial institutions such as the World Bank play?
- Would Annex-I countries' assistance for CDM replace their foreign aid?

Clearly, some of the issues noted above also apply to international trading of emissions permits. For example, establishment of rules for compliance, monitoring and enforcement (to prevent underreporting of emissions, for instance) seems essential for the success of emissions trading. In turn, this necessitates the establishment of an independent coordinating authority to facilitate international trades in permits by reducing barriers to trade, such as imperfect information about costs and benefits of trading and transaction costs involved in buying and selling permits.

This section has emphasized the important role of flexibility mechanisms in bringing together various parties of differing interests and approaches to reach an international agreement to stabilize global emissions of greenhouse gases over the next decade. It has also highlighted some of the many major issues that render the progress of the Kyoto protocol uncertain. Nevertheless, at this stage of negotiations, the key to the success of the protocol seems to lie in an agreement among the negotiators to start a period of experimentation with the CDM and trading permits in order to demonstrate the benefits of these flexibility mechanisms to the participating parties and also to allow the mechanisms and markets to become established and countries to become more familiar with their operations.

Bilateral and International Agreements

The following paragraphs provide information on several important Bilateral and International Agreements in the sphere of Global climate change and Emission reduction/trading related issues across countries/ Territories/ Regions for pollution control and sustainable development:

US/Mexico

The United States and Mexico have a long history of cooperation on environmental issues. Most recently, the two countries jointly proposed the Border 2012 program, which will focus on decreasing pollution and lowering the risks of exposure to air pollution, pesticides and other chemicals.

- US-Mexico Border 2012

US/Canada

The United States and Canada made an historic commitment to address transboundary air pollution issues in the 1991 U.S.-Canada Air Quality Agreement. To date, acid rain and ozone have been the primary focus of cooperation under the Agreement. In 2003, the two countries renewed their commitment to build on these trans-border air quality improvements and develop new cooperative projects for the years ahead.

- Agreement Between the Government of the United States of America and the Government of Canada on Air Quality;
- Canada-United States Ozone Amendment to the 1991 Air Quality Agreement.

US/China

- Memorandum of Understanding between the EPA and the State Environmental Protection Administration of the People's Republic of China on Scientific and Technical Cooperation in the Field of Environment. Strategy for Clean Air and Energy Cooperation between The Environmental Projection Agency of the United States of America and The State Environmental Protection Administration of the People's Republic of China

US/India

Memorandum of Understanding between U.S. EPA and the Ministry of Environment and Forests of the Government of India, (2002)

US/Japan

U.S.-Japan Agreement on Cooperation in Environmental Protection, (1975)

US/Taiwan

U.S.-Taiwan Bilateral Agreement on Environmental Protection

US/Thailand

U.S.-Thailand Agreement in the Area of Environmental Protection, (1999)

US/Western Europe

Spain-U.S. Joint Statement on Environmental Cooperation, (2003)
Italy-U.S. Joint Statement on Environmental Cooperation, (2002)

International Agreements and Partnership Initiatives

United Nations Economic Commission for Europe—Long Range Trans-boundary Air Pollution (LRTAP)

The Convention on Long-range Transboundary Air Pollution takes cooperative action on air pollution in the European and North American region covered by the United Nations Economic Commission for Europe (UNECE). Established in 1979, the Convention is a landmark international agreement that coordinates efforts on research, monitoring and the development of emission reduction strategies on regional air pollution and its effects. It was the first international agreement to recognize both the environmental and human health problems caused by the flow of air pollution across political borders and the need for regional solutions (http://www.unece.org/env/lrtap/).

Since 1979 the Convention on Long-range Trans-boundary Air Pollution has addressed some of the major environmental problems of the UNECE region through scientific collaboration and policy negotiation. The Convention has been extended by eight protocols that identify specific measures to be taken by Parties to cut their emissions of air pollutants. The Convention, which now has 51 Parties, identifies the Executive Secretary of UNECE as its secretariat. The aim of the Convention is that Parties shall endeavour to limit and, as far as possible, gradually reduce and prevent air pollution including long-range trans-boundary air pollution. Parties develop policies and strategies to combat the discharge of air pollutants through exchanges of information, consultation, research and monitoring. Currently, the Convention's priority activities include review and possible revision of its most recent protocols, implementation of the Convention and its protocols across the entire UNECE region (with special focus on Eastern Europe, the Caucasus and Central Asia and South-East Europe) and sharing its knowledge and information with other regions of the world.

United Nations Environment Programme–Partnership for Clean Fuels and Vehicles

The Partnership for Clean Fuels and Vehicles was established at the World Summit on Sustainable Development in September 2002 to reduce vehicular air pollution in developing countries through the promotion of clean fuels and vehicles.

Partnership for Clean Indoor Air

The Partnership for Clean Indoor Air was launched at the World Summit on Sustainable Development in Johannesburg in September 2002. The Partnership addresses the increased environmental health risk faced by more than two billion people who burn traditional biomass fuels indoors for cooking and heating. This voluntary Partnership brings together governments, public and private organizations, multilateral institutions, industry, and others to increase the use of affordable, reliable, clean, efficient, and safe home cooking and heating practices.

Sustainable Development Partnerships

This site provides information on U.S. efforts to work with other governments, the private sector, civil society and other organizations to plan and implement voluntary partnerships that promote economic growth, social development and environmental stewardship.

United Nations Framework Conventions on Climate Change (UNFCCC)

Regional Initiatives

Clean Air Initiative in Latin America, Sub-Saharan Africa, and Asia (CAI)

The Clean Air Initiative (CAI) advances innovative ways to improve air quality in cities by sharing knowledge and experiences through partnerships in selected regions of the world. CAI partners and participants foster actions to improve air quality in cities. We bring together a range of cross-cutting expertise in urban development, transport, energy reform, environmental management and environmental health.

Global Climate Change: South Asian Regional Initiatives (SARI)

The USAID South Asia Regional Initiative for Energy (SARI/Energy) is an eight country program that promotes regional energy security through energy market development, cross-border energy trade, and increased access to clean energy. SARI/Energy countries include: Afghanistan, Pakistan, India, Nepal, Bhutan, Bangladesh, Sri Lanka and the Maldives. SARI/Energy has been in operation since 2000; Afghanistan and Pakistan joined the program in 2004 (http://www.usaid.gov/our_work/environment/climate/docs/cp/sari.pdf). Rapid economic growth is driving energy demand in South Asia at approximately 6-8% a year. South Asian national electricity grids are powered by a mix of domestic coal, natural gas and hydropower, with some wind power. But inefficiencies and the slow pace of reform result in persistent and costly electrical power outages throughout the region. Petroleum consumption in the industrial and transportation sectors is increasing South Asia's dependence on imported oil and natural gas.

At present, there is little exchange of energy across borders in South Asia, or with Central Asia. However, within South Asia there are significant but disparately located energy reserves such as low quality coal, hydropower, and a significant un-harnessed wind and solar resources. The neighboring Central Asian region offers potential low cost sources of hydro and thermal electricity. These South Asian energy resource and demand trends are coupled with a growing Indian capacity for world class engineering and construction among both the parastatals and the private sector. However, there remains a broad lack of strong institutional, legal and commercial structures to facilitate sustainable energy sector investment and growth. Encouragingly, there is a nascent trend toward harnessing the strengths of the marketplace, primarily in India, to capture energy sector cost, finance and investment opportunities. Coupled with the opening of regional and trans-regional relations and trade, and integrated with global markets, this trend could be the most significant step toward energy security that South Asia has seen for decades.

Sector-Specific Climate Change Activities (SARI/ Energy Program)

The South Asia Regional Initiative for Energy Cooperation and Development (SARI/Energy) encourages regional cooperation in energy development and trade in clean energy resources among South Asian countries, including Bangladesh, Bhutan, India, the Maldives, Nepal, and Sri Lanka. The program was developed jointly by USAID and the U.S. Department of State to improve regional stability. The lack of market-driven commercial energy supplies in the region was identified as a fundamental constraint to increased energy security; economic growth and associated improvements in environmental quality; human health; and reduced GHG emissions in the energy sector. SARI/Energy forms the centerpiece effort to strengthen public-private sector partnerships and economic cooperation between countries in the region. SARI/Energy enables energy professionals from the region to share experiences and lessons learned and to catalyze the long-term process of rationalizing commercial energy supply and its regional distribution. USAID's support helps establish the groundwork and sets the stage for policies and agreements for greater cross-border energy cooperation and trade.

SARI/Energy Programs Overview

SARI/Energy targets improvement of policies and agreements for cross-border cooperation and trade in sustainable energy. The program does this through the implementation of programs designed to achieve the following results:

- Development of private sector energy plans and investments in infrastructure that facilitates regional energy exchange
- Development or modification of policies, laws, and regulations, with private sector and civil society concurrence, to facilitate regional energy trade and cooperation.
- Increased capacity of regulatory bodies to support energy exchange
- Regional best practices and standards that lead to energy efficiency and improved commercial practices
- Regional adaptation of best practices that lead to improved rural energy services

- Increased private sector participation in and civil society support for sustainable energy development

As an example of increased private sector participation, SARI/Energy has implemented nine technical assistance/training activities that focus on the development of competitive energy markets that promote market-based energy prices, decrease fossil fuel subsidies, allow open access to independent providers, establish open and transparent bidding practices, and use independent regulatory bodies. Experience in implementing sector reform (including policies, legislation, regulations, and institutional reform), both within the region and with other regions of the world, is actively and productively shared by SARI/Energy program participants through many partner activities.

For example, Bangladesh, Nepal, and Sri Lanka are following India's lead in establishing regulatory bodies; India and Nepal are learning from Bangladesh's highly successful rural electrification cooperative model; and Sri Lanka is sharing its model power purchase agreement with Nepal. SARI/Energy activities have catalyzed the formation of Energy Service Companies (ESCOs) in Nepal and Sri Lanka, improved demand side management practices in India and the Maldives, and initiated efforts to establish a regional set of harmonized energy efficiency standards and labels for appliances in South Asia. Outreach efforts include the launch of the Energy South Asia magazine, the SARI/Energy Web site (www.sari-energy.org), a media partnership program, and support for the South Asia Energy Journalists Forum.

The Future: The next phase of the program proposes to expand the target audience beyond a primarily technical audience in order to identify and strengthen internal change agents for reform, including the media, eminent academicians, other influential individuals, industry leaders, nongovernmental organizations, parliamentarians, and politicians. With South Asia still lagging behind the rest of the world in terms of intraregional energy trade, the untapped potential to develop and trade clean energy resources in an efficient manner is enormous.

Role of the World Bank in Scaling up: GEF and Carbon Finance

The World Bank Group has been active in scaling up the activities towards a low-carbon economy by developing the carbon market (the Bank currently manages 9 carbon funds with a total investment of $2 billion), utilizing GEF resources, and mobilizing private sector capital. Further improvements in World Bank and other International Financial Institutions' financial instruments include scale-up of umbrella carbon facilities, ESMAP window for clean energy project development, MIGA insurance for carbon credits, expanded use of contingent grants and guarantees, bundling programmatic Clean Development Mechanism, the European Bank for Reconstruction and Development's energy efficiency facility, the Inter-American Development Bank's sustainable energy initiative and the Asian Development Bank's Asia Pacific Energy Efficiency Fund.

The World Bank Investment Funds

By exercising access and strategic control over the Climate Investment Funds (CIFs) through 2012 the World Bank taken on a tangible and significant role in devising more climate-conscious development projects. The CIFs, which fall under the Strategic Framework on Development and Climate Change (SFDCC), are the most recent financing mechanisms approved by the Bank's Board to support its increased engagement in climate change. In September 2008, donors from ten countries pledged $6.1 billion for the CIFs, with the majority coming from the US ($2 billion), the UK ($1.5 billion) and Japan (up to $1.2 billion). The CIFs include two funds, the Clean Technology Fund and the Strategic Climate Fund (http://www.bicusa.org/en/Issue.Background.48.aspx).

The Clean Technology Fund (CTF) provides finance for low carbon energy projects or energy technologies that reduce emissions. It will not limit the types of technologies eligible for financing to new renewables (like solar, wind, small hydro power), but instead keeps the door open to support for "clean coal" and large hydroelectric dams. According to the Bank, "clean coal" represents "highly cost effective opportunities for significant GHG emissions reductions and/or there is potential for developing readiness for carbon capture and storage." Thus, the CTF supports technologies that reduce the carbon intensity of development, but not necessarily overall GHG emissions. Critics argue that the CTF supports a "business-as-usual" approach, rather than a real transition, to clean energy development.

The Strategic Climate Fund (SCF) will be more broad and flexible and will support a variety of programs that tackle climate change. The primary program of the SCF is the Pilot Program for Climate Resilience (PPCR—which replaces the previously proposed Adaptation Pilot Fund). While the CIFs will provide a substantial sum of money to address climate change, they have not been accepted without controversy. Some question whether or not CIFs will draw money away from the parallel UNFCCC structures that are already in place and whether their creation was necessary. Furthermore, many developing country governments believe that the CIF framework lacks the transparency and equitable governance structure of the UNFCCC.

Another controversial World Bank fund is the Forest Carbon Partnership Facility (FCPF), which was launched in December 2007 in order to assist countries in their "REDD" efforts (Reducing Emissions from Deforestation and Degradation). The FCPF, part of the Bank's Carbon Finance Unit (CFU), is comprised of two structures: a readiness mechanism and a carbon finance mechanism. Its goal is to provide an incentive structure for the maintenance of forests and to promote their sustainable use as a means of combating deforestation and the emissions that accompany the process. As part of the readiness mechanism, individual countries develop R-PINS, Readiness Plan Idea Notes to address how to best develop strategies to meet REDD benchmarks. In their November 2008 report entitled "Cutting Corners: World Bank's forest and carbon fund fails forests and peoples," FERN and the Forest People's Programme highlighted several areas of concern with regard to the FCPF mechanism, with lack of consultation outside of the government and international NGO realm at the top of the list. The report also focused attention on issues of human rights, land tenure rights, an inconsistent view of the various drivers of deforestation, FCPF consistency with WB safeguard policies and the lack of a monitoring mechanism for the social impacts of the REDD plans.

Climate Change and the World Bank's Track Record

The following is a brief timeline of World Bank commitments with regard to climate change:

- *The Global Environment Facility, 1991*: A resolution by the World Bank's Board of Executive Directors in 1991 led to the establishment of the Global Environment Facility (GEF), which was designated as the financial mechanism for the U.N. Framework Convention on Climate Change in 1992. Since then, the WBG has administered the GEF trust fund and has been the GEF's primary implementing agency for investment projects meant to address climate change (note the GEF was set up to also specifically address biodiversity and desertification).
- *World Bank Group Energy Sector Strategy, 2001*: The World Bank's 2001 Energy Sector Strategy specified the four main goals of energy supply to be helping the poor directly, improving macroeconomic and fiscal balances, promoting good governance and private sector development and protecting the environment. It further delineated 2010 targets for greenhouse gas emissions and energy efficiency in developing and transition countries. The Energy Sector Strategy also supported various policy measures with regards to fossil fuels.
- *Extractive Industries Review/Bonn Commitment, 2004*: In the Management's Response to the Extractive Industries Review (2004) and at the International Renewable Energies Conference in Bonn, June 2004, the World Bank Group announced a commitment to scaling up lending for new renewable energy and energy efficiency by at least 20% annually over five years (FY05-FY09), and leading a Renewable Energy and Energy Efficiency Financing and Policy Network for developing countries.
- *Clean Energy Investment Framework, 2006*: In 2006, responding to a request from the G8, the Bank developed the Clean Energy Investment

Framework (CEIF) intended to help scale up investments in clean energy and integrate climate change into development assistance. The CEIF set out four primary World Bank strategic activities: Promoting transition to a low-carbon economy—especially in Brazil, China, India, Mexico, and South Africa—by increasing analytical, knowledge, and investment support; accelerating investments that help increase supplies of clean energy; improving access to affordable energy for the poor, particularly in Africa; and assisting developing countries with adaptation to the impacts of climate change through analytical work and development of risk-management tools.

- *Strategic Framework on Development and Climate Change, October 2008*: At the October 2008 annual meetings, the Bank's Development Committee approved the successor to the CEIF, the *Strategic Framework on Development and Climate Change* (SFDCC), which spells out a much broader role for the Bank in climate change issues. The SFDCC provides the IFC, MIGA, IDA, IBRD, and other entities of the Bank Group objectives, guiding principles, areas of focus, and major initiatives to guide operational response for the next three years.

The Framework is based on six action areas, each addressing both adaptation and mitigation:

1. Support climate actions in country-led development processes;
2. Mobilize additional concessional and innovative finance;
3. Facilitate the development of market-based financing mechanisms;
4. Leverage private sector resources;
5. Support accelerated development and deployment of new technologies; and
6. Step-up policy research, knowledge and capacity building.

An extensive multi-stakeholder global consultation was undertaken between April and September of 2008 to draft the Framework in an inclusive and transparent way. The following are summaries for Phase I and Phase II of the consultations:

Phase I—Consultations on Concept and Issues Paper (WB website)
Phase II—Development and Climate Change (WB website)

Some Additional Examples

At the 2004 International Conference on Renewable Energies in Bonn, Germany, the World Bank Group committed to a target of a 20 percent average annual growth in new-renewable energy (i.e. solar, wind, biomass, geothermal, as well as hydro up to 10 MW in capacity) and energy efficiency commitments between fiscal years 2005 to 2009. The WBG has succeeded in meeting its Bonn Commitment during both FY05 and FY06, committing over $1.14 billion dollars for new-renewable energy and energy efficiency. This represents more than double the Bonn commitment for this period. Overall, World Bank Group support for RE and EE has exceeded US$10 billion (for over 100 projects in more than 28 countries).

Some Highlights

The World Bank Group's work on renewable energy and energy efficiency includes the China Renewable Energy Scale-up Program, which helps to finance RE investments, build capacity, and implement China's Renewable Energy Law and the Serbia Energy Efficiency Project, which improves heating efficiency in public buildings such as schools and hospitals, and makes heating more affordable and clean. Additional project examples include the Nepal Biogas Program, which supports the provision of clean cooking fuels and the Power Sector Priority Investments Project (PSPIP) in Timor-Leste which has distributed more than 50,000 compact fluorescent light bulbs (CFLs) free of charge to electricity consumers who have acquired pre-payment meters as a means to improve energy efficiency and reduce costs. On Monday we launched together with the Mexican government the Mexican carbon fund (FOMECAR) with the objective expanding the country's opportunities in this growing market. Recently as an example of large scale operation it was concluded an agreement for US$1 billion in China to buy HFC.

However, a review of currently available resources and instruments from the International Financial Institutions and the public and private sectors concluded that on their own they cannot lead to a meaningful transition to a low carbon economy. The issue is one of scale.

- The Global Environment Facility—the GEF—is the largest source of multilateral grant financing for low carbon technologies, and its work is vital. However, it would need increased funding to achieve significant and sustained market penetration of near commercial energy efficiency and renewable energy technologies. Scaling up the GEF's current focus on removing barriers for such technologies would need increased funding by a factor of 3. In addition, if the GEF were to expand its support to the capital investment needs of new, low greenhouse-gas emitting technologies, its resources would have to be scaled up considerably more—by at least a factor of 10.
- Existing carbon funds can be scaled up for targeted investments and their applications improved; the establishment of the Umbrella Carbon Facility has demonstrated the potential of such instruments for scaling up and increasing the efficiency of delivery of emission reductions from large projects. However, the lack of a global regulatory framework for emissions reductions after 2012 prevents investments with long lead times and a long term positive impact on a transition to a low-carbon economy.

With the mechanism of the Global Environmental Facility (GEF) created to integrate Global climate change with the viable energy production and business development a series of aims are being pursued:

- Policy barriers in the host countries of projects and internationally have to be removed.
- The attractiveness of foreign markets for investors increases significantly if sound energy prices are paid for the production of energy. Sound in this context addressing to features firstly the environmental soundness of energy production and secondly the soundness of prices which are charged from the customer. Removing inappropriate subsidies on the way.
- By these measures a first step is taken to create a business environment. In a second step the capacity of the national authorities are a crucial part of providing a good and successful business environment for both the national user of energy and technologies and the international provider of the two.
- The financial and technical aspects are checked by project feasibility studies for potential project sites.
- In general technology transfer between the more and the less developed countries has to be increased, leading to more know-how in newly industrialized and developing countries.
- There has to be a certain amount of know-how to be present to enable the staff in the developing countries to run the equipment. In a second step the foreign deliverer of equipment will use the advantages of producing directly in the market.
- This increases the need for skilled and trained workers. Thereby enabling these countries to increase their domestic production capability.
- In regard of Climate Protection Projects the target is set to reduce cost for clean energy, which includes clean coal technologies as well as renewable energy sources.

The Issue of Scale

Both the GEF and the carbon market would need to be expanded for greater impact in the transition to a low-carbon economy through the development of markets for energy efficiency and renewable energy technologies. To do so, a much larger-scale "buy down" of incremental costs will help to move to cleaner technologies. A well-functioning carbon market would allow the "buy-down" of such costs via its current financial mechanisms. To be able to function and perform well, these markets need a long-term, stable framework. This could stimulate the carbon market with a flow of funds to developing countries of tens of billions of dollars per year, but the right incentives need to be created.

There is a window of opportunity to link the expansion of energy investments in countries such as China and India with strategies to reduce local and regional pollution, and simultaneously lower carbon emissions. The World

Bank Group is committed to pursuing efforts to better respond to the needs of borrowing countries—in particular, within the scope of this Investment Framework—by reducing non-financial transaction costs. This could be done by making the pricing of existing instruments more transparent and competitive to provide incentives and resources to countries to pursue clean energy alternatives.

UNEP Guidelines for Funding CDM Projects

One of the challenges facing Clean Development Mechanism (CDM) projects today is their limited ability to secure financing for the underlying greenhouse gas emission reduction activities, particularly in the least developed countries. Among the key reasons for this is the fact that most financial intermediaries in the CDM host countries have limited or no knowledge of the CDM Modalities and Procedures. Moreover, approaches, tools and skills for CDM project appraisal are lacking or are asymmetrical to the skills in comparable institutions in developed countries. Consequently, developing country financial institutions are unable to properly evaluate the risks and rewards associated with investing or lending to developers undertaking CDM projects, and therefore have, by-and-large, refrained from financing these projects. In addition, some potential project proponents lack experience in structuring arrangements for financing a project (http://www.cd4cdm.org/Publications/FinanceCDMprojectsGuidebook.pdf).

Generation of Revenues for Developing Countries by GHG Reductions

In the Kyoto Protocol, 38 industrialized countries and countries in transition were allocated an emissions target for the period 2008-12. The targets were differentiated between countries and range from –28 percent (Luxembourg) to +10 percent (Iceland) compared to 1990 levels. They refer to a basket of six GHGs and are denominated in CO_2 equivalent. A tonne of methane has an 'exchange rate' of 21, a tonne of the industrial gas HFC-23 as high as 11,700. (CO_2 primarily originates from burning of fossil fuels, and methane from rice cultivation and enteric fermentation in livestock among others. HFC-23 is used in the refrigeration sector).

An unprecedented feature of the Protocol is the possibility of using international market mechanisms to achieve the emissions targets. There are three "Kyoto Mechanisms" that allow buying and selling of emission permits: international emissions trading, Joint Implementation and the Clean Development Mechanism (CDM). CDM allows developing countries to implement projects that reduce GHGs and thereby generate emission permits (Certified Emission Reductions or CERs). CERs can be credited by industrialized countries toward their emission targets.

Like Canada, the EU and Japan will not be able to achieve their Kyoto targets by domestic action alone—there will be a considerable demand for CERs. By end 2004, more than $1,000 million have been committed by governments and private companies to acquire CERs. New tenders are announced almost every month. A tonne of CO_2 equivalent currently trades at $ 4-5 and the price trend is rising. So the global CER demand stands at 250 million tonnes and is likely to double within the next year as there remain large emission gaps in important countries such as Italy and Spain. The overall emissions gap for the period 2008-12 will surpass 3 billion tonnes of CO_2 equivalent. However, Russia, Ukraine and the countries of Eastern Europe have a huge surplus of emissions permits, colloquially called "hot air," due to their economic breakdown in the 1990s and could supply enough to cover the whole shortfall. Therefore, the CDM market heavily depends on the behavior of the hot air countries.

To ensure that CERs are not just "paper tonnes," CDM projects have to undergo independent scrutiny by officially accredited certifiers, the so-called "Operational Entities." An international body, the CDM Executive Board, has to register each project. There is a prescribed format to be filled by each CDM project developer; the promoter has to calculate emissions reductions according to complex rules and make sure that key parameters are monitored throughout the lifetime of the project.

International Adaptation Finance: The Need for an Innovative and Strategic Approach

At present, all international adaptation funding instruments—except the recently operationalised Kyoto Protocol Adaptation Fund - are replenished through ODA-type bilateral donations. The level of international funding for adaptation in developing countries is woefully inadequate to meet projected needs. The current bilateral donation instruments are unlikely to ever be able to generate the required levels of funding, especially as it is meant to be additional to ODA (viz. experience with the Monterrey 0.7% GNI commitment for ODA). Moreover, adaptation funding is seen by most developing countries not as a matter of 'donations' but as one of costs imposed by developed countries, and as such as debt incurred by them. Accordingly, neither of the traditional ODA funding modes (grants or concessionary loans), are seen to be appropriate payment modes. Funding is expected, and must be 'acceptable', in the sense of being not only appropriate, but new and additional, predictable, equitable, and adequate (http://www.oxfordenergy.org/pdfs/EV42.pdf). The study conducted by Benito Müller (2008) at the Oxford Institute for Energy Studies highlights the following salient findings on the question of using innovative and strategic approach on International Adaptation Finance mechanism:

Innovative Sourcing: Further innovative financing mechanisms apart from the CDM Adaptation Levy are needed to fill the adaptation 'funding chasm.'

The only way to provide funding for developing country adaptation which is acceptable, in the above-mentioned sense, is through international levies on emissions from international maritime transport and aviation/air travel and/or through international auctioning of assigned amount units (i.e. an adaptation levy on the proceeds of international emissions trading).

Strategic Allocation: Internationally, funds for adaptation need to be allocated on a strategic basis and not involve international micro-management at the project level. The strategic allocation of international adaptation funds should not attempt to re-invent the wheel. It should use the existing international bodies and initiatives to allocate funding streams, and not try to duplicate them under a 'climate change banner.' Domestically, as mentioned above, there is a need to enhance 'absorptive capacity' not only at the project level, but more importantly—following the Paris Declaration - at the level of domestic policy ('adaptation mainstreaming').

Good Governance: The governance of the recently operationalised Kyoto Protocol Adaptation Fund represents a milestone in the evolution of international funding mechanisms, since for the first time developing countries have genuine ownership of such an instrument. In the case of adaptation funding, developing country ownership and public transparency of decision making is not only desirable but a prerequisite for success, particularly in the context of mainstreaming activities. Given this, the Adaptation Fund should be the main instrument for the purpose of raising and managing of international adaptation finance for developing countries.

International Finance for Adaptation: An Overview

Any view on international finance for developing country adaptation - strategic or not—must deal with the question of how much adaptation to climate change impacts in developing countries does/will cost, and how much is/will be available under current conditions. While closing the expected funding gap—or, as some have called it 'adaptation funding chasm'– is not the only strategic adaptation finance issue, it is a core issue—if there are no funds, using them appropriately becomes rather like counting angels on pinheads. This does not mean that providing adequate funds alone will be sufficient to achieve adequate adaptation: there are equally difficult problems to be overcome at the potential recipient end, often characterized in terms of '(insufficient) absorptive capacity'.

The following section summarizes the various funding issues linked to International Finance for Adaptation:

1. Current estimates of funding needs for adaptation in developing countries

- Although existing estimates of adaptation funding needs are still very vague, they all indicate a level of funding needed for developing countries to adapt to the impacts of climate change in the tens of billions /$ per annum.

- Many developing countries presently do not have the relevant 'absorptive capacity', i.e. the capacity to carry out the adaptation measures needed, even if the funding were available.
- Most will have to suffer adverse impacts of climate change that could be avoided under an improved adaptation regime.
- The responsibility for these adverse impacts—whether due to a lack of funding or of absorptive capacity - will be squarely put on industrialized countries.

How much does and will climate change adaptation cost in developing countries? While the question seems to be straightforward enough, answers—not surprisingly, given the complexity and variety of adaptation—are very difficult. The World Bank guesstimates current needs of 9-41 billion US dollars ($). Although the base data for these figures have been disputed,

Table 4.3 taken from the original source—is nonetheless interesting because it illustrates just how thin on the ground these estimates had to be because of high uncertainties associated with the methodology (guesstimates of global percentages).

All these figures are very rough estimates, based on certain 'top-down' methodologies such as the World Bank estimates of the cost of 'climate proofing' current investment flows. They are also not generally comparable, since they are about different types of adaptation needs, such as:

- Climate proofing Official Development Assistance (ODA) and other 'business as usual' investment flows;
- Climate proofing of existing infrastructure;
- Additional investments necessary because of climate change (e.g. dams, dykes etc.);
- Costs on community level (community based adaptation, capacity building by NGOs etc.); and
- 'Mainstreaming' adaptation into poverty reduction strategies and other relevant government policies

As mentioned before, adaptation in developing countries is not just an issue of funding, but also of sufficient 'absorptive capacity.' Even if the funding were available, most developing countries would, at present, not have the capacity to spend it, to carry out all the adaptation needed. The result would be an avoidable increase in what has euphemistically become known as 'residual impacts,' that is impacts that that happen despite the existing impact reduction measures (i.e. by mitigation and adaptation combined). As reported in the recent Stern Review on the Economics of Climate Change, the cost of these impacts could in future amount to two-orders of magnitude hundreds of times) more than these adaptation funding needs estimates.

Table 4.3: World Bank Preliminary Estimates of Annual Adaptation Needs Developing Countries

Investment type	Amount	Climate sensitive	Adaptation cost (%)	Costs (2000 $)
ODA and Concessional Finance	$100 bn	40%	10-20%	$4bn-$8bn
Foreign Direct Investment	$160 bn	10%	10-20%	$2bn-$3bn
Total International costs				**$6bn-$11bn**
Cross Domestic Investment	$1500 bn	2-10%	10-20%	$3bn-30bn
Total adaptation costs				**$9bn-$41bn**

Source: Table K.1, World Bank (2006a)

2. *Current International Adaptation Funding*

- At present, all international funding instruments except the recently operationalised Kyoto Protocol Adaptation Fund are replenished through ODA-type bilateral voluntary contributions/donations.
- The current level of international funding (even including the Adaptation Fund) is woefully inadequate to meet projected costs.
- The current bilateral donation instruments are unlikely to ever be able to generate the required levels of funding, especially as it is meant to be additional to ODA (viz. experience with the Monterrey 0.7% GNI commitment for ODA).

As summarized in Table 4.4, the total of multilateral fiscal payments (i.e. non-reimbursable payments from general domestic fiscal revenue) for international adaptation funding—be it through the Special Climate Change Fund (SCCF), the Least Developed Country Fund (LDCF), and the GEF Special Priority to Adaptation—until 2012 is presently estimated at around $200m. Assuming that (most of) the money received has already been spent; this leaves an average of around $13m per annum over the next five years. Clearly these figures are orders of magnitude too small to cover even the most conservative estimates of the adaptation funding needs of developing countries. This—and the fact that the situation is similar with respect to the expected developing country mitigation costs that would have to be met to reduce global greenhouse gas emissions to current levels by 2030—has led to a number of proposals for a commitment by the richer countries of the world to increase their bilateral financial flows accordingly. Thus China has suggested) that developed countries should commit to an additional 0.5% of GDP for climate change payments to developing countries—additional to the 0.7% Monterrey consensus ODA target (i.e. $260 billion 2007)—which currently would amount to an additional annual contribution of around $185 billion, albeit not all for adaptation.

While it would be a giant step forward in North-South relations if any such legally binding bilateral funding targets could be agreed and while it is arguably the only situation in which developing countries might consider taking on legally binding mitigation targets, current bilateral ODA flows show that it would be very difficult to deliver. The OECD Development Assistance Committee (DAC) has been rather upbeat about the performance of its member countries with regard to the 0.7% of GNI Monterrey ODA target. The fact is, however, that only small economies have reached or surpassed the 0.7% target (Fig. 4.5). Any substantive increase in ODA would have to come from the G7 countries; since they amount for the lion's share of all ODA (73% see Fig. 4.5). The problem is that at 0.23%, their performance was actually below the overall DAC average of 0.28%, which itself is not very encouraging, particularly in the context of falling trend over last couple of years.

Table 4.4: Multilateral Adaptation Funding

Investment	Pledged	Received
SCCF(1)	$7.8m	$49.3m
LDCF(2)	$57.1m	$49.3m
SPA(3)		$50.0m
Total Grants	**$64.9m**	**$148.6m**

Notes: (1) World Bank (2007) adaptation only; (2) World Bank (2007) total (incl. mitigation); (3) UNDP (2007).

The 'Domestic Revenue Problem': However it stands to reason, in light of the historic evidence, and on more general grounds that this or any other sudden significant increase in fiscal (bilateral/multilateral) funding is unlikely to materialize in any of the large DAC members due to a general psychological and political 'domestic revenue problem': money that is raised domestically, particularly through domestic taxation, is regarded to be nationally owned. Indeed, in the case of taxation, individual tax payers often see themselves as direct owners of the revenue raised (Fig. 4.5). And the sums involved in ODA or any other tax expenditure are generally perceived in absolute terms, which can create problems, particularly if they are in competition with other (domestic) expenditures. While it may be acceptable to send one or two hospitals' worth of tax-payers money abroad, this may no longer be the case for the equivalent of 100 hospitals. This problem is psychological because it is very much a matter of how the sums in question are perceived. It is political because the source of the revenue - particularly in the case of taxation—is also the source of political power (voters, businesses), which means that politicians may be at times less than forthcoming in showing the leadership required to overcome the problem.

Given then the less than satisfactory, not to say disappointing performance of the larger world economies (G7) with regard to achieving the Monterrey target, the question has to be: how could one raise the additional funding required to meet the developing world's adaptation needs?

Fig. 4.5: 2007 ODA as share of Gross National Income (GNI)

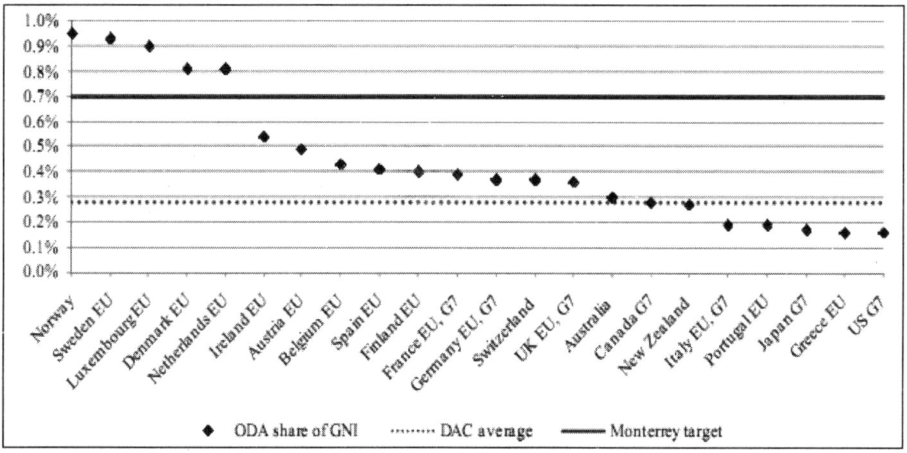

Source: OECD (2008)

3. *Need for Innovations in Raising Funds*

- Further 'innovative' financing mechanisms in addition to the CDM adaptation levy are needed to fill the adaptation 'funding chasm'.
- The only way to provide funding for developing country adaptation which is acceptable, in the sense of being new and additional, predictable, appropriate, equitable, and adequate is through international levies on bunker fuel activities and/or through international auctioning of assigned amount units (i.e. an adaptation levy on the proceeds of international emissions trading).

The developing country demand for 'new and additional' funds from developed countries, as well as for their 'adequacy' and 'predictability' is by no means new. It has been articulated again and again, not least in Article 4.3 of the UNFCCC. And, as has again been highlighted in the largely critical reactions to the recent World Bank proposal for a Pilot Adaptation Fund (now "Pilot Programme on Climate Resilience"), the funding of developing country adaptation in the form of loans—no matter how concessional—is generally rejected as not appropriate not only by developing countries:

Given the patterns of differentiated (historic) responsibilities, the costs for developing country adaptation are seen as debts to be borne by the still largely responsible industrialized world, and debts cannot be repaid by loans, or even by 'grants'—if that notion is interpreted in terms of the provider of the funds doing the recipient a favour. Moreover, given this pattern of differentiated responsibilities, there are also very strongly held views on the importance of an equitable distribution of the burden of such funding. Due to the stepped-up pace of the current UN climate change negotiations, there are host of new initiatives with regard to international adaptation finance, albeit not all of them completely satisfactory with respect these demands.

4. Conventional Funding

Foreign public sector investments/payments—whether bilateral or multilateral—have traditionally been in the form subscribed to by ODA, namely grants or (concessional) loans financed through the general budget of the donor country—i.e. based on revenue from conventional instruments such as income tax, cooperation tax, customs and excise duties, etc.

For the present purposes, this is referred to as 'conventional funding,' as opposed to funding which is raised through new, in particular, carbon-based instruments such as the auctioning of emission permits in the context of an emission trading scheme, which, as such, is genuinely additional to the conventional revenue and accordingly referred to as being 'unconventional'. Another key characteristic of these traditional payments—apart from their fiscal origin—is their fundamentally discretionary nature. Even though their overall level may be fixed and enshrined in the domestic law of the donor country, it is inevitably up to the 'donor' to decide who should get it. In short, developing countries are not taken to have a 'right' or 'entitlement' to such donations. This is something which does not fit easily with how funding for adaptation is perceived.

The World Bank Pilot Programme for Climate Resilience (PPCR)

The Pilot Program for Climate Resilience (PPCR) is part of the Strategic Climate Fund (SCF), a multi-donor Trust Fund within the World Bank's Climate Investment Funds. The overall objective of the program is to provide incentives for scaled-up action and transformational change in integrating consideration of climate resilience in national development planning consistent with poverty reduction and sustainable development goals. Within the framework of the SCF, targeted programs with dedicated funding (known as 'SCF Programs') can be established to provide financing to pilot new development approaches or scaled-up activities aimed at a specific climate change challenge or sectoral response. Resources will be mobilized and pledged to specific SCF Programs to be financed within the SCF.

A pilot program for climate resilience is proposed to provide incentives for scaled-up action and transformational change in integrating consideration of climate resilience in national development planning consistent with poverty reduction and sustainable development goals. This program is known as the Pilot Program for Climate Resilience (PPCR). The Pilot Program for Climate Resilience (PPCR) is designed to provide programmatic finance for national climate resilient national development plans. The PPCR aims to provide transformational and scaled-up support for both the development and implementation of such plans. Furthermore, its purpose is to provide lessons over the next few years that might be taken up by countries, the development community, and the future climate change regime, including the Adaptation Fund. This experience will be gained through scaled-up interventions covering the full range of sectors and sources of financing, and with sufficient resources to move quickly from planning to action. The PPCR will build upon National Adaptation Programs of Action (NAPAs), will be implemented in a manner consistent with the Paris Declaration of Aid Effectiveness, and will complement existing adaptation funds which continue to serve essential roles in tackling climate change.

The PPCR objectives are laid out in the PPCR design document (PPCR/ SC.1/CRP.1) and are summarized here (http://www.climatefundsupdate.org/ listing/pilot-program-for-climate-resilience):

(a) Pilot and demonstrate approaches for integration of climate risk and resilience into development policies and planning;
(b) Strengthen capacities at the national levels to integrate climate resilience into development planning;
(c) Scale-up and leverage climate resilient investment, building on other ongoing initiatives;
(d) Enable learning-by-doing and sharing of lessons at country, regional and global levels.

The PPCR program is based on pilot programs to be carried out in a set of selected countries. Following completion of the country selection process, the Program will be carried out in two Phases. Phase 1 will initiate a series of tasks in each respective country, including facilitation of a cross-sectoral dialogue process to arrive at a common vision of climate resilience in the medium and long-term, and formulation of a strategic approach for climate resilience. During Phase 1, to support this strategic approach an underlying investment program, a Strategic Program for Climate Resilience, will be developed. Endorsement by the PPCR-SC of the Strategic Program for further development marks the transition to Phase 2. Phase 2 will focus on implementing the Strategic Program through actions such as support to policy reform, institutional capacity building, and scaling-up other investments in key sectors. On 30th January 2009, the World Bank announced that the following countries have been invited to participate in the PPCR: Bangladesh; Bolivia; Cambodia; Mozambique; Nepal; Nigeria; Tajikistan; and Zambia.

The decision of the PPCR Sub-Committee was based on recommendations of an independent Expert Group (see section below on the decision-making structure). The criteria used include the level of vulnerability to climate change hazards and risks, country preparedness to move towards climate resilient development plans, and country distribution across regions and types of hazards. One additional country from the Middle East and North Africa region, as well as two regional groupings of countries (Caribbean and Pacific), remain under consideration for PPCR pilot programs pending further inputs from the independent Expert Group. Click here to read the full press release on this decision. Of the abovementioned countries that were selected, Bangladesh, Bolivia, Nepal, Niger and Zambia submitted final acceptance of the offer; templates have been approved by sub-committee. Mozambique just submitted final acceptance. The Sub-Committee is waiting for final templates from Tajikistan and Cambodia.

At the latest PPCR Sub-Committee meeting in May 2009, the Sub-Committee decided to invite the participation of Yemen in the PPCR. It noted that among the countries recommended, Yemen is the only country that is both an LDC and has an active MDB country investment program. It further notes that the participation of Yemen will provide an opportunity to address climate risks associated with increased precipitation in arid zones.

The Sub-Committee also noted that the Expert Group assigned priority to Djibouti based on its vulnerability to sea-level rise. The Sub-Committee was informed that the MDBs are currently working with Djibouti to strengthen analysis of its vulnerabilities to climate impacts and measures for climate resilience. Recognizing the small population size of Djibouti and its limited absorptive capacity, the Sub-Committee invited the MDBs to come back to it with a more in-depth review of the on-going work performed by the MDBs and other development partners in Djibouti and to provide a recommendation as to whether such work can usefully be supplemented by the PPCR. The Sub-Committee requested a proposal for regional pilots in the Caribbean and the Pacific to be circulated by late July. Specifically, the Sub-Committee requested that priority consideration be given to how the PPCR can best supplement the work of the MDBs in Haiti.

At the time of writing, the PPCR was earmarked to receive £250 million from the UK mostly in the form of concessional loans. Although the exact figure of how much of this funding will be in the form of grants is still uncertain, the latest figure available was that up to £13 million would be available as grants under the Strategic Climate Fund, the 'mother fund' of the PPCR. This means £50 million per annum of concessional loans, and just over £2 million per annum in grants. The Chinese +0.5 Percent of GDP Proposal In their submission replying to the UNFCCC Secretariat's call for views regarding the work programme under the Bali Road Map, China states that strategies and mechanisms for adapting to climate change shall be developed to support adoption actions in developing countries and associated financing, technology and capacity building shall be provided. Sufficient financing shall be provided by developed countries to address climate change. In addition to existing ODA, developed countries shall annually provide financial support of no less than 0.5% of their total GDP to support actions by developing countries to address climate change in developing countries. This would currently amount to $185 billion per annum. Unfortunately, nothing more is specified with regards to how much of that money would be spent on adaptation in developing countries, but since it is unlikely that adaptation is going to have a larger share than mitigation or technology transfer, $46 billion (25%) would seem to be reasonable upper limit for adaptation.

The Mexican Multilateral Climate Change Fund (MCCF) Proposal

In Bali, Mexico put forward a proposal for a Multilateral Climate Change Fund of predictable size, with contributions from developed and some developing countries matters for expanding global mitigation efforts. The size of the expected revenue is not explicitly specified. Initially, the MCCF is to be filled with some tens of billions of dollars annually, while ultimately, it is meant to be able to meet a significant part of the additional flows of finance required by developing countries to meet the costs of reducing their emissions, estimated to be in the region of $95billion in 2030. Assuming an adaptation levy of 2 percent, this would mean annual adaptation revenue of between $0.2 billion and $1.9 billion (in 2030).

According to the proposal, the amount in the MCCF must be predictable and scaleable, thereby transcending both the 'official development assistance' model and simple voluntary contributions for specific ends. This predictability will result from objective and equitable criteria negotiated multilaterally to determine national contributions as functions of indicators such as (a) green-house gas emissions, (b) population, (c) ability to pay. In other words, industrialized countries are meant to be bound by contribution commitments—the relative share of which to be determined by a Responsibility-and-Capability (R&C) indicator. While the default for the source of these (mandatory) contributions appears to be the general country budget, the proposal leaves open the possibility of also using revenues from carbon auctioning for these purposes. As concerns the nature of the envisaged funding, the Mexican proposal is thus what might be called a 'conventional/unconventional hybrid.' As concerns the envisaged developing country contributions, Least Developed Countries (http://www.oxfordenergy.org/pdfs/EV42.pdf) are, from the outset, expected to have a certain quota of the revenue at their disposal without being themselves expected to contribute.

Other ('emerging') developing country economies—presumably defined in terms of the R&C index—will, however, be expected to provide some contributions, against the insurance that they would then have the right to access amounts substantially larger than their own contributions. The amount exceeding the level of their contribution would represent a positive incentive to expand the scale of the committed effort. In contrast to industrialized countries - with their binding commitments—developing countries which, in accordance with the criteria adopted, should contribute to the MCCF but which choose not to do so would be exempted from both the obligations and the benefits of the Fund, without being subject to any kind of penalty. Creating and operating the MCCF must not be detrimental for any developing country. Finally, all contributing nations, whether developed or developing, will participate in the governance structure that will be established for the MCCF, a structure that will also be open to representatives of all beneficiaries.

Bi- and Multilateral Carbon Auction Levy Funding under IPREA

The US International Climate Change Adaptation and National Security Fund Title XIII ("International Partnerships to Reduce Emissions and Adapt"-IPREA) of the Lieberman-Warner Climate Security Act of 2008 proposed an "International Climate Change Adaptation and National Security Fund" ('the Fund') to be established in the US Treasury, with the aim of financing an International Climate Change Adaptation and National Security Program (the Programme') from 2012 till 2050. This fund was proposed under the Lieberman-Warner Climate Security Act of 2008. The bill would establish a country-wide cap-and-trade system, with 26.5% of emissions allowances auctioned in 2012, steadily ramping up to 69.5% by 2031. A portion of auction revenue (from 1% in 2012, increasing gradually to 7% by 2050) would be directed toward a newly created International Climate Change Adaptation and National Security Fund. The bill highlights the needs of 'most vulnerable developing countries,' although spending would be bound by US security-based priorities (http://www.odi.org.uk/ccef/resources/reports/s0198_oecd_adaptation.pdf).

The first purpose of the Programme is to protect the economic and national security of the United States where such interest can be advanced by minimizing, averting, or increasing resilience to potentially destabilizing global climate change impacts. To this end, the Programme shall support investments, capacity building activities and other assistance, to reduce vulnerability and promote community level resilience related to climate change and its impacts in the most vulnerable developing countries, impacts that affect economic livelihoods, result in increases in refugees and internally displaced persons, or otherwise increase social, economic, political, cultural or environmental vulnerability. In order to raise revenue for the Fund, the Administrator of the US Environmental Protection Agency (EPA) would auction a percentage of the annual emission allowances of the proposed US emission trading scheme, starting with 1% in 2012, and raising gradually to 7% in 2050. This would amount to about $1 billion in 2012, increasing to around $2 billion by 2020 and $6 billion by 2030. Up to 60% of the funding can go to international funds, provided they are created pursuant to the United Nations Framework Convention on Climate Change or an agreement negotiated under the Convention and fulfil certain additional requirements, most of which clearly satisfied by the Adaptation Fund.

Germany's International Climate Initiative (ICI)

Since early 2008, the German Federal Environment Ministry (BMU) has raised funds by auctioning 9% of its nationally allocated carbon allowances for the second phase (2008-12) of the ETS. Rather than giving away those permits to industry, the German government is auctioning the permits to generate revenue. Of the €800 million expected annual revenue from the auctions, €400 million will go to climate initiatives, €120 million of which will be allocated internationally to developing countries, and half of this amount will be allocated to adaptation and forest protection (http://www.odi.org.uk/ccef/resources/reports/s0198_oecd_adaptation.pdf). Germany's ICI is in addition to a much larger sum of money already spent bilaterally on adaptation.

The Norwegian proposal to auction AAUs

Norway has designed a proposal to finance adaptation through auctioning a portion of emission permits. The auction would occur at the international level before the AAUs are allocated to national registries, and would be auctioned by an appropriate international institution. The resulting revenue would then be placed in a fund to be used for adaptation. This proposal recommends the levy be placed at the point of issuance of allowances rather than on the market *transaction* of allowances (http://www.odi.org.uk/ccef/resources/reports/s0198_oecd_adaptation.pdf).

Combating Global Climate Change–Clinton Climate Initiative (CCI)

The William J. Clinton Foundation launched the Clinton Climate Initiative (CCI) to create and advance solutions to the core issues driving climate change. CCI takes a holistic approach, addressing the major sources of greenhouse gas emissions and the people, policies, and practices that impact them. Working with governments and businesses around the world, CCI focuses on three strategic program areas: increasing energy efficiency in cities, catalyzing the large-scale supply of clean energy, and working to stop deforestation. A relatively young initiative, CCI is delivering the following measurable and significant results in the fight against global warming and adverse effects of Climate change, mainly caused by the GHGs (http://www.clintonfoundation.org/what-we-do/clinton-climate-initiative/):

- 40 of the world's largest cities are working with us to reduce greenhouse gas emissions through a variety of large scale projects, including building retrofits, public transit improvements, and advanced waste management systems.
- CCI has brought together many of the world's largest energy service companies, banks, and building owners to cut energy consumption in buildings in more than 30 cities. Cities including London, Houston, Johannesburg, Melbourne, and Seoul are already doing projects to retrofit more than 300 municipal buildings in total. Private building owners including GE Real Estate; Merchandise Mart; and a number of significant owners and developers abroad have begun commercial retrofit projects. More than 200 million square feet of project work is already underway.
- CCI and the American College and University Presidents Climate Commitment (ACUPCC) are working together to retrofit colleges and

universities across the United States. We have already assisted more than 20 universities in developing tenders and beginning the process of building retrofits across their campuses.

- CCI has been working closely with the New York City Housing Authority (NYCHA), the largest public housing authority in the country, to develop its multi-phase energy-conservation and building retrofit program. Already, we have completed a project to install nearly 10,000 compact fluorescent light bulbs at the largest public housing development in the country, estimated to reduce greenhouse gases by 1,400 tons per year and overall electricity costs by 17%, saving NYCHA approximately $367,000 annually.
- CCI have negotiated discounted pricing agreements with more than 25 manufacturers of energy-efficient products, including building materials and systems, clean technology vehicles, and street and traffic lights. The pricing discounts on these deals range from 5-15% for commodity items and from 15-70% for non-commodity items. More than 1,100 cities worldwide now have access to these commodities through CCI's purchasing alliance.

Future of Carbon Funding: IFC

According to International Finance Corporation the term "carbon fund" or "carbon vehicle" refers to a collective investment scheme that receives money from investors and uses this money to buy carbon credits from, or invest into greenhouse gas (GHG) emissions reduction projects, generally CDM and JI. After a certain defined period, the carbon fund will then give investors carbon credits and/or cash in return. The emission reduction projects can be sourced from different developers, countries, and technologies. Carbon funds can be initiated and managed by development banks, governmental agencies, private financial institutions, or consultancies. In terms of investors, ICF identifies the following types of shareholder structures: government(s), private firm(s), or a mix of both (multi-shareholder).

Carbon funds have become an important player in the carbon markets over the past few years and have experienced considerable growth both in terms of the number of vehicles and the amount of capital raised. However, the amount of capital secured or targeted by funds created over the last two years appears to be slowing down. Indeed, the lack of clarity on the post-2012 climate policy framework at international and domestic levels, increasing competition on the buyers' side, and the decline in the number of large and low risk Clean Development Mechanism (CDM) projects create a challenging and uncertain environment for new entrants in the carbon finance community. This situation will likely be exacerbated by the credit crisis, which could reduce the amount of capital available to invest in carbon credit projects and increase the cost of capital available. But current financial difficulties can also create opportunities for some of the existing carbon funds that can benefit from a competitive advantage *vis-à-vis* other investors. Therefore, the future of the carbon fund market is complex and uncertain. However, faced with these increasing challenges, whether carbon vehicles will come out as winners or losers will mainly depend on their investment strategy and their flexibility to adapt.

Climate change has been a core competency of ICF since the mid-1980s. Globally, around 300 of our employees have experience advising corporate clients to help develop carbon strategies or working with our government clients to devise and implement climate change policy. Over the past two decades, ICF International has earned a reputation in the field of climate change for our analytical rigor, in-depth market expertise, and technical integrity through hundreds of assignments undertaken for (IFC, 2009):

- *Leading companies*, including over 60 companies among the Global FT 500 (the world's largest companies by market capitalization), and climate-friendly technology suppliers that have recognized climate change as an issue which will help to define their competitive advantage.
- *Financial institutions* on Wall Street and in Europe with an interest in the valuation and reputation impacts of climate change
- *Multilateral institutions*, such as the European Commission, Global Environment Facility, Organization for Economic Cooperation and Development, International Energy Agency, regional development banks, and United Nations agencies
- *National and local governments* in more than 50 countries including the United Kingdom, Ireland, Canada, Australia, New Zealand, Ukraine, Ukraine, and the United States.

Based on market activity during the first half of 2008, the global carbon market is expected to have grown by more than 50 percent by the end of the year compared to 2007. The total traded value could exceed $100 billion (€68 billion) in 2008 versus $64 billion (€44 billion) last year. So far, ICF has identified 84 carbon vehicles that have amassed altogether a total capital of €8,908 million. These numbers are convergent with those presented in the latest edition of *Carbon Funds Directory* by Environmental Finance Publications (Carbon Finance 2008). According to the report, the operational carbon vehicles have committed to amass $12.87 billion (€8.77 billion), while those actively raising capital were estimated to bring an additional $5.49 billion (€3.74 billion) (http://www.icfi.com/Markets/Energy/doc_files/carbon-procurement-guide.pdf).

Impact of the CDM/JI Market

The boom of investments in CDM has primarily been driven by the price differential between CDM credits—Certified Emission Reductions (CERs)—and European Union Allowances (EUAs), and between primary and secondary markets for carbon credits. In this market, carbon funds have represented a good investment option as other ways of originating carbon credits can be more risky (e.g., direct project development or financing of emission reduction projects).

Presently, the price differential continues to act as a driver for purchasing carbon credits compared to allowances, but the spread has reduced considerably. Also, a number of other factors are now affecting the CDM/JI market, thereby reducing the attractiveness of this type of investment. This year has seen downward revisions of CDM/JI credits supply forecasts. Increased scrutiny at validation, registration, and verification stages combined with a capacity crunch within Designated Operational Entities (DOEs) and the CDM Executive Board (CDM EB) has lead to a significant increase in the time required to register a project. In addition, the "low-hanging fruit" projects have been grasped: CDM and JI project opportunities are smaller, riskier, and costlier now, making it challenging to keep up with expected growth of the credit supply.

Future Repercussions for Carbon Funds

In the course of a detailed analysis, ICF has reached the following conclusions concerning the evolution of carbon vehicles and the carbon market in general (IFC, 2009):

- The financial crisis is going to have a larger impact on newly created carbon funds. Carbon funds launched recently will probably suffer from the overall lack of financial resources and the increased cost of capital. This might prevent some of them from actually being created or from reaching their target size.
- The creation of new funds will likely slow down. Although the number of carbon funds in the market has grown steadily over the last few years, the secured capital under their management has increased at a much lower rate than in the past. Taking into consideration the turmoil in the financial markets and the lack of clarity about the post-Kyoto international climate regime, investors are likely to wait for clear market signals before launching a new vehicle.
- Compliance buyers and banks adopting a more conservative approach will leave opportunities for carbon funds to seize. In the current context, certain compliance buyers and financial institutions will be less inclined to deal with increasingly small projects and companies potentially short of cash and/or with a lower credit rating. This will unleash opportunities for carbon funds that have the ability to take on more risks and/or provide pre-payment.
- New cap-and-trade schemes, provided they come into effect, might result in funds being set up to serve buyers beyond Europe. Independently from the post-Kyoto negotiations, several countries have announced that they plan to introduce their own cap-and-trade schemes that will allow, to some extent, carbon credits as a compliance

option. Once these new programs become operational, it is likely that the demand for international credits will surge, opening opportunities for new carbon vehicles probably managed by local agents and serving these specific—and increasingly fragmented—markets.

- Direct financing will likely play a larger role in carbon fund strategies. Direct financing has been gaining importance steadily as investors and developers have faced growing difficulties in finding robust projects that yield large quantities of credits on an off-take basis at attractive prices. The carbon market needs more investors that are willing to contribute equity at an early stage of project development. Carbon funds with adequate investment rules are placed ideally to fill such a gap.
- Carbon funds are starting to diversify their investment portfolio. In the face of uncertainties around the future CDM/JI market, carbon funds have started to consider sourcing other types of credits such as EUAs or Voluntary Emission Reductions (VERs), as well as investing in clean technologies outside the scope of the CDM/JI. Through such actions, funds are opening themselves up to a wider range of sectors and investors. Direct financing through equity investment in emission reducing projects will also contribute to revenue diversification.

CHAPTER 5

Carbon Management: Carbon Neutrality and Carbon Literacy

The Principle of Carbon and Climate Neutrality

Carbon neutral is a term used to describe any organization, entity, or process that has a net greenhouse gas (GHG) emissions level of zero (Dautremont-Smith *et al.* 2007). Since carbon neutrality only requires a net greenhouse gas emissions level of zero, organizations do not need to eliminate all carbon pollution to become carbon neutral. *Net emissions* differ from *gross emissions* in that gross emissions are the sum of all emissions released by the individual or entity, whereas net emissions are equivalent to the gross emissions minus any carbon offsets.

A *carbon offset* is any activity that reduces carbon emissions so as to exactly compensate for a carbon emitting activity elsewhere (Dautremont-Smith *et al.* 2007). If net emissions were greater than zero, the entity would be considered a net emitter of carbon. If they were less than zero, then the entity would be a net reducer of carbon. If the net emissions level was zero, then the entity would be carbon neutral.

Being carbon neutral, or having a net zero carbon footprint, refers to achieving *net-zero carbon emissions* by balancing a measured amount of carbon released with an equivalent amount sequestered or offset. The carbon neutral concept may be extended to include other greenhouse gases (GHG) measured in terms of their carbon dioxide equivalence—the impact a GHG has on the atmosphere expressed in the equivalent amount of CO_2. The term climate neutral is used to reflect the fact that it is not just carbon dioxide (CO_2), that is driving climate change, even if it is the most abundant, but also encompasses other greenhouse gases regulated by the Kyoto Protocol, namely: methane (CH_4), nitrous oxide (N_2O), hydrofluorocarbons (HFC), perfluorocarbons (PFC), and sulphur hexafluoride (SF6) (http://en.wikipedia.org/wiki/Carbon_neutral).

While a *zero carbon economy* may be possible in the future, present technology, infrastructure, and the availability of alternatives to carbon emitting devises make it impossible to continue the status quo without carbon pollution. For example, it would be impossible for many individuals, businesses, and other organizations to stop using fossil fuel-consuming transportation and continue with their basic operations. However, these same entities could achieve carbon neutrality without needing to wait for alternatives to fossil-fuel powered transportation to become widely available. These organizations could reduce or eliminate emissions where possible and offset carbon emissions where reduction or elimination of emissions is not an option. Organizations may also choose to offset emissions when it costs less to purchase offsets than to reduce emissions.

The concept of carbon neutrality is fairly straightforward in theory, but in practice the details are complex and in some quarters they are controversial. *Carbon neutral* means that one has calculated and accounted for all of carbon climate change impact; one has investigated reducing the production units/ emitter's carbon footprint and you also happen to purchase carbon credits to offset your unavoidable greenhouse gas emissions (http:// www.noco2.com.au/). Carbon neutrality offers individuals, businesses and other institutions the opportunity to take the personal responsibility for the global warming implications of their lifestyle. In response to a rapid increase in public awareness about global warming in the recent past, many corporations are focusing on how they can manage their carbon impacts. Emissions trading related businesses are doing this for several reasons:

- To establish their public positioning on this issue;
- To undertake practical measures to reduce energy costs leading to reduced emissions and future carbon risk;
- To satisfy their staff, who are increasingly looking to their employers to show they are taking responsible action to reduce emissions.

Many organizations in India are aiming to define their carbon footprint, to reduce this footprint by reducing the emissions from the business and, increasingly, to reduce their impact further by purchasing 'green energy' (from renewable sources) or carbon offsets. Some companies, wishing to go further, are becoming 'carbon neutral', where they completely offset their carbon footprint to theoretically have zero carbon impact on the environment. These companies can capture the Indian market by playing the significant role as the retail seller as well as buyer of the carbon credit.

Science has identified greenhouse gas (GHG) emissions as the main cause for climate change, with carbon dioxide (CO_2) being the predominant compound. Therefore GHG—in particular CO_2 emissions - should be avoided or reduced from the outset wherever possible. Unfortunately, this cannot always be achieved. But, even where emissions cannot be avoided, there is an option to contribute to climate protection – through *climate neutrality pathways.*

The *Climate Neutral Protocol* describes the quality standards used for the *'climate neutral label'*, which incidentally was developed in a Climate Partnership Project called 'Hessian Climate Partners'. The project was funded by the Hessian Ministry of the Environment. The protocol was written by 3C Consulting GmbH in close cooperation with TÜV SÜD, a global leader in the market for climate protection services. Other partners include: Deutsche Bank AG, Deutsche Gesellschaft für Technische Zusammenarbeit (GTZ) GmbH, Deutsche Post World Net AG, Frankfurter Rundschau, HEAG mobilo GmbH, KfW Bankengruppe, Neckermann.de GmbH and T-Com.

Climate Neutrality

Climate neutrality, as defined by the United Nations, signifies an organization or event that has a zero net contribution to the atmosphere of all greenhouse gases identified in the Kyoto Protocol. The set of measures introduced to avoid and reduce greenhouse-gas emissions affect all activities directly controlled by the organization, including travel, purchase of goods and services, and daily behaviour of staff. Climate neutrality also recognizes offsetting as an option to achieve full neutrality (UN Climate Neutral Strategy EMG/AM.07/11).

Climate knowledge is the foundation for the development of an effective response to the climate change challenge. The UN system plays a central role in this area, bringing together global resources for observation and analysis of climate change trends. It is committed to reinforcing its efforts to provide sound and unbiased scientific information and climate services to enable evidence-based policy and decision making at all levels (http://www.un.org/climatechange/pdfs/Acting%20on%20Climate%20Change.pdf).

As a unique international mechanism with universal membership that combines convening power, normative work, standard-setting, policy development and operational activities, the UN system in its entirety has embarked on an action oriented and coordinated effort to support the international community to rise to the climate change challenge. The initiatives taken by the Chief Executives Board for Coordination (CEB) of the United Nations that commenced in early 2007 has already established a clear framework for action with focus areas for the achievement of early results. Agencies contribute within the framework of their respective areas of expertise, mandates issued by their governing bodies and resources made available to them. Work also continues in a number of cross-cutting areas that support this effort. The overall objective is to maximize existing synergies, eliminate duplication and overlap, and optimize the impact of the collective effort of the UN system. At its April 2008 session, the CEB decided to concentrate at this stage on five focus areas, which were identified in response to the UNFCCC negotiation process and in pursuance of the broader mandates and capacities in the UN system. To ensure better coordination, convening responsibilities were assigned to UN system entities with a large volume of activities in the respective focus areas and some additional cross-cutting ones (http://www.un.org/climatechange/pdfs/Acting%20on%20Climate%20Change.pdf).

Steps Involved in Climate Neutral Activities

The damaging effect of greenhouse gases is global; therefore it is irrelevant in terms of climate protection where emissions are being caused or where they are being avoided. Unavoidable emissions at location 'A' can be compensated by initiating additional climate protection projects at location 'B', and thereby become 'climate neutralized'. This principle allows for services, events or even companies to be recognized as climate neutral and to be awarded the climate neutral label.

The following Five Steps are considered vital to develop and implement Climate Neutral Activities: each of these steps is of course based on transparent quality standards (Climate Neutral Network Offset Protocol, 2005).

- Explore where emissions can be avoided or reduced Step 2
- Calculate the unavoidable emissions to be compensated for
- Select a compensation project
- Purchase high quality emissions reduction credits
- Retire the certificates

Step 1: Explore where emissions can be avoided or reduced

For the sake of credibility exploring all options for reduction or avoidance of greenhouse gas emissions should always be the first step in any effort towards climate neutrality. We recommend introducing an environmental management system to systematically identify and realize potential options for avoiding or reducing emissions.

Step 2: Calculate the unavoidable emissions to be compensated for

The Kyoto Protocol defines six different greenhouse gases. Emissions are calculated based on data and methods from up-to-date and independent sources (e.g. the German Federal Environment Agency, the United Nations Environment Programme) or – in the case of industrial products such as cars – directly from manufacturers. Every factor influencing emissions is accounted for, and the system borders to the calculation are clearly presented, so that a conservative and comprehensible calculation is guaranteed.

Step 3: Select a compensation project

The credibility of a climate neutral activity depends on a number of factors: It should, of course, be part of a comprehensive climate protection strategy, and it should also reflect the core competencies of a company or organization. But first and foremost it is the ecological integrity of compensation projects selected to offset the activity which make or break the credibility of a measure. In this regard, the following quality standards should be observed:

- *Verification:* The project's emission reductions have to be quantifiable. Projects are monitored and verified on an ex-post basis by independent third parties.
- *Additionality:* A climate protection project is eligible for emissions compensation only if it depends on the sale of emissions reduction credits for it's financing, and only if it reduces actual emissions when compared to a realistic reference scenario (additionality). The additionality of compensation projects is audited by independent third parties.
- *Project types:* Projects using renewable resources and projects to increase energy efficiency are particularly suited to achieve climate neutrality. According to rules set by the UN certain project types, such as nuclear energy projects, are not eligible as climate compensation projects. Moreover there should be a preference for projects which are likely to bring an additional social benefit to their region.

Step 4: ***Purchase high quality emissions reduction credits***

The emissions reduction credits are purchased directly from the developers of the compensation projects via transparent purchase agreements.

Step 5: ***Retire the certificates***

For every ton of calculated greenhouse gases to be climate neutralized, one reduction certificate worth one ton of CO_2 equivalent will be retired. Fractions will be rounded up to a full ton. Every certificate will be verifiably retired after it was used for climate neutral activities in order to make sure that it will not be used more than once to compensate for emissions. Certified Emission Reductions and Emission Reduction Units will be retired into the official state-run register created for this purpose. Verified Emission Reductions are retired in a transparent process verified annually by an independent Certified Accountant.

Climate Neutral "Gold Standard"

Climate Cool products, services, and enterprises are those that reduce or offset the greenhouse gas emissions with which they are associated to achieve a net zero impact on the earth's climate. A principal goal of the Network is to completely offset the greenhouse gases generated at each stage of the life-cycle of a product or service: the sourcing of its materials; its manufacturing or production; and its distribution, use, and ultimate end-of-life disposition. Companies or institutions that offset all of the gases resulting from the full spectrum of their internal operations can also receive *'Climate Cool enterprise certification'*. Global-warming gases can be offset in a variety of ways ranging from investing in technologies that dramatically reduce carbon emissions, such as renewable energy, highly fuel-efficient vehicles or energy-efficient lighting in public schools, to the planting of trees, which absorb global-warming gases.

The Network's mission is to support companies in bringing Climate Cool innovations to market in ways which merit the *Network's "gold standard" endorsement* and create tangible competitive benefits for the companies that lead in creating this new Climate Cool market arena. The bases for this gold standard are the Network's Design Principles, below, drafted in consultation with leading environmental groups such as NRDC, the Nature Conservancy and Rocky Mountain Institute. The Design Principles are intended to guide companies in the development of a portfolio of projects that is scientifically sound, highly credible, and that merits broad stakeholder support (http://climateneutralnetwork.org/standards.php).

The Popularity of the Gold Standard

Projects with attractive sustainable development attributes are able to attract healthy price premia for CERs from projects that use the Gold Standard or similar standards certifying strong development attributes. The Gold Standard has emerged as an attractive choice for compliance buyers (and some voluntary market buyers) of CDM and JI projects that also demonstrate strong local benefits. Unlike pure voluntary market plays, where several standards try to capture the development dimension of projects, the Gold Standard has long been the only label in existence on the compliance market to do so. The label helps to place a value on the additional benefits delivered by CDM projects. The Climate, Community and Biodiversity Alliance Standard (CCBS), also a project design standard, could join the Gold Standard and help forestry projects to grab premium.

So far, the Gold Standard has been applied only to renewable energy and energy efficiency projects and just twelve CDM projects have received the Gold Standard certification. Gold Standard CERs are in high demand, from compliance buyers as well as from buyers on the voluntary market and the highest priced-primary carbon transaction in our 2007 data is a Gold Standard project that transacted at €16, a healthy premium to regular primary CERs.

Climate Neutral Network

Incorporated in 1999, the *Climate Neutral Network* (CNN), a small nonprofit organization based in Portland, Oregon, has built an impressive coalition of companies, nonprofit organizations, government agencies and academics committed to bringing climate neutral products and services to market. CNN defines Climate Cool™ products and services as those "determined to have little or no effect on the Earth's climate (http://climateneutralnetwork.org/background.php)." This is achieved by reducing and offsetting the greenhouse gases associated with a product, service, or event. One way to do this is to significantly reduce energy use and emissions of global warming gases within the companies' own facilities. After making these internal reductions, Climate Cool™companies invest in energy reductions to offset their remaining greenhouse gas emissions and achieve a net zero impact on the Earth's climate. There are a tremendous variety of offset options to reduce CO_2 emissions; for example, installing solar panels that convert sunlight into electricity, which reduces the need to burn fossil fuels; or planting trees, which absorb global warming gases. Products or companies that eliminate all of their emissions may be considered "climate neutral" and are eligible to apply for a *'Climate Cool™ certification'* and use of the Climate Cool trademark. The idea for the Climate Neutral Network emerged from an inquiry by a small group of colleagues about how much it would cost to offset the emissions created by a typical automobile burning one gallon of gas.

Climate Neutral Design Principles

The *Climate Neutral Design Principles* provide the bedrock upon which companies' investments and reductions will be founded in order to earn consumers' trust and credibility. Accordingly, the interested companies may choose to submit their draft Climate Cool strategies to the Network's Environmental Review Group for confidential review and certification. The Climate Neutral Design Principles entail the following pursuits:

I. The Climate Neutral Design Principles of Market Innovations (www.pi.energy.gov/enhancingGHGregistry/comments/.../hall2.doc) are designed to achieve the following objectives:

A. *Create new powerful choices for consumers, employees and companies:* Individuals and organizations can choose to invest in *Climate Neutral* products and services to offset their climate impacts as an optional, voluntary investment
   - these transactions can be facilitated at any point of purchase—through any companies, NGO's or other organizations with whom they are affiliated

B. *Create new* Climate Neutral *products: Climate Neutral* products or services are determined to completely offset the greenhouse gases generated across each stage of their life-cycle—including the sourcing of its materials, its manufacturing or production, its distribution, use, and ultimate end-of-life disposition.
   - *Climate Neutral* warranties would apply to products such as cars or appliances (which consume a lot of energy /fuel during use or have ambiguous lifetime durations) to cover the majority of emissions from both manufacturing and use (e.g. 60,000 miles for cars)

C. *Empower new* Climate Neutral *Enterprises:* a company can choose to become a *Climate Neutral* enterprise by offsetting the climate impacts of its internal operations across critical areas, including manufacturing, administration, transport of product to customers etc
   - a Network "metrics system" paper is available to outline and help estimate these enterprise emissions
   - companies can also offset individual components of its internal operations—approaching *Climate Neutral* certification through a

series of incremental steps—to create a series of "ramp-up" options which nonetheless maintain the integrity of the brand.

D. *Maintain Credibility and Integrity of the Climate Neutral Brand:* products or projects which would, in the view of the Network's board, undermine the credibility of the initiative or the integrity of the *Climate Neutral* brand—due to inconsistencies with the purpose of the trademark/organization for example – may be refused the *Climate Neutral* mark.
   - the board reserves the right to refuse accreditation to any application that it concludes is not in the best interests of the brand. "*Climate Neutral* hummers", for example, could pose such integrity problems.

E. *Reward early movers:* given pending legislation regarding early action crediting, the Network reserves the right to refine these principles further at a later date to ensure that companies' early *Climate Neutral* commitments would not be undermined by any future regulatory requirements.
   - a *Climate Neutral* enterprise might therefore be able to draw upon its *Climate Neutral* investments to assure its compliance with new regulatory systems; *Climate Neutral* investments made way in excess of these compliance requirements would not be available to enable other companies to increase their emissions since they would be permanently retired (see II-F below)
   - a *Climate Neutral* enterprise should therefore be able to draw upon its *Climate Neutral* investments to assure compliance with new regulatory systems since our intentions are to go beyond compliance, provided future regulatory systems are designed accordingly
   - reductions could be registered to protect companies' baselines in any future allocation system—but not to obtain saleable credits
   - although the Network seeks to promote future regulatory compliance as a result of companies *Climate Neutral* commitments, it cannot guarantee such compliance nor be held liable in any way for requirements or liabilities resulting from future regulatory systems

F. *Encourage tax benefits:* the Network supports a proactive tax treatment for *Climate Neutral* investments such that companies or individuals could benefits from appropriate tax deductions as a result of their investments – for example, by retiring them through donations to a not-for-profit—since this could increase the value of the benefits received.
   - Companies seeking *Climate Neutral* certification understand, however, that the Network's cannot be liable for any future IRS rulings or interpretations in this regard.

II. Climate Neutral Portfolios are Designed to:

A. *Be built upon a commitment toward internal energy efficiency gains:* as part of their *Climate Neutral* proposal, companies should be able to demonstrate an outstanding performance in reducing greenhouse gas emissions internally—before purchasing offsets—either as part of their new *Climate Neutral* portfolio or by pointing to past, current or future on-site reduction activities
   - outstanding performance for internal reductions could be benchmarked by reference to companies outlined in books such as Romm's *Cool Companies* (Island Press) or Lovins' *Factor Four*
   - performance demonstrated via existing programs, policies promoting energy efficiency or via new internal reductions planned as part of their current climate neutral application

B. *Reflect the principle sources of greenhouse gas emissions in the solutions they propose.* For example, if currently 60% of GHG emissions are globally attributable to fossil fuel, then, as a general rule, at least 60% of the portfolio needs to be focused on CO_2 reductions from fossil/energy sources—whether generated by on-site reductions or through offset purchases.
   - similarly, only 10-20% to focus on forest sequestration
   - a diversified *Climate Neutral* portfolio can be built over a period of five years, based on an up-front management plan—to enable companies to achieve this diversity and still invest in sizeable projects
   - *Climate Neutral* certification is designed to ensure that reductions and offsets meet these criteria. However, companies are individually responsible for the performance of these *Climate Neutral* portfolios and the Network is in no way liable for any non-performing assets or liabilities arising

C. *Create multiple benefits*: companies are encouraged to invest in offsets which confer multiple GHG savings to compound cost, community and climate benefits (e.g. methane capture which is then used to generate electricity since this has multiple benefits—methane and CO_2 displacement for fuels generation for local schools' use)

D. *Foster experimental learning and innovation:* companies can voluntarily decide to invest up to 5% of their *Climate Neutral* portfolios in experimental offset areas (transport, land use planning etc. where carbon tons are "softer," less quantifiable), in order to encourage on-going learning and innovation at the leading edge of the offset arena

E. *Balance geographic interests:* to reduce investment risk, provide educational opportunities, and support the domestic infrastructure, companies are encouraged to dedicate a substantial majority – at least two-thirds – of their offset investments to domestic investments. The stronger the domestic investment, the more credible the program.

- exceptions can be made to reflect, for example, the European emphasis for domestic reductions that the UK might wish to establish, on an individual case-by-case basis

F. *Avoid double counting:* offsets should be retired (since companies could otherwise trade permits and compromise Climate Neutral status by enabling more carbon to be emitted elsewhere), and cannot increase the US assigned amount.
   - *Climate Neutral* offsets are retired in order for the enterprise or product to receive the *Climate Neutral* mark in perpetuity. The offsets cannot be traded but rather will create long-term permanent trusts to secure these climate gains for future generations
   - Any project registered under JI/CDM cannot be used for *Climate Neutral* certification purposes since they would increase the US assigned amount.

G. *Be tradable only at the margin*: *Climate Neutral* investments could only be traded at the margin (rather than retired) if companies had voluntarily or otherwise invested in excess of their appropriate Climate Neutral requirements

H. *Offset on more than a one-to-one ratio*: portfolios need to offset the carbon on more than a one-to-one ratio in order, at a minimum, to provide for attenuation/leakage
   - discount factors for different offset projects should take account of cost/precision tradeoffs in the measuring and monitoring process

I. *Be responsive and flexible to accommodate the variations in offsets' risks & returns:* carbon gains can be front-loaded subject to appropriate discount factors to reward (1) investments which offer greater security, and (2) investments in advanced technologies which tackle the underlying consumption challenge to accelerate the introduction of products which dramatically reduce average climate emissions.
   - given the present state of the market, all offsets can be front-loaded, subject to varying discount rates; "pay as you go" approaches are preferred as the market matures
   - discount factors should take account of quality variations with respect to through-time effects
   - Ideally, advanced technology investments (including renewables, technology buy-down funds etc) might comprise, on a voluntary basis, at least 10% of the portfolio (onsite or offsite)

III. Climate Neutral Offset Investment Projects should:

A. *Qualify every offset investment according to the following criteria*: each offset investment should be:
   1. Transparent and verifiable
   2. Additional
   3. Measure, monitor, and verify reductions

4. Minimize and compensate for potential leakage
5. Establish contract enforcement and liability mechanisms (e.g. trusts, guarantees etc.)
6. Be technically sound

B. *Reflect other desirable criteria in the overall portfolio of offset investments*: desirable criteria for at least some offset projects within the portfolio include:
   1. Design for persistence and resilience—permanence
   2. Promote collateral benefits (including promotion of local acceptance, community benefits and participation)
   3. Promote innovation
   4. Be replicable

Climate-Neutral Initiatives by WMO

Large International meetings such as the World Climate Conference-3 are major sources of greenhouse-gas emissions reduction. Air travel, local transport, accommodation, meals and the conference venue itself contribute emissions. The *World Meteorological Organization* (WMO) aims at reducing its climate impact through climate-neutral initiatives and is making a special effort to keep the greenhouse-gas emissions as low as possible. In addition, WMO encourages participants to contribute, individually and collectively, to the climate-neutral effort by offsetting their carbon footprints.

WMO, inter alia, adopts a *'Green strategy'*, which includes the use of electronic means for communication (viz. announcements and flyers sent via e-mail, Web, video- and teleconferencing, online registration). By choosing an easy-to-reach conference location, WMO encourages the use of public transport and sustainable transportation options such as bicycles. WMO also promotes videoconferencing as a means to participate remotely in certain cases. WMO chooses to provide and disseminate conference material electronically. A limited number of handouts are printed on environmentally friendly paper. Recycling services are provided at the conference site. Participants are encouraged to dress according to the season, in order to keep energy consumption through air-conditioning and heating at its minimum. The WMO green meeting strategy also includes the selection of food/goods suppliers taking into account environmental concerns (http://www.wmo.int/wcc3/WCC3_green_en.php).

A climate neutral program can consist of tracking, reducing and offsetting your business flights, which is usually the largest source of unavoidable emissions in a service-oriented business. As a manufacturer, one can determine and offset the carbon footprint of the various manufactured products and services and use appropriate *'Climate Neutral Labels.'* One could also extend your product line with a climate neutral rate to give one's customers a climate-friendly option. First Climate's marketing experts can design a wide range of communication tools, such as online carbon calculators, micro web-sites, climate card games or information brochures to positively reinforce your climate neutral campaign.

Ambitious businesses have the opportunity to turn their company into a climate leader by implementing a carbon management system and offset their entire carbon footprint. A well-implemented program anticipates climate risks and turns them into business opportunities. Such programs often go hand-in-hand with leaner manufacturing, process optimization, customer relationship management or awareness-raising and education campaigns. First, Climate can support you along the entire way: assessing your initial carbon position, implementing a carbon management system, offsetting the unavoidable emissions with suited emission reduction projects, tracking the carbon performance and communicating your climate message to your stakeholders (http://www.firstclimate.com/en/climate-neutral/climate-neutrality.html).

Climate Science Literacy

Climate Science Literacy is an understanding of your influence on climate and climate's influence on you and society. A climate-literate person:

- understands the essential principles of Earth's climate system,
- knows how to assess scientifically credible information about climate,
- communicates about climate and climate change in a meaningful way, and
- is able to make informed and responsible decisions with regard to actions that may affect climate.

No single person is expected to understand every detail about all of the fundamental climate science literacy concepts. Full comprehension of these interconnected concepts will require a Systems-thinking approach, meaning the ability to understand complex interconnections among all of the components of the climate system. Moreover, as climate science progresses and as efforts to educate the people about climate's influence on them and their influence on the climate system mature, public understanding will continue to grow.

Climate is an ideal interdisciplinary theme for lifelong learning about the scientific process and the ways in which humans affect and are affected by the Earth's systems (http://downloads.climatescience.gov/Literacy/Climate%20Literacy%20Booklet%20Low-Res.pdf). This rich topic can be approached at many levels, from comparing the daily weather with long-term records to exploring abstract representations of climate in computer models to examining how climate change impacts human and ecosystem health. Learners of all ages can use data from their own experiments, data collected by satellites and other observation systems, or records from a range of physical, chemical, biological, geographical, social, economic, and historical sources to explore the impacts of climate and potential adaptation and mitigation strategies.

Climate Literacy Crisis

Since 1997, a social scientist - Jon Krosnick, of Stanford University, has tracked Americans' awareness, understanding, and opinions about global warming through periodic random surveys. Between 1997 and 2006, he observed a steady increase in public concern about the issue, which spiked sharply upward between periods 2005-06. He reported, "A vast majority, 85 percent, believes global warming probably is occurring, up slightly from 80 percent in a 1998 poll. But fewer than four in 10 are very sure of it, a level of uncertainty that reflects broad and continued belief that scientists themselves disagree on whether or not it's happening." He notes that, in 2006, 64 percent of Americans perceived "a lot of disagreement" among scientists about whether global warming is happening (Krosnick, 2006). Krosnick's findings are independently supported by Georgetown University Professor Thomas Brewer, who synthesized the results from a variety of surveys conducted between 1989 and 2006 (by ABC/Washington Post, Gallup, ORCA, PIPA, and PPIC) to measure public awareness, understanding and opinions about climate change.

Collectively those surveys reveal a consensus among Americans that global warming is happening, and that they are worried a "great deal" or "fair amount" about it (Brewer, 2006). And yet a section of climate scientists notes that there is a disparity between Americans' concerns about global warming and their perceptions of uncertainty and disagreement among climate scientists. In fact, there is no such disagreement among climate scientists. Earth's average temperature has risen by at least 0.5°C over the last century, climate modelers predict the globe will continue warming through the course of the 21st century, and human emissions of greenhouse gases are major cause of the warming trend—each of these statements are empirically observed, reproducible facts that are not questioned the peer reviewed climate science literature (IPCC, 2007).

Journalists are trained to "balance" their reports by presenting both (or all) sides of a given issue. Though noble in its original intent, this practice, ironically, sometimes introduces the potential for *bias* in reports about climate change research. A recent review of the climate science literature reveals no evidence of controversy among climate scientists about whether the globe has warmed by at least 0.5°C in the last century, nor whether humans are substantial reason for the warming due to the increase in greenhouse gas emissions (Corbett and Durfee, 2004; Boykoff and Boykoff, 2004). Whereas there may be little or no disagreement among the science community about the fact that the globe is warming and humans are largely the cause of it, journalists continue to seek out dissenting or alternative views to preserve their notion (or at least the appearance of) balance in their reports. Such journalistic practice tends to distort climate science in the minds of non-scientists by fomenting the false impression that there is more uncertainty among scientists than there really is (Corbett and Durfee, 2004; Feynman, 1986).

Importance of Climate Science Literacy

During the 20th century, Earth's globally averaged surface temperature rose by approximately 1.08°F (0.6°C). Additional warming of more than 0.25°F (0.14°C) has been measured since 2000. Though the total increase may seem small, it likely represents an extraordinarily rapid rate of change compared to changes in the previous 10,000 years. Over the 21st century, climate scientists expect Earth's temperature to continue increasing, very likely more than it did during the 20th century. Two anticipated results are rising global sea level and increasing frequency and intensity of heat waves, droughts, and floods. These changes will affect almost every aspect of human society, including economic prosperity, human and environmental health, and national security.

Scientific observations and climate model results indicate that human activities are now the primary cause of most of the ongoing increase in Earth's globally averaged surface temperature. Climate change will bring economic and environmental challenges as well as opportunities, and citizens who have an understanding of climate science will be better prepared to respond to both. Society needs citizens who understand the climate system and know how to apply that knowledge in their careers and in their engagement as active members of their communities. Climate change will continue to be a significant element of public discourse. Understanding the essential principles of climate science will enable all people to assess news stories and contribute to their everyday conversations as *informed citizens*.

Guiding Principle of Carbon Literacy

The seven Essential Principles of Climate Literacy are framed by the *'Climate Literacy Handbook: Guiding Principle for informed climate decision'*, which relates to the crucially important social, economic, and environmental solutions required in meeting the climate challenges we and future generations face. On the basis of suggested guidelines humans can take actions to reduce climate change and its impacts in the following manner (http://www.eoearth.org/article/Climate_Literacy_Handbook:_Guiding_Principle_for_informed_climate_decision):

A. Climate information can be used to reduce vulnerabilities or enhance the resilience of communities and ecosystems affected by climate change. Continuing to improve scientific understanding of the climate system and the quality of reports to policy and decision-makers is crucial.
B. Reducing human vulnerability to the impacts of climate change

depends not only upon our ability to understand climate science, but also upon our ability to integrate that knowledge into human society. Decisions that involve Earth's climate must be made with an understanding of the complex inter-connections among the physical and biological components of the Earth system as well as the consequences of such decisions on social, economic, and cultural systems.

C. The impacts of climate change may affect the security of nations. Reduced availability of water, food, and land can lead to competition and conflict among humans, potentially resulting in large groups of climate refugees.

D. Humans may be able to mitigate climate change or lessen its severity by reducing greenhouse gas concentrations through processes that move carbon out of the atmosphere or reduce greenhouse gas emissions.

E. A combination of strategies is needed to reduce greenhouse gas emissions. The most immediate strategy is conservation of oil, gas, and coal, which we rely on as fuels for most of our transportation, heating, cooling, agriculture, and electricity. Short-term strategies involve switching from carbon-intensive to renewable energy sources, which also requires building new infrastructure for alternative energy sources. Long-term strategies involve innovative research and a fundamental change in the way humans use energy.

F. Humans can adapt to climate change by reducing their vulnerability to its impacts. Actions such as moving to higher ground to avoid rising sea levels, planting new crops that will thrive under new climate conditions, or using new building technologies represent adaptation strategies. Adaptation often requires financial investment in new or enhanced research, technology, and infrastructure.

G. Actions taken by individuals, communities, states, and countries all influence climates. Practices and policies followed in homes, schools, businesses, and governments can affect climate. Climate-related decisions made by one generation can provide opportunities as well as limit the range of possibilities open to the next generation. Steps toward reducing the impact of climate change may influence the present generation by providing other benefits such as improved public health infrastructure and sustainable built environments.

Carbon Neutrality

Carbon neutral, also called *carbon neutrality* is a term used to describe the action of organizations, businesses and individuals taking action to remove as much carbon dioxide from the atmosphere as each put in to it. The overall goal of carbon neutrality is to achieve a zero carbon footprint. For example, a business may plant trees in different places around the world to offset the electricity the business uses. This practice is often called *carbon offset* or *offsetting* (http://www.webopedia.com/TERM/C/carbon_neutral.html).

Being carbon neutral involves calculating your total climate-damaging carbon emissions, reducing them where possible, and then balancing your remaining emissions, often by purchasing a carbon offset: paying to plant new trees or investing in "green" technologies such as solar and wind power. The rise of carbon neutral reflects the growing importance of the green movement in the United States. In a CBS News/New York Times Poll in May 2006, 66% of respondents agreed that global warming is a problem that's causing a serious impact now. 2006 also saw the launch of a new (and naturally, carbon neutral) magazine about eco-living, Editor in chief of the New Oxford American Dictionary 2e, said "The increasing use of the word carbon neutral reflects not just the greening of our culture, but the greening of our language. When you see first graders trying to make their classrooms carbon neutral, you know the word has become main-stream (http://blog.oup.com/2006/11/carbon_neutral_/)."

The concept of 'carbon neutrality' is becoming an increasingly popular way for organizations and individuals to demonstrate their concern over climate change. However, the term has been used to describe a variety of different policies, ranging from operational activities (e.g. running a Government department) to the planning of a new housing growth area (http://www.sd-commission.org.uk/pages/carbon-neutrality-and-offsetting.html). Going carbon neutral is a way to reduce the impacts associated to the greenhouse gas (GHG) emissions created during the business or production cycle. Given that climate change is a worldwide phenomenon, the concept of carbon neutrality is based on the principle that a GHG emissions reduction achieved elsewhere has the same positive effect as a reduction made locally. When releasing GHG emissions into the atmosphere, a company can effectively neutralize their global warming impact by purchasing carbon offsets. Offsets are emissions reductions achieved by projects elsewhere, such as energy efficiency investments, wind farms, or solar installations. By purchasing these reductions, a firm's emissions levels and their net climate impact are reduced. Offset purchases will form the final component of a carbon neutral program, supplementing measures taken to reduce the organization carbon footprint, including those related to internal abatement and green power purchases.

Carbon Neutrality Strategy

Many of the world's leading companies are taking steps to voluntarily reach carbon neutrality by reducing their own direct and indirect emissions and offsetting their remaining emissions by developing a portfolio mix of carbon offsets from international markets. Motives to develop a robust corporate or product carbon-neutral strategy may be based on one or more of

Fig. 5.1: Impact of Climate Neutrality Approach on Environmental Quality

Source: http://climateneutralnetwork.org/standards.php

these elements:

Regulatory: By voluntarily measuring and assigning costs to carbon emissions, a company can prepare for a future carbon-constrained economy in which GHG emissions are regulated and/or taxed (http://www.icfi.com/markets/climate-change/doc_files/carbon-neutrality.pdf).

Financial: Effectively managing GHG emissions can help companies generate long-term value by cutting costs, reducing risk, and increasing revenues. Going carbon neutral can be a central part of this management strategy and can help to establish a framework for identifying and pursuing cost-effective emissions reduction and savings opportunities by resorting to the following broad approaches:

Marketing: Demonstrating environmental leadership on the corporate, product, or service level can create strong brand presence, increase customer loyalty, and appeal to new clients and the general public. First movers in the area of carbon neutrality are likely to gain an edge over business-as-usual competitors, particularly as environmental awareness and public opinion on climate change grows.

Corporate Social Responsibility: Neutralizing GHG emissions is a powerful way to show stakeholders (e.g., customers, shareholders, community, nongovernmental organizations) that a company is taking responsibility for emissions and addressing climate change. By going carbon neutral, a firm can have a significant positive impact on world climate.

Strategies and Goals: Companies typically share the same general objectives in pursuing a carbon neutrality strategy; however, well-defined goals can facilitate development of a more tailored strategy. Carbon neutrality can be applied to an entire entity, a facility, a process, or even a product. Within a manufacturing perspective, these strategies should address the trend of environmental procurement criteria, increasingly a component of large corporations' policies. A product's environmental footprint is fast becoming a determining factor as corporations address sustainability down their supply chain. Consumer awareness and desire to mitigate individual environmental impacts also are playing a significant role in the strategy. Specific product lines can go carbon neutral with either the manufacturer or the consumer paying the cost of offsetting carbon emissions through the purchase of the product.

Entity-Based Offsetting: From a corporate perspective, having a strong basis for the offsetting strategy means that an organization should have a clear understanding of the following:

- The projected evolution of business-as-usual GHG emissions in the coming years;
- The volume of GHG emissions reductions achieved per annum through internal reduction programs;
- The timescale for achieving the carbon neutrality strategy; and
- The scope and geographic coverage of carbon neutrality.

With these data, a company can determine how GHGs should be offset through purchases of green energy, and through purchases of renewable certificates and voluntary and regulatory carbon credits. Determination of these volumes is depicted in the following graph, in which each internal or external emissions reduction option is represented by a "wedge."

Product-Specific Offsetting: Companies can choose to offset the emissions associated with a specific product line in one of two ways. The first, and presently most common, approach is to offset the use phase of the product, typically with products that consume energy to perform a function. The energy consumption for an individual unit, over a specific time period, is quantified, as are the associated GHG emissions. Offsets then are purchased to neutralize those emissions. An alternative method examines the emissions associated with the life cycle of the product. Life-cycle analysis quantifies the emissions from raw material extraction, through manufacturing, transportation, use, and eventual disposal. This is considered the most advanced method for quantifying GHG emissions and provides the most comprehensive carbon footprint.

A Carbon Neutral Company

A carbon neutral company is one where net CO_2 equivalent emissions from its activities are zero. In principle, a product or service can also be made carbon neutral. The basic options for achieving carbon neutrality are likely to be a combination of reducing emissions, and - since it will be very difficult to reduce emissions to zero—investing in carbon offset projects. These are projects that enable greenhouse gases to be absorbed or introduce efficiencies or renewable energy to reduce the need to burn fossil fuels. Carbon neutrality means that emissions are reduced; and that remaining emissions are offset (creating or investing in another activity that absorbs the same amount of emissions) (Fig. 5.2). Becoming carbon neutral is part of the process towards being carbon positive. A carbon positive company will be fully engaged in driving the growth of the low-carbon economy, and will shape its own business model, and its products and services to maximize its contribution to reducing CO_2 emissions and promote technologies which support quality of life with low CO_2 emissions.

Steps to Achieve Carbon Neutrality by A Company

The most interesting questions arise when a company asks itself how far it will go, and how fast and how much money it should spend, and what combination of strategies it should adopt. What balance of internal vs. external measures should a company take? Here are the few basic suggested steps to follow to achieve Carbon Neutrality:

STEP 1: Define your Carbon Footprint

The first step in a carbon management process is to determine the extent of your carbon footprint, i.e. your total carbon impact on the environment. Many companies start by calculating their direct operational emissions from electricity, gas and transport use. But there is more to it than that – What about emissions from inputs into the business such as paper, emissions from waste generated, or energy from outsourced activities? If emissions from these types of activities are not included, then it is quite possible that important carbon impacts will be inadvertently omitted.

STEP 2: Minimize your CO_2 Footprint to increase Energy Efficiency

Most companies can achieve savings of 10-15% of their energy use (which directly relates to greenhouse gas emissions through a consistent energy management program and implementing projects which pay back in less than 3 years.

Fig. 5.2: Carbon Neutral and Carbon Positive Balances

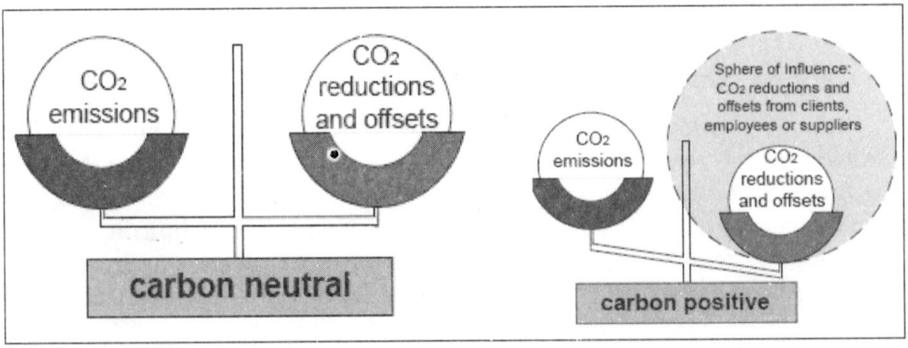

Lowering Process Emissions to generate Green Energy

Renewable energy, such as from new solar and wind power projects, does not rely upon the use of fossil fuels and thus does not increase greenhouse gases in the atmosphere.

STEP 3: Offset your Carbon Footprint

It is important to understand that carbon offsets vary in price and quality, from lower cost offsets from measures such as methane combustion to those from renewable energy. The following are the logical steps to follow for achieving Carbon Neutrality (Fig. 5.3):

Fig. 5.3: Steps of Achieving the Carbon Neutrality

Assess the Carbon Footprint
↓
Account for emission Free electricity purchases
↓
Implement the emission Reduction measures
↓
Compute the remaining Carbon emissions
↓
Purchase GHG offsets
↓
Communicate the carbon Neutrality

Carbon Literacy

Carbon literacy embraces the following attributes (http://en.wiktionary org/wiki/carbon_literacy):

1. refers or means to possess a general knowledge or awareness of the concepts, causes, and the effects of atmospheric pollution or greenhouse gases on the environment, especially in relation to the climate of earth as in climate change or
2. to have the skills necessary to limit or reduce personal and/or collective impact, i.e. carbon footprint, on the environment through more efficient and wise use of available resources and energy;
3. to have the knowledge and skills necessary to participate in personal carbon trading systems and platforms which focus on financial incentives and investments in new methods, systems, and technologies that limit or reduce greenhouse gas pollution, as in carbon capturing

or carbon sequestering, and the research and development of alternative energy sources.

World civilization is dependent on fossil fuels and is starting to recognize the consequences. However we are still predominantly carbon illiterate. Carbon dioxide, the primary greenhouse gas caused by burning fossil fuels, is invisible to all and unknown to many and there are few economic penalties for its release. The key question is how to break this vicious circle and create action. Consumers are becoming more aware but it is unrealistic to expect major change until carbon is properly priced in the economy -which requires government action. Lead Energy/ Power generating companies have the ability to break the cycle by helping raise consumer awareness, challenging government and looking for business opportunities in a carbon-constrained world. There is clearly a need for more research to be done in understanding what exactly encourages people to reduce their emissions; on the extent to which the practice of offsetting has an effect on such behaviour; and on how much it can play a role in educating people about climate change. The biggest-spending consumers in the voluntary offset market are businesses. As with individuals, the motivation for offsetting varies: some companies argue that they recognize the threat of climate change and try to act responsibly; but also some of the motivations for offsetting here will be strategic—they might be to meet Corporate Social Responsibility (CSR) obligations, to generate goodwill, or to attract the growing number of customers attracted by environmental action. The overall framework of suggested Action plan that could be used to bridge the gap between the Carbon Literate vis-à-vis Carbon Illiterate Society is shown below in Fig. 5.4.

Steps to Evaluate the Feasibility of Carbon Neutrality

The following vital steps are involved in properly evaluating the feasibility of Carbon Neutrality in a Production System with environmental/GHG

Fig. 5.4: Gap should be Reduced by Gensol's Marketing

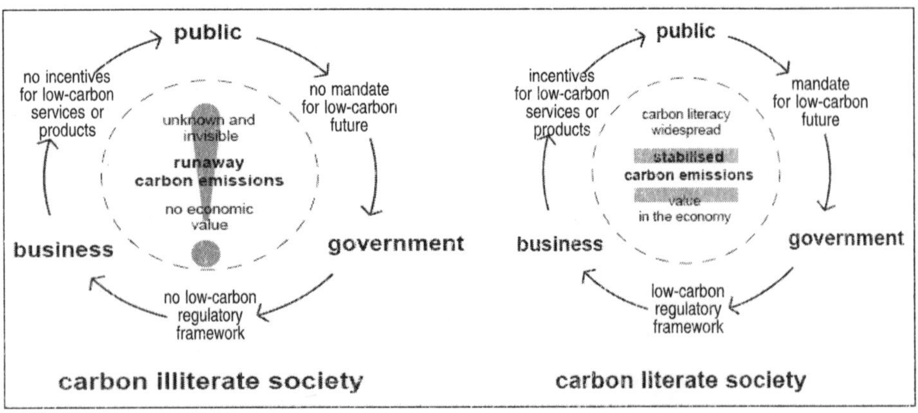

implication:

1. Conduct and create a greenhouse gas emissions inventory and establishment of an emissions baseline;
2. Identify, investigate, and evaluate options for reducing or eliminating emissions from individual sources;
3. Conduct modeling to predict and evaluation future emissions under different reduction scenarios;
4. Develop timeframe and costing models for emission reductions for each source and for selected reduction scenarios;
5. Evaluate the role of offsets in achieving emissions and develop offsetting options;
6. Prepare an implementation plan and costs for the selected approach.

Comprehensive System Approach to Carbon Literacy

No single person is expected to understand every detail about all of the fundamental climate science literacy concepts. Full comprehension of these interconnected concepts will require a systems-thinking approach, meaning the ability to understand complex interconnections among all of the components of the climate system. Moreover, as climate science progresses and as efforts to educate the people about climate's influence on them and their influence on the climate system mature, public understanding will continue to grow.

Climate is an ideal interdisciplinary theme for lifelong learning about the scientific process and the ways in which humans affect and are affected by the Earth's systems. This rich topic can be approached at many levels, from comparing the daily weather with long-term records to exploring abstract representations of climate in computer models to examining how climate change impacts human and ecosystem health. Learners of all ages can use data from their own experiments, data collected by satellites and other observation systems, or records from a range of physical, chemical, biological, geographical, social, economic, and historical sources to explore the impacts of climate and potential adaptation and mitigation strategies.

In the coming decades, scientists expect climate change to have an increasing impact on human and natural systems. In a warmer world, accessibility to food, water, raw materials, and energy are likely to change. Human health, biodiversity, economic stability, and national security are also expected to be affected by climate change. Climate model projections suggest that negative effects of climate change will significantly outweigh positive ones. The nation's ability to prepare for and adapt to new conditions may be exceeded as the rate of climate change increases.

Reducing our vulnerability to these impacts depends not only upon our ability to understand climate science and the implications of climate change, but also upon our ability to integrate and use that knowledge effectively. Changes in our economy and infrastructure as well as individual attitudes, societal values, and government policies will be required to alter the current trajectory of climate's impact on human lives. The resolve of individuals, communities, and countries to identify and implement effective management strategies for critical institutional and natural resources will be necessary to ensure the stability of both human and natural systems as temperatures rise.

A Standard Climate Science literacy document focuses primarily on the physical and biological science aspects of climate and climate change. Yet as nations and the international community seek solutions to global climate change over the coming decades, a more comprehensive, interdisciplinary approach to climate literacy—one that includes economic and social considerations—will play a vital role in knowledgeable planning, decision making, and governance. A fresh effort is needed in development within the social sciences community to produce a comprehensive companion document that will address these aspects of climate literacy. Such documents will hopefully promote informed decision-making and effective systems-level responses to climate change that reflect a fundamental understanding of climate science.

Conducting Carbon Emissions Inventory

For an organization, the term carbon footprint or greenhouse gas inventory includes the carbon emissions and other greenhouse gas (GHG) emissions generated directly from the organization's activities or use of fuels, and also indirectly from the use of electricity and from the use and disposal of materials, products and services. By measuring its carbon footprint or GHG inventory, the organization can manage and reduce emissions over time, and also use it for disclosure to stakeholders or for marketing and corporate social responsibility (CSR) purposes.

After calculating the carbon footprint, it is then possible for the organization to take active steps to manage the emissions. The organization can:

- Set emissions reduction targets
- Identify opportunities for energy efficiency and reduction of emissions
- Take action to implement emissions reduction projects
- Monitor the performance of the projects and improve accordingly

The approach to an organization's carbon footprint usually involves five steps (http://www.lowcarbonsg.com/2009/05/08/measure-your-organisations-carbon-footprint-or-greenhouse-gas-inventory/):

1. Define a consistent methodology
2. Specify the boundary and scope involved
3. Obtain the emissions data and calculate the carbon footprint
4. Verify the results with a third party
5. Disclose the carbon footprint in a report and to stakeholders

The World Business Council for Sustainable Development (WBCSD) and the World Resources Institute (WRI) have published the "Greenhouse Gas Protocol: a corporate accounting and reporting standard". The standard is the most widely accepted set of standards for both calculation of GHG emissions and deciding which carbon sources should be included in an inventory. The Campus Carbon Calculator version 5.0, an excel-based document created by an environmental non-governmental organization Clean-Air Cool-Planet was utilized. The program contains a series of spreadsheets that comprise three general modules: data input, emissions factors, and summary. Basic data is collected from the college's activities and loaded in. The calculator then uses these variables to calculate GHG emissions, and offsets based on conversion factors stored in the emissions factors module. All emission and conversions factors came from the US DOE, EPA or the USDOT. The gross emissions, net emissions, and emissions by source for all GHG emissions are converted into metric tons of carbon dioxide equivalents (MTCDE) and can be viewed in the summary module.

The Challenge of Carbon Neutrality

The challenge of the 21st century will be for human society to regain a healthy relationship with our 'living' earth. The signs are clear that right now we are failing in that task. Unless we change our ways, it is likely that comfortable civilization as we know it will disappear. Climate change is not just an environmental problem—it is a societal problem. The simple truth is that humanity is changing the atmosphere - every day we add around 70 million tonnes of invisible carbon dioxide, the very gas that we know controls the temperature of the planet. Climate change is a critical threat to the continued existence of our civilization. The climate change that we are now seeing—also known as global warming—is caused predominantly by human activities. The largest single factor is the emission of carbon dioxide (CO_2) from the use of fossil fuels. Directly or indirectly, every activity which uses fossil fuel energy will cause the generation of CO_2 emissions and thus contribute to climate change.

Rules for Transfer of Carbon Credits and Allowances

A set of common rules govern who can transfer allowances or credits and who can trade them. To understand these rules, it is important to distinguish between a "trade" of emission rights and a "transfer" and/or "acquisition" of these rights. When an emissions source that possesses an allowance or a credit conveys it to another covered source that uses it to meet its compliance obligations, it is said to "transfer" that allowance or credit to the other source. However, it is possible for the initial holder of the allowance or credit to first "trade" its rights to a broker or other entity before it is eventually transferred to the covered party that will ultimately benefit from it in terms of regulatory compliance.

To illustrate the point, assume that Facility A possesses excess allowances under a cap-and-trade system. It trades the right to these excess allowances to Broker B, who then trades them to Broker C, who trades them to Facility D. Facility D uses them to cover its emissions under the cap-and-trade system. The ultimate result of this transaction is that Facility A has "transferred" some of its allowances to Facility D. Along the way; Facility A "traded" the allowances to Broker B, who in turn traded them to Broker C, who traded them to Facility D. Another way of putting this is that "trading" is what happens when a source grants another entity some type of ownership right in the emissions asset, whereas "transferring" occurs when one source conveys the allowance or credit to another covered party that will employ it to comply with its regulatory obligations. Programs have rules as to who can engage in the "transfer" of allowances. Under the Guidelines for negotiations for Transfer of Carbon Credits a set of rules of the following kind that can limit "trade" in allowances and how they can trade them (http://www.unido.org/fileadmin/import/72285_Kyoto_Protocol_carbon_credits.pdf) can be suggested:

- *Project eligibility*: Some baseline-credit systems set limits on the types of emission reduction projects that are eligible to generate credits under the program. These limits can be geographic, such as a requirement that the project be upwind of the source that purchases the credit. They can also be temporal, such as a requirement that the project occur within a specified timeframe. Other limits may turn on the nature of the project. An example of this in the GHG area would be rules regarding whether parties can generate credits by expanding forests or other "carbon sinks."
- *Company or facility eligibility*: Program rules may also specify which companies or facilities are eligible to generate and/or purchase credits. For example, such rules may require that a facility be in full compliance with air pollution regulations before it will be allowed to purchase emission reduction credits.
- *Baseline calculation:* Baseline-credit programs often define how sources should go about calculating their baseline. For example, such rules may

require sources to calculate their baseline in terms of actual emissions or may let them use allowable emissions. Rules may also define the time period within which baseline emissions should be measured (e.g., any consecutive 12 months within the last two years).

- *Additionality*: Many baseline-credit programs require that, in order to generate credits, reductions be "additional to" those that would have occurred in the normal course of events. Program rules structure the inquiry into which reductions are truly "additional."
- *Public participation*: Some baseline-credit initiatives require public comment on the eligibility of a given project, the additionality of reductions, and/or the validity or magnitude of reductions. Program rules may set out a process for sources to follow in seeking this public participation.
- *Certification*: Most baseline-credit systems establish a process for certifying that a given emission reduction meets program requirements such as those that have just been described. A given reduction must be certified before it can generate credits for transfer or trade.
- *Enforceability*: Programs may require sources to demonstrate that their reductions are enforceable. For example, they may require that a facility incorporate the reductions into a permit or other enforceable legal document. They may also require a showing that the reduction is practically enforceable by the relevant authority.

Carbon Neutrality and Offsetting

Carbon Neutrality: The Case of SDC

The concept of 'carbon neutrality' is becoming an increasingly popular way for organizations and individuals to demonstrate their concern over climate change. However, the term has been used to describe a variety of different policies, ranging from operational activities (e.g. running a Government department) to the planning of a new housing growth area.

The SDC has engaged with stakeholders on the issue of carbon neutrality and carbon offsetting, and this has helped us to arrive at a definition which we hope will be refined over time. The SDC defines carbon neutrality as any product, activity or organization that causes no net increase in CO_2 emissions to the atmosphere under 'business-as-usual' conditions. Hence, carbon neutrality allows emissions to be netted off in some other location, a process which is usually called 'carbon offsetting' (http://www.sd-commission.org.uk/pages/carbon-neutrality-and-offsetting.html). SDC believes that carbon neutrality should be seen as a process that aims to progressively reduce emissions, rather than just a route to carbon offsetting. For organizations, it should therefore lead to the adoption of a comprehensive carbon management strategy, which attempts to quantify and reduce the lifecycle carbon emissions of the operation, service or facility in question. Lifecycle emissions may be defined as the sum of direct and indirect emissions, and would therefore require engagement both up and down the supply chain to ensure maximum impact. Further guidance on carbon neutrality and the importance of good carbon management is available from the Carbon Trust.

Carbon Offsetting: The Case of SDC

The use of carbon offsetting is gaining momentum. A useful definition of carbon offsets has been developed by the Carbon Trust:

> "Carbon offsets are generated from projects that avoid or absorb/sequester carbon dioxide, or any of the other main greenhouse gases. These projects can take various forms, including renewable power, energy efficiency, fuel switching (e.g. from oil to natural gas), reforestation, or destruction of greenhouse gases (e.g. methane, HFC 23)."

Carbon Offsetting has the following characteristics:

- "The purchase of credits from greenhouse gas emission reduction projects in one place to counter the emissions of greenhouse gases in another place"
- Growth of voluntary off-set market: worth >US$91m (Hamilton et al., 2007); ca. 15% of market are individual/ private consumers (3% of UK population; DEFRA, 2007)
- Initial research (with corporate consumers) suggests rationales for off-setting are diverse; but there are also ethical and technical criticisms.

Carbon offsets (sometimes called 'carbon credits') are available from two quite separate sources: the compliance market, which is a product of the legal instruments created to support the UNFCCC and the Kyoto Protocol, and the voluntary market, which has developed in response to consumer and business demand for carbon offsetting services.

However, it is not currently possible to do carbon offsetting in the UK, as this would lead to the double-counting of any reduction in emissions (as all reductions are already claimed by Government in helping to meet our international obligations). For organizations that want to keep funds within their organization or community, the SDC recommends they look at alternative mechanisms, such as investment in 'revolving funds' (for example, the public sector revolving fund managed by Salix Finance).

It is important to note that carbon offset projects would not be necessary if there was an universal cap on international greenhouse gas emissions. However, in the absence of a viable international framework covering all countries or sectors, there is a potential role for carbon offsetting in helping to stimulate additional low carbon investment in developing countries whilst providing low cost emissions reductions for the offset purchaser. As stated above, carbon offsetting should be done as part of a comprehensive carbon management strategy. For these reasons, the SDC cautiously supports the use of carbon offsetting in the absence of a global cap on emissions.

There are a number of sustainable development benefits that offsetting can deliver to less developed countries, such as the provision of new sources of energy and a reduction of local air pollution. This was the logic behind the establishment of the Clean Development Mechanism (CDM) as part of the Kyoto Protocol. Furthermore, carbon offsetting may raise carbon awareness (or 'carbon literacy') among individuals and businesses, helping to put a cost (albeit a small one) on carbon-emitting activities. The SDC's work in this area, through the Sustainable Consumption Roundtable (in conjunction with the National Consumer Council), suggests that actions such as carbon offsetting could act as a 'practical catalyst', helping to drive other actions towards more sustainable behaviour. It is also important that carbon offsets deliver verifiable reductions in greenhouse gas emissions in a way that is consistent with the principles of sustainable development. This means that organizations and individuals need to exercise caution when purchasing carbon offsets to ensure they are of a sufficiently high standard. Initiatives such as the Gold Standard for CDM and voluntary market carbon offsets are particularly welcome, and an increasing number of carbon offset providers are offering credits that are independently certified.

The Challenge of Standardizing Carbon Offsets

'Tradable GHG reductions' are often purchased to offset or *"neutralize"* a company's carbon dioxide emissions, and are therefore commonly referred to as *"carbon offsets."* Carbon offsets are a unique commodity. In essence, they are a claim or property right to a specific, verified reduction in GHG emissions. Carbon offsets originate from projects or activities (i.e., "offset projects") explicitly designed to reduce GHG emissions. In purchasing and retiring an offset, a company in effect makes the claim that it has foregone reducing its own emissions, but is instead paying someone else to achieve the same net result.

Carbon offsets are intangible, and so their integrity depends on assurances about how well they meet certain defining criteria. It is commonly agreed that a carbon offset commodity must represent GHG emission reductions that are:

- *Real*—Quantified GHG reductions should represent actual emission reductions and not artifacts of (incomplete) accounting.
- *Surplus*—GHG reductions should be a response to the buyer's promise to purchase them, not reductions that would have happened anyway under "business as usual" (i.e., they should be "additional").
- *Verifiable*—GHG reductions should result from projects whose performance can be readily monitored and verified.
- *Permanent*—GHG reductions (or removals, in the case of sequestration) should be permanent, and/or be backed by guarantees if they are reversed, i.e., re-emitted to the atmosphere.
- *Enforceable*—GHG reductions should be backed by contracts or legal

instruments that define their creation, provide for transparency, and ensure exclusive ownership.

The challenge, of course, is elaborating these basic criteria into an actual standard that ensures a uniform commodity. What constitutes a sufficient level of assurance that offset reductions are real, surplus, verifiable, permanent, and enforceable? What requirements can provide confidence that a ton of GHG reductions from one project is the same as a ton from another?

In fact, three related sets of *"standards"* are necessary to fully define a carbon offset commodity:

1. *Procedural and technical standards:* These are standards related to the validation, monitoring, and verification of offset projects, as well as the certification and crediting of GHG reductions. Procedural and technical standards ensure that offsets are verifiable.
2. *Contractual standards:* These are standards for the establishment and transfer of property rights related to GHG reductions, and for information disclosure. They can include terms for payment and delivery, allocation of risk, and compensation where GHG removals or reductions are reversed or not realized. Contractual standards are necessary to avoid double-issuance and double-counting of GHG reductions, and ensure that offsets are enforceable.
3. *Accounting standards:* These are standards related to the actual quantification of GHG reductions. Accounting standards specify methods for defining quantification boundaries, estimating baseline emissions, and correcting for unintended changes in emissions (i.e. *"leakage"*). Accounting standards also encompass methods for demonstrating that a project would not have happened anyway (i.e., "additionality"). Finally, they may specify methods for treating non-permanent GHG removals on equal footing with permanent GHG reductions. Accounting standards are a first-order requirement for ensuring that "a ton is a ton" and ensure that offsets are real, surplus, and permanent.

In addition to defining these various standards, many carbon offset programs will impose eligibility criteria for offset projects intended to ensure that they are compatible with goals beyond simply reducing GHG emissions. Eligibility criteria may exclude certain types of projects based on secondary environmental or social concerns (e.g., nuclear waste, or community displacement caused by hydro- reservoirs), or they may ensure that projects contribute to additional social, economic, and environmental objectives (e.g., "sustainable development"). While these criteria are ancillary to defining a GHG reduction, they nevertheless help to define or brand a particular carbon offset commodity, and are therefore germane to carbon offset "standards" and the linking of different markets.

Carbon Philanthropy: A New Approach to Carbon Management

The arrival of carbon emissions on the personal and corporate landscape has seen a sudden rise in the number of providers offering offset products to enable emitters to address their emissions. The voluntary offset market mirrors the essential elements of the cap and trade process for carbon emissions agreed under the Kyoto Protocol and two years ago was seen as a sensible and logical approach to excess carbon emissions. So much so, that major companies rushed to buy offsets and then make claims of "carbon neutrality", individuals felt less guilty about air travel and car use, and service providers used offsets to enhance the "green" credentials of their services (Kimber, 2008). This rise of offsets has been accompanied by greater carbon literacy amongst companies and consumers, with hard questions being asked, not just of the offset providers, but also of the relevance of offsets in the wider carbon emissions solution/ climate change challenge. At a simple level the idea of buying a solution to a major global challenge, without any reduction in emissions creation, seemed too good to be true.

There is no doubt that the deployment of voluntary offsets has helped to raise awareness of emissions values and the need to address excess carbon emissions and their effects, but for most corporate and individual consumers there is still a low level of understanding of the products available and ability to evaluate relative merits and value. This is a problem in a voluntary market, where the purchase of an offset is a voluntary donation is effectively a charitable donation. If the motive for offset purchase is to voluntarily transfer money into mitigation of the causes and effects of climate change, there is a clear argument for making donations, some of which may be to offset projects, but abandoning the notion of matching gram for gram emissions to offset products in the voluntary sector, where methodology of both calculation and projects is not always clear, nor the administrative charge of the provider.

Real Cost of Carbon: The Case of DEFRA

By using the two main international and domestic methodologies for carbon measurement, the GHG Protocol and DEFRA respectively values could be created for carbon emissions that could be converted into monetary value at the prevailing social/economic cost of carbon. DEFRA currently use the Shadow Price of Carbon and value this at £26 per ton, greater than the market price of EUAs or most voluntary offsets. This is higher than the social and economic costs of carbon used by Stern in his assessment of the economic impact of climate change. This higher value would reflect the wider cost of carbon and not just the delivery price of an offset to the purchaser, which can vary considerably. It would further provide assurance that purchasers were taking full responsibility for their emissions ((http://www.teqs.net/ DEFRAPFSresponse.pdf).

The use of a monetary value conversion for carbon would provide companies especially, with a fixed value for budgeting and assessing potential carbon cost risks to their business, but would also provide a notional financial value that could inform their corporate philanthropy programmes. Environmental charitable giving comprises less than 8% of charitable giving in the UK and yet the charities that are working to prevent climate change will be increasingly pressured as the consequences of increased emissions are felt. Increased drought, flooding, disease that may result from climate change will also put greater pressure on those charities who specialize in disaster relief and who are, all too frequently, the front line of international aid efforts, whilst money/aid pledged from governments and agencies abroad takes time to arrive. Companies using this mechanic would be able to amend their corporate philanthropy programmes to support climate change mitigation, using recognized charities whose operations and reporting requirements are supported by legislation. The secondary value of this approach, no less important from a corporate perspective is the embedding of carbon as a cost of operation, using a fixed conversion. The reporting required to support carbon-based philanthropy will enable corporate decision-making based upon the cost of services in relation to their carbon cost. Currently, carbon emissions reduction and carbon efficient purchasing are not widely enabled by corporate reporting or MIS.

This carbon philanthropy approach will not immunize companies from criticism with regard to their environmental impact, but coupled with clear emissions reporting and monetary conversion will give their investors, stakeholders, employees and customers a greater understanding of the level of responsibility taken by them. For companies who are actively engaged in carbon reduction programmes, the fixed conversion value will provide a mechanic for carbon budgeting and clear long term financial comparison of performance year on year. This carbon budgeting will, in simple terms, mirror national allocation allowances and provide a pre regulatory framework for carbon management. It will further have the merit of preparing companies for the increased regulatory environment for carbon emissions that is the likely consequence of the continued failure of voluntary activity to stem the increase in carbon emissions.

The Full-Cost of Carbon Reduction

Within carbon budgeting there will be a greater understanding of the real cost of purchasing. Currently a company purchasing air travel will receive reports from suppliers on the cost of the services purchased and from a carbon perspective will see short haul travel emissions reduction as a switch from air to rail. Whilst the per passenger emissions values per kilometre support this view, the difference in cost between peak domestic rail fares and low cost carrier flight can result in the company paying a high cost for its emissions reduction on the journey. To provide a full analysis for decision-making the cost of each journey option needs to be assessed to ensure a full understanding of the cost of carbon and its reduction to a company. This will enable companies to demonstrate the costs that they have incurred in choosing lower emitting options for power and travel, a more accurate reflection of the costs of carbon reduction to a company than the simple use of the annual emissions values as a financial cost based upon offset purchase. Investors and consumers will be able to differentiate between the cost and ROI of carbon reduction measures and to establish the difference between "business as usual" emissions supported by offset payments and the value of reduction and investment in reduction by companies. The latter is the sustainable long-term approach that will deliver environmental and corporate benefits.

For some companies, the carbon philanthropy approach will provide an opportunity for them to create their own internal funding systems based upon the SPC, to create internal corporate funds that can be used for the reduction of carbon emissions. Examples already exist of government agencies and private companies where employees vote for and bid for money from such internal funds to support projects to reduce the overall emissions of their company. This internalized approach is a voluntary and thus primarily philanthropic approach, but one which enables employee mobilization and engagement with corporate carbon reduction. The funds generated for such programmes fall into philanthropy and not operating cost, even though they may have a significant effect on both operating costs and environmental impact, which will benefit the company. The internal model for the voluntary distribution of funding could exist hand in hand with donations to external agencies that reflect the core values or a company, its geographical scope of operations or location.

Towards Achieving Carbon Neutral Milestone

Carbon emissions reduction and mitigation is a voluntary activity for the vast majority of companies and the current use of offset as a primary strategy is likely to create significant problems for companies who have elected to aim for short term "carbon neutral" goals. In 2006 and much of 2007 most consumers and investors were not sufficiently carbon literate to question these aims, although there was an uneasiness about the process of buying environmental absolution. 2008 has seen an increase in companies withdrawing from offset purchase and greater questions being raised around the business as usual emissions of "carbon neutral" companies. Reduction of emissions will be the hallmark of carbon responsible companies and the use of a transparent carbon philanthropy mechanic, supported by clear cost metrics, to reflect the voluntary financial commitments of companies, either through support for charities and/or internal emissions reduction programmes, will deliver tangible benefits that investors, employees and consumers can identify. The use of a financial mechanic for reporting carbon will be helpful preparation for a regulatory environment, support the important link between carbon efficiency and cost efficiency, thereby bringing carbon into business operations as an environmental and cost metric.

Principles and Technologies of Carbon Management

The total amount of carbon on Earth is constant, some of it existing as pure forms of the element, such as coal, but most of it existing in compounds with other elements, such as oxygen. The main 'reservoirs' for carbon in the environment are the atmosphere, the oceans, the ground and the biosphere of plants and animals, that includes humans. Carbon in the environment moves between these reservoirs by natural processes of the carbon cycle such as plant growth and decay, ocean warming and cooling. Most of this exchange is via the atmosphere where the carbon exists mainly as the gas carbon dioxide (CO_2), in a very small proportion compared to the main gases of the atmosphere. But this small proportion of carbon dioxide has a controlling effect on the greenhouse action which is warming the Earth's climate.

Our activities of burning fuels to supply energy for transport, buildings and manufacturing emit amounts of carbon dioxide that far exceed those generated by natural fires or volcanic activity. Controlling the quantities of carbon fuels we use, together with the methods used to burn them, will also control the amount of carbon dioxide emitted into the atmosphere. This 'carbon management' is therefore interdependent with energy management (http://knol.google.com/k/randall-mcmullan/energy-use-and-carbon-management/1hvmbypv7oiib/4#).

Carbon management is a key resource management and policy issue of the 21st century. The atmospheric concentration of CO_2 increased by about 25% during the 20th century and is continuing to increase in part due to the burning of fossil fuels and changes in land cover and land use. Increases in the atmospheric concentration of CO_2 and other greenhouse gases are likely to produce significant changes in global climate and accompanying changes in the energy and water cycles. These changes will have profound impacts on the Earth's ecosystems and society.

New technologies for reduction of carbon emissions and storage of carbon deep underground or in the oceans are two principal options to achieve long-term reduction in the concentration of CO_2 in the atmosphere. As these technologies are being developed and applied, a reduction in the rate of increase in atmospheric CO_2 may be possible by increasing storage of carbon in soils, above ground biomass and aquatic environments. Such carbon sinks already absorb about 50% of the carbon emitted into the atmosphere annually. Researches at NASA indicate that changes in land management practices can enhance terrestrial and oceanic carbon sinks (http://nasascience.nasa.gov/earth-science/applied-sciences/national-applications/carbon-management).

In developing a science-based carbon management regime, knowledge of the carbon cycle is fully exploited to assist operational agencies fulfill their mandates to manage carbon and to support local, regional, national, and global policy and planning for control of carbon in the environment. The application element also collaborates with academic and government laboratories and operational agencies in the development, testing, and implementation of new technologies for measuring, monitoring, and validation of carbon management practices.

Carbon management is now at the heart of tackling climate change and has rapidly become a central part of the global business environment (http://www.geos.ed.ac.uk/masters/cman_info/). Carbon Management is the process of understanding how and where an organization's commercial activities generate greenhouse gas emissions (GHG), in order to then minimize these emissions in an ongoing and financially sustainable way. It extends from both internal activities to the consumption of an organization's products, and ultimately is about incorporating an understanding of carbon data into strategic corporate decision-making.

The volume of global trade has more than doubled in the last decade, reaching six times the rate of growth in the world's GDP during that time. This phenomenon has been fueled by relatively cheap energy with little attention given, unfortunately, to the impact on climate change.

With an estimated economic damage of about US$85 for each ton of carbon dioxide, capping greenhouse gas emissions—and putting a price tag on them —became inevitable. Indeed, under the European Union (EU) emissions trading scheme (EU ETS), such a setup is already in effect for certain industries. Similar schemes are popping up across the United States in separate groups of states and in other industrial economies worldwide (Bramante *et al*, 2009).

According to the 2008 IBM Global CEO Study and a separate IBM study, corporate social responsibility (CSR)—of which carbon control is a significant part—is at the top of the agenda for chief executives. It is fast transcending its traditional compliance and philanthropy image and is becoming a revenue opportunity for businesses as customers increasingly demand transparency and accountability from the organizations with which they do business

(http://www.sdcexec.com/web/online/In-Depth/Mastering-Carbon-Management/4$10733).

Proactive Carbon Management provides:

- Legislation compliance
- Reduction in operational costs
- Streamlined operations
- Improved asset utilization
- The ability to meet growing community and consumer demands to meet corporate social Responsibilities
- Lowering negative impact on the global environment

At a national and global level, climate change has come to the fore as a key environmental sustainability issue. Many governments are taking steps to reduce GHG emissions through national policies including the introduction of mandatory emissions trading programs, voluntary programs, carbon or energy taxes, and regulations and standards on energy efficiency and emissions.

As a result, today's organizations must understand the GHG risks, to ensure long-term success in a competitive business environment, and to be prepared for future national or regional regulations. This document identifies the key business requirements in implementing carbon management for the Australian National Greenhouse and Energy Reporting (NGER) Act 2007 – which came into effect July 1, 2008. It also explains why carbon management is a critical undertaking for businesses by answering key executive questions (http://www.climatechange.gov.au/reporting/index.html):

- How can effective carbon management increase profit?
- What premium will you have to pay to address climate change if you do not make the deadline for NGER reporting?
- Are your customers and stakeholders demanding to know what action you are taking to measure and reduce your GHG emissions.
- Does your corporate social responsibility (CSR) policy include a carbon reduction strategy?

Of any social issue, the climate change debate has had the most significant impact in escalating the importance of CSR. The growing body of scientific evidence, amplified through credible communicators – particularly Al Gore16, and the Stern Review on climate change17 – has led to a number of CEOs reviewing how radically they will need to change. And an unprecedented high number are concluding that the answer is a great deal more profound than they could ever have imagined.

The Economist Intelligence Unit in 2006/2007 conducted a survey of 634 global business executives resulting in a report called 'A Change in the Climate' (http://www.sweden.se/upload/promotion_forum/Brand_Sweden/Bevis/Internationell_ranking_axplock.doc). It found that European and Asia Pacific companies are ahead of the curve in addressing carbon emissions and revealed that companies want government regulations to shape how they progress. When Asia Pacific corporation executives were asked about incorporating carbon management in a CSR strategy, the key findings found were as follows:

- Business is not keeping up with the changing public mood. Less than 10 per cent of executives admit that their organizations monitor their overall carbon footprint and just 18 per cent have a carbon reduction plan in place. Although these numbers look set to rise rapidly, nearly one-half of firms have no intention of implementing carbon reduction plans within the next three years. So, companies seeking competitive advantage will need to start adopting carbon management best practices.
- Business is reacting to reputation risk, not exploiting business opportunities.
- Companies do not expect the costs to be high.

Businesses engaged in carbon reduction are spending only about 0.6 per cent of their operating expenses in this area. By 2010, more than half expect to either have no costs or else result in a net positive impact.

- Companies starting out on carbon reduction face a steep learning curve
- Government has a crucial role to play. Government regulation is the single largest factor in shaping how companies address carbon issues.

Business Case for Carbon Management

The business case for taking action on carbon management is continuing to grow and take momentum. The challenge for Irish enterprises is to become aware of the implications of climate change for their business in order to turn potential risks into business opportunities. Proactive businesses who take action to reduce their carbon footprint can make significant gains from the associated increases in internal efficiency and from reductions in resource costs. Other benefits include enhanced reputation, stakeholder support and improved environmental compliance. Tackling climate change can therefore result in a positive outcome both for the environment and for business.

When looking at certain specific business case for taking action on carbon management, there are some key factors which should be taken into consideration:

- Cost Savings
- Corporate Social Responsibility
- Staying ahead of regulations
- Direct impact of climate change

From management's and the accountants' point of view, the financial case for carbon management will more than likely be the most important. The other benefits, outlined above are less quantifiable in financial terms. Nevertheless, these factors are extremely important and to neglect them may result in financial implications.

Cost Savings

There are many opportunities to save money through the implementation of a robust, practical and suitable carbon management plan for your business. Some cost saving opportunities will be short term while others will be medium to long-term savings which may require some initial investment. When assessing the business case, the length of pay-back time which you prepared to accept should be carefully considered.

Cost savings can be achieved through some of the following:

- Decreasing energy use (including transport) through increased operational efficiency, increased energy awareness among staff and installation of more energy efficient technology.
- More efficient use of resources leading to less costs in terms of

procurement, water charges, waste charges etc.

- Being prepared if a carbon tax is implemented which will mean reduced costs than if you had not made any adjustments to your activities.
- In the long-term, energy prices are likely to rise—securing alternative energy sources and becoming more energy efficient will mean that your organization is less susceptible to the financial implications this may bring.

The above scenarios are relatively easy to quantify and you should work with the appropriate personnel within your organization to develop your own business case model.

CO_2 Capture Technology (CCT) for Minimising Emission Hazards

Meeting the world's growing demand for affordable, secure and convenient energy while reducing Green House Gas emissions—particularly CO_2—is the main challenge facing our industry and more particularly the energy sector today. Global energy demand is expected to double over the next quarter of a century and fossil fuels will dominate the supply of energy for the next three decades. In 2008, the G8 shared a goal of reducing global greenhouse gas emissions by at least 50% by 2050, and recognized Carbon Capture and Storage (CCS) as one of the options to reduce emissions.

While CCS won't solve society's energy and environmental problems on its own, it provides mean to make use of the abundant fossil fuels needed in the energy mix which could make an important contribution over this timeframe. Currently nearly all of the CO_2 injected underground - nearly 0.5 gigatons (Gt) of CO_2 thus far – has been a consequence of CO_2 flood enhanced oil recovery; however, this amounts to a fraction of a percent of the world's CO_2 emissions from fossil fuels. For CCS to play a major role in reducing CO_2 emissions, it should be increased significantly (http://www.spe.org/events/09adb2/documents/09ADB2_brochureFINAL.pdf).

Capturing CO_2 might be applied to large point sources, such as large fossil fuel or biomass energy facilities, industries with major CO_2 emissions, natural gas processing, synthetic fuel plants and fossil fuel-based hydrogen production plants. Air capture is also possible. But carbon dioxide contains oxygen, and so capturing and storing CO_2 would slow down the oxygen cycle in the biosphere. Concentrated CO_2 from the combustion of coal in oxygen is relatively pure, and could be directly processed. In other instances, especially with air capture, a scrubbing process would be needed. The main challenge regarding CO_2 capture technology is to reduce the overall cost by lowering both the energy and the capital cost requirements. While costs and energy requirements for today's capture processes are high, the opportunities for significant reductions exist, since researchers have only recently started to address these needs. One strategy that looks extremely promising is to combine CO_2 removal with advanced coal energy conversion processes that have features which will enable low energy intensive capture (Herzog *et al*, 1997: CCT-MIT: http://sequestration.mit.edu/pdf/WhitePaper.pdf).

Broadly, three different types of technologies exist: post-combustion, pre-combustion, and oxy-fuel combustion.

- In *post-combustion capture*, the CO_2 is removed after combustion of the fossil fuel - this is the scheme that would be applied to conventional power plants. Here, carbon dioxide is captured from flue gases at power stations or other large point sources. The technology is well understood and is currently used in other industrial applications, although not at the same scale as might be required in a commercial scale power station.
- The technology for *pre-combustion* is widely applied in fertilizer, chemical, gaseous fuel (H_2, CH_4), and power production. In these cases, the fossil fuel is partially oxidized, for instance in a gasifier. The resulting syngas (CO and H_2) is shifted into CO_2 and more H_2. The resulting CO_2 can be captured from a relatively pure exhaust stream. The H_2 can now be used as fuel; the carbon dioxide is removed before combustion takes place. There are several advantages and disadvantages when compared to conventional post combustion carbon dioxide capture.
- In *oxy-fuel combustion* the fuel is burned in oxygen instead of air. To limit the resulting flame temperatures to levels common during conventional combustion, cooled flue gas is recirculated and injected into the combustion chamber. The flue gas consists of mainly carbon dioxide and water vapour, the latter of which is condensed through cooling. The result is an almost pure carbon dioxide stream that can be transported to the sequestration site and stored. Power plant processes based on oxyfuel combustion are sometimes referred to as "zero emission" cycles, because the CO_2 stored is not a fraction removed from the flue gas stream (as in the cases of pre- and post-combustion capture) but the flue gas stream itself. It should be noted, however, that a certain fraction of the CO_2 generated during combustion will inevitably end up in the condensed water. To warrant the label "zero emission" the water would thus have to be treated or disposed of appropriately. The technique is promising, but the initial air separation step demands a lot of energy.

The following flow diagram (Fig. 5.5) provides the overview of CCT Processes and Systems (http://www.greenfacts.org/en/co2-capture-storage/figtableboxes/figure-3.htm):

An alternate method, which is under development, is chemical looping combustion (CLC). Chemical looping uses a metal oxide as a solid oxygen carrier. Metal oxide particles react with a solid, liquid or gaseous fuel in a fluidized bed combustor, producing solid metal particles and a mixture of

Fig. 5.5: An overview of CCT Processes and Systems Integration Pathways

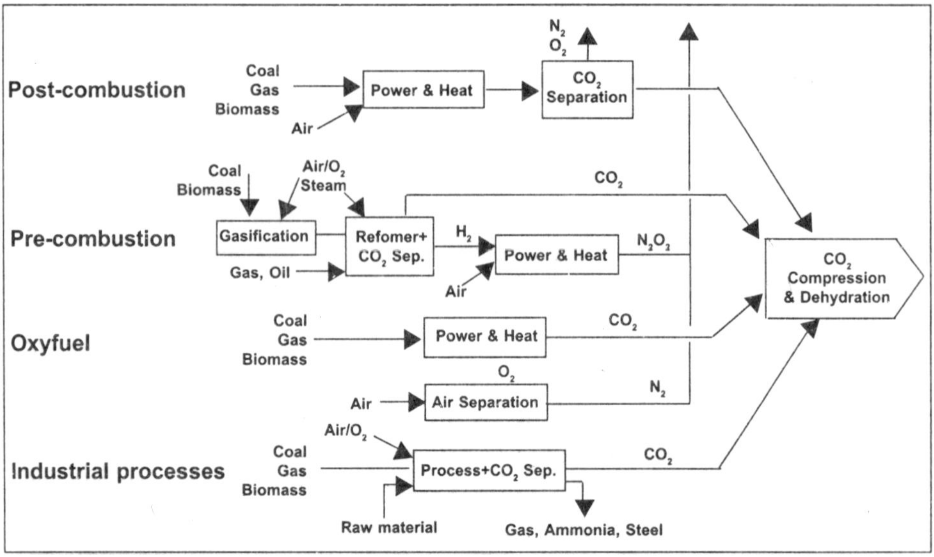

carbon dioxide and water vapour. The water vapour is condensed, leaving pure carbon dioxide which can be sequestered. The solid metal particles are circulated to another fluidized bed where they react with air, producing heat and regenerating metal oxide particles that are recirculated to the fluidized bed combustor. A variant of chemical looping is calcium looping, which uses the alternate carbonation and then calcination of a CaO based carrier as a means of capturing CO_2.

A few engineering proposals have been made for the more difficult task of capturing CO_2 directly from the air, but work in this area is still in its infancy. Global Research Technologies demonstrated a pre-prototype in 2007. Capture costs are estimated to be higher than from point sources, but may be feasible for dealing with emissions from diffuse sources like automobiles and aircraft. The theoretically required energy for air capture is only slightly more than for capture from point sources. The additional costs come from the devices that use the natural air flow. Removing CO_2 from the atmosphere is a form of geo-engineering by greenhouse gas remediation. Techniques of this type have received widespread media coverage as they offer the promise of a comprehensive solution to global warming if they can be coupled with effective carbon sequestration technologies.

It is more usual to see such techniques proposed for air capture, than for flue gas treatment. Carbon dioxide capture and storage is more commonly proposed on plants burning coal in oxygen extracted from the air, which means the CO_2 is highly concentrated and no scrubbing process is necessary. According to the Wallula Energy Resource Center in Washington State, by gasifying the coal, it is possible to capture approximately 65% of carbon dioxide embedded in coal and sequester them into the solid form.

Carbon Capture and Storage (CCS) Mechanism

Carbon capture and storage (**CCS**) is a theoretical approach to mitigating the contribution of fossil fuel emissions to global warming, based on capturing carbon dioxide (CO_2) from large point sources such as fossil fuel power plants. It can also be used to describe the scrubbing of CO_2 from ambient air as a geo-engineering technique. The carbon dioxide might then be permanently stored away from the atmosphere. The term *carbon dioxide capture and storage* has also been used to describe biological techniques such as biochar burial, which use trees, plankton, etc. to capture CO_2 from the air. However, it is more conventional to use the term *carbon capture and storage* to describe non-biological processes of capturing carbon dioxide from combustion at the source. The Carbon Capture and Storage (CCS) Mechanism showing both terrestrial and geological sequestration of carbon dioxide emissions from a coal-fired plant is given in Fig. 5.6.

Although CO_2 is injected into geological formations for various purposes, the long term storage of CO_2 is a relatively untried concept. The first integrated pilot-scale CCS power plant was to begin operating in September 2008 in the eastern German power plant Schwarze Pumpe in the hope of answering questions about technological feasibility and economic efficiency.

It has been theorized that CCS applied to a modern conventional power plant could reduce CO_2 emissions to the atmosphere by approximately 80-90% compared to a plant without CCS. The IPCC estimates that the economic potential of CCS could be between 10% and 55% of the total carbon mitigation effort until year 2100 (Section 8.3.3 of IPCC report).

Fig. 5.6: Schematic showing both Terrestrial and Geological Sequestration of Carbon Dioxide Emissions from a Coal-fired Plant

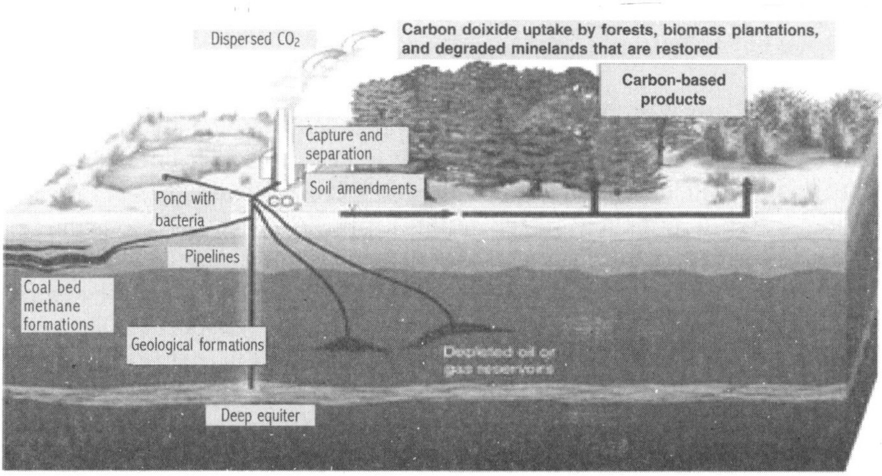

Source: http://en.wikipedia.org/wiki/Carbon_capture_and_storage

Capturing and compressing CO_2 requires much energy and would increase the fuel needs of a coal-fired plant with CCS by 25%-40%. These and other system costs are estimated to increase the cost of energy from a new power plant with CCS by 21-91%. These estimates apply to purpose-built plants near a storage location: applying the technology to preexisting plants or plants far from a storage location will be more expensive. However, recent industry reports suggest that with successful research, development and deployment (RD&D), sequestered coal-based electricity generation in 2025 will cost less than un-sequestered coal-based electricity generation today.

Storage of the CO_2 is envisaged either in deep geological formations, in Deep Ocean masses, or in the form of mineral carbonates. In the case of deep ocean storage, there is a risk of greatly increasing the problem of ocean acidification, a problem that also stems from the excess of carbon dioxide already in the atmosphere and oceans. Geological formations are currently considered the most promising sequestration sites. In its 2007 *Carbon Sequestration Atlas*, the National Energy Technology Laboratory (NETL) reported that North America has enough storage capacity at its current rate of production for more than 900 years worth of carbon dioxide. A general problem is that long term predictions about submarine or underground storage security are very difficult and uncertain and CO_2 might leak from the storage into the atmosphere.

When used on plants which use biomass, CCS is known as bio-energy with carbon capture and storage. This has the potential to be used as a carbon sequestration technique, and can thus be regarded as geo-engineering (http://www.netl.doe.gov/publications/press/2007/07016-Carbon_Sequestration_Atlas_Publish.html). The Department of Energy's Regional Carbon Sequestration Partnerships have identified the power plant and other stationary sources of more than 3.8 billion tons a year of the greenhouse gas CO_2 in the United States and Canada and companion candidate storage capacity for more than 3,500 billion tons.

Transportation Mechanism

The captured carbon dioxide next needs to be moved to a storage site. Two methods of less than 1000km, carbon dioxide gas can be pressurized so that it acts like a liquid and then moved through pipelines. Several thousands of kilometers of pipeline for this purpose already exist in the US and Canada. For very long distances, carbon dioxide could be transported as a liquid by ship – similar to the way in which liquefied natural gas is transported. There are energy costs associated with both methods, since pressurizing the carbon dioxide gas into a liquid requires energy, and surface transport has associated fuel costs. After capture, the CO_2 would have to be transported to suitable storage sites. This is done by pipeline, which is generally the cheapest form of transport. In 2008, there were approximately 5,800 km of CO_2 pipelines in the

United States, used to transport CO_2 to oil production fields where the CO_2 is injected in older fields to extract oil. The injection of CO_2 to produce oil is generally called "Enhanced Oil Recovery" or EOR. In addition, there are several pilot programs in various stages to test the long-term storage of CO_2 in non-oil producing geologic formations.

Storage of Carbon Dioxide and Natural Gas (Sequestration)

Once the carbon dioxide has been captured, it needs to be stored on a long-term basis, for many thousands of years. The volumes of gas produced make building special storage facilities uneconomical. Natural geological formations have been identified as potential storage sites, specifically in rocks under the land or under the seabed. It is difficult to bury carbon dioxide though as it is light and buoyant so will tend to float to the surface. The right storage site needs to be more than 800m deep, so that the natural pressure is high enough to keep the carbon dioxide in its liquid form.

Secondly, the right site needs a particular rock formation – a layer of porous rock that can act as a sponge and soak up the liquid carbon dioxide, topped off with a layer of non-porous 'cap rock' that can provide a 'lid' to keep the carbon dioxide inside. Disused oil and gas fields appear to be ideal, as they have natural geological 'lids' that stopped the oil rising to the surface. They are also fairly well understood—oil producers in Hungary and the USA have been injecting carbon dioxide into declining oilfields for 30 years as it helps to displace hard to reach oil and so increases the amount of oil that can be collected. This is known as Enhanced Oil Recovery (EOR) and can produce 5-15% extra oil. Un-mineable coal seams could also provide useful stores for carbon dioxide. Underground, coal is coated by a layer of methane. When you inject carbon dioxide in, the carbon dioxide will stick to the coal surface, replacing the methane, which can be captured and sold as a fuel. This technique is still under development.

Deep underground stores of salt water called saline aquifers could be another storage site for carbon dioxide. These aquifers currently appear to have little value to humans as the salt and minerals in the water makes it undrinkable. Carbon dioxide pumped into the aquifers will displace the water in the pores of the rocks and then dissolve in the water. Over a timescale of millennia, it will eventually react with the dissolved salts and minerals to create solid carbonate materials similar to lime scale. Importantly, saline aquifers are common offshore from the UK, reducing the need to transport carbon dioxide over long distances. They also have massive storage capacity—recent assessments suggest that the total UK offshore storage capacity in depleted oil and gas fields could be equivalent to about 40 years worth of the UK's total carbon dioxide emissions at 2006 rates. Storage capacity in all accessible saline aquifers is expected to be equivalent to several centuries' worth of total UK carbon dioxide emissions.

Relatively little is known about saline aquifers compared to oilfields, however, and their ability to keep carbon dioxide contained over a long period is untested. There is current research aiming to evaluate their potential however, and there are many examples worldwide where carbon dioxide has been trapped naturally in porous sediments deep below ground for tens of millions of years. Another possible way to store captured carbon dioxide would be to inject it into the deep oceans, between 800 and 4000 metres. The carbon dioxide could be released into the ocean water or the deep seabed via a fixed pipeline or from a moving ship. Below a certain depth the pressure of the seawater causes the carbon dioxide to form solid lumps that slowly dissolve into the water. The UK Government does not support this method of storage however as it is not compatible with international marine treaties, the impact on the plants and animals living in the oceans is unclear and there is a risk that some of the carbon dioxide could eventually return to the atmosphere. Most importantly, dissolving large volume of carbon dioxide in seawater increases ocean acidity, which could adversely affect ocean life. Pumping carbon dioxide from fossil fuel combustion directly into the sea is not considered to be a suitable solution to increasing carbon dioxide concentrations in the atmosphere. Various forms have been conceived for permanent storage of CO_2. These forms include gaseous storage in various deep geological formations (including saline formations and exhausted gas fields), liquid storage in the ocean, and solid storage by reaction of CO_2 with metal oxides to produce stable carbonates.

System Efficiency

Carbon capture technologies reduce the efficiency of power stations, so that more fuel has to be burned to produce the same amount of energy, producing more emissions. The fuel needs of a plant with CCS technologies increases by between 10 and 40%, depending upon the type of plant and technology used. Additional energy is also used in transporting and storing the captured carbon dioxide. As an illustration, a coal-fired power station using a wet-scrubbing post-combustion capture technique would use an extra 93kg (24%) of fuel per MW hour of electricity produced. This means that more carbon dioxide is produced by plants with CCS, although significantly less is actually released into the atmosphere. The overall impact that the technologies can have then, will depend upon maximizing the amount of carbon dioxide captured in the first place and balancing this against the extra energy capture technology and transport. Improving the efficiency of CCS technologies so that they use less energy will be vital. It is also possible that small amounts of carbon dioxide could leak from the storage site, so this needs to be taken into account for any calculations and sites will need to be selected carefully.

Technological Challenge of CCS

Carbon Capture and Storage represents a major technological challenge. There are also uncertainties around the costs and regulatory issues that need to be resolved. As yet, although key elements of the individual components (i.e. capture, transport and storage) have been demonstrated, there has been no commercial-scale CCS power station developed in any country. Demonstrating that they can work on a commercial scale will be the key to bringing the costs down and to helping make the technology commercially viable.

Geological Storage (Geo-sequestration)

Geological sequestration (geo-sequestration) is the long-term geological storage of carbon dioxide—a greenhouse gas. The process involves capturing carbon dioxide and transferring it to deep underground geological formations such as depleted oil and gas fields, deep saline reservoirs or deep un-mineable coal deposits. Within these, carbon dioxide can be safely stored, thus preventing it from entering the atmosphere (http://www.griffincoal.com.au/default.aspx?MenuID=190).This method involves injecting carbon dioxide, generally in supercritical form, directly into underground geological formations. Oil fields, gas fields, saline formations, un-minable coal seams, and saline-filled basalt formations have been suggested as storage sites. Various physical (e.g., highly impermeable cap rock) and geochemical trapping

Fig. 5.7: Long-term Storage of CO_2-GHGs through Geosequestation

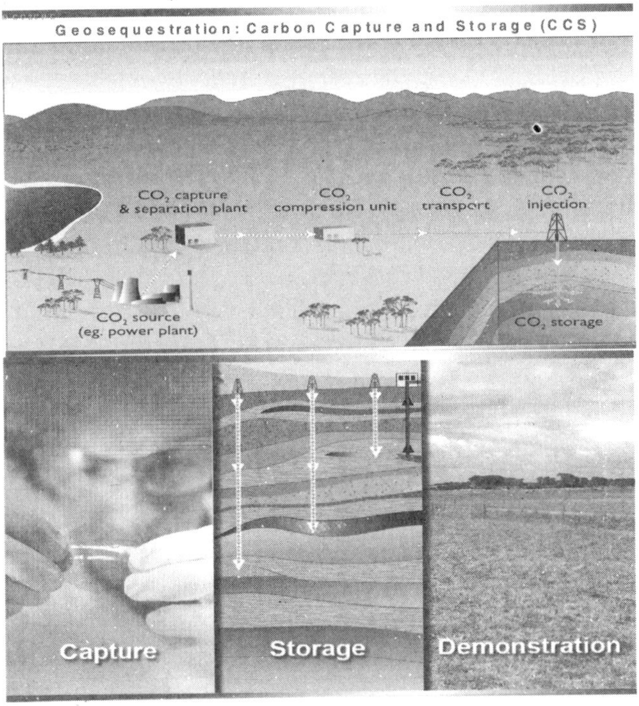

mechanisms would prevent the CO_2 from escaping to the surface. The long-term geological storage of carbon dioxide of GHGs through geosequestation mechanism is shown in Fig. 5.7.

CO_2 is sometimes injected into declining oil fields to increase oil recovery. Approximately 30 to 50 million metric tonnes of CO_2 are injected annually in the United States into declining oil fields. This option is attractive because the geology of hydrocarbon reservoirs are generally well understood and storage costs may be partly offset by the sale of additional oil that is recovered. Disadvantages of old oil fields are their geographic distribution and their limited capacity, as well as that the subsequent burning of the additional oil so recovered will offset much or all of the reduction in CO_2 emissions.

Un-minable coal seams can be used to store CO_2 because CO_2 adsorbs to the surface of coal. However, the technical feasibility depends on the permeability of the coal bed. In the process of absorption the coal releases previously absorbed methane, and the methane can be recovered (enhanced coal bed methane recovery). The sale of the methane can be used to offset a portion of the cost of the CO_2 storage. However, burning the resultant methane would produce CO_2, which would negate some of the benefit of sequestering the original CO_2.

Saline formations contain highly mineralized brines, and have so far been considered of no benefit to humans. Saline aquifers have been used for storage of chemical waste in a few cases. The main advantage of saline aquifers is their large potential storage volume and their common occurrence. The major disadvantage of saline aquifers is that relatively little is known about them, compared to oil fields. To keep the cost of storage acceptable the geophysical exploration may be limited, resulting in larger uncertainty about the aquifer structure. Unlike storage in oil fields or coal beds no side product will offset the storage cost. Leakage of CO_2 back into the atmosphere may be a problem in saline aquifer storage. However, current research shows that several *trapping mechanisms* immobilize the CO_2 underground, reducing the risk of leakage.

For well-selected, designed and managed geological storage sites, the IPCC estimates that CO_2 could be trapped for millions of years, and the sites are likely to retain over 99% of the injected CO_2 over 1,000 years. In 2009 it was reported that scientists had mapped 6,000 square miles of rock formations in the U.S. that could be used to store 500 years worth of U.S. carbon dioxide emissions.

Geo-Chemical Carbon Sequestration

Emissions of CO_2 from the use of fossil energy may be controlled by capturing CO_2 in energy production facilities and then injecting the CO_2 into deep sedimentary formations. The capture and storage of CO_2 poses two principal difficulties: (1) the capture of CO_2 from combustion products is energy-intensive, expensive, and likely applicable only to large-scale stationary processes; and (2) the buoyancy of gas-phase CO_2 in reservoirs poses inherent risks of leakage.

In contrast, CO_2 is naturally captured directly from the atmosphere by its reaction with silicate minerals to form carbonates as rocks are weathered. Unfortunately, this process, while thermodynamically favored, is very slow. If such weathering processes could be artificially accelerated, it might be possible to manage the CO_2 produced by fossil fuels while avoiding some of the difficulties of conventional CO_2 capture and storage. Geochemical immobilization can effectively eliminate the risk of CO_2 leakage. In addition, the use of geochemical processes allows the direct capture of CO_2 from the air, thus potentially lowering the cost of managing emissions from dispersed sources. Research is needed in the following broad areas (http://books.nap.edu/openbook.php?record_id=10699&page=14):

- Assessment of the reactivity, abundance, and economic availability of suitable alkaline minerals and rocks;
- Development of processes that can accelerate the carbonation reactions; and
- Assessment of the capacity, cost, and environmental impact of these processes.

Ocean Carbon Storage

Another form of carbon storage is in the oceans. Several concepts have been proposed so far which includes:

- 'Dissolution' injects CO_2 by ship or pipeline into the water column at depths of 1000 m or more, and the CO_2 subsequently dissolves.
- 'lake' deposits CO_2 directly onto the sea floor at depths greater than 3000 m, where CO_2 is denser than water and is expected to form a 'lake' that would delay dissolution of CO_2 into the environment.
- convert the CO_2 to bicarbonates (using limestone).
- Store the CO_2 in solid clathrate hydrates already existing on the ocean floor or growing more solid clathrate.

The environmental effects of oceanic storage are generally negative, but poorly understood. Large concentrations of CO_2 kills ocean organisms, but another problem is that dissolved CO_2 would eventually equilibrate with the atmosphere, so the storage would not be permanent. Also, as part of the CO_2 reacts with the water to form carbonic acid, H_2CO_3, the acidity of the ocean water increases. The resulting environmental effects on benthic life forms of the bathypelagic, abyssopelagic and hadopelagic zones are poorly understood. Even though life appears to be rather sparse in the deep ocean basins, energy and chemical effects in these deep basins could have far reaching implications. Much more work is needed here to define the extent of the potential problems.

An additional method of long term ocean based sequestration is to gather crop residue such as corn stalks or excess hay into large weighted bales of biomass and deposit it in the areas of the deep ocean basin. Dropping these residues in alluvial fans would cause the residues to be quickly buried in silt on the sea floor, sequestering the biomass for very long time spans. Alluvial fans exist in all of the world's oceans and seas where river deltas fall off the edge of the continental shelf such as the Mississippi alluvial fan in the Gulf of Mexico and the Nile alluvial fan in the Mediterranean Sea.

Global warming should be called ocean warming, as more than 80% of the added heat resides in the ocean. Clear alterations to the ocean have already been detected from observations. The magnitude and patterns of these changes are consistent with an attribution to human activities and not explained by natural variability alone. Global average land and ocean surface temperatures increased at a rate of about 0.2°C/decade over the last few decades (Hansen et al., 2006), and ocean temperatures down to 3000 m (10,000 feet) depth are also on the rise. Averages rates of sea-level rise over the last several decades were 1.8±0.5 mm/y, with an even larger rate (3.1±0.7 mm/y) over the most recent decade. Higher precipitation rates are observed at mid to high latitude and lower rates in the tropics and subtropics. Corresponding changes have been measured in surface water salinities. One of the most striking trends is the decline in Arctic sea-ice extent, particularly over the summer. September Arctic ice-cover from 2002-06 was 18% lower than pre-1980 ice-cover (http://www.arctic.noaa.gov/detect/ice-seaice.shtml, and some models predict near ice-free conditions by 2040. Recent studies of the Greenland ice sheet highlight an alarming increase in surface melting over the summer, and percolation of that melt water to the base of the ice sheet where the melt-water could lubricate ice flow and potentially greatly accelerate ice loss and sea-level rise. These new findings have not been full incorporated into projected sea-level rise estimates, which thus may be underestimated.

Over half of human carbon dioxide emissions to the atmosphere are absorbed by the ocean and land biospheres (Sarmiento and Gruber, 2002), and the excess carbon absorbed by the ocean results in increased ocean acidity. The physical and chemical mechanisms by which this occurs are well understood. Once carbon dioxide enters the ocean, it combines with water to form carbonic acid and a series of acid-base products, resulting in a lowering of pH values. The amount and distribution of human-generated carbon in the oceans are well determined from an international ocean survey conducted in the late 1980s and early 1990s (Sabine et al., 2004). The rate of ocean carbon uptake is controlled by ocean circulation. Most of the excess carbon is found in the upper few hundred meters of the ocean (upper 1200 feet) and in high-latitude regions, where cold dense waters sink into the deep ocean. Surface water pH values have already dropped by about 0.1 pH units from preindustrial levels and are expected to drop by additional 0.14-0.35 units by the end of the 21st century (Orr *et al.*, 2005).

Mineral Storage

Carbon sequestration by reacting naturally occurring Mg and Ca containing minerals with CO_2 to form carbonates has many unique advantages.

Most notable is the fact that carbonates have a lower energy state than CO_2, which is why mineral carbonation is thermodynamically favorable and occurs naturally (e.g., the weathering of rock over geologic time periods). Secondly, the raw materials such as magnesium based minerals are abundant. Finally, the produced carbonates are unarguably stable and thus re-release of CO_2 into the atmosphere is not an issue. However, conventional carbonation pathways are slow under ambient temperatures and pressures. The significant challenge being addressed by this effort is to identify an industrially and environmentally viable carbonation route that will allow mineral sequestration to be implemented with acceptable economics.

In this process, CO_2 is exothermically reacted with abundantly available metal oxides which produce stable carbonates. This process occurs naturally over many years and is responsible for much of the surface limestone. The reaction rate can be made faster, for example by reacting at higher temperatures and/or pressures, or by pre-treatment of the minerals, although this method can require additional energy. The IPCC estimates that a power plant equipped with CCS using mineral storage will need 60-180% more energy than a power plant without CCS. The following table lists principal metal oxides of Earth's Crust. Theoretically, up to 22% of this mineral mass is able to form carbonates.

Leakage

A major concern with CCS is whether leakage of stored CO_2 will compromise CCS as a climate change mitigation option. For well-selected, designed and managed geological storage sites, IPCC estimates that risks are comparable to those associated with current hydrocarbon activity. CO_2 could be trapped for millions of years, and well selected stores are likely to retain over 99% of the injected CO_2 over 1000 years. For ocean storage, the retention of CO_2 would depend on the depth; IPCC estimates 30–85% would be retained after 500 years for depths 1000–3000 m. Mineral storage is not regarded as having any risks of leakage. The IPCC recommends that limits be set to the amount of leakage that can take place. This might rule out deep ocean storage as an option.

It should also be noted that at the conditions of the deeper oceans, (about 400 bar or 40 MPa, 280 K) water–CO_2 mixing is *very* low (where carbonate formation/acidification is the rate limiting step), but the formation of water-CO_2 hydrates is favorable (a kind of solid water cage that surrounds the CO_2). To further investigate the safety of CO_2 sequestration, one can look into Norway's Sleipner gas field, as it is the oldest plant that stores CO_2 on an industrial scale. According to an environmental assessment of the gas field which was conducted after ten years of operation, the author affirmed that geo-sequestration of CO_2 was the most definite form of permanent *geological* storage of CO_2.

Available geological information shows absence of major tectonic events after the deposition of the Utsira formation. This implies that the geological environment is tectonically stable and a site suitable for carbon dioxide storage. The solubility trapping is the most permanent and secure form of geological storage. The Phase I of the Weyburn Project in Weyburn, Saskatchewan, Canada has determined that the likelihood of stored CO_2 release is less than one percent in 5,000 years. A 330-kilometre-long pipe connects the oil fields of Weyburn Sask., to a coal gassification plant in Beulah, North Dakota. Tonnes of carbon dioxide (CO_2) are transported through it daily to be injected beneath Canadian soil (Fig. 5.8). This example of carbon capture and storage is seen by many government and industry leaders as a vital way of reducing CO_2 emissions ((http://www.canadiangeographic.ca/magazine/JF08/indepth/weyburn.asp). The federal government sees it not as a quick fix, but rather a means of transforming a low emissions energy industry. CG spoke with Carolyn K. Preston, executive director of the Petroleum Technology Research Centre (PTRC), more about the CCS research her centre is spearheading in Weyburn.

CO_2 re-use

A potentially useful way of dealing with industrial sources of CO_2 is to convert it into hydrocarbons where it can be stored or reused as fuel or to make plastics. There are a number of projects investigating this possibility. Currently, bio-fuels represent the other potentially carbon-neutral jet fuel available. The steps involved in conversion of CO_2 into Hydrocarbons are as follows:

1. ***Single step methods: $CO_2 + H_2 \rightarrow$ methanol***

A proven process to produce a hydrocarbon is to make methanol. Methanol is rather easily synthesized from CO_2 and H_2. Based on this fact the idea of a methanol economy was born.

2. ***Single step methods: $CO_2 \rightarrow$ hydrocarbons***

At the department of Industrial Chemistry and Engineering of Materials at the University of Messina, Italy there is a project to develop a system which

Fig. 5.8: CCS Projects: Weyburn, Sask. Q&A: In the search for the best solutions for reducing carbon dioxide emissions, carbon capture and storage (CCS)—capturing the greenhouse gas and injecting into the ground instead of the atmosphere—is one of the most promising

Source: http://www.canadiangeographic.ca/magazine/JF08/indepth/weyburn.asp.

works like a fuel-cell in reverse, whereby a catalyst is used that enables sunlight to split water into hydrogen ions and oxygen gas. The ions cross a membrane where they react with the CO_2 to create hydrocarbons.

3. Step methods: $CO_2 \rightarrow CO \rightarrow$ Hydrocarbons

If CO_2 is heated to 2400°C, it splits into carbon monoxide and oxygen. The Fischer-Tropsch process can then be used to convert the CO into hydrocarbons. The required temperature can be achieved by using a chamber containing a mirror to focus sunlight on the gas. There are a couple of rival teams developing such chambers, at Solarec and at Sandia National Laboratory, both based in New Mexico. According to Sandia these chambers could provide enough fuel to power 100% of domestic vehicles using 5800 km^2, but unlike bio-fuels this would not take fertile land away from crops but would be land that is not being used for anything else.

Carbon dioxide scrubbing variants exist based on potassium carbonate which can be used to create liquid fuels. Although the creation of fuel from atmospheric CO_2 is not a geo-engineering technique, nor does it actually function as greenhouse gas remediation, it nevertheless is potentially very useful in the creation of a low carbon economy, as transport fuels, especially aviation fuel, are currently hard to make other than by using fossil fuels. Whilst electric car technology is widely available, and can be used with renewable energy for carbon neutral driving, there are no electric jet airliners available, nor are there likely to be in the foreseeable future.

Country-wise Performance of Successful CCS Projects

As of 2007, four industrial-scale storage projects are in operation. Sleipner is the oldest project (1996) and is located in the North Sea where Norway's Statoil Hydro strips carbon dioxide from natural gas with amine solvents and disposes of this carbon dioxide in a deep saline aquifer(http://en.wikipedia.org/wiki/Sleipner_gas_field). The carbon dioxide is a waste product of the field's natural gas production and the gas contains more (9% CO_2) than is allowed into the natural gas distribution network. Storing it underground avoids this problem and saves Statoil hundreds of millions of euro in avoided carbon taxes. Since 1996, Sleipner has stored about one million tonnes CO_2 a year. A second project in the Snøhvit gas field has been set up in the Barents Sea stores with the capacity of 700,000 tonnes per year (http://en.wikipedia.org/wiki/Carbon_capture_and_storage).

The Weyburn-Midale CO_2 Project is currently the world's largest carbon capture and storage project. Started in 2000, Weyburn is located on an oil reservoir discovered in 1954 in Weyburn, southeastern Saskatchewan, Canada (http://en.wikipedia.org/wiki/Weyburn,_Saskatchewan). The CO_2 for this project is captured at the Dakota Gasification Company plant in Beulah, North Dakota which has produced methane from coal for more than 30 years. At Weyburn, the CO_2 will also be used for enhanced oil recovery with an injection rate of about 1.5 million tonnes per year. The first phase finished in 2004, and demonstrated that CO_2 can be stored underground at the site safely and indefinitely. The second phase, expected to last until 2009, is investigating how the technology can be expanded on a larger scale (Casey, 2008).

The fourth site is In Salah, which like Sleipner and Snøhvit is a natural gas reservoir located in In Salah, Algeria. The CO_2 will be separated from the natural gas and re-injected into the subsurface at a rate of about 1.2 million tonnes per year. In July 2008, the Government of Alberta announced a $2 billion investment in three to five large-scale carbon capture and storage projects. Full Project Proposals for the projects are due March 31, 2009 and selected projects announced by June 30, 2009 (http://en.wikipedia.org/wiki/In_Salah).

A major Canadian initiative called the Alberta Saline Aquifer Project (ASAP) is a consortium of 34 companies that are developing a pilot site for commercial scale carbon capture and storage in a saline aquifer. The initial pilot will sequester 1,000 tonnes per day in 2010, while the commercial phase could see 10,000 tonnes per day as soon as 2015 (http://www.albertaasap.com/). Another Canadian initiative called the Integrated CO_2 Network (ICO_2N) is a proposed system for the capture, transport and storage of carbon dioxide (CO_2). ICO_2N members represent a group of industry participants providing a framework for carbon capture and storage development in Canada (Casey, 2008).

Based in Wallula, Washington, Wallula Energy Resource Center is proposing a coal plant that incorporates the use of technology and carbon sequestration in order to create electricity in a clean and environmentally friendly manner. The Wallula Energy Resource Center plans to use Integrated Gasification Combined Cycle (IGCC) to gasify the coal therefore capturing 65% of the coals CO_2 and sequestering the CO_2 into basalt formations underground. This proposed plant would be able to generate approximately 914 megawatts of electricity, an amount equal to half of Seattle's total power requirements. IGCC, coupled with sequestration, will enable WERC to fully comply with the CO_2 emission performance standards recently adopted by Washington State (http://wallulaenergy.com/).

In October 2007, the Bureau of Economic Geology at The University of Texas at Austin received a 10-year, $38 million subcontract to conduct the first intensively monitored, long-term project in the United States studying the feasibility of injecting a large volume of CO_2 for underground storage. The project is a research program of the Southeast Regional Carbon Sequestration Partnership (SECARB), funded by the National Energy Technology Laboratory of the U.S. Department of Energy (DOE). The SECARB partnership will demonstrate CO_2 injection rate and storage capacity in the Tuscaloosa-Woodbine geologic system that stretches from Texas to Florida. The region has the potential to store more than 200 billion tons of CO_2 from major point sources in the region, equal to about 33 years of U.S. emissions overall at present rates. Beginning in fall 2007, the project will inject CO_2 at the rate of one million tons per year, for up to 1.5 years, into brine up to 10,000 feet (3,000 m) below the land surface near the Cranfield oil field about 15 miles (25 km) east of Natchez, Mississippi. Experimental equipment will measure the ability of the subsurface to accept and retain CO_2. Currently, the United States government has approved the construction of what is touted as the world's first CCS power plant, FutureGen. On January 29, 2008, however, the Department of Energy announced it was recasting the FutureGen project and on 24 June 2008, DoE published a funding opportunity announcement seeking proposals for an IGCC project, with integrated CCS, of at least 250MW (http://en.wikipedia.org/wiki/Carbon_capture_and_storage).

Examples of carbon sequestration at an existing US coal plant can be found at utility company Luminant's pilot version at its Big Brown Steam Electric Station in Fairfield, Texas. This system is converting carbon from smokestacks into baking soda. Skyonic plans to circumvent storage problems of liquid CO_2 by storing baking soda in mines, landfills, or simply to be sold as industrial or food grade baking soda. GreenFuel Technologies Corp. is piloting and implementing algae based carbon capture, circumventing storage issues by then converting algae into fuel or feed.

In November 2008, the DOE awarded a $66.9 million, eight-year grant to a research partnership headed by Montana State University to demonstrate that underground geologic formations "can store huge volumes of carbon dioxide economically, safely and permanently." Researchers under the Big Sky Regional Carbon Sequestration Project plan to inject up to one million tons of CO_2 into sandstone beneath southwestern Wyoming.

American Clean Coal Fuels, in their Illinois Clean Fuels project, is developing a 30,000 Barrel Per Day Biomass and Coal to Liquids project in Oakland Illinois, which will market the CO_2 created at the plant for Enhanced Oil Recovery applications (http://www.cleancoalfuels.com/cleancoalfuels_projects.html). The project is expected to come online in mid-2013. By combining sequestration and biomass feedstock, the ICF project will achieve dramatic reductions in the lifecycle carbon footprint of the fuels they produce. Baard Energy, in their Ohio River Clean Fuels project, are developing a 53,000 BPD Coal and Biomass to Liquids project, which has announced plans to market the plant's CO_2 for Enhanced Oil Recovery. Rentech is developing a 29,600 barrel per day coal and biomass to liquids plant in Natchez Mississippi which will market the plant's CO_2 for enhanced oil recovery (http://www.baardenergy.com/orcf.htm). The first phase of the project is expected in 2011. DKRW is developing a 15,000-20,000 Barrel Per Day coal to liquids plant in Medicine Bow Wyoming, which will market it plant's CO_2 for enhanced oil recovery. The project is expected to begin operation in 2013. In addition to individual carbon capture and sequestration projects, there are a number of U.S. programs designed to research, develop and deploy CCS technologies on a broad scale. These include the National Energy Technology Laboratory's (NETL) Carbon Sequestration Program, regional carbon sequestration partnerships and the Carbon Sequestration Leadership Forum (CSLF- http://www.cslforum.org/).

The United Kingdom Government has launched a tender process for a CCS demonstration project. The project will use post-combustion technology on coal fired power generation at 300-400mw or equivalent. The project aims to be operational by 2014. The Government announced in June 2008 that four companies had prequalified for the following stages of the competition, BP Alternative Energy International Limited, EON UK Plc, Peel Power Limited and Scottish Power Generation Limited. BP have subsequently withdrawn from the competition claiming it could not find a power generator partner and RWE npower is seeking a judicial review of the process after it did not qualify. Doosan Babcock modified a Test Rig at Renfrew in Scotland to accommodate Oxyfuel firing on pulverized coal with recycled flue gas and demonstrate the operation of a full scale 40 MW burner for use in coal-fired boilers. Sponsors of the project include the UK Department for Business Enterprise and Regulatory Reform (BERR) and a group of industrial sponsors and university partners comprising Scottish and Southern Energy (Prime Sponsor), E.ON UK PLC, Drax Power Limited, ScottishPower, EDF Energy, Dong Energy Generation, Air Products Plc (Sponsors), and Imperial College and University of Nottingham (http://www.pandct.com/media/shownews.asp?ID=17013).

Limitations of CCS for Power Stations

One limitation of CCS is its energy penalty. The technology is expected to use between 10 and 40% of the energy produced by a power station. Wide scale adoption of CCS may erase efficiency gains of the last 50 years, and increase resource consumption by one third. However even taking the fuel penalty into account overall levels of CO_2 abatement remain high, at approximately 80-90% compared to a plant without CCS. It is theoretically possible for CCS, when combined with combustion of biomass, to result in net negative emissions, but this is not currently feasible given the lack of development of CCS technologies and the limitations of biomass production.

A second concern regards the permanence of storage schemes. It is claimed that safe and permanent storage of CO_2 cannot be guaranteed and that even very low leakage rates could undermine any climate mitigation effect. However, the IPCC conclude that the proportion of CO_2 retained in appropriately selected and managed geological reservoirs is very likely to exceed 99% over 100 years and is likely to exceed 99% over 1,000 years.

Finally, there is the issue of cost. Greenpeace claim that CCS could lead to a doubling of plant costs. However CCS may still be economically attractive in comparison to other forms of low carbon electricity generation. It is also claimed by opponents to CCS that money spent on CCS will divert investments away from other solutions to climate change.

Cost of CCS

Although the processes involved in CCS have been demonstrated in other industrial applications, no commercial scale projects which integrate these processes exist, the costs therefore remain highly uncertain. The increased energy requirements of capturing and compressing CO_2 significantly raise the operating costs of CCS-equipped power plants. In addition there is added investment or capital costs (Cassey, 2009). The process would increase the fuel requirement of a plant with CCS by about 25% for a coal-fired plant and about 15% for a gas-fired plant. The cost of this extra fuel, as well as storage and other system costs are estimated to increase the costs of energy from a power plant with CCS by 30-60%, depending on the specific circumstances. Pre-commercial CCS demonstration projects are likely to be more expensive than mature CCS technology, the total additional costs of an early large scale CCS demonstration project are estimated to be 0.5-1.1bn per project over the project lifetime (Mckinsey, 2008).

The break-up of cost implications of Energy with and without CCS is shown in the Table below http://en.wikipedia.org/wiki/Carbon_capture_and_storage).

Costs of Energy with and without CCS (2002 US$ per kWh)

	Natural gas combined cycle	Pulverized coal	Integrated gasification combined cycle
Without capture (reference plant)	0.03-0.05	0.04-0.05	0.04-0.06
With capture and geological storage	0.04-0.08	0.06-0.10	0.06-0.09
(Cost of capture and geological storage)	0.01-0.03	0.02-0.05	0.02-0.03
With capture and Enhanced oil recovery	0.04-0.07	0.05-0.08	0.04-0.08

All costs refer to costs for energy from newly built, large-scale plants. Natural gas combined cycle costs are based on natural gas prices of US$2.80–4.40 per GJ (LHV based). Energy costs for PC and IGCC are based on bituminous coal costs of US$1.00–1.50 per GJ (LHV. This is to be noted that the costs are very dependent on fuel prices (which change continuously), in addition to other factors such as capital costs. Also to note that for EOR, the savings are greater for higher oil prices. Current gas and oil prices are substantially higher than the figures used here (IPCC, 2005)

The cost of CCS depends on the cost of capture and storage which vary according to the method used. Geological storage in saline formations or depleted oil or gas fields typically cost US$0.50–8.00 per tonne of CO_2 injected, plus an additional US$0.10–0.30 for monitoring costs. However, when storage is combined with enhanced oil recovery to extract extra oil from an oil field, the storage could yield net benefits of US$10–16 per tonne of CO_2 injected (based on 2003 oil prices). This would likely negate some of the effect of the carbon capture when the oil was burnt as fuel. However, as the table above shows, the benefits do not outweigh the extra costs of capture. Comparisons of CCS with other energy sources can be found in wind energy, solar energy, and Economics of new nuclear power plants.

Environmental Effects of CCS Projects

Additional energy is required for CO_2 capture, and this means that substantially more fuel has to be used, depending on the plant type. For new supercritical pulverized coal (PC) plants using current technology, the extra energy requirements range from 24-40%, while for natural gas combined cycle (NGCC) plants the range is 11-22% and for coal-based gasification combined cycle (IGCC) systems it is 14-25% (IPCC, 2005). Obviously, fuel use and environmental problems arising from mining and extraction of coal or gas increase accordingly. Plants equipped with flue gas de-sulfurization (FGD) systems for SO_2 control require proportionally greater amounts of limestone and systems equipped with SCR systems for NO_X require proportionally greater amounts of ammonia. IPCC has provided estimates of air emissions from various CCS plant designs. While CO_2 is drastically reduced (though never completely captured), emissions of air pollutants increase significantly, generally due to the energy penalty of capture. Hence, the use of CCS entails a reduction in air quality.

Regulation of Carbon Capture and Storage

The *raison d'être* for carbon capture and storage (or sequestration) (CCS) is to enable continued use of fossil fuels in a carbon emission constrained world. Benefits include economic competitiveness, energy security and a non-disruptive transition to low-carbon energy systems. The technology is conceptually simple: carbon dioxide (CO_2) is captured from electric power plants or industrial sources, transported to the injection site, and injected deep underground for storage. CCS is at a very early stage of deployment, with only four commercial-scale operations world-wide, but it has the potential to play an important role in the portfolio of climate change mitigation technologies, supplementing the carbon emission reductions to be achieved by energy efficiency, conservation, and renewables. It is estimated that CCS could be used to achieve between 15% and 55% of the carbon emission reductions necessary to avoid dangerous levels of climate change (IPCC, 2005), and that achieving emission reduction goals will be less costly with CCS than without it (MIT, 2007). To achieve this potential would require large-scale, worldwide deployment of CCS in the electric utility industry, capturing and storing billions of tonnes of CO_2 per year (International Risk Governance Council (IRGC) Policy Report, 2007).

Large-scale CCS deployment will require the creation of a regime to manage risks and supporting policies to facilitate technology investment. Within this framework, regulatory, legal, and public perception considerations

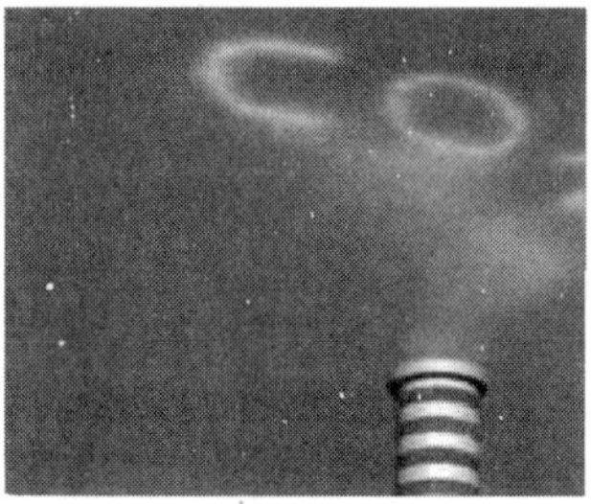

emerge as crucial factors that could either accelerate or inhibit CCS deployment. Policy makers worldwide need to work towards a system of regulation and risk governance for CCS that is globally consistent nationally coordinated, and which adequately manages local risks. This policy brief examines regulatory issues post-capture, particularly the transport and geological storage of CO_2. It identifies key areas where relevant stakeholders should collaborate internationally and proposes a model for development of national deployment and regulation, which incorporates jurisdictional specificities. The regulatory context of CCS activities as related to compliance of Kyoto Protocol is presented below (http://www.irgc.org/IMG/pdf/Policy_Brief_CCS.pdf) (Fig. 5.9).

CCS Project Life-Cycle

For practical implementation, CCS will need to be regulated as an industrial process, with regulations geared to each project stage: capture, transportation, site selection and permitting, site operations, site closure, and long-term stewardship. While all the elements of this industrial process exist, they are not yet developed to scale nor are they integrated. The structure of the future CCS industry could take a number of possible forms in terms of the relationships between CO_2 producers, CO_2 pipeline operators, and geological storage site operators. This policy brief does not cover regulatory issues related to capture but it is worth noting that, while the long-term potential for CCS lies in capturing CO_2 at fossil-fired electric power plants, significant short-term potential lies in other industrial processes that already generate a concentrated

Fig. 5.9: Regulatory Context of CCS Activities to Comply with Kyoto Protocol

	Emissions trading	Other policies
International	• Kyoto Protocol: – Joint Implementation (JI) – Clean Development Mechanism (CDM) ← AAUs	• Kyoto Protocol: – National inventory • IPCC Good Practice / Special Report • Asia-Pacific Clean Development Partnership • Carbon Sequestration Leadership Forum – regulatory considerations – international R&D co-operation
Domestic	• Mandatory emissions trading schemes: – baseline and credit (NSW GGAS) – cap and trade (EU ETS) • Voluntary/retail schemes Credit for early action? ←	• Regulatory frameworks: – eg. Australian draft regulatory guiding principles • Project development approvals • Research & Development: – pilot projects • Subsidies/other financial incentives: – direct funding

CO_2 stream, such as natural gas, ammonia, or hydrogen production. Regulation of transport and geological storage must be designed to manage CO_2 from both electric utilities and from these other industries. As each CCS project site will be unique, the characterization and management of geological and technical uncertainty—shades of grey as opposed to black and white—requires methodologies and technologies tailored to the particular circumstances and risks of each site (IRGC Policy Report, 2007). The life-cycle stages of a CCS project are illustrated in Fig. 5.10.

This model assumes that: commercial firms will operate CCS sites; those firms will continue to be responsible for some period of time post-closure; and, ownership will pass to a government entity at some point.

Site characterization: careful site selection is the single most important way to manage short- and long-term risks of CCS. Establishment of generalized CCS sitting guidelines that can be customized to local geology is an important first regulatory step that can be immediately undertaken. Such efforts are underway in Australia, the US, Canada, and throughout the EU. The site characterization phase will extend into site development. Installation of injection wells and monitoring systems will add detailed understanding of site geological features.

Site operations: regulation of site operations will centre on pipeline transport, injection, and monitoring. Most CCS projects will require pipeline transport from source to sink. Regulatory requirements regarding injection will shape the industry by specifying parameters such as injection well design, allowed injection quantities, reservoir pressure limits, purity of CO_2 stream, and financial responsibility standards.

Current regulations cover some of these, but modifications are needed to adequately manage the risks of injecting large volumes of buoyant CO_2. Additionally, monitoring must verify that sites are performing as expected. Establishment of general monitoring and verification (M&V) requirements will be useful to both industry and regulators but, given the heterogeneity of both capture systems and geology, each site's M&V regime will need to be site-specific and adaptive over time. Extensive baseline measurements before injection will be essential. Technical requirements will not be the only driver for M&V regulations, as the monitoring of results is also important for public acceptance.

Closure/Post-closure: closure requirements, established as part of the permitting process, will guide operations, decommissioning, M&V, and regulatory oversight throughout the project, because all parties will be motivated to have the site successfully meet closure requirements at the end of operations. After injection ceases, the CCS site operator should retain responsibility for a specified post-closure period to establish stability of storage. The duration of this post-closure liability period is debatable, however: durations of several years to several decades are possible and may vary across jurisdictions and across projects.

Fig. 5.10: Life-cycle Stages of a CCS Project

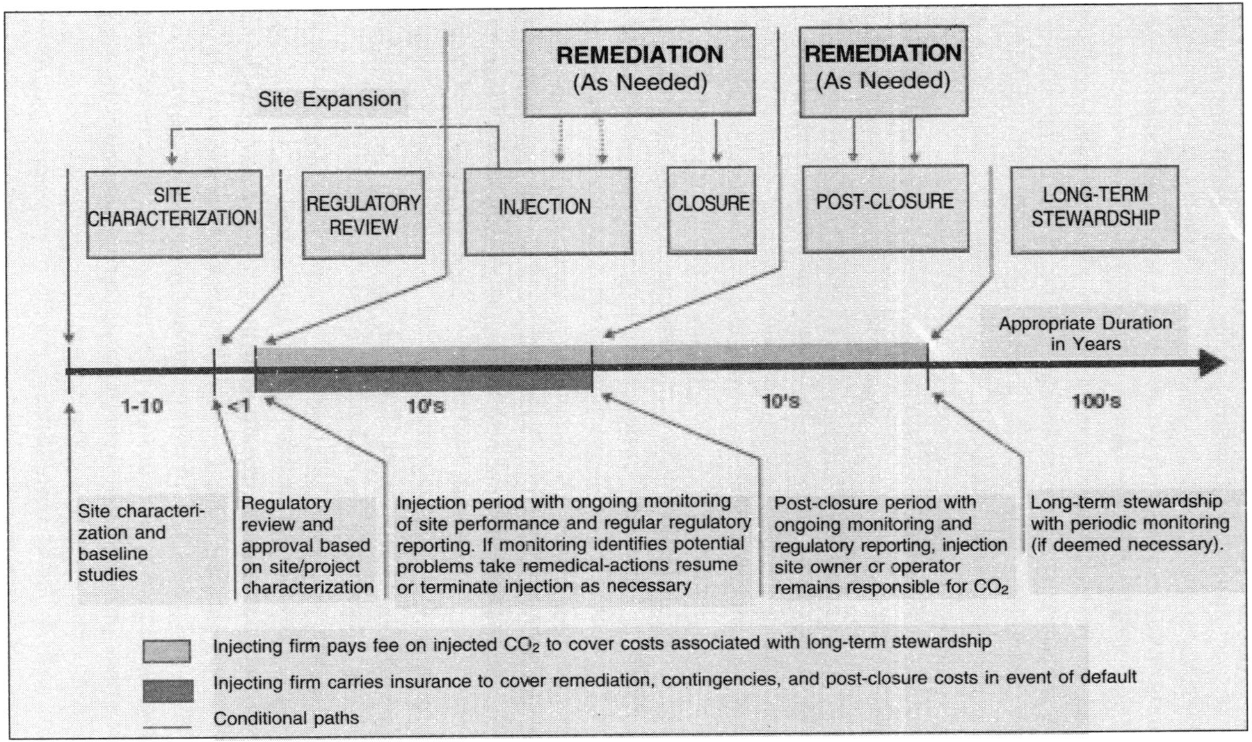

Source: IRGC-http://www.irgc.org/IMG/pdf/Policy_Brief_CCS.pdf [Illustration from Rubin et al., 2007]

Long-term stewardship: CCS requires CO_2 to remain sequestered underground for hundreds to thousands of years. Public assumption of long-term responsibility will probably be required at some point after site closure, conditional upon proof that CO_2 storage is behaving predictably, as nations are the only entities that can make credible commitments over such long storage time periods. If this model is implemented, regulations will need to specify the temporal and technical requirements to qualify for ownership transfer. See Section 5.4 for more discussion on long-term liability and responsibility. Balancing the needs of all actors through project cycle A comprehensive CCS regulatory framework must balance competing needs and interests of local, national and international publics, CO_2 generators, CO_2 pipeline operators, geological storage site developers, financial and insurance institutions supporting the project, government agencies setting safety and environmental requirements, and national and international agencies managing climate regimes.

The public: public priorities will vary by context. The global public's interest in CCS is to avoid dangerous climate change. National and state level publics will be concerned with economic competitiveness, including the cost of electricity, and with cost and effectiveness of regulatory agencies. The local public will focus on health, safety and environmental concerns, as well as property rights and property value issues, especially for on-shore sites. Off-shore sites, depending on the country, may attract less public interest; however, environmental issues will remain a public concern.

CO_2 generators: CO_2 generators will need secure repositories for CO_2 coupled to a reliable pipeline system linking sources to sinks. Clear definition of CO_2 ownership transference through the industrial process from capture to injection to storage will be important, as will a stable climate policy regime (with clear targets, and additional financial certainty – e.g. a known floor to the long-term carbon price) to make CCS economically viable.

CO_2 pipeline operators: CO_2 pipeline operators need a profitable business model to justify investment in this vital infrastructure.

Geological storage site developers: project developers need CCS to be legal and profitable. They must satisfy the rules established by the four key bodies that will govern CCS: government agencies setting safety and environmental regulations; national or international agencies administering climate regimes; insurance institutions participating in liability coverage; and, financial institutions supporting the project.

Local and national regulators: protection of human health and the environment is the primary objective of local and national regulators. They will also strive to minimize the cost of regulation to both the public and industry, and equitably balance the risks of CCS between public and private actors.

Climate regime administrators: a climate regime will need to accurately measure CO_2 emissions avoided. Development of harmonized greenhouse gas

(GHG) accounting procedures, as well as a minimum standard for international CCS site operating procedures (site selection, injection, and monitoring) will be necessary to enable international carbon trading and ensure that the value of emissions allowances is not eroded by leakage from CCS sites.

Insurance companies: the ability of insurers and reinsures to assess risk will depend on which activities they are asked to cover and the limits on liability provided under national, state, or provincial law. This analysis will depend on available data and site experience.

Financial underwriting companies: CCS projects will not be possible without financing. Financial institutions will require that CCS be profitable. This calls for incentive structures to render CCS economically viable and special incentives to encourage "first of a kind" projects. It will probably also entail long-term contracts with CO_2 generators, and mechanisms to ensure continued viability of CCS plants. Finally, no investment can occur before legal operation is assured. On-shore and off-shore regulation and guidance must explicitly address CO_2 and associated substances to remove legal uncertainty, must clarify ownership rights and responsibility of injected CO_2, and, in some jurisdictions, must clarify ownership of subsurface pore space and ownership of mineral rights affected by CO_2 storage.

Coalbed Methane Technology (CBM)

Methane and coal are formed together during coalification, a process in which plant biomass is converted by biological and geological forces into coal. Methane is stored in coal seams and the surrounding strata and released during coal mining. Deeper coal seams contain much larger amounts of methane than shallow seams. Small amounts of methane are also released during the processing, transport, and storage of coal. Although the volume of methane contribution to the total GHG emissions is three times smaller than that of carbon dioxide, at the same time, methane is a particularly strong GHG; its greenhouse potential is 21 times higher than that of CO_2. Over the last two centuries, methane concentrations in the atmosphere have more than doubled, largely due to human-related activities, including rice production, waste disposal, cattle ranching, large scale extraction and transportation of oil and natural gas, and mining.

Global Potential of Coalbed Methane

On a global basis, coalbed methane now contributes more than 1TCF (trillion cubic feet) of gas per annum. Coalbed methane accounts for between 3 percent and 4 percent of all gas production in the U.S.A.

Coalbed Methane resource appraisal, drilling and production testing are presently underway in at least a dozen other countries, and the proportion of non-US production can be expected to soar during the next 10 years. China, the world's largest coal producer, has generated particular interest. Its coalbed methane resources are estimated to range from between 1,000 and 2,800 TCF, many times larger than its conventional gas potential. Chinese coal mine operations already extract gas from within the underground workings. In 1990, 110 mines recovered 15 BCF of coalbed methane. Although modest, this production has demonstrated to the Chinese authorities the viability of the resource. As a result a number of areas have now been joint-ventured with foreign companies with initial drilling showing promising results.

India's coalbed methane resource potential has been estimated at 280 TCF, again far surpassing its conventional gas endowment (http://www.geocities.com/pemnq/cbm.html). Partnerships between U.S. and Indian companies are presently drilling and evaluating several large coalbed methane exploration permits. On the basis in initial investigations, Bangladesh, too, is regarded as having the potential to produce moderate to large volumes of coalbed methane in close proximity to major markets. In the Philippines, interest in coalbed methane is being sparked by the fact that high gas prices would compensate for the several southern African countries with extensive coal deposits. Drilling success in Zimbabwe has demonstrated the prospect of commercial coalbed methane development in the country.

In several parts of Eastern Europe, a number of coalbed methane ventures involving foreign investment and applications of U.S. technology are reported to have met with mixed success. Productions tests in some regions are said to favour commercial production, which is anticipated one negotiation with regulatory authorities have been resolved.

The tempo of coalbed methane exploration has increased significantly in Australia since the entry of major international coalbed methane producers, such as Amoco and Conoco, and the involvement of major utilities such as Pacific Power and AGL, together with a number of smaller groups. These efforts are backed by a major research effort which is modifying US technologies to suit local conditions.

The optimum linkage of probable process technologies to Coal Bed Methane operations is not simple. The proportion of the carbon in the entering fossil fuel (normally coal) that may be sequestered is a function of the efficacy of the capturing technique employed. For the base case when 80% of the carbon dioxide is recovered in each process stage, the maximum fraction of the fossil fuel carbon sequestered is not likely to exceed about two- thirds of that in the incoming coal feed. This value is considerably less in the liquid fuel co- product cases when the emissions released on ultimate consumption of the methanol in vehicles are included (http://pages.ca.inter.net/~jhwalsh/wcbm1.html)

Characteristics of Coalbed Methane

In conventional sandstone and limestone reservoirs, gas occurs in a free or dissolved phase, but in CBM reservoirs, gas exists almost exclusively in a condensed, liquid-like state. Coal has high micro-porosity with large internal surface areas and, therefore, adsorbs and retains large amounts of gas. CBM reservoirs occur at various depths, but most exploration focuses on reservoirs above 6000 ft depth. Shallow CBM reservoirs (<3000 ft) contain bacteriogenic gas. The generation of gas per ton of coal in shallow reservoirs is often low (<50scft/ton), but the thickness and extent of some coal deposits have resulted in gigantic accumulations. This is observed in the Powder River Basin, USA with a resource between 24 and 34 Tcf of CBM. Deep CBM reservoirs (>3000 ft) contain methane generated by the cracking of kerogen and oil in coal to gas. Generation of thermogenic gas increases exponentially as vitrinite reflectance (coal rank) increases. Therefore, deep CBM reservoirs can be attractive exploration targets due to their great generation potential and low water saturation. Over-pressured CBM reservoirs occur at various depths of burial; their source is thermogenic gas that has migrated up-dip from deep coal beds. The influx of meteoric water through permeable coal beds controls gas migration. This can lead to accumulations of gas at permeable barriers or the flushing of gas to the surface. The infiltration of meteoric water can also lead to the generation of bacterial gas in shallow coal beds.

Methane recovery from coal seams provides a number of benefits. The frequent accidents occurring in mines are largely caused by the explosion of coal mine methane, which accumulates in the mining process. This methane can be recovered and sold. Methane recovered from the mining sites will not only significantly reduce greenhouse gas emissions, but could also be a substitute for fuel, which is currently being imported from outside Ukraine.

Methane emissions from coal mines can be reduced by recovering and using methane from underground mines and by oxidating of methane from ventilation air. Recent technologies make both ways easily feasible. Methane recovery technologies include vertical wells drilled from the surface or boreholes drilled from inside of the mines. Depending on gas quality, methane recovered from underground mines may be sold to natural gas companies, used to generate electricity, used on-site as fuel for drying coal, or sold to nearby industrial or commercial facilities.

Origin of Coalbed Methane

Coalbed methane (CBM), or coalbed natural gas, is an unconventional natural gas in coal that formed as: (1) thermogenic gas in high rank coals (bituminous to anthracite) through the process of coalification at high temperatures, (2) biogenic gas in low rank coals (sub-bituminous) generated by bacterial activity, and (3) secondary biogenic gas accumulating in high rank coal (Rice, 1993; Faiz and Hendry, 2006; Flores, 2008). CBM is comprised of 88-98% CH_4, which is generated in significant amounts (100–300 m^3/g of coal) during the coalification process (Rice, 1993), with minor amounts of CO_2 and N_2. Thus, the abundance of CH_4 popularized the CBM terminology. Heavy hydrocarbons such as ethane, butane, propane, and pentane, are mainly associated with the thermogenic gas. Average desorbed gas content in bituminous coal (9–19 m^3/t) is higher than in sub-bituminous coal (1–4 m^3/t) (Creedy *et al.*, 2001; Stricker *et al.*, 2006).

Storage and Flow of Gas and Water

Gases, such as CH_4 and CO_2 molecules and N_2, are attracted as monolayers by Van der Waals, which is a weak force or bond, and adsorbed on surfaces of coal fractures or cleats and pores such as micropore (<2 nm), mesopore (2-50 nm), and macropore (>50nm) (Clarkson and Bustin, 1996) and held by groundwater hydrostatic pressure. The pore sizes may control the amount of adsorbed CH_4, CO_2, and N_2 with molecular diameters of 0.38, 0.33, and 0.36 nm, respectively. Average adsorption ratio of CH_4: CO_2 is about 1:14 in lignite, 1:10 in sub-bituminous coal, 1:3 in bituminous coal, and 1:2 in semi-anthracite–anthracite coal which is why coals have excellent potential for CO_2 sequestration coupled with enhanced recovery of CBM especially in low rank coals. Although molecular sizes of CH_4 and CO_2 appear similar, differences in molecular diameters and shapes may control replacement of one molecule for the other. Moreover, hydrostatic pressures must be maintained to hold gas in the coal, otherwise CO_2 sequestration by molecular replacement of CH_4 is not possible.

Coal stores 6–7 times as much gas as conventional reservoirs of equal rock volume due to the large internal surface area. About 98% of the gas is stored in pores in the coal matrix and about 2% in coal fractures (Mavor and Nelson, 1997). Also, gas occurs as solute in groundwater in the coal. Thus, a small amount of "free" gas from the fracture system and gas solute is released during the initial dewatering of CBM wells to reduce hydrostatic pressure in coal reservoirs. Gas diffusion occurs from pores through the coal matrix, with the gas migrating into cleats where two-phase transport of gas and water is governed by Darcy's Law. Gas flow in coal is controlled by permeability or connectivity of open fractures, which in turn is mainly controlled by rank. Because fracture spacing is less in bituminous coal than in sub-bituminous coal (Law, 1993), permeability in sub-bituminous coal is 1–1,000 md compared to 1–15 md in bituminous coal (Creedy *et al.*, 2001).

Continuous dewatering, from a few days to months depending on the volume water in the coal reservoir, creates local (i.e., borehole) and regional (i.e., several boreholes) depressions of low hydrostatic pressure, which then causes gas and water to flow or drain from high-pressure areas. Gas and water are separated in the borehole and both are collected at the wellhead (http://www.scitopics.com/Coalbed_Methane_Gas_of_the_Past_Present_and_Future.html).

Indian Scenario in CBM

India's Directorate General of Hydrocarbon (DGH) has identified nine blocks in four states—Gujarat, Jharkhand, Madhya Pradesh and West Bengal

Fig. 5.11A: Flow Chart of Gas and Water Transport from Coal Matrix through Cleat Fractures to the Borehole during Dewatering of a Coal Reservoir

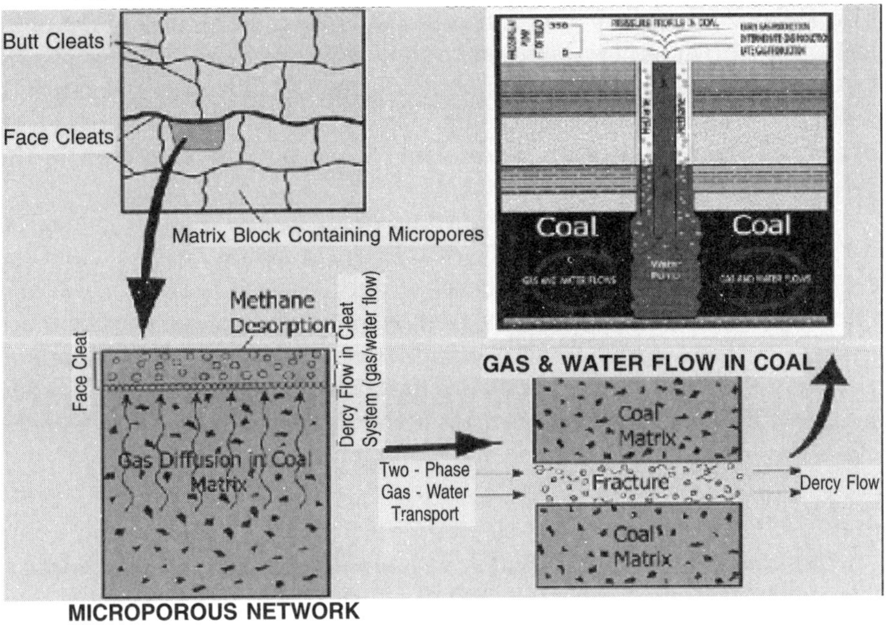

Figure 5.11B: Manging Surface Discharge, Reinjection, and Reuse of Co-produced Water from Coalbed Methane Wells.

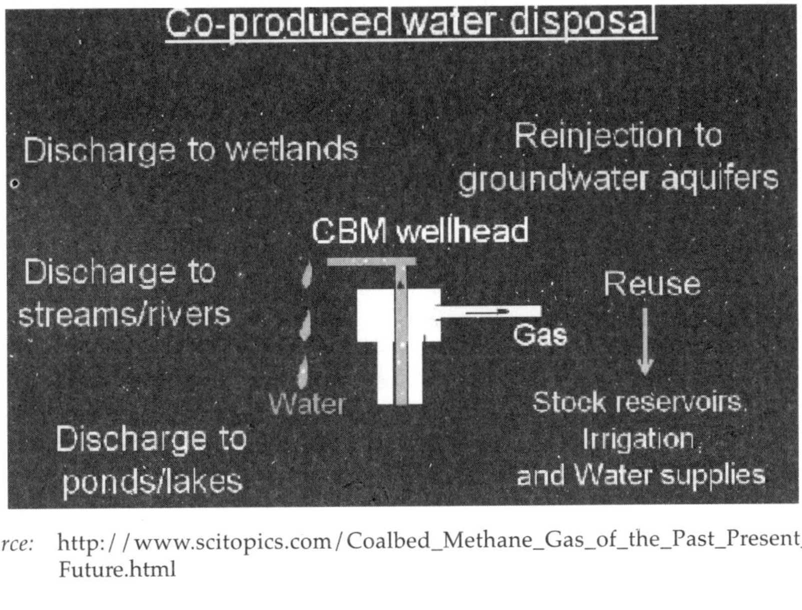

Source: http://www.scitopics.com/Coalbed_Methane_Gas_of_the_Past_Present_and_Future.html

for exploration and production of CBM that will be offered for bidding. The government is also planning to come out with the third round of exploration blocks for oil and gas under the New Exploration Licensing Policy by the end of December 2001, exactly after a year of announcing the second round of 25 blocks. The roadshows for the second round, which includes eight deepwater blocks in the west coast for the first time, and eight shallow water blocks both on east and west coast and nine blocks on land, has been very successful with government confident of an investment inflow of over $2 billion in the upstream exploration and production.

The coal-bearing formations of India occur in two distinct geological horizons: the Lower Gondwana (Permian) belts of Peninsular India and the Tertiary sediments (Eocene-Oligocene) of north-eastern India, Gujarat and Jammu-Kashmir. Entrapped within them across a wide range of subsurface depths, is methane gas whose recovery has been recognized as a viable option to enhance the natural gas potential of the country. Accordingly, a Coal Bed Methane (CBM) Exploitation Project has been formulated by the DGH with the help of the Ministry of Petroleum and Natural Gas.

CBM Reserves in the USA

The contiguous United States is estimated to have CBNG in-place resources of 700 trillion cubic feet (Tcf), of which 100 Tcf may be economically recoverable. The most prolific basins exist in the western United States, but eastern areas of the Nation also have notable reserves of CBNG. Other areas that have significant CBNG potential include Alaska and the Illinois Basin.

U.S. CBNG proved reserves and production have grown nearly every year since 1989. In 2003, CBNG accounted for 18.7 Tcf, or 10%, of the Nation's proved dry gas reserves, with 1.6 Tcf being produced, or >8% of dry gas production.

Fig. 5.12: The United States Coal Bed Natural Gas Resources

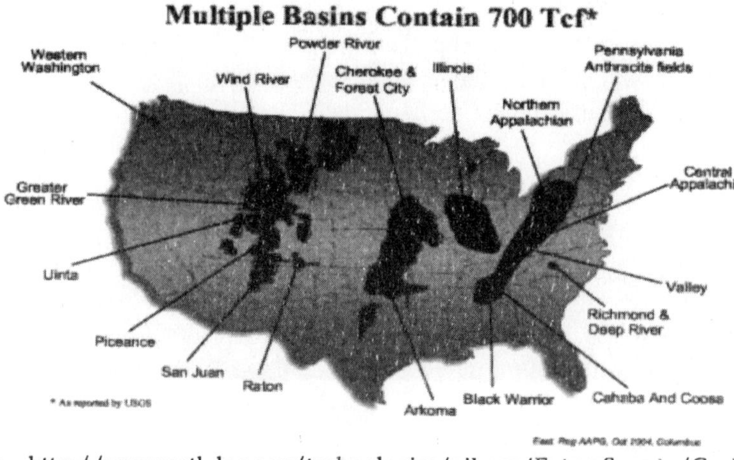

Source: http://www.netl.doe.gov/technologies/oil-gas/FutureSupply/CoalBedNG/CoalBed_NG.html

CBNG produced in Colorado, New Mexico, and Wyoming totaled nearly 1.3 Tcf during 2003, which represents 80% of total CBNG production. Other notable producing areas include the Central Appalachian and Warrior basins in the eastern United States and the Uinta and Raton basins in the Rocky Mountain region. The majority of future CBNG production is expected from western basins (http://www.netl.doe.gov/technologies/oil-gas/FutureSupply/CoalBedNG/CoalBed_NG.html).

The Reality Check on Carbon Neutrality

In response to a rapid increase in public awareness about global warming in the last 12 months, many corporations are focusing on how they can manage their carbon impacts. Businesses are doing this for several reasons:

- To establish their public positioning on this issue;
- To undertake practical measures to reduce energy costs leading to reduced emissions and future carbon risk;
- To satisfy their staff, who is increasingly looking to their employers to show they are taking responsible action to reduce emissions.

Many organizations are aiming to define their carbon footprint, to reduce this footprint by reducing the emissions from the business and, increasingly, to reduce their impact further by purchasing 'green energy' (from renewable sources) or carbon offsets. Some companies, wishing to go further, are becoming 'carbon neutral', where they completely offset their carbon footprint to theoretically have zero carbon impact on the environment.

The challenge in Australia for companies, analysts and shareholders is to ensure that there is a common understanding of what it means to be 'carbon neutral', and agreed standards for each of the elements involved in achieving that status. Without this clarity, there will be uncertainty about the validity of 'carbon neutral' status. This will undermine the confidence in the term 'carbon neutral' and the market positioning value for companies that invest to become genuinely 'carbon neutral'. Furthermore, companies that do not use a rigorous approach could be inadvertently misleading their shareholders and the public, as well as committing to future financial obligations and other risks which they do not fully understand.

According to Kevin Smith *et al* (2007) in their Report on 'The Carbon Neutral Myth: Offsets indulges on your Carbon Sins' (Transnational Institute, The Netherlands www.carbontradewatch.org www.tni.org), the sale of offset indulgences is a dead-end detour off the path of action required in the face of climate change. There is an urgent need to return to political organizing for a wider, societal transition to a low carbon economy, while simultaneously taking direct responsibility for reducing our personal emissions. Offset schemes are shifting the focus of action about climate change onto lifestyles, detracting from the local participation and movement building that is critical to the realization of genuine social change. It is hoped that the rising awareness of the shortcomings of offset credits will contribute to a reformation of the climate change debate.

Carbon Neutrality Pilot Project: A Case of Calvin College

The students of Calvin College's Engineering 333 and Biology 354 classes were given the task of determining "What would it take to make Calvin College Carbon Neutral?" At first glance, it would not appear that classes on Advanced Thermal System Design and Investigations in Plant Ecology would have much in common. However, the biology and engineering students comprising these classes banded together to tackle the problem. Coming to a satisfactory conclusion would require (http://www.calvin.edu/admin/provost/environmental/):

- Creating an inventory of Green House Gas (GHG) emissions and sequestration potential
- Coming up with detailed solutions to achieve Carbon Neutrality
- Working out a practical schedule to implement those solutions, and
- Developing a realistic plan to finance the process of achieving Carbon Neutrality Approach

In order to tackle this semester-long project, the students subdivided into five groups, each targeting a specific area to inventory GHG emissions and sequestration. The five groups, taking a cue from Calvin College's Statement of Sustainability, investigated the following five areas:

- Energy Use and Purchasing
- Land Use and Waste Water Management
- Recycling and Solid Waste Management
- Construction and Renovation
- Transportation

While creating the inventory, each group also researched potential solutions to the problem of GHG emissions. For the sake of consistency, all groups reported results in Metric Tons Carbon Dioxide Emitted per year, abbreviated MTCE. During the course of the project, it became clear that some areas of study contributed far more to Calvin's carbon footprint than others. The two groups working on land use and waste water management and recycling and solid waste management wrapped up their investigations and focused on the financial aspect of the project. All groups had been encouraged to brainstorm solutions, both frivolous and realistic. The students working on Finance filtered through these suggestions, selecting projects and objectives that were both feasible and marketable. Once these had been selected, a financial plan; which took into account inflation, the time-value of money, and the college's total budget, was drawn up.

Fig 5.13: Results of the Inventory

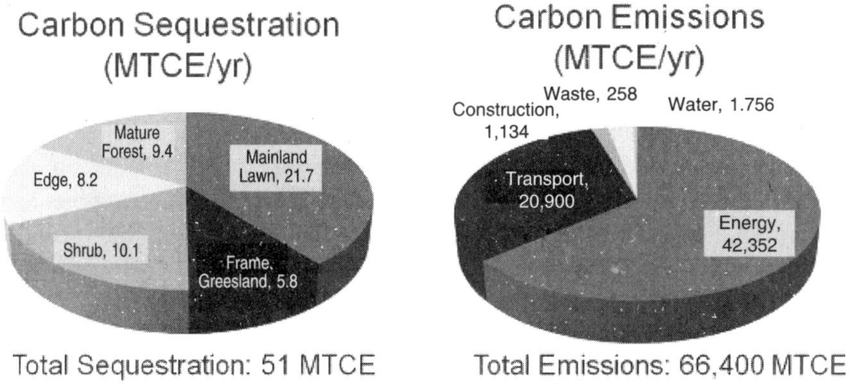

Results

Achieving Carbon Neutrality is possible but quite a challenge. And it will remain that way as long as we burn fossil fuel to produce heat and electricity and to power our vehicles. Figure 5.13 shows the results of the Emissions and Sequestration inventory (http://www.calvin.edu/admin/provost/environmental/)

Because Calvin's sequestration potential is so much lower than it emissions, reducing Calvin's Carbon footprint to zero required creativity. Ultimately, it is not currently feasible to become Carbon neutral without looking off campus. Calvin can reduce its emissions by reducing consumption, increasing efficiency, using renewable power and finally making up the difference by purchasing Carbon offsets from a third party. Details on methods to reduce consumption and increase efficiency on campus can be found in the following appendices.

The main thrust of the proposed plan to achieve Carbon Neutrality centers around the installation of wind turbines on property owned by Calvin. These wind turbines would generate clean, renewable energy and reduce Calvin's dependence on electricity generated by burning fossil fuels. The rest of Calvin's footprint would be countered by purchasing Carbon Offsets from a third party such as TerraPass.com. Other projects, such as Calvin-owned bikes, would play into the overall plan, doing a small part to decrease emissions and playing a significant role in raising public awareness. This plan outlined here could be achieved by dedicating 1.16% of Calvin's $86 million budget each year for 11 years toward a Green Energy Fund. Any monetary savings resulting from campus improvements would be deposited back into this Green Energy Fund to provide capital for future carbon reduction projects.

Carbon Neutral Design Project

Never before have had architectural educators and professionals faced the magnitude and urgency of todays ecological challenges resulting from global warming and climate change. To meet these challenges, the Society of *Building Science Educators (SBSE)* has initiated the *Carbon Neutral Design (CND) Project* to create and disseminate the resources and tools needed to integrate carbon neutral and zero-energy design into professional architecture programs and practice ((http://www.architecture.uwaterloo.ca/faculty_projects/terri/carbon-aia/introduction.html).This project is a direct response to the *2010 Imperative* (which is a call for architectural educators to address carbon neutral design and fossil fuel reduction in the design studio, to improve ecological literacy for design students, and to integrate related issues in the design and operations of university facilities and campuses) as well as the *Architecture 2030 Challenge* (the realization of carbon neutral architecture by the year 2030).

The Carbon Neutral Design (CND) Project is a joint effort by the AIA, members of AIA Committee on the Environment (COTE), and the Society of Building Science Educators (SBSE) to produce educational and resource materials for carbon neutral design. SBSE, a non-profit educational organization with members from the academy, governmental agencies and publishers, has for over twenty years served as an incubator for new approaches to teaching environmentally responsive architectural design. This partnership with the academy has capitalized on the expertise of building science educators and practitioners to produce this peer reviewed web-ready content on carbon neutral design. The CND Project acknowledges that there are identifiable starting positions and best practices that are required to succeed in reducing the energy demands of a building to allow it to reach a state of Carbon Neutrality.

This, the initial phase of the Carbon Neutral Design Project will focus on reducing the carbon emissions for the operating energy of the building. Operating energy has been selected over embodied energy as it represents the major contribution to the carbon emissions of a building, and is the focus of the 2030 Challenge. Practices, protocols and tools to reduce the carbon associated with embodied energy will form the focus of subsequent research associated with this project. Funding for the project has been provided by the American Institute of Architects, the Brico Fund and a private donor.

Co-sponsoring Organizations- SBSE Carbon Neutral Design Project

The Society of Building Science Educators (SBSE) is working with the following organizations to make the CND Project a reality:

- Architecture 2030
- The Association of Collegiate Schools of Architecture (ACSA)
- The American Institute of Architects Educator Practitioner Network (AIA EPN)
- The American Institute of Architects Committee on the Environment (AIA COTE)

- The American Institute of Architects Sustainability Discussion Group (SDiG)
- The United States Green Building Council Formal Education Committee (USGBC)
- BuildingGreen.com

Carbon Neutrality: A Case of Newcastle Building Society

In a recent competition among staff members Newcastle Building Society asked individuals to list the number of things they could do to save energy. Simon Booth, a finance manager at Newcastle, who has pledged to save 20% of all the energy he uses, was chosen as the lucky winner of a mountain bike. The prize draw, to promote the ways staff can save energy in the workplace as well as at home, is the latest initiative in the Society's target to become the largest business in the North East to commit to a Carbon Neutral future. The draw took place at Newcastle Building Society's head office by Chief Executive, Colin Seccombe, and Steve Hunter, Director of the Energy Saving Trust advice centre- who provided the second prize of a solar radio. Newcastle Building Society, which has more than 1,000 staff and assets under administration of over £4.4bn, recently announced its ambitious programme to minimize and balance out its carbon dioxide (CO_2) emissions, working with CarbonNeutral Newcastle. The process of going Carbon Neutral involves a carbon footprint, energy audit, staff awareness and green travel initiatives. Unavoidable emissions can be offset through investment in local insulation and renewable energy schemes ((http://www.newcastle.co.uk/content/docs/2007-press/04-12-2007.pdf).

AVIVA's Program on CSR on Carbon Literacy—A Case Study

Aviva's environmental programme includes reducing its carbon emissions and offsetting all residual emissions by supporting innovative projects across the world. Aviva, the world's fifth largest insurance group, is the market leader for life and pension products and general insurance in the UK, and is the leading provider of life and pensions to Europe. Its main activities are long-term savings, fund management and general insurance. In the UK, Aviva's main operations are Norwich Union, RAC and the asset management business Morley. Internationally the company has a presence across Europe, North America and Asia-Pacific. Globally it has 59,000 employees serving more than 40 million customers. Aviva's vision for CSR is acting responsibly for the long term in how it does business to help meet its ambition of being the world's most trusted savings, investment and insurance provider (AVIVA CSR Report, 2008).

Aviva has a CSR policy which is global and applies throughout the Aviva Group. The policy governs its approach to all its activities. It is based on the underlying company philosophy that a responsible approach to developing relationships between companies and the communities they serve – global or local – is vital to delivering business success. Aviva started reporting on its environmental performance in 1999 and is committed to continuous improvement. Aviva reports fewer than seven elements of CSR: standards of business conduct; customers; people; diversity; suppliers; community and the environment. This case study focuses on Aviva's environmental programme, particularly its work to reduce its carbon emissions and offset all residual emissions, making the company the first insurer to carbon neutralize its worldwide operations.

The Drivers for Carbon Neutrality

Aviva is only too aware of the strategic implications of climate change for the financial services sector and therefore managing its impact on the environment is central to their CSR programme. The decision to become the first insurer to carbon neutralizes its worldwide operations in December 2006 was driven by a number of factors relating to the Group's desire to:

- Be responsible and minimize its contribution to causing climate change and encourage others to do the same;
- Maintain its forward thinking reputation and create a market differentiator;
- Create competitive advantage to strengthen opportunities to access new markets (e.g. socially responsible investment) or new products with environmental focus (e.g. discounts on hybrid cars);
- Increase employee engagement and trust by motivating staff to take action to reduce energy use and improve energy efficiency in the office and, in turn, at home; and
- Identify business risks associated with climate change by taking a proactive approach.

Prior to the company's announcement of carbon neutrality in December 2006, it had already made good progress in reducing its carbon dioxide output, cutting its building and travel related CO_2 by 54% (2002-05). Aviva manages its environmental footprint by resorting to the following management policies:

Energy-efficient property management – Aviva takes advantage of new build and refurbishment projects to realize environmental savings. For example, Aviva's Irish business, Hibernian, has a new environmentally efficient HQ building with low energy lighting, recyclable furniture and carpets, and plenty of natural light.

Energy consumption – 55% of the electricity Aviva uses globally comes from zero emissions.

All sources and all businesses are encouraged to reduce energy consumption, e.g. 'switch off' campaigns or by providing shuttle buses for employees' commute to work from public transport.

Improving waste management – Across the Group Aviva recycles 59% of their waste.

Paper usage – Aviva is constantly looking for ways to use less paper (e.g. via double-sided printing and electronic archiving) and to use paper with recycled content. Print stations have replaced 400 printers and copiers in Hibernian's HQ with just 46, saving paper and power.

Cutting down travel – Aviva tries to cut down on business travel wherever it can. Initiatives include a CEO-sponsored 'tele-presence' project in the UK which involves staff using teleconferencing rather than travelling to meetings. Aviva also has in-house car-pooling schemes. In 2007, the focus on carbon management was strengthened and has been made a corporate objective.

Carbon offsetting – the offsetting process began with an audit of the Group operations' carbon emissions throughout 25 countries. An independent third party has verified the final figure and the company approached two carbon brokers to buy the equivalent carbon credits – amounts of carbon that can be created and purchased. The criteria for selecting projects to source carbon credits was developed internally. One challenge was the wide ranging understanding of climate change in the different countries in which it operates. Another challenge in reducing emissions and offsetting the remaining emissions is the business growth and recent acquisitions, for example of the RAC in the UK, mean that its worldwide carbon footprint has nearly doubled.

Maintaining Momentum

In 2007, Aviva will offset its remaining emissions on a retrospective basis starting with emissions generated in 2006 – some 125,000 tonnes in total. The offsetting will compensate for its carbon output caused by the consumption of non-renewable sourced electricity, gas and oil from buildings and business travel, including air, car and train, across all its global operations.

To offset 2006 carbon emissions, six projects have been selected. The projects have a global focus and are a blend of social and commercial projects. These projects include a biogas project in Sri Lanka that is generating energy from cow manure through biogas technology as a free sustainable source of power. This process replaces wood fuel and saves 3,000 tonnes of CO_2 equivalent.

Business Benefits

Core to Aviva's business is making provisions for future events. The climate change strategy is about making provisions for the future effects of climate change through adapting and creating new products and services to meet this challenge; this is the main business benefit to the group.

Carbon Management and Corporate Social Responsibility

Often a company has not fully explored the type and content of the carbon credits available and subsequently purchased credits that have less additional benefits. This sometimes leaves a carbon management initiative running independently from the overall Corporate Social Responsibility (CSR) strategy, creating a rather 'sterile' methodology of carbon abatement. More and more companies are now demonstrating their 'carbon consciousnesses by seeking out inventive and broad viewed carbon offsets. One approach that is becoming more and more desirable is to invest in smaller scale, high quality projects that demonstrate social or biodiversity benefits in addition to carbon mitigation. Some organizations look to lower the impact of their business on the environment by including projects that have a strong social focus, such as including disadvantaged people in the completion of the carbon projects themselves or by combining agro-forestry projects that open up Fair Trade commercial opportunities to the farmers.

It is this emphasis on complementary benefits that cements carbon offsetting into the CSR arena and provides another area of advantageous marketing for the more savvy company. It provides reassurance that your investment is providing a wide range of conscientiousness benefits in addition to compensating for the greenhouse gases you emit. Whatever offsetting portfolio one opt for, one should be clear on how it fits with CSR strategy and be confident of both the footprint calculations and the quality of the carbon credits ((http://www.bluegreencarbon.com/library/CSR.pdf).

The overall success of a carbon management plan relies on the co-operation and involvement of everyone within an organization including senior management. This commitment is essential to driving such a complex and serious subject and showing staff that they are serious about the initiative. Genuine management commitment will ensure that the necessary resources are devoted and employees are more likely to become involved. Carbon management can be time and resource intensive so without management support, direction and an understanding of the issue, it can be very difficult to make real change.

The following section gives some ideas for putting this case forward; however the case for each business is different.

Carbon Management Information Centre

Natural gas provides the least-cost option for major reductions in carbon emissions compared to electric and oil equipment on a full fuel cycle ("source-to-site") basis. This fact is generally not recognized by policymakers, regulators, customers, and environmental groups. Large increases in U.S. carbon emissions are being driven in the residential and commercial sectors by electricity usage. The public perception is often that natural gas is part of the carbon problem, not part of the solution. There are many opportunities for achieving least-cost carbon reductions with natural gas and for using renewable solar thermal and biogas energy solutions.

GTI (energyutilization@gastechnology.org) has launched a Carbon Management Information Center (CMIC) to help address natural gas industry issues and opportunities in the evolving arena of carbon emission controls. The Carbon Management Information Center (CMIC) was established to serve the following objectives in the area of Carbon Management (http://www.gastechnology.org/webroot/app/xn/xd.aspx?it=enweb&xd=15GTISolutions/CMIC/CMIC_home.xml):

- Serve as a clearinghouse for relevant carbon management information.
- Develop, where necessary, credible information products and functional tools to meet the needs of investors and their customers.
- Help investors inform policy makers, public utility commissions, trade allies, and customers about the significant environmental, energy efficiency, and cost savings of direct natural gas use.

The Carbon Management Information Center (CMIC) serves the gas industry, its customers, and other stakeholders by developing resources and analytical tools that:

- Clearly and fairly evaluate opportunities for natural gas to improve total energy efficiency, reduce greenhouse gas emissions, and lower energy costs for consumers.
- Identify opportunities to capture financial value through carbon emission reduction and energy efficiency programs.
- Provide a clear, concise, and technically-defensible message to policymakers, regulatory authorities, public interest groups and others in reducing the nation's energy consumption and carbon emissions.

Global Carbon Management Software Market, 2008-12

Companies, around the world, are implementing carbon management software to calculate the amount of carbon emissions, let out through direct or indirect means. The software enables organizations in tracking and reporting carbon emissions, along with measuring the impact of carbon emissions on environment. It also provides an analysis of Green house gas inventory and carbon business processes, carbon footprint, and requirements for emissions compliance. Moreover, carbon management is becoming significant for an organization on account of increasing regulation and consumer pressure. Thus, Global carbon management is becoming a part of business process to facilitate reduction in carbon emissions through compliance with environmental regulations. Moreover, demand for carbon management solutions is expected to increase from those companies which are targeting to achieve energy efficiency. These solutions manages internal emissions by aggregating data emission from multiple sources, such as electricity meters, water meters, transportation fuel consumption, business air travel etc. The data obtained is then used to perform calculations by application of accounting capabilities, reporting functions, and business intelligence tools of carbon management software.

The report forecasts the size of Global Carbon Management Software

Market over the period 2008-12, and segments the market into various geographic regions. Further, it discusses the key market trends, drivers and challenges of Global Carbon Management Software Market, and highlights some of the vendor initiatives in this market. TechNavio Insights is a set of reports based on TechNavio - a market intelligence platform for the IT industry. It builds on the intelligence available within TechNavio, and leverages on the custom research experience of the ' Technology Navigators'. TechNavio is built on years of experience of Infiniti Research in deep dive custom research and consulting for over 30 Fortune 500 companies and numerous large and mid-sized companies (http://www.the-infoshop.com/report/infi92090-carbon-m-softw.html).

Carbon Management Geographic Information System (GIS)

This project is developing tools to provide U.S. Department of Energy (DOE) research and development programs and project managers with the capability for real-time display and analysis of carbon dioxide (CO_2) sources, potential sequestration sinks, and other data, such as transportation corridors, within a spatial database. This type of program can assist decision makers by providing visual access to high quality, current, consistent data obtained from distributed datasets. The main tool being used is a geographic information system (GIS) to model and analyze the data and to display their spatial relationships. The Massachusetts Institute of Technology's (MIT) Carbon Management (CM) system is employing GIS tools to support decision making within the CM system. MIT is using GIS software to prepare a user-friendly model. Various social, economic, regulatory, and political aspects of sequestration are also being considered as part of the project (http://www.netl.doe.gov/publications/factsheets/project/Proj285.pdf).

MIT is taking a top-down approach to analyzing the potential for CO_2 capture and storage in the United States. To avoid duplication of effort while conducting this project, MIT is working closely with the ongoing National Carbon Sequestration Database and Geographic Information System (NATCARB) project. The primary use of the Carbon Management GIS will be as a systems analysis tool that can be used on a local, regional, or national scale. One of the options for mitigating CO_2 emissions from power plants and other point sources is sequestration in geologic formations, whether solely for containment purposes, for enhanced oil recovery (EOR), or for enhanced gas recovery (EGR). However, to minimize transportation and other costs, sources and sinks should be in close proximity to each other. The software being developed by MIT in this project will permit rapid visualization of the relationship between CO_2 sources and potential sequestration sites. It will ultima' ly aid DOE in the development of meaningful and economically feasible sequestration demonstration projects. Such projects are essential if sequestration is to become a technically, economically, environmentally, and socially acceptable option that supports the continued use of our abundant domestic coal resources to help meet the near- and long-term energy needs of the United States.

CDP and IBM Partnership Program on CSR in Carbon Management

Defining carbon leadership

Carbon leaders have decided that climate change is an important issue and that their organizations will do their part to reduce and minimize their carbon dioxide and GHG emissions. They have set targets for emissions reduction; they have tools in place to collect and manage information; and they publicly disclose their findings and commitments and may share their data with one or more organizations such as Carbon Disclosure Project (CDP).

Such organizations appear to have realized that the cost of carbon management is far outweighed by the benefits and opportunities it creates. The CDP and IBM are working in partnership to conduct research into how organizations manage carbon information. It was set out to explore how leading organizations are capturing and managing carbon data. The CDP also wanted to find out more about the journey of establishing a carbon information management system and process. 11 FT500 organizations representing a broad spectrum of industries participated in this study to help gain greater insight into this emerging and previously unexplored area.

It was hoped that some experience of carbon leaders would have been captured in a Best Practice Guide for companies wanting to improve their carbon management. Behind the carbon leaders, are many thousands of other organizations which may be termed, variously, *'carbon novices'*, *'carbon agnostics' or 'carbon unversed'* according to their experience and outlook. All of them can learn from the *'carbon leaders'* (CDP-IBM- http://www-05.ibm.com/innovation/uk/green/pdf/joint_cdp_and_ibm_study.pdf).

Stakeholders are Diverse and Exert Different Pressures

Carbon leaders recognize a range of stakeholders with an interest in developing and using carbon information. Internal to the organization, key stakeholders are those engaged at all levels – from executive management to the wider employee base – in the data collection process and the drive for change. Additionally, specific functions, including procurement, energy management, audit and external communications, are closely involved. For six years, companies globally have been reporting their carbon emissions data and climate change strategies to the CDP, a collaboration of 385 investors, holding $57 trillion in assets under management.

CDP and IBM are partnering together for the first time to investigate how some of the world's leading companies on climate change are actually collecting this data and how they are addressing issues of carbon management across their business. Through interviews with leading companies across a range of sectors, CDP and IBM have uncovered some fascinating insights into best practice and the key drivers in successful carbon management.

This report serves as an important guide to any company wishing to improve its understanding of what's involved in developing and implementing a successful carbon management strategy. From setting measurement criteria, to gaining buy-in from senior and board-level management, to a highly informative best practice matrix, it will help guide you through the 'do's' and 'don'ts' of carbon management. Whatever is the company's position on the path to carbon management, by sharing best practice from some of the leaders in this field, this report will provide valuable insight to the carbon leaders, as much as to the carbon novices.

Carbon Management Role in Future

CDP supplied a sample of companies with several years of experience in carbon data management. All companies are respondents to it. IBM provided research professionals to guide a discussion with the 11 carbon leaders. The interviewees were senior professionals and executives, primarily in corporate and social responsibility roles. Findings are intended to provide a qualitative insight into how carbon leaders capture and manage carbon data and may form a useful basis for further, more quantitative work across a wider range of organizations.

Carbon Footprints

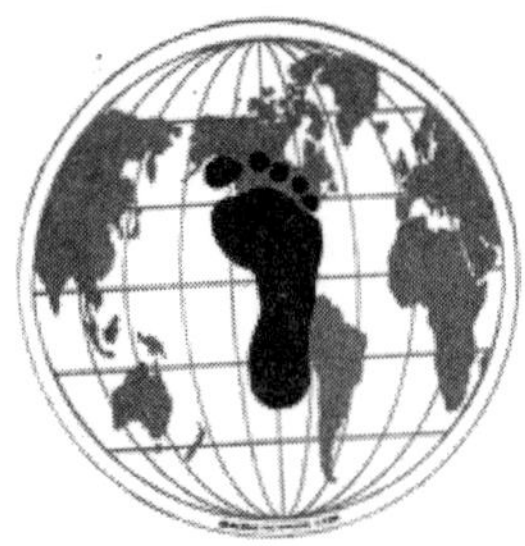

A *carbon footprint* is "the total set of GHG (greenhouse gas) emissions caused directly and indirectly by an individual, organization, event or product" (UK Carbon Trust, 2008). An individual, nation, or organization's carbon footprint is measured by undertaking a GHG emissions assessment. Once the size of a carbon footprint is known, a strategy can be devised to reduce it. Carbon offsets, or the mitigation of carbon emissions through the development of alternative projects such as solar or wind energy or reforestation, represent one way of managing a carbon footprint. The concept and name of the carbon footprint originates from the ecological footprint discussion. The carbon footprint is a subset of the ecological footprint

Carbon Footprinting and Labelling

The word 'Carbon', when used within the phrase 'carbon footprint' is often used as a short-cut term to describe the main greenhouse gases (e.g. Carbon and Nitrous Oxide – N_2O) in terms of carbon dioxide – CO_2, Methane – CH_4 equivalents. However, on occasion it is also used to refer only to CO_2 so it is important to clarify what is meant when the term is used (http://en.wikipedia.org/wiki/Carbon_footprint).

'Carbon dioxide equivalents' (CO_2e) is the internationally recognized way of expressing the amount of global warming of a particular greenhouse gas in terms of the amount of CO_2 required to achieve the same warming effect over 100 years. Methane has 23 times the global warming potential of Carbon Dioxide, whilst Nitrous Oxide has approximately 300 times the potential. So for example, 1kg of methane has the global warming potential of 23kg of CO_2.

Carbon Trust's original label

working with the Carbon Trust

100g

CO2

Source: Carbon Trust.

'Global warming' is most often used to refer to the greenhouse gas effect caused by human activities. It must be recognized however that there is a natural greenhouse effect.

A 'carbon footprint' is the total emissions of greenhouse gases (in carbon equivalents) from whichever source is being measured – be it at an individual, organization or product level.

'Embedded carbon' is the term used to describe the way in which the carbon footprint of a product, as measured by a full lifecycle assessment from 'cradle to grave', can be represented in terms of kg of CO_2 per kg of product.

The term 'carbon label' is used to refer to the communication of embedded carbon to consumers. The Carbon Trust has adopted a label which gives a product's embedded carbon emissions in terms of grams of CO_2e for the particular pack weight.

Why do Companies Calculate Carbon Footprints?

There is near universal consensus that greenhouse gases produced by human activity lead significantly to global warming, and its consequences. The early incentive for some businesses to understand their footprints came in the form of the EU's Emissions Trading Scheme. As companies could sell permits they did not need, they had a real incentive to understand their carbon impact (footprint) and to reduce it.

More recently, there is increasing public understanding of carbon footprints and their impact and an increased desire to buy 'low carbon' products and service. Indeed, Carbon Trust research reports that 66 per cent of the UK population is more likely to buy products and services with a low carbon footprint. Many companies which have undertaken some form of carbon footprint of their operations have identified 'hotspots' (e.g. associated with energy use) and have been able to address these, saving both money and carbon emissions.

The drivers for food and grocery companies to understand their carbon footprints in the future are likely to be both consumer demand for lower carbon products and regulation. In the UK, within the Climate Change Act there is specific reference to introducing a mandatory scheme called the Carbon Reduction Commitment for greenhouse gas emissions which is likely to cover significant parts of the food chain.

How to Calculate Carbon Footprints?

The traditional method of carbon management was for each company in a supply chain to focus simply on their own carbon footprint, from the operations they directly undertake. Other companies in the chain may also undertake an analysis, but possibly using a different methodology, so it was difficult to form a complete picture of the overall carbon footprint of a product.

Today, the generally accepted method used in calculating a carbon footprint is based on 'Life Cycle Assessment' (LCA) (http://en.wikipedia.org/wiki/Life_cycle_assessment). An LCA evaluates quantitatively the environmental impact of a product or service in each stage of its life - from 'cradle to grave'. So for example, an LCA of a food product sold through a retail outlet would look at the following stages (Fig. 5.14):

Taking a full supply approach has a number of advantages over the traditional approach:

- the analysis covers the entire supply chain for a product, and so companies must collaborate on providing information and undertaking the LCA;
- the carbon savings from a full supply chain assessment can come from internal efficiency improvements or from external process change. In this sense, there are many similarities between full supply chain LCA's and 'Value Chain Analyses' conducted by the Food Chain Centre.

There is an international standard which sets the necessary principles and framework, and requirements and guidelines for a LCA (ISO 14040/14044). The Carbon Trust have developed a draft methodology based on the ISO standard

Fig. 5.14: LCA of a Food Product Sold through a Retail Outlet

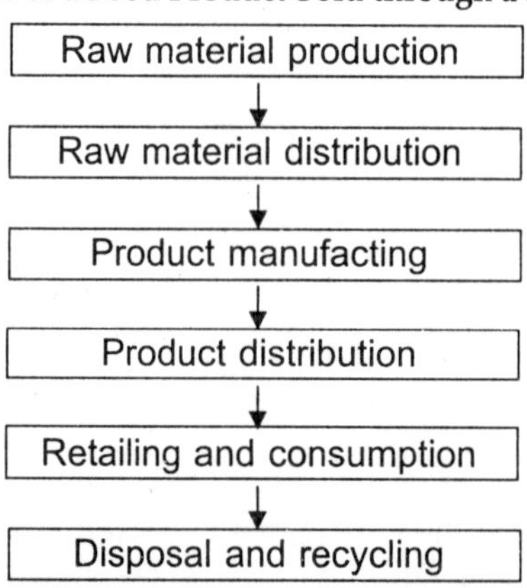

and is currently using it in a number of pilot projects with food and grocery (and other) companies.

For companies that go through the process and wish to, they are able to use the Carbon Trust's consumer facing label, which gives the weight of CO_2e per pack. There are rules for carbon reductions which allow the continued use of the label.

Carbon Trust's current label

Source: Carbon Trust

Age-related Carbon Footprint

A number of studies have calculated the carbon footprint of organizations and nations. One UK study by the Stockholm Environment Institute (Gary Haq *et al.*, 2007) examined age-related carbon emissions based on expenditure and consumption. The study found that on average people aged 50-65 years have a higher carbon footprint than any other age group. Individuals aged 50-55 years old have a carbon footprint of approximately 13.5 tonnes/ capita per year, compared to the UK average of 12 tonnes.

Carbon Emission Level

A carbon emission label or carbon label describes the carbon dioxide emissions embodied in a product. The world's first carbon label, which shows this carbon footprint embodied in a product in bringing it to the shelf, was introduced in the UK in 2006 by the Carbon Trust. Examples of products featuring their carbon footprint are Walkers Crisps, innocent Smoothies and Boots plc shampoos, which all featured the label from March 2007. By 2008 many other products featured the carbon reduction label including Tesco on an initial 20 own-range products such as orange juice and washing powder. HBOS feature it on their online bank account. The Carbon Trust label also requires companies to commit to reduce the embodied carbon in the labeled product or they lose the right to feature the label. An independent panel is currently verifying the process alongside Defra and the British Standards Institute BSI and a new standard PAS2050 is due to be introduced in mid-2008.

Other versions now exist, one of them, the Carbon Counted version of the carbon label, which started in January 2007, uses a live carbon supply chain to determine the amount of carbon dioxide emitted to bring a product to market. This third party certified system, based on an open standard, eliminates the need for heavy auditing and guess work associated with values determined when using isolated accounting methods. This also addresses how to consistently and fairly apply the smaller details such as the heating, cooling, lighting etc. in the shops the products are sold in.

Another label initiative started in spring 2008 in Switzerland. The independent association climatop labels the most climate friendly products with their label "approved by climatop". In contrast to the label of Carbon Trust, this label does not indicate the carbon footprint of a specific product; it labels those products out of a comparable group of products with a remarkably lower carbon charge. As a rule of thumb, products have to be at least 20% better as other products from the same category. Therefore life cycle assessments or the products are calculated by independent offices, and the calculations are reviewed by a third party. Beside the fact that it has to be proven that those products have a lower climate charge, the products also have to fulfil several environmental and social standards. Examples of labeled products can be found at the Swiss retailer Migros, such as an organic fair trade sugar from Paraguay, recycling kitchen towels or laundry detergents.

California State representative Ira Ruskin sponsored a carbon labeling bill—the Carbon Labeling Act of 2009—in the California state legislature, which has been voted out of the Assembly Committee on Natural Resources. The act would require the State Air Resources Board to develop and implement a program for the voluntary assessment, verification, and standardized labeling of the carbon footprint of consumer products sold in the state.

Carbon Labeling Act of 2008

Carbon Labeling Act of 2008 in California aims to "implement a program for the voluntary assessment, verification, and standardized labeling of the carbon footprint, as defined, of consumer products sold in this state." According to the Act, the California Air Resources Board (CARB) will "establish a methodology for determining and communicating the carbon footprint of a consumer product. If feasible, the state board shall establish standards and methodologies for determining and communicating to consumers on a product label whether a product has a lower carbon footprint than the average comparable product available in the state" (http://www.learnaboutcarbon.net/standards/carbon-footprint-standards-agreements-legislation/carbon-labeling-act-2008).

Who does carbon footprints?

- Carbon Footprint Basics
- Carbon Labeling Act of 2008
- PAS 2050
- PCF Pilot Project

Carbon footprints can apply to businesses, organizations, or individuals. World-wide, many are using carbon footprint information to manage energy efficiencies and environmental impacts of products, services, or lifestyles

Carbon Footprint—Assessment of CO_2 Labeling Programmes

Product Carbon Footprint in Europe

More and more companies are recognizing the opportunities and challenges climate change poses to their business. They are confronted with rising expectations from stakeholders to act on climate change. They increasingly understand that taking actions on climate change may lead to efficiency gains, lower risk exposure, and that it can improve consumer recognition. The goods and services a company provides have an inherent carbon profile, which is determined by the greenhouse gas emissions that are incurred along the entire supply chain. If measured and quantified, these emissions, on a product-by-product basis, can potentially provide a significant indicator for the carbon performance of a company. The overall emissions of a company in comparison to other companies are difficult to interpret. In contrast, emissions associated with the manufacture and provision of a certain product could become a benchmark criterion and hence allow for credible product differentiation. However, the assessment of Product Carbon Footprints (PCF) is not trivial and further steps are necessary to develop a uniform methodology for this purpose (http://ecf.pik-potsdam.de/Processes/carbon-footprint).

European CO_2-Labelling of Products and Services

The first initiatives in the area are shaping to address the related questions and gain valuable practical experiences in measuring and communicating the carbon footprint of individual products. The British Carbon Trust has announced a pilot carbon labeling scheme in March 2007 for the UK with a number of companies successfully adding a carbon label to products, available off the shelves already. In parallel to the British labeling scheme, the German "PCF Pilotprojekt Deutschland" was successfully launched after a series of two business-led dialogue forums in Berlin. Seven major players from various industries (food and drinks, fast moving consumer goods, retail, textile, telecommunications) are currently measuring PCF of some of their products. They aim to identify suitable tools to empower the consumer to take carbon conscious decisions (http://www.pcf-projekt.de/main/news/). Just recently Swiss Migros announced the labeling of a washing power and French Casino the labeling of 3000 products. However, competing standards may lead to confusion, lowered efficiency and eventually discourage companies' involvement. Therefore, a central objective of the different initiatives is the need to define one single common standard, which may lead, one day, to a European carbon label.

Dialogue Forum Low Carbon Economy

Dialogue Forum Low Carbon Economy (DKM) is an enterprise-led platform for stakeholders to share experience and build synergies in the development of low carbon market tools. Presently DKM is focused on establishing the transparency consumers need to identify and compare low carbon products and services. DKM#3 is organized by European Climate Forum and kindly supported by sponsors (http://ecf.pik-potsdam.de/Processes/carbon-footprint).

Zero-Emission Technologies

Zero-emission refers to an engine, motor, or other energy source, that emits no waste products that pollutes the environment or disrupts the climate. Vehicles and other mobile machinery used for transport (over land, sea, air, and rail) and for other uses (agricultural, mobile power generation, etc) contribute heavily to climate change and pollution, so zero emission engines are an area of active research. These technologies almost in all cases include an electric engine powered by an energy source compact enough to be installed in the vehicle. These sources include hydrogen fuel cells, batteries, super capacitors, and flywheel energy storage devices. In some cases, such as compressed air engines, the engine may be mechanical rather than electrical. This mechanical engine is then powered by a passive energy source like compressed air, or a combustible non-polluting gas like hydrogen. These engines can be used in all vehicles, from cars to boats to propeller airplanes. For boats, energy sources such as nuclear power and solar panels can also be a viable option, in addition to traditional sails and turbo sails (http://en.wikipedia.org/wiki/Zero_emission).

Zero Emission Energy Production: Current Technology for Power Plants

Renewable energy sources, including solar power, wind power, tidal power and hydroelectric power are often referred to as zero emission power, since their operation produce very few, or zero, emissions. On the other hand, the construction of such plants may involve emissions if the materials are created or assembled using equipment that produces emissions. For example, the production of high purity silicon for photovoltaic cells consumes large quantities of carbon, contributing to CO_2 emissions, and this is also true for the production of steel and concrete for wind turbines and hydroelectric dams.

Similarly, Nuclear power plants do not emit significant amounts of pollutants during their operation, but the extraction of Uranium ore and construction of waste repositories usually involves machinery powered by gasoline. The construction of the plant itself also requires materials similar to those for renewable energy sources, though because of nuclear power plants large output per station, less so per amount of energy produced. All in all, emissions due to the operation, fuel production and waste management from nuclear power plants are similar in magnitude to those of the renewable. However, though nuclear power plants do not emit large amounts of pollution during normal operation, they do produce large quantities of radioactive waste, which has to be stored for several centuries before it's radioactivity has decreased to that of the uranium ore. There is also concern that flawed designs or operation can lead to nuclear accidents, emitting radioactive pollutants, as happened during the Chernobyl disaster.

Potential Future Technologies

Many technological improvements have been suggested to reduce emissions from various energy sources. Carbon dioxide could be captured from fossil plants and stored underground rather than released into the atmosphere (a technology known as carbon capture and storage). Improved efficiency for solar cells, or cells made from novel materials, could offset the emissions associated with silicon production. New reactor technology could enable nuclear reactors to produce orders of magnitude more energy without increased emissions, and recycled or new materials could reduce the emissions associated with construction of wind turbines and hydroelectric dams.

Zero-Emission Technologies for Fossil Fuels

Technology Status Report

A wide array of R&D efforts that address zero-emission technologies for fossil-fuels are underway across the globe through sponsorship by country-specific R&D programs, international organizations and industry. The May 2002 publication of the 'International Energy Agency, Zero Emissions Technologies for Fossil Fuels: Technology Status Report' has an appendix that summarizes some of the zero emissions projects by project sponsor. The appendix is by no means a comprehensive listing of the R&D underway in this area and it is in the process of updating and enhancing this information (IEA-http://www.earthscape.org/r1/ES17006/TSRAppendix2003.pdf).

Zero-Emission AGV Technology Developed by Numexia and TTS

Numexia has been working alongside TTS Port Equipment AB to develop its revolutionary technology in creating the new TTS Cassette AGV (C-AGV) range. Swiss transport and power technology expert Numexia SA has adapted its ground-breaking contactless energy transfer technology to pioneer a zero-emission energy solution for Automated Guided Vehicles (AGVs) in use at container terminals. Numexia has been working alongside TTS Port Equipment AB to develop its revolutionary technology in creating the new TTS Cassette AGV (C-AGV) range (http://www.tts-marine.com/Pressroom/Press-releases/ZERO—Emission-AGV-technology-developed-by-Numexia-and-TTS-/).

Intensive research and testing on AGVs carried out on the Numexia virtual terminal, has enabled the companies to develop and test the energy chain technology, which achieves optimal efficiency with zero emissions. Numexia has developed a range of energy transfer systems for different applications, such as the high-performance TTS C-AGV platform, which can carry container loads of up to 61 tons, while providing unprecedented manoeuvrability and flexibility.

The contact-less energy transfer technology contains ground-based and vehicle-based segments. The two key components to the ground based system are the power electronics element and coils, which enables the C-AGV to receive energy both under the quay crane and the stacking cranes areas. In addition to the ground based system, the vehicle based system employs the same technology and uses super capacitors to store the energy, which is then used by the specially designed electric wheel motors. In comparison with other work in this field around the world, Numexia's technology offers very high efficiency, high power capability, a low electromagnetic field, as well as safety and reliability. The C-AGV range is built on the combined expertise of TTS and Numexia, enabling it to meet zero emission requirements while providing unprecedented manoeuvrability and performance.

Long-term US Strategies to Cut GHG Emissions

The initiatives outlined the USCCAP (United States Climate Action Partnership) Action Plan will continue to reduce emission levels from projected levels beyond the year 2000 (http://www.gcrio.org/USCCAP/longterm.html). However, this plan is unlikely to stabilize emissions at 1990 levels in the longer term under reasonable assumptions regarding economic growth, the diffusion of existing technologies, and new technology development. Therefore, the Administration will develop policies to address the longer term trends in greenhouse gas emissions. These policies must address technologies of energy supply and use, and condition markets for the long-term transition away from activities, fuels and technologies that generate large emissions of greenhouse gases.

The policies contained in the Action Plan are directed primarily at creating effective markets for investments in existing or nearly commercially-available technology that reduce greenhouse gas emissions. The core of a long term strategy must ensure that a constant stream of improved technology is available and that market conditions are favorable to their adoption. The Action Plan is likely to stimulate a modest acceleration in technology development, but this impact is not readily quantified. Such gains will lay the foundation for the development of technologies that could contribute to significant reductions in greenhouse gas emissions in both the United States and abroad. But a long term economic and technology development strategy must quickly be developed in order for progress to continue on greenhouse gas emission reductions into the next century.

Research and development into the technologies that could contribute to greenhouse gas emission reductions will be a critical part of the long-term effort. Research priorities to reduce energy demand include advanced building systems, transportation equipment and systems, and manufacturing technology to reduce energy and material requirements. Research priorities for low-carbon energy supply technologies could include sustainable biomass energy systems, advanced natural gas turbine and fuel cell technologies, cogeneration systems, energy storage systems, renewable energy technologies, hydrogen fuel systems, and continued research into nuclear safety and waste disposal options that could maintain the option of commercial nuclear power.

The Administration will begin a thorough evaluation of budget, technology, and economic policies that directly or indirectly affect future greenhouse gas emission trends. The mitigation of greenhouse gases in the U.S. and abroad should become a fundamental guiding principle of economic, energy, environmental, and international policies. A key component of the efforts will be an extensive consideration of transportation sector options. Much of the anticipated growth in greenhouse gas emissions after 2000 will be in the transportation sector. Even with actions taken under this plan, we expect that overall use of automobiles and trucks will increase as vehicles are added to the nation's fleet and as vehicles are driven greater distances. In order to continue emission reduction trends beyond the turn of the century, additional cost-effective measures will be needed to reduce greenhouse gas emissions of individual vehicles and to increase America's transportation options, so that personal mobility can be enhanced while private vehicles can be driven less.

On September 29, 1993, the Administration announced a bold initiative aimed at dramatically reducing the impact of automobiles on the environment. This historic partnership with the major U.S. automakers has the goal of producing a new generation of world-competitive automobiles that are up to three times more fuel efficient than today's models — and therefore would emit one-third the CO2 of comparable cars today —- while meeting all forthcoming standards for safety and conventional air pollutants. If successful, this highly efficient new generation of automobiles could begin to come into widespread production by 2010. This initiative holds out enormous long-term progress for the health of both the global environment and the U.S. automobile industry. But over the next 15 years, additional measures will be required to curb the rising trend of greenhouse gas emissions from the transportation sector.

The US Policy under the Annual Action Plan of USCCAP underlines the following future policies towards Long-Term US Strategies to cut GHG Emissions:

- The White House National Economic Council, the Office on Environmental Policy and the Office of Science and Technology Policy to co-chair a long-term strategy working group to examine all budget, technology, R&D, regulatory and economic policies that could impact greenhouse gas emission levels beyond the year 2000. The task force will make initial recommendations by the end of 1994.
- The National Economic Council, the Office on Environmental Policy, and the Office of Science and Technology Policy to co-chair a process, to be completed not later than one year from issuance of this plan, to develop measures to significantly reduce greenhouse gas emissions from personal motor vehicles, including cars and light trucks.
- The process will involve the Departments of Transportation and

Energy, the Environmental Protection Agency, and all relevant departments and agencies in the Federal government. The Administration will also strongly encourage the full participation of the automobile industry, state and local government, the environmental community, and others with potential solutions to offer.

- The Administration will look at the full range of options under existing authority as well as alternative methods to cost-effectively reduce total greenhouse gas emissions attributable to personal motor vehicle use, while meeting or exceeding applicable vehicle safety and clean air requirements.
- The goal will be to identify and implement strategies that yield significant greenhouse gas emission reductions from personal motor vehicles, including cars and light trucks. Included in the issues to be examined will be a combination of regulatory and non-regulatory measures to substantially improve new vehicle fuel efficiency.

Cutting Carbon Emissions on Profit: Opportunities for the U.S.

This report on 'Cutting Carbon Emissions at a Profit: Opportunities for the U.S.' identifies and corrects shortcomings in recent modeling studies on the economics of reducing greenhouse gas emissions in the U.S (Krause *et al* 2007: http://stephenschneider.stanford.edu/Publications/PDF_Papers/Krause2001.pdf). The major assessments of the Kyoto Protocol —by the U.S Energy Information Administration, the Clinton White House Council of Economic Advisers, the U.S. Department of Energy Inter-Laboratory Working Group, and the Stanford Energy Modeling Forum — are found to be seriously incomplete. Each study is shown to omit one or several of four major cost-reducing policy options, resulting in cost estimates that are far too pessimistic. The study on the subject is the first to integrate all cost-cutting policy options into a coherent least-cost policy framework. Three domestic policies — a national carbon cap and permit trading program, productivity-enhancing market reforms and technology programs, and recycling of permit auction revenues into economically advantageous tax cuts — are combined with international emission allowance trading.

In analyzing this integrated least-cost approach, the present study introduces no new models. This reassessment of the Report led to the following principal findings of the Report:

1. The U.S. could meet the emission reduction targets set forth in the Kyoto Protocol by 2010 and exceed them by 2020 while increasing economic output from baseline growth projections.
2. In 2010, an integrated least-cost strategy would produce an annual net output gain of about $50-60 billion/yr or roughly 0.5 percent of GDP. By 2020, this gain grows to $120 billion/yr or 1 percent of GDP. On a cumulative net present value basis, the U.S. would gain $250 billion by 2010 and $600 billion by 2020.

3. Most of these economic gains can be achieved through a purely domestic no-regrets strategy. International trading adds some further benefits, but these are not decisive for a positive economic outcome.
4. A strong synergy exists between a national energy policy aimed at safeguarding the economy and a least-cost policy aimed at slowing climate change. By reducing consumption of oil and natural gas relative to rising business-as-usual trends, a climate policy would help protect the U.S. against energy price shocks.
5. Net economic benefits can be realized in the early years of implementation and continue to grow over time. As energy-using equipment and capital stocks turn over, market, organizational, and institutional reforms have the effect of speeding up and completing the penetration of currently available, highly cost-effective energy efficiency technologies that require little or no time-consuming research, demonstration, and commercialization.
6. Potential economic savings from energy productivity gains far exceed the costs of technology R&D programs. Together with expanded markets under a climate protection policy, these have the effect of accelerating cost reductions for renewable energy sources and other low-carbon technology options.
7. Postponing least-cost emissions reduction policies or relying on incomplete, one-sided policy strategies would result in lost opportunities for the U.S. economy of $50-150 billion/yr in 2010.
8. In the context of an integrated least-cost strategy, credits for carbon sinks and constraints on the use of the Kyoto flexibility mechanisms are of only minor economic significance.
9. An integrated least-cost approach would not only be more effectively in insulating U.S. industries from competitiveness problems than a global emissions trading approach applied in isolation; it would actually improve U.S. competitiveness. Productivity gains and tax shifts would reduce production costs and export prices in most industries below baseline levels rather than merely limiting increases in costs and prices.
10. The perception that emission reduction targets such as those of the Kyoto Protocol are unavoidably costly or unfair is the result of outdated modeling assessments. Integrated economic analysis such as that contained in this report is needed as an input for future climate negotiations.

UNEP's Carbon Neutral Network to Decarbonize Economies

The United Nations Environment Programme (UNEP) is providing leadership and encourages partnership in caring for the environment by inspiring, informing and enabling nations and peoples to improve their quality of life of future generations. UNEP's basic aim is to provide coherence and strengthen the varied environmental activities taking place throughout the world by the systematic functions of United Nations. UNEP was conceived as a catalyzing agency for the entire United Nations family to help focus on environmental issues, monitor trends and facilitate coordinated international action to safeguard the environment. It has been described as the environmental conscience of the United Nations system given its mandate to motivate and inspire, raise environmental awareness and increase action, and to coordinate the environmental work of all the UN organizations and agencies. One of the most important strategies for implementing the work programme will be the identification of strategic partnerships with Major Groups. This has been done with the realization that there are growing knowledgeable and increasingly active NGO communities, private sector and other Major Groups on environmental and natural resources management issues. The involvement of these groups will receive increasing attention (http://www.karmayog.org/startanngo/startanngo_16320.htm).

Under the initiative the UNEP in cooperation with UN Environmental Group has established 'Climate Neutral Network (CN Net)' with an aim to federate the nations, cities, local authorities, communities and companies which would work to be 'carbon neutral' by bring their emission to zero level. Four countries, four cities and five companies announced their joining with the initiative by the UNEP's Governing Council (GC)/Global Ministerial Environmental Forum (GMEF) meeting held in 2007. The four countries to be partner of the CN Net are Costa Rica, Iceland, New Zealand and Norway while the four cities are Arendal of Norway, Rizhao of China, Vancouver of Canada and Vaxjo of Sweden (http://www.climatemediapartnership.org/spip.php?article405).The five companies became the first to join the CN Net are Cooperative Financial Services, largest solar installation company of UK, Interface Inc., a commercial interiors company of USA, Brazilian multinational cosmetic company Natura, Nedbank, a leading bank of South Africa and Singapore's largest power company Senoko Power.

CN Net would assist in building confidence through demonstrable action at national and local level for transition to a low, even zero carbon future in which the carbon neutrality would be a reality if inspiring and practical actions could be federated around the world. UNEP network would be truly information exchange network and over the coming months, intergovernmental bodies, organizations, civil society groups and individuals from all sectors of society would be invited to join the network.

Going Carbon Neutral: A Guidance Document for Carbon Neutrality within CARE

Human-related emissions of carbon dioxide (CO_2) and other greenhouse gases are triggering climate change. Many non-governmental organizations (NGOs), especially those concerned with environmental conservation, have long recognized the need to take direct action. Citing concerns about safeguarding institutional legitimacy and credibility, as well as a desire to "do the right thing," many have sought to minimize their greenhouse gas emissions. Participants of the October 2007 CARE & Carbon Workshop resolved to take immediate steps to minimize emissions and move towards and eventually achieve neutrality through multiple benefit offset projects. This report provides CARE with the information it needs to debate and develop appropriate internal emissions reductions strategies. As a non-profit organization, a leader in civil society and an advocate for policy change, CARE faces a unique set of ethical and administrative challenges.

Some governments, major corporations and NGOs are already pursuing the goal of carbon neutrality by offsetting their reduced carbon footprints. However, critics see carbon offsets as the modern day equivalent of "indulgences" sold to absolve us of our climatic sins. They worry that carbon offsets may become the equivalent of SUV-for-tree swaps, giving consumers no motivation for changing unsustainable habits. This report challenges such pessimistic – and reductionist – points of view. By shifting the debate from *whether* to pursue carbon neutrality to practical details about *how* to develop an effective plan, CARE can encourage a response to inadequacies in the voluntary market based on prescriptions for reform rather than rejection. Indeed, by engaging with the voluntary market as a purchaser – as well as provider of carbon credits – CARE can strengthen carbon markets and broaden the distribution of offset-derived benefits to impoverished communities.

The study concludes with the following major resolutions of participants in the October 2007 CARE & Carbon Workshop (http://www.careclimatechange.org/files/GoingCarbonNeutral-ACAREGuidanceDocument.pdf):

1. CARE seeks to aggressively minimize its carbon footprint through win-win and cost neutral measures.
2. The savings from cost-saving neutrality measures should be reinvested in efforts to further reduce CARE's carbon footprint.
3. CARE aims to offset remaining emissions through premium credits yielding multiple benefits for impoverished communities.
4. Achieving carbon neutrality within CARE may be best achieved through tailored reduction plans designed and implemented by individual field and country offices as well as CARE International Member headquarters and regional management units. Better-resourced entities can provide support to field and country offices, as needed.

Addressing Climate Change, Energy Security and Sustainable Development Issues: Indian Scenario

The Ministry of Power (Bureau of Energy Efficiency, 2007) brought out a detailed status position paper on "Addressing Energy Security and Climate Change" in October, 2007 in a Report Form, which will bring about integration and various initiatives institutionalized through specialized institutions, such as the Ministry of New & Renewable Energy, the Bureau of Energy Efficiency, and the Technology Information, Forecasting & Assessment Council, with specific mandates to promote climate friendly technologies. The National Environment Policy, 2006, provides the basis for the integration of environmental considerations in the policies of various sectors. The Policy Statement for Abatement of Pollution, 1992, stressed the prevention of pollution at the source based on the "polluter pays" principle. The Forest Policy, 1988, highlights environmental protection through preservation and restoration of the ecological balance. Sustainable development has become part of all climate change policy discussions at the global level, particularly due to adoption of Agenda 21 and the various Conventions resulting from the UNCED-1992 (http://www.un.org/geninfo/bp/enviro.html). The generally accepted and used definition as given by the Brundtland Commission in 1983 (http://simple.wikipedia.org/wiki/Brundtland_Commission#cite_note-0) is 'development that meets the needs of the present without compromising the ability of future generations to meet their own needs'.

Greenhouse Gas (GHG) Emissions Reduction Strategy

India emits the fifth most carbon of any country in the world. At 253 million metric tons, only the U.S., China, Russia, and Japan surpassed its level of carbon emissions in 1998. Carbon emissions have grown nine-fold over the past forty years. Going forward, carbon emissions are forecast to grow 3.2% per annum until 2020. To put this in perspective, carbon emissions levels are estimated to increase by 3.9% for China and by 1.3% for the United States. India is a non-Annex I country under the United Nations Framework Convention on Climate Change, and as such, is not required to drastically reduce its carbon emissions. Climate change and other sustainable development policies are often but not always synergistic. However, sustainable development has become an integrating concept embracing economic, social and environmental issues (Lélé, 2001).

Three critical components in promoting sustainable development are economic growth, social equity and environmental sustainability. Switching to more sustainable development paths can make a major contribution to climate change mitigation, but implementation may require overcoming multiple barriers. Climate change and other sustainable development policies often benefit each other, though not always. There is a growing understanding of the possibilities to choose and implement mitigation options in several sectors to create synergies and avoid conflicts with other aspects of sustainable development. Making development more sustainable by changing development paths can make a major contribution to climate change mitigation, but implementation may require resources to overcome multiple barriers (http://

www.global-greenhouse-warming.com/sustainable-development-policies.html).

Energy Carbon Nexus

Energy consumption and climate change are highly correlated. Leading enterprises increasingly understand the links between greenhouse gas (GHG) emissions and energy consumption. To put it simply, if we reduce energy use, we will reduce GHG emissions. Companies need short-term solutions to optimize energy usage and better understand their GHG emissions. In the long term, they want to optimize the usage and financial impact of energy and reduce GHG emissions. Energy is a key contributor to carbon and other GHG emissions. Recent energy prices have seen extreme volatility because of supply-and-demand imbalances and financial speculation. Most energy experts anticipate a return to higher energy pricing when the current economic downturn ends. The ability to manage the acquisition, use, and cost of energy will be critical to combating climate change and to promoting sustainable, profitable growth (http://www.sapsustainabilityreport.com/solutions/energyAndCarbon.html?sct=challenge).

Competitive pressure, consumer interest, regulation, energy costs, market opportunities and future risk make greenhouse gas ("GHG," "carbon") management a board-level issue. Companies and other organizations are being challenged to create an effective, fiscally responsible, and sustainable response (http://www.environcorp.com/services/article.php?id=3110). Commercial and industrial businesses around the world are under pressure to cut energy usage and spending, go "green," and take measures to address climate change. The drivers for change come from many sources such as those from:

- Consumers who place increasing value on good environmental stewardship and reflect it in their buying preferences.
- Retailers who demand their suppliers embrace their own particular brand of green strategy.
- Governments that globally mandate change.
- Investors who rightly demand to know the green credentials of the companies in which they are investing.

Energy for Sustainable Development

Energy is a critical instrument for Socio-economic and sustainable development (Suarez, 1995). Energy is central to sustainable development and poverty reduction efforts since it affects all aspects of development - social, economic, and environmental - including livelihoods, access to water, agricultural productivity, health, population levels, education, and gender-related issues. And yet, some 2.4 billion people in developing countries lack modern fuels for cooking and heating and approximately 1.6 billion people do not have access to electricity. The current energy infrastructure is unsustainable, vulnerable to natural disasters, and woefully insufficient in many developing countries.

Government of India should therefore provide incentives, within legal and policy frameworks, for public, private, and community level action towards greater fuel efficiency and the use of appropriate technologies. Countries with such incentive systems can serve as examples of utility based incentives for Renewable Energy Technologies (RETs), tax incentives, fuel displacement levies and feed-in laws, extension of loans to renewable energy promoters and consumers shifting to RETs, market mechanisms, and pricing policies such as net metering facilities and tax-exemption on public bus purchases. Such incentives have the potential of encouraging private sector players to promote the use of renewable energy, including wind, solar, and hydropower. Developed countries should further international cooperation to transfer renewable energy technologies and capacity building to developing countries.

The Government of India should also implement existing commitments with relevance to energy for sustainable development, including those made in Agenda 21, the WSSD, and the Johannesburg Plan of Implementation (WSSD, 2002). It is critical for countries to implement fully and expeditiously the UN Framework Convention on Climate Change, its Kyoto Protocol, and related climate change agreements, policies, and partnerships. Government should involve civil society in multi-stakeholder participatory processes during the design, implementation, and monitoring of energy programmes (http://www.unep.org/civil_society/GCSF/contributions/statement_to_GC9.pdf).

Harmonizing Emerging Global Markets and Energy Security

Approaches to ensuring energy supply security in the 21st century should differ from past approaches that concentrated on oil substitution. Besides sustainable growth challenges, new approaches need to tackle the new energy security issues raised by market liberalization. The enhanced role of markets is tied closely to the process of globalization. Globalization, which is still gaining momentum, has encouraged competition and strengthened markets and regional and international trade, particularly for crude oil and oil products, natural gas, and energy services. Globalization is bringing new opportunities for energy security, such as better access to markets and services and the transfer of technologies that are helping to reduce the cost of energy exploration and expand proven reserves. International trade in energy resources (including Emissions trading) and services is vital for energy security.

The idea of sustainable development is gaining acceptance on the official level as well as among the public. Sustainable development demands environmental preservation. Energy production and utilization, particularly in the case of fossil fuels, can be major sources of environmental degradation. These detrimental environmental impacts have a direct bearing on the future of energy—in terms of fuels and the extent of their use—and on energy security. The United Nations Framework Convention on Climate Change, adopted at the Rio Earth Summit in 1992 (Johnson, 1993), and the Kyoto Protocol, signed by more than 160 countries in 1997 (UNFCC, 1997), call for major reductions of greenhouse gas emissions, which are caused mainly by energy use. In the G-8 Meeting of the Major Economic Forum (MEF) held on 9 July, 2009 in L'Aquila, Italy, the leaders of the largest economies agreed to a goal of achieving at least a 50% reduction in global greenhouse gas emissions by 2050. As global emissions have risen by around a fifth since 1990, a later baseline would represent a higher final emissions goal. It was recognized that the 80% emissions reduction goal would require robust mid-term reductions, although no intermediate targets were set (http://www.world-nuclear-news.org/EE-G8_endorses_emissions_cuts_and_nuclear-0907096.html).

The year 2009 is a crucial year for climate change. At a pivotal UN conference in Copenhagen at the end of 2009, governments of the world need to agree the continuation of the global climate treaty, building on and strengthening the Kyoto Protocol. The G8 leaders meeting in Italy this July have availed the opportunity to set the course for the Copenhagen conference to succeed. In doing this they realized the triple advantage of action against climate change (http://us-cdn.creamermedia.co.za/assets/articles/attachments/22098_wwfbinaryitem12911.pdf):

- Action now will start to clean up our climate act, keeping the option open for the planet to avoid catastrophic climate change.
- Action now will set the world on a sustainable energy path, which in the short run can help stabilize the economy, and in a few decades can provide ample energy for the whole planet.

Action now can help develop a low carbon economy that helps avoiding millions of climate refugees and massive cost for later adaptation to climate impacts. Fulfilling the commitments as agreed and at the schedules approved would greatly affect the use of energy resources and could compromise global economic progress. There is a large gap between the commitments and the means for implementation. Targets agreed upon by negotiators were not necessarily implemented by legislators or other policymakers. Implementation of such targets is hindered not only by cost but also by the need to maintain energy security. All indications are that fossil fuels will continue to dominate global energy resources for at least the first decades of the 21st century. Moreover, the demand for energy services will continue to increase. Most of the growth will be in developing countries, which can ill afford the high cost of containment measures. It is therefore essential to find means to contain energy-related emissions without compromising energy security.

Bibliography

Agrawal, S.K. (2006), Accounting and Taxation Aspects of Carbon Trading, October 2006, *The Chartered Accountant*, pp. 509-513.

Ammann, Caspar; *et al.* (2007), "Solar influence on climate during the past millennium: Results from transient simulations with the NCAR Climate Simulation Model" (PDF). Proceedings of the National Academy of Sciences of the United States of America 104 (10): 3713–3718. doi:10.1073/pnas.0605064103. PMID 17360418. http://www.pnas.org/cgi/reprint/104/10/3713.pdf. "Simulations with only natural forcing components included yield an early 20th century peak warming of ~ 0.2 °C (~ 1950 AD), which is reduced to about half by the end of the century because of increased volcanism.".

Azmi, B.N. Al; Nassehi, V. and Khan, A.R. (2008), Impact of emissions from power stations on the ambient air quality of selected urban areas in Kuwait, *American J. Environ. Sci.*, Dec., 2008.

APCEL (2007) Asia-Pacific Centre for Environmental Law: The Kyoto Protocol and the International Carbon Market: Perspectives for a Post]Kyoto Agreement, Faculty of Law, National University of Singapore, August 29, 2007

Archer, David (2005). "Fate of fossil fuel CO_2 in geologic time" (PDF). Journal of Geophysical Research 110 (C9): C09S05.1–C09S05.6. doi: 10.1029/2004JC002625. http://geosci.uchicago.edu/~archer/reprints/archer.2005.fate_co2.pdf.

Australian National Greenhouse and Energy Reporting (NGER) Act 2007 -http://www.climatechange.gov.au/reporting/index.html

Bearden, David M. (1999), 98-563: Air Quality and Emissions Trading—An Overview of Current Issues, CRS Report for Congress, Updated August 16, 1999.

Bhagwati, J.N. (1958), "Immiserizing Growth: A Geometrical Note." *The Review of Economic Studies* 25(3): 201-205.

Boykoff, M.T. and J.M. Boykoff (2004), Balance as Bias: Warming and the US Prestige Press. *Global Environmental Change*, 14; pp. 125-136.

Bradford, David (2002), Improving on Kyoto: Greenhouse Gas Control. Princeton University, New Jersey.

Bramante, Jim Sanjeev Nagrath and Karen Butner (2009), Mastering Carbon Management Balancing tradeoffs to optimize supply chain efficiencies: http://www.sdcexec.com/web/online/In-Depth/Mastering-Carbon-Management/4$10733

British High Commission and IT Power, (2007), Developing successful CDM Partnership, The UK Trade and Investment, The British Council: http://www.berr.gov.uk/files/file50420.pdf

Bureau of Energy Efficiency—The Ministry of Power (2007), "Addressing Energy Security and Climate Change" in October, 2007, Govt. of India, New Delhi.

Brundtland Commission-1983: http://simple.wikipedia.org/wiki/Brundtland_Commission#cite_note-0

Canada National Round Table International on the Environment and the Economy DPR 2001-2002: http://www.tbs-sct.gc.ca/rma/dpr/01-02/NRTEE/NRTEE0102dpr_e.asp

Canada, National Round Table on the Environment, Domestic Emissions Trading: Glossary of Terms, Ottawa: The Roundtable, [2001]. http://www.nrtee-

Capoor, K., and P. Ambrosi, (2006), *State and Trends of the Carbon Market 2008*. Washington, D.C.: World Bank.

Capoor, K., and P. Ambrosi, (2007), *State and Trends of the Carbon Market 2007*. Washington, D.C.: World Bank.

Capoor, K., and P. Ambrosi, (2008), *State of the Carbon Market 2006: A Focus on Africa*. Washington, D.C.: World Bank.

CCT-MIT: http://sequestration.mit.edu/pdf/WhitePaper.pdf

Carbon currency-Carbon Credit: http://opencarbonworld.com/wiki/carbon-credit.html

Clinton Climate Initiative-CCI: http://www.clintonfoundation.org/what-we-do/clinton-climate-initiative/

Carbon Emissions Offset Directory; Eco-Business Links, 2007.

Carbon-offsetting: All credit to them—Climate Change, Environment - Independent.co.uk

Carbon Offsetting-SDC: http://www.sd-commission.org.uk/pages/carbon-neutrality-and-offsetting.html

Cassey, Allan (2008), *Carbon Cemetery*, Canadian Geographic Magazine, Jan/Feb 2008, p. 61.

Cassey, Allan (2008), New Scientist No2645, 1st March 2008.

Cassey, Allan (2008), http://www.nytimes.com/2008/02/19/science/19carb.html?_r=1

Cassey, Allan, Science, 27 February 2009, Vol 323, p 1158, "Stimulus Gives DOE Billions for Carbon-Capture Project"

CDM & Carbon Funds: http://www.itpi.co.in/Resources/Clean_development_mechanisms/Business%20Plan%20for%20Establishing%20Indian%20Carbon%20Fund%200407.pdf

CDM Methodology Panel : http://cdm.unfccc.int/methodologies/PAmethodologies/Clarifications and http://cdm. unfccc.int/Reference/Guidclarif

CDM Policy Framework-Canada: http://www.co2offsetresearch.org/policy/Canada.html

CES-KULeuven (2008), DWTC/SSTC Final Report – VITO- MARKAL: A Model to Support Greenhouse Reduction Policies (MARKAL-http://www.belspo.be/belspo/home/publ/pub_ostc/CG2131/rappCG22syn_en.pdf)

Clarkson, C.R., Bustin, R.M. (1996), Variation in micropore capacity and size distribution with composition in bituminous coal of the western Canadian sedimentary basin, Fuel, 75, 1483–1498.

Clean-Air Canada: http://www.cleanaircanada.org/

Climate Change 2007: Impacts, Adaptation and Vulnerability (http://environment.about.com/od/globalwarming/a/ipcc_report_two.htm

Climate Change Convention as reported by the United Nations (http://unfccc.int/files/essential_background/background_publications_htmlpdf/application/pdf _/ghg table_06.pdf

Climate Neutral Network Offset Protocol (2005): www.climateneutralnetwork.org-http://climateneutralnetwork.org/downloads/CNN_OffsetProtocol.pdf

Copenhegen Treaty: 2009- http://us-cdn.creamermedia.co.za/assets/articles/attachments/22098_wwfbinaryitem12911.pdf

Coase, R.H. (1960), "The Problem of Social Cost," *Journal of Law and Economics* 3: 1-44.

Corbett, J.B. and J.L. Durfee (2004), Testing Public (UN) Certainty of Science: Media Representations of Global Warming. Science Communication, 26, 2, pp. 129-151.

Creedy, D.P., Garner, K., Holloway, S., Ren, T.X., Armstrong, W., 2001. A review of the worldwide status of coalbed methane extraction and utilization. British Department of Trade and Industry's Cleaner Coal Technology Transfer Programme, Report No, COAL R210 DTI Pub URN 01/1040, 78p

Criqui, P., S. Mima, and L. Viguier (1999), "Marginal Abatement Costs of CO_2 Emission Reductions, Geographical Flexibility and Concrete Ceilings: an Assessment Using the POLES Model." Energy Policy 27: 585-601.

Dales, J.H. (1968), "Land, Water and Ownership," Canadian Journal of Economics 1: 791-804

Dautremont-Smith J., Cortese A. D., Dyer G., and Walton J. (2007), Implementation Guide Information and Resources for Participating Institutions v. 1.0., American College and University Presidents Climate Commitment (ACUPCC): http://www.presidentsclimatecommitment.org/pdf/ACUPCC_IG_Final.pdf

Dieter Helm New College, Oxford, August 10th 2003.(http://www.dieterhelm .co.uk/publications/MARKALModelComments.pdf)

Dixon, Lloyd; Isaac Porche, Jonathan Kulick (2002), Driving Emissions to Zero: Are the Benefits of California's Zero Emission Vehicle Program Worth the Costs? RAND Corporation, ISBN 0-8330-3212-7

Dlugokencky, E.J., *et al.*, 1998, Continuing decline in the growth rate of the atmospheric methane burden. Nature, 393, 447–450.

Drake, Frances Global Warming: The Science of Climate Change: http://www.infibeam.com/Books/info/Frances-Drake/Global-Warming-The-Science-of-Climate-Change/0340653027.html

DTI (2006), The Energy Challenge", Department of Trade and Industry, HM Government, July, 2006: http://www.berr.gov.uk/files/file31890.pdf

DTI-MARKAL (2003), Brief Comments on the MARKAL Model and its Use by the DTI

Economic Intelligence Unit: http://www.sweden.se/upload/promotion_forum/Brand_Sweden/Bevis/Internationell_ranking_axplock.doc

EcoSecurities (2003), The Role of Emissions Trading in Asian Clean Energy Finance, EcoSecurities Ltd.

Environmental Finance Solutions, Presented at PowerGenAsia Ho Chi Minh City, Vietnam, 23 September, 2003.

Ecosystem Marketplace, State of the Voluntary Carbon Markets, 2007.

EIA's Annual Energy Outlook 2007 (AEO, 2007).

Ellis & Bosi (1999), Multi-Project Baselines: http://ies.lbl.gov/MultiProjectBaselines

Encyclopedia Britannica, Volume 27.

Energy Conservation Act-1983: http://www.energymanagertraining.com/new_index.php

Environment Canada (2008a), Turning the Corner: Regulatory Framework for Industrial Greenhouse Gas Emissions, Environment Canada. Retrieved July 15 2008 from Environment Canada

Environment Canada (2008b), Turning the Corner: Canada's Offset System for Greenhouse Gases. Environment Canada

Environment Canada (2008c), Turning the Corner: Canada's Offset System for Greenhouse Gases, Guide for Protocol Developers, Environment Canada. DRAFT August 2008

EPA Office of Air and Radiation (1998), *Acid Rain Program Compliance Report*, August 1999. The report is available electronically from EPA's web site at http://www.epa.gov/acidrain.

EPA (2006): GHG Emissions down in 2006: http://www.environmentalleader.com/2008/04/15/epa-ghg-emissions-down-in-2006/

Faiz, M., Hendry, P. (2006), Significance of microbial activity in Australian coal bed methane reservoirs—a Review, Bulletin of Canadian Petroleum Geology 54, 261–272.

Feynman, R. (1986), Personal observations on the reliability of the Shuttle, Report of the Presidential Commission on the Space Shuttle Challenger Accident, National Aeronautics and Space Administration (NASA), http://science.ksc.nasa.gov/shuttle/missions/ 51-l/docs/rogers-commission/Appendix-F.txt

Fischer, Carolyn, and Richard Newell (2005), Environmental and Technology Policies for Climate Change and Renewable Energy, Discussion Paper 04-05. Washington, DC: Resources for the Future.

Flores, R.M. (1998), Coalbed methane: from hazard to resource, *International Journal of Coal Geology*, 35, 3–26

Flores, R.M. (Ed.), (2008), Microbes, Methanogenesis, and Microbial Gas in Coal, *International Journal of Coal Geology* 76, p. 192.

FutureGen Power Plant, U.S. Climate Change Technology Program – Technology Options for the near and Long Term, November 2003, p. 61: http://www.climatetechnology.gov/library/2003/tech-options/tech-options-2-1-1.pdf

G-8 Summit, Germany: http://en.wikipedia.org/wiki/33rd_G8_summit

German Advisory Council on Global Change (http://www.wbgu.de/wbgu_sn2003_engl.html

GHG Assessment Handbook, 1988- The World Bank.

Ghosh, Sujit (2008), Carbon Credit Trading & Tax, Financial Express, 20 May, 2008.

Gielen, D., and T. Kram (2000), The role of Kyoto mechanisms: results from MARKAL analysis, CN-Policy Studies, Amsterdam.

Gillenwater, Michael (2008). "Redefining RECs—Part 1: Untangling attributes and offsets". *Energy Policy* 36: 2109, doi:10.1016/j.enpol.2008.02.036.

Gillenwater, Michael (2008). "Redefining RECs (Part 2): Untangling certificates and emission markets", *Energy Policy* 36: 2120, doi:10.1016/j.enpol.2008.02.019.

Gillenwater, Michael; Derik Broekhoff, Mark Trexler, Jasmine Hyman, Rob Fowler (2007), "Policing the voluntary carbon market". *Nature Reports Climate Change* **6** (0711): 85–87. Doi: 10.1038/climate.2007.58.

Greenhouse gases" - http://www.lenntech.com/greenhouse-effect/greenhouse-gases.htm#ixzz0GDPksrV7&A

G-8 Emissions Reduction Target: http://www.world-nuclear-news.org/EE-G8_endorses_emissions_cuts_and_nuclear-0907096.html)

G-8 Summit, L'Aquila, Italy (http://www.world-nuclear-news.org/EE-G8_endorses_emissions_cuts_and_nuclear-0907096.html)

GTZ-India CDM Project: http://www.gtz.de/en/weltweit/asien-pazifik/indien/9521.htm

Hamilton, J.D (1983), 'Oil and the macro-economy since World War II', Journal of Political Economy, 91, 228–248.

Handbook for the International Treaties for the Protection of the Ozone Layer, 2000: http://www.unep.org/ozone or http://www.unep.ch/ozone

Hegerl, Gabriele C.; *et al.* (2007). "Understanding and Attributing Climate Change" (PDF). Climate Change 2007: The Physical Science Basis. Contribution of Working Group I to the Fourth Assessment.

Herman, E Ott (1994), Emissions Trading in Kyoto Protocol; Finished and unfinished business, Wuppertal Institute for Climate, Environment and Energy, Linkage Journal, Vol 3, No. 4, 1994: http://www.iisd.ca/journal/ott.html

Hidenori Niizawa,Tatsuyoshi Saijo and Akinobu Yasumoto (2003), Proposal of Upstream Emissions Trading in Japan, OECD Global Forum for Sustainable Development: Emission Trading, OECD HQ, Paris, 17-18 March,2003: http://www.isr.gov.au/content/controlfiles/display_details.cfm?objectid=6D35F6EF-3AFE-443B-

Herzog, Howard; Elisabeth Drake, Eric Adams (1997), CO_2 Capture, Reuse, and Storage Technologies for Mitigating Global Climate Change- A White Paper; DOE Order No, DE-AF22-96PC01257, Energy Laboratory, Massachusetts Institute of Technology, USA.

Hossein Farzin, Y. (2008), Kyoto Greenhouse Gas Emissions Reduction Targets: Economic Issues and Prospects-Department of Agricultural and Resource Economics at UC Davis: http://www.agecon.ucdavis.edu/extension/update/articles/fall2000_2.pdf

IEA (2001), International Emission Trading From Concept to Reality, IEA, 2009

IPCC-Report of the Intergovernmental Panel on Climate Change, http://ipcc-wg1.ucar.edu/wg1/Report/AR4WG1_Print_Ch09.pdf, "Recent estimates indicate a relatively small combined effect of natural forcings on the global mean temperature evolution of the seconds half of the 20th century, with a small net cooling from the combined effects of solar and volcanic forcings."

ICAP (2007), International Carbon Action Partnership-Political Declaration, Lisbon, 29 October 2007: http://www.icapcarbonaction.com/declaration.htm

I.C.C.T (2007), Air Pollution and Greenhouse Gas Emissions from Ocean-going Ships: Impacts, Mitigation Options and Opportunities for Managing Growth, ICCT, March, 2007: http://www.theicct.org/documents/MarineES_Final_Web.pdf

IETA, (2002), Accounting for carbon under the UK Emissions Trading Scheme, IETA, 2002.

IFC (2009), Carbon Finance Outlook, January, 2009- http://www.icfi.com/Markets/Energy/doc_files/carbon-procurement-guide.pdf

IFRS—First time Adoption of International Financial Reporting Standards (July, 2002) http://en.wikipedia.org/wiki/International_Financial_Reporting_Standards

India's CERs production: http://www.cdmindia.com/gtz/publications/INDIA-%20Worlds%20Leading%20CDM%20Destination.pdf

Intergovernmental Panel on Climate Change (2001), "Atmospheric Chemistry and Greenhouse Gases", Climate Change 2001: The Scientific Basis. Cambridge, UK: Cambridge University Press. http://www.grida.no/publications/other/ipcc_tar/?src=/CLIMATE/IPCC_TAR/WG1/127.htm

International Energy Agency (2008) World Energy Outlook (Part C, The role of energy in climate policy).

International Greenhouse Partnerships (IGP) Program- http://envfor.nic.in/cc/cdm/information.htm

IPCC (1990): Climate Change: The IPCC Scientific Assessment. J. T. Houghton, G. J. Jenkins, and J. J. Ephraums (eds.), Cambridge University Press, Cambridge, UK, 365 pp.

IPCC (2001) Report, Intergovernmental Panel on Climate Change.

IPCC (2001), "Summary for Policymakers" (PDF), Climate Change 2001: The Scientific Basis. Contribution of Working Group I to the Third Assessment Report of the Intergovernmental Panel on Climate Change. IPCC. http://www.grida.no/climate/ipcc_tar/wg1/pdf/WG1_TAR-FRONT.pdf. Retrieved on 21 April 2009.

IPCC (2007) Intergovernmental Panel on Climate Change, http://www.ipcc.ch/ipccreports/assessments-reports.htm

IPCC (2007), "Chapter 1: Historical Overview of Climate Change Science" (PDF), *IPCC WG1, AR4 Report*. IPCC, p97 (PDF page 5 of 36), http://ipcc-wg1.ucar.edu/wg1/Report/AR4WG1,_Print_Ch01.pdf, Retrieved on 21 April 2009.

IPCC (2007), "Summary for Policymakers" (PDF), Climate Change 2007: The Physical Science Basis. Contribution of Working Group I to the Fourth Assessment Report of the Intergovernmental Panel on Climate Change. IPCC. http://ipcc-wg1.ucar.edu/wg1/Report/AR4WG1_Print_SPM.pdf.

IPCC (2007), Intergovernmental Panel on Climate Change, http://www.ipcc.ch/ipccreports/assessments-reports.htm

IPCC (2008), International Rules for Greenhouse Gas Emissions Trading: Defining the principles, Modalities, rules and guidelines for verification, reporting and accountability, UNCTAD, 2008.

IPCC Database on Greenhouse Gas Emission Factors (EFDB)-http:// www.ipcc-nggip-iges.or.jp/EFDB/main.php

IPCC Fourth Assessment (2007), Mitigation of Climate Change, (Summary for Policymakers), http://www.ipcc.ch/SPM040507.pdf

IPCC, (1996): Climate Change 1995: Economic and Social Dimensions of Climate Change. eds JP Bruce, H Lee, and EF Haites, Intergovernmental Panel on Climate Change, Cambridge University Press, Cambridge, U.K.

IPCC, 2001: Climate Change 2001: The Scientific Basis. The Contribution of Working Group I to the Third Assessment Report of the Intergovernmental Panel on Climate Change, 944. eds JT Houghton, Y. Ding, DJ Griggs, M Noguer, P. van der Linden and D. Xiaosu, Intergovernmental Panel on Climate Change, Cambridge University Press, Cambridge, U.K.

International Risk Governance Council (IRGC, 2007), Policy Report on Regulation of Carbon Capture Storage, 2007.

Johnson, S.P. (1993), The Earth Summit: The United Nations Conference on Environment and Development (UNCED), London, Graham & Trotman/ Martinus Nijhoff

Keppler F, Hamilton JT, Brass M, Röckmann T (2006), "Methane emissions from terrestrial plants under aerobic conditions", *Nature* 439: 187–191.

Kerr, Stephen, G. (2008), Accounting Policy and Carbon credits, Journal of Business & Economics Research – August 2008, Vol. 6, No. 8.

Key Elements of Offsets Standards; Broekhoff, 2007.

Kevin Smith et al (2007, Report on 'The Carbon Neutral Myth: Offsets indulges on your Carbon Sins' (Transnational Institute, The Netherlands www.carbontradewatch.org www.tni.org

Kimber, Hugo (2008), Carbon Philanthropy: Supporting a New Approach to Carbon Management and Reduction: http://www.thecarbonconsultancy.co.uk/pdfs/ Carbon-Philanthropy.pdf

Knight, F.H. (1924), "Some Fallacies in the Interpretation of Social Cost," *Quarterly Journal of Economics* 38: 582-606.

Kopp, Ray (2005), So Kyoto Is Going Into Effect—How Do We Make It Work? Washington, DC: Resources for the Future; Available at http://www.rff.org/ rff/News/Features/So-Kyoto-Is-Going-Into-Effect.cfm. (Accessed February 16, 2005)

Krause *et al*, (2007), http://stephenschneider.stanford.edu/Publications/PDF_ Papers/Krause2001.pdf

Krosnick, J.A. *et al*. (2006), The Origins and Consequences of Democratic Citizens' Policy Agendas: A Study of Popular Concern about Global Warming, Climatic Change. Issue Volume 77. pp. 7-43 http://www.climate.noaa.gov/education/ pdfs /GWNationalSeriousness.pdf

Larry Pakerr, Congressional Research Service Updated October 1, 2002: Global Climate Change: Market-Based Strategies to Reduce Greenhouse Gases.

Lélé, S.M. (1991), 'Sustainable development: A critical review', World Development, 19(6): 607–606.

Likens, G.E., W.C. Keene, J.M. Miller and J.N. Galloway, 1987, Chemistry of precipitation from a remote, terrestrial site in Australia, J. Geophys. Res. 92(D11):13,299-13,314.

Lipsey, R.G., and K. Lancaster (1956), "The General Theory of Second Best" *The Review of Economic Studies* 24(1): 11-32.

MarK Van Wees, (2002): http://www.capsd.nl/pages/documents/CDMcurrentdevelopments.pdf

MARKAL Model-http://www.law.monash.edu.au/regstudies/synopsis.scenarios-sensitivities-on-long-term-carbon-reductions-using-the-uk-markal-and-markalmacro-energy-system-models.pdf

Marshall, Bibler, C.J., J.S., Pilcher, R.C. (1998), Status of worldwide coal mine methane emissions and use.

In: Flores, R.M. (Ed.), Coalbed Methane: from Coal-Mine Outbursts to a Gas Resource, *International Journal of Coal Geology* 35, 283–310.

McKibbin, Warwick J., M. Ross, R. Shackleton, and P. Wilcoxen (1999), Emissions Trading, Capital Flows and the Kyoto Protocol, Energy Journal Special Issue, 287-334.

CCS—Assessing the Economics, Mckinsey, 2008 http://www.mckinsey.com/clientservice/ccsi/pdf/CCS_Assessing_the_Economics.pdf

McKinsey and Company (2009), Pathways to a Low Carbon Economy, Version 2 of the Global Greenhouse Gas Abatement Curve.

MCX—Carbon Credits Contract: http://www.mcxindia.com/PressRelease/MCX_PR_21jan08.pdf

Michaelowa, A. 2008: "Demand and Supply on the International Carbon Markets." background Paper to World Bank CDM in China (in press), June 2008.

Müller, Benito (2008), International Adaptation Finance: The Need for an Innovative and Strategic Approach, Oxford Institute for Energy Studies, EV 42; June 2008: http://www.oxfordenergy.org/pdfs/EV42.pdf

Murphy, C. (2002), 'Hog Wild for Pollution Trading' Fortune,http://www.fortune.com/indext.jhtml?channel=print_article.jhtml&doc_id=209008

Murphy Deborah, John Drexhage and Peter Wooders (2009), International Carbon Market Mechanisms in a Post-2012 Climate Change Agreement, IISD- May 2009: http://www.iisd.org/pdf/2009/international_carbon_market_mechanisms.pdf

New Energy Finance (2008), Carbon Market to break $100 billion in 2008, New Energy Finance Note October, 2008.

Newbery, D.M. (1992), "Should carbon taxes be additional to other transport fuel taxes?" The *Energy Journal* 13(2): 49-60.

Newbery, D.M. (2001), Harmonizing Energy Taxes in the EU, paper prepared for conference Tax Policy in the European Union held in OCFEB, Erasmus University, 17-19 October.

Orr, J.C., V.J. Fabry, O. Aumont et al. (2005), Anthropogenic ocean acidification over the twenty-first century and its impact on marine calcifying organisms, Nature, 437, 681-686, doi:10.1038/nature04095.

Ozone Secretariat, UNEP, Handbook for the International Treaties for the Protection of the Ozone Layer, pp.367, 2000, (ISBN: 92- 807-1867-3)

Pew Center (June 2002), Climate Change Activities in the United States, Pew Center on Global Climate Change, Washington, DC.

Philibert, Cédric (2005), Approaches for Future International Co-operation. Paris: OECD.

Policy Directions to 2050: http://www.wbcsd.org/plugins/DocSearch/details.asp?type=DocDet&ObjectId=MjM1MTE

PowerGen Europe (2006), Emissions management strategies and emission trading and potential barrier: The faster way to satisfy EU ETS compliance requirements, Dipl.-Ing. Wilfried Sicking, SIEMENS AG, Power Generation, Erlangen, SPPA-M3000 Emissions Management.

Ramanathan, V., (1975), Greenhouse effect due to chlorofluorocarbons: climatic implications. Science 190, 50–52

Ramanathan, V., Cicerone, R.J., Singh, H.B., Kiehl, J.T., (1985), Trace gas trends and their potential role in climate change, J. Geophys. Res. 90, 5547–5566.

Regional Greenhouse Gas Initiative (RGGI) (http://www.rff.org/rff/News/Features/Regional-Greenhouse-Gas-Initiative.cfm

Revised IPCC Guidelines for National GHG Inventories, 1996.

Ricardo Bayon, et al (2007), Voluntary Carbon Markets: An International Guide to What They Are and How They Work, London: Earthscan, 2007, p. 4.

Robin Clarke, A. *(1996)*, Pilot Greenhouse Gas Trading System: The Legal Issues, Geneva: United Nations, 1996, p. 16.

Rochon, Emily *et al* (2008), False Hope: Why carbon capture and storage won't save the climate Greenpeace, May 2008, p. 5.

Ron Bass, Jones & Stokes (2008), Evaluating Greenhouse Gases and Climate Change Impacts under NEPA (http://www.icfi.com/docs/GHG-CC-NEPA.pdf)

Royal Society (2005), "Joint science academies' statement: Global response to climate change". http://royalsociety.org/displaypagedoc.asp?id=20742. Retrieved on 19 April 2009

Sabine, C.L., R.A. Feely, N. Gruber *et al.* (2004), The oceanic sink for anthropogenic CO_2, Science, 305, 367-371.

Sarmiento, J. L. and N. Gruber (2002), Sinks for Anthropogenic Carbon, *Physics Today*, August, 30-36.

Schipper, L., Figueroa, M.J., Price, L., and Espey, M. (1993), "Mind the Gap: The Vicious Circle of measuring automobile fuel use". Energy Policy (October, 1993).

Othmar Schwank, Othmer; Madeleine Guyer (2009), CDM in China: Contribution to National and International Goals of Climate Policy, INFRAS, Research and Consulting, CH-8045 Zurich January 2009.

S.M. Enzler: Characteristics of GHGs: http://www.lenntech.com/greenhouse-effect/greenhouse-gases.htm

S.M.I.C. Report (1971), In: Wilson, C.M., Matthews,W.H. (Eds.), Report of the Study of Man's Impact on Climate (SMIC). M.I.T. Press, Cambridge, Mass, London, England.

Sergey Paltsev, John M. Reilly, Henry D. Jacoby, A. Denny Ellerman and Kok Hou Tay (2003), Emissions Trading to Reduce Greenhouse Gas Emissions in the United States: The McCain-Lieberman Proposal, Global Science Policy Change, MIT, Report No. 97, June, 2003.

South Coast Air Quality Management District, *RECLAIM Audit Report for the 1997 Compliance Year,* March 12, 1999, p. 25.

Stephens, J. (2006), Growing interest in carbon capture and storage (CCS) for climate change mitigation.

Sustainability: Science, Practice, & Policy 2(2):4–13. http://ejournal.nbii.org/archives/vol2iss2/0604-016.stephens.html, Published online November 29, 2006.

Silverstein, K. (2002), 'Emissions Trading: Sending Pollution Up in Smoke, Fortune. www.scientech.com

Solomon, S., *et al.* (2009), "Irreversible Climate change due to carbon dioxide emissions". *Proceedings of the National Academy of Sciences* 106 (6): 1704–1709. doi: 10.1073/pnas.0812721106.

Sorrell, S., Emissions Trading After Kyoto. Introduction to Environmental Economics of Science and Technology Policy Research, 2004 (http://www.sussex.ac.uk/Users/prpp4/lec8.ppt)

The UK Government's Quality Assurance Scheme for Carbon Offsetting UK Government DEFRA, 9 February 2009.

The World Bank (2007), State and Trends of the Carbon Market, August, 2007: http://web.worldbank.org/WBSITE/EXTERNAL/NEWS/0, content MDK: 21319781~pagePK: 64257043~piPK: 437376~theSitePK: 4607, 00.html

Tietenberg, T. (1994), Implementation Issues for Global Tradable Carbon Entitlements, International Environmental Economics: Theories, Models and Applications to Climate Change, International Trade and Acidification. E. C. van Ierland. Amsterdam, Elsevier.

Tietenberg, T. (2000), Environmental and Natural Resource Economics, Fifth Edition, Addison Wesley Longman, Inc., Reading, MA.

Tom Tietenberg, Michael Grubb, Axel Michaelowa, Byron Swift and ZhongXiang Zhang trnee.ca/eng/programs/ArchivedPrograms/Emission%2DTrading/overview_glossary.htm

UN Climate Neutral Strategy EMG/AM.07/11: http://www.unemg.org/climateneutralun/Portals/24/Documents/Meetings/CNUN/UNClimateNeutralStrategy.pdf

UNCED (1992):http://www.un.org/geninfo/bp/enviro.html

UNEP Guidelines for Funding CDM Projects-http://www.cd4cdm.org/Publications/FinanceCDMprojectsGuidebook.pdf

UNEP, Synthesis of the Reports of the Scientific, Environmental Effects and Technology and Economic Assessment Panels of the Montreal Protocol. A Decade of Assessments for Decision Makers Regarding the Protection of the Ozone Layer: 1989-1998, United Nations Environment Programme, pp. 161, 1999. (ISBN: 92-807- 1733-2).

UNFCCC (1997)—Kyoto Protocol: The United Nations Framework Convention on Climate Change (UNFCCC or FCCC), 1997: http://en.wikipedia.org/wiki/United_Nations_Framework_Convention_on_Climate_Change

UNFCCC (2005), Report of the Convergence of the Parties Serving as the Meeting of the Parties to the Kyoto Protocol, Proceeding of United Nations Framework Convention on Climate Change, Document Number FCCC/KP/CMP/2005/8/Add.2, 2005.

UNFCCC/CCNUCC-CDM – Executive Board, Version 02 / November 2003

UNFCCC, (2009), *Emissions trading and the project-based mechanisms: draft conclusions proposed by the Chair*. Ad Hoc Working Group on Further Commitments under the Kyoto Protocol, Seventh session, Bonn, March 29 to April. FCCC/KP/AWG/2009/L.2.

UNFCCC Summit, 2009 Copenhagenhttp://us-cdn.creamermedia.co.za/assets/articles/attachments/22098_wwfbinaryitem12911.pdf

UNIDO Knowledge Network for Industrial Technology Transfer-www.unido.org/knitt/index.html

United States Environmental Protection Agency, Carbon Sequestration in Agriculture and Forestry.

US International Climate Change Adaptation and National Security Fund Title XIII ("International Partnerships to Reduce Emissions and Adapt"-IPREA) of the Lieberman-Warner Climate Security Act of 2008.

URC 2008b (UNEP- Risoe Centre 2008): CDM supply report, June, 2008.

USEPA (1994), EPA 400-F-92-014 U.S. Environmental Protection Agency, Office of the Mobile Sources, Fact Sheet-12, August, 1994 (http://www.epa.gov/OMS/consumer/12-miles.pdf)

USPEA Emission Inventory Improvement Program: http://www.epa.gov/ttn/chief/eiip/techreport/volume08/index.html

W.A. Pizer, W.A (2007), Evolution of Global Climate Change Agreement, pizer@rff.org.: http://www.rff.org/documents/RFF-DP-07-03.pdf).

Wang,W.-C., Yung, Y.L., Lacis, A.A., Mo, T., Hansen, J.E., 1976. Greenhouse effects due to man-made perturbation of trace gases. Science 194, 685–690. Doi: 10.1126/ Science.194.4266.685.

Warren Bell and John Drexhage (2005), Climate Change and the International Carbon Market, International Institute for Sustainable Development, USA (http://www.iisd.org)

Weyant, J.P. (1999), "The Costs of the Kyoto Protocol: A Multi-Model Evaluation." The Energy Journal, Special Issue.

Weyant, John P (2004), Introduction and Overview, Energy Economics 26(4):502–515.

Weyant, John P. and Jennifer Hill (1999), The Costs of the Kyoto Protocol: A Multi-Model Evaluation, Introduction and Overview. Energy Journal Special Issue, vii–xliv.

World Bank, State and Trends of the Carbon Market, 2007.

World Summit on Sustainable Development (WSSD, 2002)-Development imperatives: http://www.gdrc.org/uem/energy/renewable-energy-agenda-2004.html

World Bank (2009), CDM in China: Moving towards a value added approach, Beijing/ Washington (in press).

Yamartino, R.J; D. G. Smith; S. A. Bremer; D. Heinold; D. Lamich (2008), Impact of Aircraft Emissions on Air Quality in the Vicinity of Airports. Volume II, An Updated Model Assessment of Aircraft Generated Air Pollution at LAX, JFK, and ORD: http://www.stormingmedia.us/50/5001/A500190.html

Weblinks

A Change in the Climate': http://www.sweden.se/upload/promotion_forum/Brand_Sweden/Bevis/Internationell_ranking_axplock.doc

Acid rain Effect-http://www.essortment.com/all/acidraineffect_rqmz.htm
Acid rain on aquatic life- http://www.epa.gov/acidrain/effects/surface_water.html
Acid rain on Human Health- http://www.epa.gov/acidrain/effects/health.html
Acid Rains: http://www.geocities.com/CapeCanaveral/Hall/9111/DOCS.HTML
Acid Rains:
http://www.google.co.in/search?hl=en&q=Acid+Rains&btnG=Google+Search&meta=&aq=f&oq
Air Quality : http://ncseonline.org/nle/crsreports/air/air-27.cfm
Asia Carbon Group and the Asian Carbon Exchange: http://www.asiacarbon.com/aboutus.htm
AVIVA CSR Report, 2008: http://files.the-group.net/library/aviva/csrreport2008/pdfs/csr08_elements_communities.pdf
Blue-Green Carbon: CSR: http://www.bluegreencarbon.com/library/CSR.pdf
British Antarctic Survey: http://www.sciencedaily.com/releases/2007/07/070726104756.htm
Butner-http://www.recercat.net/handle/2072/1750
Carbon Credit Capital (CCC): (http://www.carboncreditcapital.com/
Carbon Credit Exchange trading market (http://www.cmawebline.org/documents/JAMAR-v6.1- %20Carbon%20Strat%20Mang%20Accounting.pdf
CCT Processes & Systems (http://www.greenfacts.org/en/co2-capture-storage/figtableboxes/figure-3.htm
Carbon Credit Offsets : http://en.wikipedia.org/wiki/Carbon_offset
Carbon Credit/ carbon offset: http://en.wikipedia.org/wiki/Carbon_credit
Carbon dioxide world emission- http://infoweb.magi.com/~dwalsh/wfsesr.html
CDM-IBM- http://www-05.ibm.com/innovation/uk/green/pdf/joint_cdp_and_ibm_study.pdf).
Carbon Emissions Offset Directory; Eco-Business Links, 2007.
Carbon & Energy Management: http://www.environcorp.com/services/article.php?id=3110
Carbon Finance: http://en.wikipedia.org/wiki/Carbon_finance
Carbon Footprint:http://www.lowcarbonsg.com/2009/05/08/measure-your-organisations-carbon-footprint-or-greenhouse-gas-inventory/
Carbon footprints: (http://www.ecosecurities.com/Home/EcoSecurities_the_carbon_market/The_carbon_market/default.aspx
Carbon Foot-printing: http://en.wikipedia.org/wiki/Carbon_footprint
Carbon Labelling Act-2008-California: http://www.learnaboutcarbon.net/standards/carbon-footprint-standards-agreements-legislation/carbon-labeling-act-2008
Carbon Literacy- http://en.wiktionary.org/wiki/carbon_literacy
Carbon Neutral Company: http://en.wikipedia.org/wiki/The_Carbon_Neutral_Company
Carbon Neutral design Project: http://www.architecture.uwaterloo.ca/faculty_projects/terri/carbon-aia/introduction.html
Carbon Neutral: http://blog.oup.com/2006/11/carbon_neutral_/
Carbon Neutral: http://www.firstclimate.com/en/climate-neutral/climate-neutrality.html

Carbon Neutrality: http://www.icfi.com/markets/climate-change/doc_files/carbon-neutrality.pdf

Carbon Offset: http://www.webopedia.com/TERM/C/carbon_neutral.html

Carbon Offsets: http://www.odi.org.uk/fpeg/publications/policybriefs/forestrybriefings/0707_forestrybriefing13_carbonoffsets.pdf, Retrieved on 2007

Carbon Offsetting: http://www.sd-commission.org.uk/pages/carbon-neutrality-and-offsetting.html

Carbon Tax: http://en.wikipedia.org/wiki/Carbon_tax

Carbon Trading Ethics: http://economictimes.indiatimes.com/Opinion/The-ethics-of-carbon-trading/articleshow/2612592.cms

Carbon-offsetting: All credit to them - Climate Change, Environment - Independent.co.uk

CarbonPositive: http://www.carbonpositive.net/

Carbon Management-NASA: http://nasascience.nasa.gov/earth-science/applied-sciences/national-applications/carbon-management

Carbon Management: http://www.geos.ed.ac.uk/masters/cman_info/

Carbon Software Market: http://www.the-infoshop.com/report/infi92090-carbon-m-softw.html

CD4CDM: http://www.cd4cdm.org/

CDM & Carbon Funds:
http://www.itpi.co.in/Resources/Clean_development_mechanisms/Business%20Plan%20for%20Establishing%20Indian%20Carbon%20Fund%200407.pdf

CDM Executive Board: http://cdmindia.nic.in/cdm_india.htm

CDM Methodology Panel : http://cdm.unfccc.int/methodologies/PAmethodologies/Clarifications and http://cdm. unfccc.int/Reference/Guidclarif

CDM Policy Framework-Canada: http://www.co2offsetresearch.org/policy/Canada.html

CDM Potential in Malaysia: http://cdm.eib.org.my/subindex.php?menu=7&submenu=43

CDM-India: http://cdmindia.com/India%20CDM%20Potential.pdf

CDM-India:http://www.cdmindia.com/gtz/publications/INDIA-%20Worlds%20Leading%20CDM%20Destination.pdf

Chicago Climate Exchange (CCX-2007)- http://www.chicagoclimatex.com/

Chicago Climate Exchange: http://www.chicagoclimatex.com/

Clean-Air Canada: http://www.cleanaircanada.org/

Climate Change 2007: Impacts, Adaptation and Vulnerability (http://environment.about.com/od/globalwarming/a/ipcc_report_two.htm

Climate Change Convention as reported by the United Nations (http://unfccc.int/files/essential_background/background_publications_htmlpdf/application/pdf _/ghg table_06.pdf

Climate change: http://www.bibalex.org/Libraries/Presentation/Static/Climate_Change.pdf

CCP-Leakage: http://www.canadiangeographic.ca/magazine/JF08/indepth/weyburn.asp

CCP-Carbon Capture & Storage: http://en.wikipedia.org/wiki/Carbon_capture_and_storage
Carbon Labeling Program: http://ecf.pik-potsdam.de/Processes/carbon-footprint
Climate Literacy: "The Essential Principles of Climate Sciences" U.S. Global Change Research Program (USGCRP): http://www.globalchange.gov/resources/educators/climate-literacy
Climate Literacy: http://downloads.climatescience.gov/Literacy/Climate%20Literacy%20Booklet%20Low-Res.pdf.
Climate Literacy: The Essential Principles of Climate Sciences- A Guide for Individuals & Community:
Climate Neutral Design Principle-_www.pi.energy.gov/enhancingGHGregistry/comments/.../hall2.doc
Climate Neutral Gold Standard: http://climateneutralnetwork.org/standards.php
Climate Neutral Market Innovations: www.pi.energy.gov/enhancingGHGregistry/comments/.../hall2.doc
Climate Neutrality-http://en.wikipedia.org/wiki/Carbon_neutra
Coalbed Methane: http://www.geocities.com/pemnq/cbm.html
CBM in USA: http://www.netl.doe.gov/technologies/oil-gas/FutureSupply/CoalBedNG/CoalBed_NG.html
CBM: http://pages.ca.inter.net/~jhwalsh/wcbm1.html
DEFRA: http://www.teqs.net/DEFRAPFSresponse.pdf
DEFRA-TEQ Study-2008-http://www.powerswitch.org.uk/forum/viewtopic.php?t=7692&sid=1e73d770929f7484721da4f4cf31da26
Dieter Helm New College, Oxford, August 10th 2003: http://www.dieterhelm.co.uk/publications/MARKALModelComments.pdf)
Domestic energy market: http://www.infoplease.com/ipa/A0004686.html
DTI-MARKAL (2003), Brief Comments on the MARKAL Model and its Use by the DTI
Ecology and Environment, Inc. (EEI): http://www.google.com/finance?q=NASDAQ:EEI
Economics of Emission Trading: http://en.wikipedia.org/w/index.php?title=Emissions_trading&action=edit&section=4
Ecosystem Marketplace, State of the Voluntary Carbon Markets- 2007
Effects of Acid rain on Surface water: http://www.epa.gov/acidrain/effects/surface_water.html
Effects of Acid rains: http://www.ace.mmu.ac.uk/eae/acid_rain/older/Impacts.html
Effects of Global Warming on Society: www.comnett.net/~wit/food.html
Emission & Sequestration Inventory: http://www.calvin.edu/admin/provost/environmental/
Emissions-Thermal Power plant: http://www.enzenglobal.com/pdf_downloads/environmental_impact.pdf
Emission Trading: http://en.wikipedia.org/wiki/Emissions_trading
Emissions Trading: http://www.btinternet.com/~foster.enviro/web/resources/glossary.htm

Emissions Trading-Beijing Olympics: http://in.news.yahoo.com/139/20090726/981/tsc-emission-control-during-beijing-olym.html

Emissions Marketing Association: http://www.environmentalmarkets.org/index.ww

Emissions-Climate change: http://europa.eu.int/comm/environment/climat/emission.htm

Energy Conservation Act-1983: http://www.energymanagertraining.com/new_index.php

Emission Accounting-Australian Model: http://www.climatechange.gov.au/ncas/publications/strategic/index.html

Energy Market: (http://www.infoplease.com/ipa/A0004686.html

Energy Carbon Nexus: http://www.sapsustainabilityreport.com/solutions/energyAndCarbon.html?sct=challenge

EU-ETS : www.cdmcapacity.org/glossary.html

Emissions Trading in China: http://www.ecoworld.com/business/china-to-adopt-emissions-trading.html

European Climate Exchange: http://www.europeanclimateexchange.com/index_flash.php

European Union Emissions Trading Scheme (EU ETS): http://europa.eu.int/comm/environment/climat/emission.htm

Feedstock Diversity in Energy Production: www.doc.mmu.ac.uk/aric/gcc/cell.html#pos6

Flooding & Global warming: Clorofluorocarbons-http://www.scientificamerican.com/article.cfm?id=chlorofluorocarbons-cfcs

G-8 Summit, Germany: http://en.wikipedia.org/wiki/33rd_G8_summit

German Advisory Council on Global Change (http://www.wbgu.de/wbgu_sn2003_engl.html

GHG Assessment Handbook, 1988- The World Bank

GHG in Earth's atmosphere: www.abcnews.com/sections/us/global106.html

GHG Protocols: www.ghgprotocol.org

GIS-MIT: http://www.netl.doe.gov/publications/factsheets/project/Proj285.pdf

Global Carbon Market: http://www.wbcsd.org/DocRoot/wrHqIUtkoNq4sC2wodX8/linkages.pdf

Global Environmental Concerns: http://www.bee-india.nic.in/GuideBooks/1Ch9.pdf

Global warming & Greenhouse Effect: www.eecs.umich.edu/mathscience/funexperiments/agesubject/lessons/images/diagrampage.html

Global warming on Society: www.comnett.net/~wit/food.html

Global warming Potential: http://www.epa.gov/climatechange/emissions/downloads/ghg_gwp.pdf

Global warming: http://en.wikipedia.org/wiki/Global_warming

Global warming: www.epa.gov/oppeoeel/globalwarming/impacts/health/index.html

Greenhouse Effect: http://www.mpcfaculty.net/mark_bishop/greenhouse.htm

Greenhouse Effect: http://www.sciencelearn.org.nz/contexts/icy_ecosystems/science_ideas_and_concepts/greenhouse_effect

Greenhouse Effect: www.eecs.umich.edu/mathscience/funexperiments/agesubject/lessons/images/diagrampage.html

Greenhouse Gas-Credit Aggregation Pool (GG-CAP): http://www.natsource.com/

Greenhouse gases" - http://www.lenntech.com/greenhouse-effect/greenhouse-gases.htm#ixzz0GDPksrV7&A

Greenhouse Gases: http://r0.unctad.org/ghg/download/other/pilot_ghg.pdf

GTZ-India CDM Project: http://www.gtz.de/en/weltweit/asien-pazifik/indien/9521.htm

Handbook for the International Treaties for the Protection of the Ozone Layer, 2000:

MWGGA: http://en.wikipedia.org/wiki/Midwestern_Greenhouse_Gas_Accord

http://www.mcxindia.com/PressRelease/MCX_PR_21jan08.pdf 979905B5A6B8E39D

http://www.unep.org/ozone or http://www.unep.ch/ozone

http://cdm.unfccc.int/methodologies/ PA methodologies

http://ecobusinesslinks.com/carbon_offset_wind_credits_carbon_reduction.htm, Retrieved on 2007

http://infoweb.magi.com/~dwalsh/wfsesr.html

http://papers.ssrn.com/sol3/papers.cfm?abstract_id=452400

http://www.architecture.uwaterloo.ca/faculty_projects/terri/carbon-aia/introduction.html

http://www.decisioncraft.com/energy/papers/ecc/re/recc.pdf

http://www.ec.gc.ca/doc/media/m_124/report_eng.pdf.

http://www.gefweb.org/

http://www.geocities.com/CapeCanaveral/Hall/9111/DOCS.HTML

http://www.hunton.com/practices/practice_detail.aspx?gr_H4ID=1087

http://www.iisd.ca/vol12/enb1276e.html

http://www.isr.gov.au/content/controlfiles/display_details.cfm?objectid=6D35F6EF-3AFE-443B-

http://www.nature.com/climate/2007/0711/full/climate.2007.58.html.

http://www.newcastle.co.uk/content/docs/2007-press/04-12-2007.pdf

http://www.rsacarbonlimited.org/uploads/documents/CarbonLimited_InterimRecommendations_37.pdf

http://www.teap.org/

http://www.unmfs.org/

Ice berg melting & rise of sea level -http://www.britannica.com/EBchecked/topic/281212/iceberg/273740/Climatic-impacts-of-icebergs

ICEcap Ltd.: http://www.icecapltd.com/index.asp

ICX: http://www.chicagoclimatex.com/content.jsf?id=1622

IEA-http://www.earthscape.org/r1/ES17006/TSRAppendix2003.pdf

IEA's Energy Policy: http://www.iea.org/Textbase/subjectqueries/keyresult.asp?KEYWORD_ID=4106

IETA, (2002), Accounting for carbon under the UK Emissions Trading Scheme, IETA, 2002.

IFC (2009), Carbon Finance Outlook, January, 2009: http://www.icfi.com/Markets/Energy/doc_files/carbon-procurement-guide.pdf

IFC Carbon Fund-http://www.icfi.com/Markets/Energy/doc_files/carbon-procurement-guide.pdf
IFC-International Carbon Finance: http://www.ifc.org/carbonfinance
IFRS- First time Adoption of International Financial Reporting Standards (July, 2002): http://en.wikipedia.org/wiki/International_Financial_Reporting_Standards
IGP: http://www.econ4env.co.za/archives/climate/connolly.pdf
Impact of Acid rain: http://www.ace.mmu.ac.uk/eae/acid_rain/older/Impacts.html
India's CERs production: http://www.cdmindia.com/gtz/publications/INDIA-%20Worlds%20Leading%20CDM%20Destination.pdf
International Emissions Trading Association: http://www.ieta.org/ieta/www/pages/index.php
International Energy Agency (2008) World Energy Outlook (Part C, The role of energy in climate policy).
International Finance Corporation (IFC): http://www.ifc.org/ifcext/enviro.nsf/AttachmentsByTitle/p_2005SustReport_15/$FILE/ClimateChange.pdf
International Greenhouse Partnerships (IGP) Program-
IPCC 2001: http://www.epa.gov/climatechange/emissions/downloads/ghg_gwp.pdf
IPCC climate change info (http://www.ipcc.ch/present/graphics.htm)
IPCC Database on Greenhouse Gas Emission Factors (EFDB)-http: // www.ipcc-nggip-iges.or.jp/EFDB/main.php
IPCC Fourth Assessment (2007), Impacts, Adaptation and Vulnerability.(Summary for Policymakers). http://www.ipcc.ch/SPM13apr07.pdf
IPCC Fourth Assessment (2007), Mitigation of Climate Change, (Summary for Policymakers)
IPCC: http://www.ipcc.ch/SPM040507.pdf
Life Cycle Assessment (LCA):http://en.wikipedia.org/wiki/Life_cycle_assessment)
LRTAP- http://www.unece.org/env/lrtap/
Kyoto Protocol Agreement: http://dsp-psd.pwgsc.gc.ca/Collection-R/LoPBdP/MR/mr150-e.htm
MARKAL Model-http://www.law.monash.edu.au/regstudies/synopsis.scenarios-sensitivities-on-long-term-carbon-reductions-using-the-uk-markal-and-markalmacro-energy-system-models.pdf
MCX - Carbon Credits Contract
Melting Iceberg in Antartic-http://www.nature-blog.com/2008/07/arctic-ice-melting-at-record-speed.html
Methane Concentration in Rice Field: www.envirolink.org/orgs/edf/sitemap.html
Montréal Protocol:http://www.undp.org/seed/eap/montreal/index.htm
New Clues to Ozone Depletion: http://www.sciencedaily.com/releases/2007/07/070726104756.htm).
Ozone Depletion -Environmental Indicators-http://www.epa.gov/Ozone/science/indicat/index.html
Ozone Depletion- http://en.wikipedia.org/wiki/Ozone_depletion
Ozone depletion:
Ozone depletion: http://en.wikipedia.org/wiki/Ozone_depletion

Ozone depletion-Environmental Indicators: http://www.epa.gov/Ozone/science/indicat/index.html

Ozone Hole- http://www.ucsusa.org/global_warming/science_and_impacts/science/the-science-of-ozone.html

Ozone hole : http://www.unep.org/ozone

Ozone hole: http://www.atm.ch.cam.ac.uk/tour/part3.html

Ozone hole: http://www.ucsusa.org/global_warming/science_and_impacts/science/the-science-of-ozone.html

Ozone-UNEP : http://www.unep.ch/ozone

PCF Pilotprojekt Deutschland: http://www.pcf-projekt.de/main/news/

Policy Directions to 2050: http://www.wbcsd.org/plugins/DocSearch/details.asp?type=DocDet&ObjectId=MjM1MTE

Regional Carbon Sequestration Partnerships (RCSP)-http://www.netl.doe.gov/publications/press/2007/07016-Carbon_Sequestration_Atlas_Publish.html

Regional Greenhouse Gas Initiative (RGGI) (http://www.rff.org/rff/News/Features/Regional-Greenhouse-Gas-Initiative.cfm

Renewable Energy Certificate System (RECS): http://www.recs.org/

Revised IPCC Guidelines for National GHG Inventories-1996

Rise of Sea level: www.envirolink.org/orgs/edf/sitemap.html

SARI/Energy Web site (www.sari-energy.org)

SARI-http://www.usaid.gov/our_work/environment/climate/docs/cp/sari.pdf

Science of Ozone Hole- http://www.atm.ch.cam.ac.uk/tour/part3.html

SDC- http://www.sd-commission.org.uk/publications/downloads/carbon_neutral_scotland_workshop_report.pdf

Second Version: March 2009; www.climatescience.gov

Suarez, C.E. (1995), 'Energy needs for sustainable development: Energy as an instrument for socio-economic development', in T.B. Johansson and J.Goldemberg (eds.), Energy for Sustainable Development: A Policy Agenda, New York, United Nations Development Programme, pp. 18–27.

Sustainable Development: http://www.global-greenhouse-warming.com/sustainable-development-policies.html

Tax Planning: http://taxpolicy.ird.govt.nz/publications/files/emissions.pdf

The World Bank : http://www-esd.worldbank.org/mp

The World Climate Report: www.abc.com/sections/us/global106.html

Tradable Entitlements: http://www.unctad.org/en/docs/poditctedm9.en.pdf

UK Carbon Trust-2008: http://www.climatechangecorp.com/content.asp?ContentID=5007

UN Climate Neutral Strategy EMG/AM.07/11: http://www.unemg.org/climateneutralun/Portals/24/Documents/Meetings/CNUN/UNClimateNeutralStrategy.pdf

UNEP Decarbonization: http://www.climatemediapartnership.org/spip.php?article405

UNEP Guidelines for Funding CDM Projects-http://www.cd4cdm.org/Publications/FinanceCDMprojectsGuidebook.pdf

UNFCCC/CCNUCC-CDM – Executive Board, Version 02 / November 2003

UNIDO Knowledge Network for Industrial Technology Transfer-www.unido.org/knitt/index.html

UNIDO : http://www.unido.org/

United Nations Development Programme – Millennium Development Goals (MDG) Carbon Facility: http://www.undp.org/mdgcarbonfacility/docs/BookletMDGCarbonFacility.pdf

UNEP: http://www.unep.org/civil_society/GCSF/contributions/statement_to_GC9.pdf

UNEP- http://www.karmayog.org/startanngo/startanngo_16320.htm

United States Environmental Protection Agency, Carbon Sequestration in Agriculture and Forestry

USPEA Emission Inventory Improvement Program:

http://www.epa.gov/ttn/chief/eiip/techreport/volume08/index.html

WB-CIF: http://www.bicusa.org/en/Issue.Background.48.aspx

WCC-3 and climate neutrality: http://www.wmo.int/wcc3/WCC3_green_en.php

World Bank - List of Carbon Funds: http://carbonfinance.org/Router.cfm?Page=Funds&ItemID=24670

World Bank Carbon Finance Unit: http://carbonfinance.org/Router.cfm?Page=Home&ItemID=24675

World Bank, State and Trends of the Carbon Market- 2007

World Climate Report (http://www.co2andclimate.org/climate/previous_issues/vol3/v3n20/feature1.htm)

www.abcnews.com/sections/us/global106.html

www.aloha.net~jhanson/page70.htm

www.doc.mmu.ac.uk/aric/gcc/cell.html#pos6

www.envirolink.org/orgs/edf/sitemap.html

http://www.un.org/climatechange/pdfs Acting%20on%20Climate%20Change.pdf

http://www.unido.org/fileadmin/import/72285_Kyoto_Protocol_carbon_credits.pdf

Zero Emission-http://www.tts-marine.com/Pressroom/Press-releases/ZERO—Emission-AGV-technology-developed-by-Numexia-and-TTS-/

Zoro-EmissionTechnologies: http://en.wikipedia.org/wiki/Zero_emission#Potential_future_technologies

Economics of Kyoto Protocol: http://www.ieta.org/ieta/www/pages/getfile.php?docID=129

Glossary

A

A/R CDM Project activity (A/R): An A/R CDM project activity is an afforestation or reforestation measure, operation or action that aims at achieving net anthropogenic GHG removals by sinks. The Kyoto Protocol and the CDM modalities and procedures use the term "project activity" as opposed to "project". An A/R CDM project activity could, therefore, be identical with or a component or aspect of a project undertaken or planned.

A/R: Afforestation and reforestation: Term given to the class of projects devoted to the planting of trees on unforested land for carbon emissions reduction and other environmental benefits.

AAU: Assigned Amount Unit. Allowances for carbon emissions allocated to developed countries up to their target level under the Kyoto Protocol. These allowances are tradable under Kyoto's international emission trading mechanisms in place from 2008 to 2012. Each AAU equates to one tonne of CO_2e.

ABARE: The Australian Bureau of Agricultural and Resource Economics, is a government economic research agency which provides independent research and forecasting on numerous specialized areas, including the quality of the Australian environment.

Abatement: Adjustments within communities and ecosystems to cope with changing climate conditions. Examples include; the construction of flood walls to protect property from stronger storms and heavier precipitation, or the planting of agricultural crops and trees more suited to warmer temperatures and drier soil conditions.

Absorption of Radiation: The uptake of *radiation* by a solid body, liquid or gas. The absorbed energy may be transferred or re-emitted.

ACCC: Australian Competition and Consumer Commission. An independent Commonwealth statutory authority formed in 1995 to administer the Trade Practices Act 1974 and other acts. The ACCC promotes competition and fair trade in the market place to benefit consumers, business and the community. Its primary responsibility is to ensure that individuals and businesses comply with the Commonwealth competition, fair trading and consumer protection laws.

Acceptance/Approval: These instruments have the same legal effect as ratification and are used when, at national level, constitutional law does not require a treaty to be ratified by the head of state.

Accession: Accession is the act whereby a state accepts the offer or the opportunity to become a party to a treaty already negotiated and signed by other states.

Acid Rain: Also known as "acid deposition." Acidic aerosols in the *atmosphere* are removed from the atmosphere by wet deposition (rain, snow, fog) or dry deposition (particles sticking to vegetation). Acidic aerosols are present in the atmosphere primarily due to discharges of gaseous sulfur oxides (*sulfur dioxide*) and *nitrogen oxides* from both *anthropogenic* and natural sources. In the atmosphere these gases combine with water to form acids.

Actual net greenhouse gas removals by sinks (A/R - SSC A/R): Actual net greenhouse gas (GHG) removals by sinks" is the sum of the verifiable changes in carbon stocks in the carbon pools within the project boundary, minus the increase in emissions of the GHGs measured in CO_2 equivalents by the sources that are increased as a result of the implementation of the afforestation or reforestation (A/R) project activity within the project boundary, attributable to the A/R CDM project activity.

Adaptation -Adaptation is the process of preparing for and adjusting to the climate changes that will occur. For example, adaptation actions may moderate the harm from extreme weather events or may exploit the beneficial opportunities of changing rainfall patterns. Adaptation to climate change refers to adjustment in natural or human systems in response to actual or expected climatic stimuli or their effects, which moderates harm or exploits beneficial opportunities. Various types of adaptation can be distinguished, including anticipatory and reactive adaptation, private and public adaptation, and autonomous and planned adaptation.

Adaptive Capacity: is the ability of a system to adjust to climate change (including climate variability and extremes) to moderate potential damages, to take advantage of opportunities, or to cope with the consequences. Also see the page explaining the link between adaptation and mitigation.

Additional (SSC A/R): A small-scale afforestation or reforestation project activity under the CDM is additional if the actual net greenhouse gas removals by sinks are increased above the sum of the changes in carbon stocks in the carbon pools within the project boundary that would have occurred in the absence of the registered small-scale afforestation or reforestation project activity under the CDM.

Additionality: A key eligibility test for projects designed to generate carbon credits under the Kyoto Protocol's CDM and JI programmes, as well as other carbon offset schemes. This test dictates that projects are only eligible for carbon credits if the resulting emission reductions weren't going to happen anyway, ie. They are additional to what would have occurred without the carbon credit incentive.

Advection Fog: Fog formed when wind flows over a surface with a different temperature.

Aerator: An aerator is a device that breaks the flow from your shower head (or other water supply) into a finer spray of droplets - so you get just as wet while using less water.

Aerosol: A collection of airborne solid or liquid particles, with a typical size between 0.01 and 10 micrometers (μm) and residing in the atmosphere for at least several hours. Aerosols may be of either natural or anthropogenic origin. Aerosols may influence climate in two ways: directly through scattering and absorbing radiation, and indirectly through acting as condensation nuclei for cloud formation or modifying the optical properties and lifetime of clouds. The term has also come to be associated, erroneously, with the propellant used in "aerosol sprays."

Afforestation (A/R - SSC A/R): "Afforestation" is the direct human-induced conversion of land that has not been forested for a period of at least 50 years to forested land through planting, seeding and/or the human-induced promotion of natural seed sources.

AGO: The Australian Greenhouse Office has now been incorporated into the Department of Climate Change.

AGO Factors and Methods Workbook: The workbook has now been updated and replaced by the National Greenhouse Accounts Factors. For more information, see here.

AGO GF: 'Australian Greenhouse Office Greenhouse Friendly' Abatement Program was launched in 2001 and now forms part of the Australian Government's Greenhouse Challenge Plus programme. It provides businesses and consumers with the opportunity to sell and purchase greenhouse neutral products and services and broaden the basis for investment in additional greenhouse gas abatement. For more information, see here.

Agro-ecology is the science of sustainable agriculture; the methods of agro-ecology have as their goal achieving sustainability of agricultural systems balanced in all spheres. This includes the socio-economic and the ecological or environmental.

Air pollution: Air pollution usually refers to the presence of any chemical or particulate that alters the normal make-up of our atmosphere, causing direct threats to human health (such as breathing difficulties) or longer-term damage through its effects on our planet's ecosystem. Pollutants include smoke and dust, nitrogen oxides, methane, and the fumes from aerosol sprays and other solvents. Industrial processes and transport are major contributors to air pollution, but it can also be caused by natural processes such as forest fires and volcanoes. See also acid rain.

Albedo: Refers to the ratio of light from the sun that is reflected by the Earth's surface to the light received by it. Unreflected light is converted to infrared radiation (i.e., heat), which causes atmospheric warming (see "radiative forcing"). Thus, surfaces with a high albedo (e.g., snow and ice) generally contribute to cooling, whereas surfaces with a low albedo (e.g., forests) generally contribute to warming. Changes in land use that significantly alter the characteristics of land surfaces can therefore influence the climate through changes in albedo.

Algal Bloom is the explosive growth of blue green algae that deprives aquatic life of oxygen. Algal blooms can be toxic to animals and humans.

Alliance of Small Island States (AOSIS): The group of Pacific and Caribbean nations who call for relatively fast action by developed nations to reduce *greenhouse gas emissions*. The AOSIS countries fear the effects of rising sea levels and increased storm activity predicted to accompany *global warming*. Its plan is to hold *Annex I Parties* to a 20 percent reduction in *carbon dioxide* emissions by the year 2005.

Alliance of Small Island States (AOSIS): A coalition of some 43 low-lying and small island countries, most of which are members of the G77 that are particularly vulnerable to the potential adverse consequences of climate change such as sea-level rise, coral bleaching, and increased frequency and intensity of tropical storms.

Allianz: Glossary of Emissions Trading: Deutsche Emissionshandelsstelle im Bundesamt – German Emissions Trading Authority (www.dehst.de)

Allocation -Under an emissions trading scheme, permits can initially either be given away for free or auctioned. The Australian. Government is currently considering giving away some permits to energy-intensive, trade-exposed industries and to facilities seriously disadvantaged by the introduction of emissions trading. They propose auctioning the remaining permits.

Allowance: The quantity of carbon dioxide emissions which a company or country is permitted to emit, as a result of regulation

Alternative Energy: Energy derived from nontraditional sources (e.g., compressed natural gas, solar, hydroelectric, wind).

Alternative Fuels Substitutes: Substances for use with traditional oil-derived motor vehicle fuels, like petrol (gasoline) and diesel. This includes mixtures of alcohol-based fuels with petrol, methanol, ethanol, compressed natural gas, and others.

Anaerobic digester: An industrial system that converts waste products into useful energy while reducing greenhouse gas emissions.

Ancillary Benefits: Complementary benefits of a climate policy including improvements in local air quality and reduced reliance of imported fossil fuels.

Annex 1 or Annex B: In practice, Annex 1 of the Convention and Annex B of the Protocol are used almost interchangeably. However, strictly speaking, it is the Annex 1 countries that can invest in JI/CDM projects as well as host JI projects, and non-Annex 1 countries that can host CDM projects, even though it is the Annex B countries that have the emission reduction obligations under the Protocol. Note that Belarus and Turkey are listed in Annex 1 but not Annex B; and that Croatia, Liechtenstein, Monaco and Slovenia are listed in Annex B but not Annex 1. CDP refers to Annex B countries throughout.

Annex A: A list in the Kyoto Protocol of the six greenhouse gases and the sources of emissions covered under the Kyoto Protocol. See also "Basket of Gases."

Annex B Countries: Annex B of the Kyoto Protocol contains a list of the industrial nations committed to regulating their greenhouse gas emissions in the period between 2008 and 2012. These are: Australia, Austria, Belgium, Bulgaria, Canada, Croatia, Czech Republic, Denmark, European Economic Community (including UK), Estonia, Finland, France, Germany, Greece, Hungary, Iceland, Italy, Japan, Latvia, Liechtenstein, Lithuania, Luxembourg, Monaco, The Netherlands, New Zealand, Norway, Poland, Portugal, Romania, Russian Federation, Slovakia, Slovenia, Spain, Sweden, Switzerland, Ukraine, United States. One Annex B countries - the United States - has not ratified the Kyoto agreement and therefore do not face any emission reduction obligations under Kyoto at this time. The Kyoto Protocol offers three 'flexibility mechanisms' to allow countries to meet its targets.

Annex I Countries/Parties: Group of countries included in Annex I (as amended in 1998) to the United Nations Framework Convention on Climate Change, including all the developed countries in the Organization of Economic Co-operation and Development, and economies in transition. By default, the other countries are referred to as Non-Annex I countries. Under Articles 4.2 (a) and 4.2 (b) of the Convention, Annex I countries commit themselves specifically to the aim of returning individually or jointly to their 1990 levels of greenhouse gas emissions by the year 2000.

Annex I Parties: The 40 countries plus the European Economic Community listed in Annex I of the UNFCCC that agreed to try to limit their GHG emissions: Australia, Austria, Belarus, Belgium, Bulgaria, Canada, Croatia, Czech Republic, Denmark, European Economic Community, Estonia, Finland, France, Germany, Greece, Hungary, Iceland, Italy, Japan, Latvia, Liechtenstein, Lithuania, Luxembourg, Monaco, The Netherlands, New Zealand, Norway, Poland, Portugal, Romania, Russian Federation, Slovakia, Slovenia, Spain, Sweden, Switzerland, Turkey, Ukraine, United States.

Annex I, or Annex B: The signatory nations to the Kyoto Protocol that are subject to caps on their emissions of greenhouse gases and committed to reduction targets – countries with developed economies. Annex I refers to the 36 countries identified for reduction in the UNFCCC while the Annex B is an adjusted list of 39 countries identified under the more recent Kyoto Protocol. Annex B countries have their reduction targets formally stated.

Annex II: A subset of Annex 1/B, Annex II countries is signatory nations to the UNFCCC which are also members of the OECD - the most industrialized economies. They have extra obligations to help developing nations combat climate change via technology transfer and financial help.

Anthropogenic Climate Change: Anthropogenic means "human made". So in the context of climate change it refers to greenhouse gases, or emissions that are produced as the result of human activities.

Anthropogenic climate impacts: Changes due to human activity rather than nature; viz. factory emissions or deforestation, rather than a smoking volcano.

Anthropogenic emissions: Greenhouse-gas emissions resulting from human activities. Anthropogenic sources of greenhouse gases include industry, agriculture, mining, transportation, construction, and deforestation.

Anthropogenic: (Merriam-Webster) -"of, relating to, or resulting from the influence of human beings on nature"; *as in* "most of the observed increase in globally averaged temperatures since the mid-twentieth century is very likely due to the observed increase in anthropogenic (man-made) greenhouse gas concentrations"—Summary for Policymakers (Intergovernmental Panel on Climate Change)

AP6: In July 2005, the U.S joined five Asia-Pacific nations - Australia, China, India, Japan and Korea - in launching the Asia-Pacific Partnership on Clean Development and Climate - known as AP6.

AP7 (formerly AP6): Asia-Pacific Partnership on Clean Development and Climate sets up a non-binding partnership between the governments of Australia, People's Republic of China, India, Japan, South Korea, Canada and the United States. The partnership sets out a voluntary framework for action, which encourages cooperation between member parties and the private sector to develop and implement clean energy technologies.

Appendix A (SSC - SSC A/R): Refers to the Appendix A of the simplified modalities and procedures for small-scale CDM project activities which provides for a simplified project design document. The latest form for simplified project design document for small-scale CDM project activities is available in the section 'Project design documents/forms' of the UNFCCC CDM website: http://cdm.unfccc.int/Reference.

Appendix B (SSC - SSC A/R): The Appendix B of the simplified modalities and procedures for small-scale CDM project activities corresponds to the indicative list of simplified baseline and monitoring methodologies for selected small-scale CDM project activity categories. This list is updated and modified by decisions by the Executive Board based on inputs by project participants. The latest version of the list is available on the UNFCCC CDM website under the section on approved small-scale methodologies.

Appendix C (SSC - SSC A/R): The Appendix- C of the simplified modalities and procedures for small-scale CDM project activities corresponds to the procedures for determining the occurrence of debundling.

Approval by Parties involved (All types): A written approval constitutes the authorization by a designated national authority (DNA) of specific entity (ies)' participation as project proponents in the specific CDM project activity. The approval covers the requirements of paragraphs 33 and 40 (a) and (f) of the CDM modalities and procedures.

Argon – (Ar) constitutes 1.3 percent of the atmosphere by weight and 0.94 percent by volume. Argon is isolated on a large scale by the fractional distillation of liquid air. It is used in gas-filled electric light bulbs, radio tubes, and Geiger counters. It also is widely utilized as an inert atmosphere for arc-welding metals, such as aluminium and stainless steel; for the production and fabrication of metals, such as titanium, zirconium, and uranium; and for growing crystals of semiconductors, such as silicon and germanium.

Asia-Pacific Partnership on Clean Development and Climate (APP, AP6): The Asia-Pacific climate pact is a rival international climate change agreement to the Kyoto Protocol. Its initiators in 2005 were the United States and Australia, the only two industrialized nations not to have ratified the Kyoto treaty at that time (Australia since has ratified in 2007). The group also includes China, India, Japan, South Korea and now Canada. APP rejects Kyoto-style emission reduction targets in favour of encouraging business to invest in clean fossil-fuel technology and renewable energy.

Assigned Amount: In the Kyoto Protocol, the permitted emissions, in CO_2 equivalents, during a commitment period. It is calculated using the Quantified Emission Limitation and Reduction Commitment (QELRC), together with rules specifying how and what emissions are to be counted.

Assigned amount (AA) and assigned amount units (AAUs): The assigned amount is the total amount of greenhouse gas that each country is allowed to emit during the first commitment period of the Kyoto Protocol. This total amount is then broken down into measurable units.

Assurance: The process of an independent third party checking the methodology, data and calculation processes to ensure they are robust.

ASX100: Australia's 100 largest companies based on market capitalization.

Atmosphere: The gaseous envelope surrounding the Earth. The dry atmosphere consists almost entirely of nitrogen (78.1% volume mixing ratio) and oxygen (20.9% volume mixing ratio), together with a number of trace gases, such as argon (0.93% volume mixing ratio), helium, radioactively active greenhouse gases such as carbon dioxide (0.035% volume mixing ratio), and ozone. In addition the atmosphere contains water vapor, whose amount is highly variable but typically 1% volume mixing ratio. The atmosphere also contains clouds and aerosols. The layer nearest the Earth is the *troposphere*, which reaches up to an altitude of about 8 km (about 5 miles) in the Polar Regions and up to 17 km (nearly 11 miles) above the equator. The *stratosphere*, which reaches to an altitude of about 50 km (31 miles) lies atop the troposphere. The mesosphere which extends up to 80-90 km is atop the stratosphere, and finally, the thermosphere, or ionosphere, gradually diminishes and forms a fuzzy border with outer space. There is relatively little mixing of gases between layers.

Atmospheric Lifetime: The lifetime of a greenhouse gas refers to the approximate amount of time it would take for the anthropogenic increment to an atmospheric pollutant concentration to return to its natural level (assuming emissions cease) as a result of either being converted to another chemical compound or being taken out of the atmosphere via a sink. This time depends on the pollutant's sources and sinks as well as its reactivity. The lifetime of a pollutant is often considered in conjunction with the mixing of pollutants in the atmosphere; a long lifetime will allow the pollutant to mix throughout the atmosphere. Average lifetimes can vary from about a week (sulfate aerosols) to more than a century (chlorofluorocarbons (CFCs), carbon dioxide).

Attachment A to Appendix B" (SSC): The attachment A to Appendix B corresponds to list of barriers project participants shall use in order to demonstrate that a small-scale project activity would not have occurred otherwise (i.e. is additional). The latest version of the attachment A to Appendix B can be found on the UNFCCC CDM website under the section on approved small-scale methodologies (please refer to http://cdm.unfccc.int/methodologies/SSCmethodologies/approved). A simplified baseline and monitoring methodology listed in Appendix B to the simplified modalities and procedures for small-scale CDM project activities may be used for a small-scale CDM project activity if the project participants are able to demonstrate to a designated operational entity that the project activity would otherwise not be implemented due to the existence of one or more of the barriers listed in attachment A of appendix B. Where specified in appendix B for a project category, quantitative evidence that the project activity would otherwise not be implemented may be provided instead of a demonstration based on the barriers listed in attachment A to appendix B.

Auctioning: A method of allocating permits; Australia's trading scheme is likely to involve a mix of free allocation of some permits and auctioning the remainder.

Australian Emissions Trading Scheme (AETS): Australian Government policy is to introduce a cap and trade scheme for the stationary energy, industrial processes, fugitives, transports and waste sectors by 2010. The centerpiece of the AETS is a carbon emissions market. A price on carbon is needed to address the market failure of currently unpriced greenhouse gas emissions.

Australian Standard 4978 : This standard provides a common approach to determining how much carbon a given area of forest can be expected to remove from the atmosphere, taking account of the species of tree, how densely the trees are planted and a number of other factors.

Australia's National Greenhouse Accounts: A comprehensive set of reports outlining Australia's greenhouse gas emissions. Further information is available at: http://www.greenhouse.gov.au/inventory/

B

Banking and borrowing: The ability under an emissions trading scheme to save emission permits issued in one for use in later years (banking), or bring forward some of a future year's permit allocation for use in the current year (borrowing).

Banking: The possibility of carrying over emissions allowances from one compliance period to the following period is known as banking. In Germany, it is not possible to bank emissions allowances from the 2005-2007 compliance periods for use in the 2008-2012 compliance period; unused 2005-2007 emissions allowances become invalid after April 30, 2008. Banking will, however, be possible in subsequent compliance periods. Banking is most worthwhile if increased prices for emissions allowances are expected, or to hedge against future price changes.

Base Year: Targets for reducing GHG emissions are often defined in relation to a base year. In the Kyoto Protocol, 1990 is the base year for most countries for the major GHGs; 1995 can be used as the base year for some of the minor GHGs.

Baseline - approved methodology (P - A/R): A baseline methodology approved by the Executive Board is publicly available along with relevant guidance on the UNFCCC CDM website (http://unfccc.int/cdm).

Baseline - new methodology (P - A/R): Project participants may propose a new baseline methodology established in a transparent and conservative manner. In developing a new baseline methodology, the first step is to identify the most appropriate approach for the project activity and then an applicable methodology. Project participants shall submit a proposal for a new methodology to a designated operational entity by forwarding a completed "Proposed New Baseline and Monitoring Methodologies (CDM-NM)" and the Project Design Document (CDM-PDD) with sections A to C, including relevant annexes, completed in order to demonstrate the application of the proposed new methodology to a proposed project activity.

Baseline and credit: A type of emissions trading scheme where firms are encouraged to reduce their greenhouse gas emissions below a projected "business as usual" path of increasing emissions. Any reductions below that future path earn credits for the difference which can be sold to other emitters struggling to contain increases to baseline levels.

Baseline and monitoring - simplified A/R approved methodology (SSC A/R): A simplified baseline and monitoring methodology approved by the Executive Board for selected small-scale afforestation and reforestation (A/R) CDM project activity categories is publicly available on the UNFCCC CDM website (http://unfccc.int/cdm).

Baseline and monitoring methodology (SSC A/R): A methodology is an application of a baseline approach. The most likely baseline scenario of the small-scale A/R CDM project activity is considered to be the land-use prior to the implementation of the project activity, whichever are the case grasslands or croplands. Project activities implemented on settlements or wetlands are presently not included in SSC.

Baseline approach (P - SSC): A baseline approach is the basis for a baseline methodology. The Executive Board agreed that the three approaches identified in sub-paragraphs 48 (a) to (c) of the CDM modalities and procedures be the only ones applicable to CDM project activities.

Baseline approach for A/R CDM project activities (A/R): A baseline approach is the basis for a baseline methodology. The Executive Board agreed that the three approaches identified in sub-paragraphs 22 (a) to (d) of the CDM A/R modalities and procedures shall be the only ones applicable to A/R CDM project activities.

Baseline approach for SSC A/R CDM project activities (SSC A/R): A baseline approach is the basis for a baseline methodology. The most likely baseline scenario of the small-scale A/R CDM project activity is considered to be the land-use prior to the implementation of the project activity, whichever is the case grasslands or croplands. Project activities implemented on settlements or wetlands are presently not included in SSC-AR-CDM.

Baseline Emissions: The emissions that would occur without policy intervention (in a business-as-usual scenario). Baseline estimates are needed to determine the effectiveness of emissions reduction programs (often called mitigation strategies).

Baseline for small-scale CDM project activities - approved methodology (SSC): A baseline methodology approved by the Executive Board is included in an indicative list of simplified baseline and monitoring methodologies for selected small-scale CDM project activity categories (contained in the Appendix B to the simplified modalities and procedures for small-scale CDM project activities) and is publicly available along with relevant guidance on the UNFCCC CDM website (http://cdm.unfccc.int/methodologies/SSCmethodologies/approved)

Baseline methodology (A/R): A methodology is an application of an approach as defined in paragraph 22 of the CDM A/R modalities and procedures, to an individual A/R CDM project activity, for the determination of the baseline scenario. A baseline methodology should reflect aspects such as environmental conditions and past land uses and land use changes. No methodology is excluded a priori so that project participants have the opportunity to propose a methodology.

Baseline methodology (P - SSC): A methodology is an application of an approach as defined in paragraph 48 of the CDM modalities and procedures, to an individual project activity, reflecting aspects such as sector and region. No methodology is excluded a priori so that project participants have the opportunity to propose any methodology. In considering paragraph 48, the Executive Board agreed that, in the two cases below, the following applies:

Baseline net greenhouse gas removals by sinks (A/R - SSC A/R): "Baseline net GHG removals by sinks" is the sum of the changes in carbon stocks in the carbon pools within the project boundary that would have occurred in the absence of the A/R CDM project activity.

Baseline scenario (P - SSC): The baseline for a CDM project activity is the scenario that reasonably represents the anthropogenic emissions by sources of greenhouse gases (GHG) that would occur in the absence of the proposed project activity. A baseline shall cover emissions from all gases, sectors and source categories listed in Annex A (of the Kyoto Protocol) within the project boundary. A baseline shall be deemed to reasonably represent the anthropogenic emissions by sources that would occur in the absence of the proposed project activity if it is derived using a baseline methodology referred to in paragraphs 37 and 38 of the CDM modalities and procedures. Different scenarios may be elaborated as potential evolutions of the situation existing before the proposed CDM project activity. The continuation of a current activity could be one of them; implementing the proposed project activity may be another; and many others could be envisaged. Baseline methodologies shall require a narrative description of all reasonable baseline scenarios. To elaborate the different scenarios, different elements shall be taken into consideration, including related guidance issued by the Executive Board.

Baseline scenario for A/R CDM project activities (A/R): The baseline scenario for an A/R CDM project activity is the scenario that reasonably represents the sum of the changes in carbon stocks in the carbon pools within the project boundary that would occur in the absence of the A/R CDM project activity. A baseline scenario shall be derived using a baseline methodology referred to in paragraphs 12 and 13 of the CDM A/R modalities and procedures. A baseline shall cover all carbon pools within the project boundary but project participants may choose not to account for one or more carbon pools if they provide transparent and verifiable information indicating that the choice will not increase the expected net anthropogenic GHG removals by sinks. Different baseline scenarios may be elaborated as potential projections of the situation existing before the proposed A/R CDM project activity. The continuation of an existing activity could be one of them; the implementation of the proposed A/R CDM project activity may be another; and many others could be envisaged. Baseline methodologies shall require a narrative description of all reasonable baseline scenarios. To elaborate the different scenarios, different elements shall be taken into consideration, including related guidance issued by the Executive Board. For instance, the project participants shall take into account national / sectoral policies and circumstances, ongoing technological improvements, past land uses and land-use changes, investment barriers, etc. (see paragraph b (vii) of Appendix C to decision 3/CMP.1 and paragraphs 20 (e) and 22 of decision 5/CMP.1).

Baseline scenario for SSC A/R CDM project activities (SSC A/R): The baseline for a proposed small-scale afforestation or reforestation project activity under the CDM is the scenario that reasonably represents the sum of the changes in carbon stocks in the carbon pools within the project boundary that would have occurred in the absence of the proposed project activity. A baseline shall be deemed to reasonably represent the sum of the changes in carbon stocks in the carbon pools within the project boundary that would occur in the absence of the proposed small-scale afforestation or reforestation project activity under the CDM if it is derived using a baseline methodology referred to in Appendix B of as "CDM-SSC-A/R modalities and procedures" of decision 6/CMP.1. A baseline shall cover all carbon pools as considered for small scale A/R CDM project activities within the project boundary.

Baselines: The baseline estimates of population, GDP, energy use and hence resultant greenhouse gas emissions without climate policies, determine how big a reduction is required, and also what the impacts of climate change without policy will be.

Basket of Gases: This refers to the group six of greenhouse gases regulated under the Kyoto Protocol. They are listed in Annex A of the Kyoto Protocol and include: carbon dioxide (CO_2), methane (CH_4), nitrous oxide (N_2O), hydrofluorocarbons (HFCs), perfluorocarbons (PFCs), and sulphur hexafluoride (SF_6).

Benchmarks: Instruments for determining what allowance should be allocated to each installation type and for recognizing Early Action. Standard emissions values for installation types and industries are determined, for example using BAT (Best Available Technology) or average values, and these standard values are then used to calculate what emissions allowances should be allocated.

Berlin Mandate: A ruling negotiated at the first *Conference of the Parties* (CoP 1), which took place in March, 1995, concluding that the present commitments under the Framework Convention on Climate Change are not adequate. Under the Framework Convention, developed countries pledged to take measures aimed at returning their *greenhouse gas emissions* to 1990 levels by the year 2000. The Berlin Mandate establishes a process that would enable the Parties to take appropriate action for the period beyond 2000, including a strengthening of developed country commitments, through the adoption of a protocol or other legal instruments.

Biochar: Carbon-rich charcoal created when plant matter is heated in an oxygen-free environment. Carbon that would otherwise combine with oxygen, burn, and be emitted to the air is contained in the charcoal, which can be used to fertilise soils or make biofuels.

Biodegradable: Organic matter that can break down or decompose rapidly under natural conditions and processes is referred to as biodegradable. Garden and food waste, animal waste, and most paper products, as well as plastics derived from vegetable content, will biodegrade, but not plastic carrier bags and polystyrene cups, for example.

Biodiesel: Bio diesel is fuel generated from vegetable oil that can be used pure or blended with regular diesel (diesel produced by refining crude oil) in conventional, unmodified diesel engines. It is not the same as waste vegetable oil, otherwise known as 'unwashed biodiesel', which requires engine modification.

Biodiversity: Biodiversity (a contraction of the phrase biological diversity) means the variety of life on earth, or within one particular ecosystem, in terms of the number of distinct biological species present. Tropical rainforests, for example, support a huge variety of species, so are highly biodiverse, while Polar Regions are far less so. Scientists worry that deforestation, such as that associated with planting crops for biofuels, animal feed and human consumption is reducing biodiversity. See also monoculture.

Bio-ethanol: Fuel from crops such as sugar cane, corn or palm oil.

Biofuel: Biofuel is a general term for fuel, including biodiesel, that is derived from biomass - living or recently dead organic matter. In general it is made from sugar, starch, vegetable oils or animal fats. Examples include bio-ethanol from energy crops such sugar cane, corn, palm oil, and rape seed. Biofuels are renewable fuels made from plants that can be used to supplement or replace the fossil fuels petroleum and diesel used for transport. The two main biofuels are ethanol and biodiesel. Ethanol is produced from the fermentation of sugar or starch in crops such as corn and sugar cane. Biodiesel is made from vegetable oils in crops such as soybean, or from animal fats. Depending on the processes used to make biofuels, greenhouse emissions from cars and fuel-powered machinery can be substantially reduced by their use.

Biogeochemical Cycle: Movements through the Earth system of key chemical constituents essential to life, such as carbon, nitrogen, oxygen, and phosphorus.

Biogeochemical Cycle: The chemical interactions that take place among the atmosphere, *biosphere, hydrosphere,* and *geosphere.*

Biomass (All types): Biomass means non-fossilized and biodegradable organic material originating from plants, animals and micro-organisms. This shall also include products, by-products, residues and waste from agriculture, forestry and related industries as well as the non-fossilized and biodegradable organic fractions of industrial and municipal wastes. Biomass also includes gases and liquids recovered from the decomposition of non-fossilized and biodegradable organic material. Biomass residues mean biomass by-products, residues and waste streams from agriculture, forestry and related industries.

Biomass boiler/stove: A biomass boiler is a central heating system or a stove powered by biomass (usually wood) rather than fossil fuels form of pellets or wood.

Biomass Energy: Energy produced by combusting renewable biomass materials such as wood. The *carbon dioxide* emitted from burning biomass will not increase total atmospheric carbon dioxide if this consumption is done on a sustainable basis (i.e., if in a given period of time, regrowth of biomass takes up as much carbon dioxide as is released from biomass combustion). Biomass energy is often suggested as a replacement for *fossil fuel combustion* which has large *greenhouse gas emissions.*

Bioplastics: Instead of petroleum, bio-renewable materials such as starch from corn or whey from cheese-making can be used to make plastics. Industry uses microbes or their enzymes to convert biomass to feedstocks, building blocks for biodegradable plastics, industrial solvents and specialty lubricants.

Biosphere: The part of the Earth system comprising all ecosystems and living organisms, in the atmosphere, on land (terrestrial biosphere) or in the oceans (marine biosphere), including derived dead organic matter, such as litter, soil organic matter and oceanic detritus.

Black Carbon Aerosols: Particles of carbon in the atmosphere produced by inefficient combustion of fossil fuels or biomass. Black carbon aerosols absorb light from the sun, shading and cooling the Earth's surface, but contribute to significant warming of the atmosphere (see "radiative forcing").

Black Carbon: Operationally defined species based on measurement of light absorption and chemical reactivity and/or thermal stability; consists of soot, charcoal, and/or possible light-absorbing refractory organic matter.

Borehole: Any exploratory hole drilled into the Earth or ice to gather geophysical data. Climate researchers often take ice core samples, a type of borehole, to predict atmospheric composition in earlier years.

Broker: A broker is an intermediary, who buys and sells carbon offsets on behalf of clients.

Bryd-Hagel Resolution: In June 1997, anticipating the December 1997 meeting in Kyoto, Senator Robert C. Byrd (D-WV) introduced, with Sen. Chuck Hagel (R-NE) and 44 other cosponsors, a resolution stating that the impending Kyoto Protocol (or any subsequent international climate change agreement) should not - "(A) mandate new commitments to limit or reduce GHG emissions for the Annex I Parties [i.e. industrialized countries], unless the protocol or other agreement also mandates new specific scheduled commitments to limit or reduce GHG emissions for Developing Country Parties within the same compliance period, or (B) would result in serious harm to the economy of the United States..."

Bubble: An option in the Kyoto Protocol that allows a group of countries to meet their targets jointly by aggregating their total emissions. The member states of the European Union are utilizing this option.

Building regulations: Potentially a way to reduce energy use in buildings (for heating, cooling, lighting) which typically accounts for around 40% of CO_2 emissions in developed countries. Tend to be resisted by the building industry because they raise costs.

Bundle (SSC A/R): Bringing together of several small-scale CDM project activities, to form a single CDM project activity or portfolio without the loss of distinctive characteristics of each component project activity and with the total bundle not exceeding the limits stipulated in paragraph 6(c) of decision 17/CP.7 with the aim of lowering transaction costs per unit. A bundle of small-scale afforestation and reforestation project activities satisfies the conditions for bundling and the overall monitoring plan for the bundled small-scale afforestation and reforestation project activities is appropriate

Bundled offsets: Bundled offsets are emissions that do not come from one single project but are a collection of offsets from various projects. The bundles are sometimes from a mixture of accreditation schemes and types.

Burden: The total mass of a gaseous substance in the atmosphere that causes worry and concern.

C

Call option: A call option is a financial contract between two parties, the buyer and the seller of an abatement option. It is the option to buy abatement at a specified time in the future. The buyer of the option has the right, but not the obligation to buy an agreed quantity of abatement from the seller of the option at a certain time for a certain price (the strike price). The seller (or "writer") is obligated to sell the commodity (e.g. carbon offsets). The buyer pays a fee (called a premium) for this right.

Cap: Upper emissions limit or emissions goal for emissions trading at the national level and targets for all macro-sectors (energy production; industry; trade, commerce and services; transport and households) and particularly for what is known as the emissions trading segment, that is the parts of industry that participate in emissions trading in accordance with the guideline. The overall goal is based on Burden Sharing, whereas the targets for the macro-sectors are based on the national climate protection program, the contribution the individual sectors make to carbon dioxide emissions as a whole, and the German Government's climate policy priorities.

Cap and Trade: Cap and Trade is a market-based policy tool for protecting human health and the environment. A cap and trade program first sets an aggressive cap, or maximum limit, on emissions. Sources covered by the program then receive authorizations to emit in the form of emissions allowances, with the total amount of allowances limited by the cap. Each source can design its own compliance strategy to meet the overall reduction requirement, including sale or purchase of allowances, installation of pollution controls, implementation of efficiency measures, among other options. Individual control requirements are not specified under a cap and trade program, but each emissions source must surrender allowances equal to its actual emissions in order to comply. Sources must also completely and accurately measure and report all emissions in a timely manner to guarantee that the overall cap is achieved.

Capital Costs: Costs associated with the capital or investment expenditures on land, plant, equipment and inventories. Unlike labor and operating costs, capital costs are independent of the level of output. (IPCC)

Capital Stocks: The accumulation of machines and structures that are available to an economy at any point in time to prune goods or render services. These activities usually require a quantity of energy that is determined largely by the rate at which that machine or structure is used. (IPCC)

Capture and Storage (CCS): The aim is to prevent CO_2 entering the atmosphere by capturing and storing it in, for example, old oil and gas fields.

Car clubs: Not, in this instance, a club for car enthusiasts, but an organisation providing car rental to members at very short notice and usually for short periods, such as an hour or two.

Car pooling: Car pooling - or lift sharing - is a way of reducing CO_2 emissions from private transport, especially commuter travel, by sharing journeys. It can be arranged informally among friends and colleagues or, increasingly, through dedicated websites. Its aim is to have fewer cars on the road with more people in each.

Carbon: Carbon is the fourth most common chemical element in the universe, and carbon compounds - in other words, carbon chemically combined with other elements - are the basis of all known life forms on earth. Pure carbon appears in many apparently diverse forms, from diamond to graphite to charcoal, but it is much more commonly found in substances such as coal, oil, natural gas, wood and peat that we use for fuel. When we burn these substances to provide energy - either directly in our homes as heat, or in power stations to produce electricity - the combustion process produces 'oxides' of carbon, including the gas CO_2.

Carbon (Natural) Sequestration: Carbon (natural) sequestration refers to the long-term storage of carbon in forests and vegetation, soils or in the oceans.

Carbon and Environment Daily: Glossary of Environmental Management.

Carbon Calculation: A method of calculating the amount of greenhouse gas emissions produced in a certain time frame from an activity. Carbon calculation can be done for anything from one car journey to a whole business or industry. There is a number of different calculation methods used in the carbon offset market.

Carbon capture: Carbon capture, carbon sequestration, or CCS (carbon capture and storage) are all terms to describe relatively new technologies designed to let major producers of CO_2 emissions, such as fossil fuel-burning power stations, prevent the CO_2 they create being released into the atmosphere. Instead it is stored by being injected into underground or undersea geological formations. Some CCS technology is already in operation on a limited scale in other countries but its use is not yet widespread.

Carbon capture and storage (CCS): the process of capturing greenhouse gas pollution from coal or gas power plants and storing it underground instead of releasing it into the atmosphere. Technology used to capture emissions of carbon dioxide, often from coal-fired electricity generators, for transport and long-term storage. Captured carbon dioxide can be stored in a variety of geological or ocean sites and can reduce emissions by up to 70-80% from a power plant.

Carbon Mitigation Strategies: Carbon Mitigation Strategies Energy efficiency and conservation Efficient vehicles? car size, power Reduce vehicle use ? mass transit, urban design Efficient buildings ? Solar heating, ventilation Efficient baseload coal plants Renewable electricity and fuels Wind power Photo Voltaic (Solar) power H_2 fuel cell car Biomass fuel for fossil fuel Fuel shift Gas baseload power for coal power Nuclear fission for coal power CO_2 capture and storage CO_2 capture at baseload power plant ? technology already in use for H_2 Capture CO_2 at H_2 plant Capture CO_2 at coal-to-synfuels plants Geological storage Forests and agriculture Reduce deforestation, afforestation and reforestation Conservation tillage .

Carbon credit: A generic term to assign a value to a reduction or offset of greenhouse gas emissions. A carbon credit is usually equivalent to one tonne of carbon dioxide equivalent (CO_2-e). A carbon credit can be used by a business or individual to reduce their carbon footprint by investing in an activity that has reduced or sequestered greenhouse gases at another site. Also, an Entitlement Certificates issued by the UNFCCC to implementers of approved CDM projects; each represents one metric ton of carbon dioxide or equivalent Greenhouse Gas.

Carbon Cycle: All parts (reservoirs) and fluxes of carbon. The cycle is usually thought of as four main reservoirs of carbon interconnected by pathways of exchange. The reservoirs are the atmosphere, terrestrial biosphere (usually includes freshwater systems), oceans, and sediments (includes fossil fuels). The annual movements of carbon, the carbon exchanges between reservoirs, occur because of various chemical, physical, geological, and biological processes. The ocean contains the largest pool of carbon near the surface of the Earth, but most of that pool is not involved with rapid exchange with the atmosphere.

Carbon Dioxide (CO_2): The *greenhouse gas* whose concentration is being most affected directly by human activities. CO_2 also serves as the reference to compare all other greenhouse gases. The major source of CO_2 *emissions* is *fossil fuel combustion*. CO_2 emissions are also a product of forest clearing, *biomass* burning, and non-energy production processes such as cement production. Atmospheric concentrations of CO_2 have been increasing at a rate of about 0.5% per year and are now about 30% above pre-industrial levels.The major source of CO_2 emissions is fossil fuel combustion. CO_2 emissions are also a product of forest clearing, biomass burning, and non-energy production processes such as cement production. Atmospheric concentrations of CO_2 have been increasing at a rate of about 0.5% per year and are now about 30% above pre-industrial levels (EPA). Human activities such as fossil fuel combustion and deforestation have increased atmospheric concentrations of CO_2 by approximately 30 percent since the industrial revolution. CO_2 is the standard used to determine the "global warming potentials" (GWPs) of other gases. CO_2 has been assigned a 100-year GWP of 1 (i.e., the warming effects over a 100-year time frame relative to other gases).

Carbon Dioxide equivalent (CO_2e): The universal unit of measurement used to indicate the global warming potential (GWP) of each of the six greenhouse gases (carbon dioxide, methane, nitrous oxide, hydrofluorocarbons (HFCs), perfluorocarbons (PFCs) and sulfur hexafluoride). CO_2e is used to evaluate the impacts of releasing (or avoiding the release of) different greenhouse gases.

Carbon Dioxide Fertilization: The enhancement of the growth of plants as a result of increased atmospheric CO_2 concentration. Depending on their mechanism of photosynthesis, certain types of plants are more sensitive to changes in atmospheric CO_2 concentration.

Carbon Equivalent (CE): A metric measure used to compare the emissions of the different greenhouse gases based upon their global warming potential (GWP). Greenhouse gas emissions in the U.S. are most commonly expressed as "million metric tons of carbon equivalents" (MMTCE). Global warming potentials are used to convert greenhouse gases to carbon dioxide equivalents. Carbon dioxide equivalents can then be converted to carbon equivalents by multiplying the carbon dioxide equivalents by 12/44 (the ratio of the molecular weight of carbon to carbon dioxide). Thus, the formula to derive carbon equivalents is: MMTCE = (million metric tons of a gas) * (GWP of the gas) * (12/44) (EPA)

Carbon Flux: Shifts or flows of carbon over time from one pool to another.

Carbon 'footprint': A carbon footprint is "the total set of GHG (greenhouse gas) emissions caused directly and indirectly by an individual, organization, event or product" (UK Carbon Trust 2008). An individual, nation, or organization's carbon footprint is measured by undertaking a GHG emissions assessment. Once the size of a carbon footprint is known, a strategy can be devised to reduce it.

Carbon Innovators Network: EPA Victoria set up this forum for organizations that want to address the strategic and practical issues of greenhouse gas emissions. The network aims to stimulate debate and innovation in carbon management, provide the support and tools organizations need to develop sound carbon management strategies, and transform climate change from a business cost to a business opportunity.

Carbon Intensity: This is a measure of how much fossil fuel is required to produce a certain amount of economic output. Thus carbon intensity is not the amount of carbon emitted, but a number pro-rated by the Gross Domestic Product (GDP) or purchasing power parity (PPP). For example, if GDP and emissions both rose 3% in a given year, the carbon intensity would remain unchanged despite the fact that the actual emissions had risen. On a global basis, carbon intensity fell by 13% from 1990 to 2000, despite the fact that total emissions grew. In China, carbon intensity fell by 47% at the same time that CO_2 emissions rose by 39%. Carbon intensity measures the dependence of current business operations on carbon emissions.

Carbon Management Principles: A set of Principles developed by EPA Victoria to guide carbon management strategies. They are, in order: Measure, Set objectives, Avoid, Reduce, Switch, Sequester, Assess and Offset. Find out more about the carbon management principles here.

Carbon neutral: A business or a process is described as carbon neutral if it doesn't add to the net amount of CO_2 in the atmosphere. This can be achieved by emitting no CO_2 to begin with - by using only renewable energy, say—or by 'offsetting' emissions (a controversial issue) which means compensating for emissions by another action which might reduce atmospheric CO_2, such as planting trees. In practice, it is impossible for a person to live in an entirely carbon neutral way because even if you cut out energy consumption derived from fossil fuels, most products and services people rely on will have embodied emissions. As the organization or product will typically have caused some greenhouse gas emissions, it is usually necessary to use carbon offsets to achieve neutrality. Carbon offsets are emissions reductions that have been made elsewhere and which are then sold to the entity that seeks to reduce its impact. In order to become carbon neutral it is important to have a very accurate calculation of the amount of emissions which need to be offset – requiring calculation of a carbon footprint.

Carbon offset: A carbon offset is an investment in a project or activity that reduces greenhouse gas emissions offsite or away from the polluting activity. The offset is used to compensate for greenhouse gas emissions from a business's activities. Businesses may seek to offset the emissions of their business as a whole, for specific activities such as air travel or for lifecycle emissions associated with a business product or service. Businesses can then purchase 'offsets' or 'credits' from emission reduction or carbon sequestration activities occurring at another location to offset their calculated emissions. Carbon offset products typically involve projects which invest in forestry activities, energy efficiency projects or the avoidance of methane emissions. Offsets are usually measured in tonnes of carbon dioxide equivalents, and can be sourced from a variety of projects that reduce or sequester emissions of one or more of the six key greenhouse gases.

Carbon offsetting: where an investment is made in a project that will lead to the prevention or removal of carbon dioxide from the atmosphere (for example, planting trees or building renewable energy power stations to avoid the construction of coal ones).

Carbon Pollution Reduction Scheme (CPRS): Australia's cap and trade scheme which will come in to effect in 2010. The CPRS will place a limit, or cap, on the amount of carbon pollution industry can emit and allow trading of carbon credits. It will concentrate on the biggest polluters, by placing obligations on around 1000 Australian companies in total. However it will affect all Australians through indirect price increases.

Carbon pools (A/R - SSC A/R): Carbon pools are: above-ground biomass, belowground biomass, litter, dead wood and soil organic carbon. Project participants may choose not to account for one or more carbon pools if they provide transparent and verifiable information that indicates that the choice will not increase the expected net anthropogenic GHG removals by sinks.

Carbon positive: An individual, family or organization that is responsible for taking more greenhouse gases out of the atmosphere than it emits is said to be "carbon positive". This requires minimizing one's own emissions and more than offsetting remaining emissions by paying for activities such as forest planting or investing in renewable energy. A carbon positive company would actively drive new carbon markets and develop climate-friendly products or services, effectively achieving negative emissions in a positive way.

Carbon price: An economic value placed on the emission of greenhouse gases into the atmosphere from human activity. This price is designed to create a disincentive for emissions and incentive to avoid them. A carbon price takes the form of either a carbon tax or an emissions trading scheme.

Carbon Sequestration: Carbon sequestration refers to the capture and long-term storage of carbon in forests and soils or in the oceans. For example trees and other plants sequester carbon dioxide from the atmosphere as they grow, through the process of photosynthesis. Carbon Sequestration also refers to the uptake and storage of carbon. Trees and plants, for example, absorb carbon dioxide, release the oxygen and store the carbon. Fossil fuels were at one time biomass and continue to store the carbon until burned. (EPA)

Carbon sinks: Natural and potentially man-made features on the Earth's surface where carbon dioxide is removed from the atmosphere. The major natural sinks are forests and oceans which have processes that absorb CO_2. Carbon sinks are vital to fighting global warming because they counteract sources of carbon emissions, such as industry and transport.

Carbon tariff: Import duty levied by countries with greenhouse gas emission caps in place on carbon-intensive goods from countries without such controls in place. The intention is to protect the competiveness of local industries whose goods have higher prices than their imported rivals because they reflect the cost of carbon.

Carbon taxes: A surcharge, levy or cost on the carbon content of a product or activity which aims to discourage the use of fossil fuels and thereby reduce carbon dioxide emissions. Australian governments do not currently support the introduction of a carbon tax.

Carbon Trading: A term used to describe an ETS involving greenhouse gas emissions.

Carbon Trends: an Annual Update of the Global Carbon Budget and Trends (Global Carbon Project, 2008)—analysis supported by references, slide presentations, additional emission figures, data files, and a description of data sources and calculations.

Cavity walls: A wall comprising two layers of brick or block separated by a cavity (gap). This air space acts as an insulator, but does so more effectively if filled with an insulating material, such as plastic foam or natural materials like wool or recycled newspaper.

CCB Standards: The standard focuses exclusively on bio-sequestration (forestry) projects and emphasizes the social and environmental benefits of such projects. CCBS is a project design standard and offers rules and guidance for project design and development.

CCPO—Climate Change Project Office: The Climate Change Project Office (CCPO) is jointly funded by the Department of Trade and Industry (DTI) and the Department for Environment, Food and Rural Affairs (Defra). They are a UK Government advisory office set up to assist UK businesses who wish to pursue opportunities arising from the Kyoto Protocol. They provide advice and support for projects which reduce greenhouse gas emissions and which could be eligible for tradable emission reduction credits.

CCX—Chicago Climate Exchange: The Chicago Climate Exchange is the Worlds first and North America's only marketplace for voluntary legally binding emissions reductions with emissions trading and offsets for all six greenhouse gases.

CCX-Transaction Fee: Fee charged by CCX and deducted from each enrollee's payment received from the sale of XFOs through the CCX Trading Platform.

CDM—Clean Development Mechanism: The Clean Development Mechanism (CDM) is an arrangement under the Kyoto Protocol allowing industrialized countries with a greenhouse gas reduction commitment (so-called Annex 1 countries) to invest in emission reducing projects in developing countries as an alternative to what are generally considered as more costly emission reductions in their own countries. The CDM is supervised by the CDM Executive Board (CDM EB) and is under the guidance of the Conference of the Parties (COP/MOP) of the United Nations Framework Convention on Climate Change (UNFCCC).

Certification (P-SSC): Certification is the written assurance by the designated operational entity that, during a specified time period, a project activity achieved the reductions in anthropogenic emissions by sources of greenhouse gases (GHG) as verified.

Certified Emission Reductions (CER): CERs are emissions certificates issued by bodies of the UN Framework Convention on Climate Change and the Kyoto Protocol for the successful completion of Clean Development Mechanism (CDM) climate protection projects. A CER can be sold or counted toward Annex I countries' emissions commitments. Reductions must be additional to any that would otherwise occur.

CFCs—Chlorofluorocarbons and Related Compounds: This family of anthropogenic compounds includes chlorofluorcarbons (CFCs), bromofluorcarbons (halons), methyl chloroform, carbon tetrachloride, methyl bromide, and hydrochlorofluorcarbons (HCFCs). These compounds have been shown to deplete stratospheric ozone, and therefore are typically referred to as ozone depleting substances. The most ozone-depleting of these compounds are being phased out under the Montreal Protocol.

CFI: Carbon Financial Instrument: The name of the futures contract through which parcels of emission permits are traded on the European Climate Exchange and the Chicago Climate Exchange. Each CFI consists of 100 permits (mandatory EUAs in Europe and voluntary allowances and offsets on the Chicago market) covering the emission of 100 tonnes of CO_2.

CFL bulbs: CFLs are compact fluorescent lamps. They operate on the same principle as fluorescent lights but the tube is folded into a more compact design so they are more versatile and can be used in devices designed to take traditional incandescent bulbs. They use about 80% less electricity than traditional bulbs to produce the same light and last considerably longer. Their mercury content means disposal, especially of broken bulbs, requires extra care.

CFL: Compact fluorescent lamps use less power and have a longer rated life than incandescent light bulbs and therefore use less energy across their lifecycle.

Chlorofluorocarbons (CFCs): CFCs are synthetic industrial gases composed of chlorine, fluorine, and carbon. They have been used as refrigerants, aerosol propellants, cleaning solvents and in the manufacture of plastic foam. There are no natural sources of CFCs. CFCs have an atmospheric lifetime of decades to centuries, and they have 100-year "global warming potentials" thousands of times that of CO_2, depending on the gas. In addition to being greenhouse gases, CFCs also contribute to ozone depletion in the stratosphere and are controlled under the Montreal Protocol.

Chlorofluorocarbons and Related Compounds: This family of *anthropogenic* compounds includes chlorofluorcarbons (CFCs), bromofluorcarbons (*halons*), methyl chloroform, carbon tetrachloride, methyl bromide, and hydrochlorofluorcarbons (HCFCs). These compounds have been shown to deplete stratospheric *ozone*, and therefore are typically referred to as *ozone depleting substances*. The most ozone-depleting of these compounds are being phased out under the *Montreal Protocol*.

CHP—combined heat and power: CHP—also referred to as cogeneration - is a system that recovers the heat produced as a byproduct of electricity generatior instead of simply venting it via cooling towers. This recovered heat can then be used for domestic or industrial heating close to the plant.

Clean Coal: Refers to types or aspects of coal-fired electricity production, which is purported to have a reduced impact on the environment. For example flue gases can be treated with steam with the purpose of removing sulfur dioxide, and returned so as to make the carbon dioxide in the flue gas economically recoverable.

Clean Coal Technologies - Clean coal technologies are technologies designed to enhance both the efficiency and the environmental acceptability of coal extraction, preparation and use. These technologies reduce emissions, reduce waste, and increase the amount of energy gained from each tonne of coal. When combined with carbon capture and storage, clean coal technologies have been found to be capable of achieving cuts of 75% in greenhouse gas emissions.

Clean Development Mechanism (CDM) - A mechanism established by Article 12 of the Kyoto Protocol for project-based emission reduction activities in developing countries. The CDM's two main objectives are to address the sustainable development needs of the host country, and increase the opportunities for meeting emissions reduction commitments. Such projects produce CDM credits known as Certified Emission Reduction certificates.

Climate: Climate in a narrow sense is usually defined as the "average weather," or more rigorously, as the statistical description in terms of the mean and variability of relevant quantities over a period of time ranging from months to thousands of years. The classical period is 3 decades, as defined by the World Meteorological Organization (WMO). These quantities are most often surface variables such as temperature, precipitation, and wind. Climate in a wider sense is the state, including a statistical description, of the climate system.

Climate change: Climate change refers to the change in the state of the climate that can be identified (e.g. by using statistical tests) by changes in the mean and/or variability in typical rainfall, temperature, humidity, cyclones and other extreme weather events. The United Nations Framework Convention of Climate Change (see *United Nations Framework Convention of Climate Change (UNFCC)*) defines climate change as 'a change of climate which is attributed directly or indirectly to human activity that alters the composition of the global atmosphere and which is in addition to natural climate variability over comparable time periods'.

Climate Change Action Plan: Unveiled in October, 1993 by President Clinton, the CCAP is the U.S. plan for meeting its pledge to reduce greenhouse gas emissions under the terms of the Framework Convention on Climate Change (FCCC). The goal of the CCAP is to reduce U.S. emissions of anthropogenic greenhouse gases to 1990 levels by the year 2000. The CCAP, which consists of some 50 voluntary federal programs that span all sectors of the economy, uses a win-win approach by helping program partners save energy, save money, and gain access to clean technology while also reducing greenhouse gas emissions.

Climate Change Commitment: The Term introduced by researchers of the National Center of Atmospheric Research (NCAR) in Boulder, Colorado, to clarify the seriousness of climate change to non-climatologists. It means the amount of climate change that will inevitably occur in the coming century as a result of human behavior in the 20th century.

Climate Cooling: A fall in the average atmospheric temperature.

Climate Feedback: An interaction mechanism between processes in the climate system is called a climate feedback, when the result of an initial process triggers changes in a second process that in turn influences the initial one. A positive feedback intensifies the original process, and a negative feedback reduces it.

Climate Forecast: The result of an attempt to produce a most likely description or estimate of the actual evolution of the climate in the long term.

Climate Lag: The delay that occurs in climate change as a result of some factor that changes only very slowly. For example, the effects of releasing more carbon dioxide into the atmosphere may not be known for some time because a large fraction is dissolved in the ocean and only released to the atmosphere many years later. (EPA)

Climate Model: A quantitative way of representing the interactions of the atmosphere, oceans, land surface, and ice. Models can range from relatively simple to quite comprehensive.

Climate Modeling: The simulation of the climate using computer-based models.

Climate Projection: A description of the response of the climate system to emission or concentration scenarios of greenhouse gases and aerosols, or radiative forcing scenarios, often based upon simulations by climate models. Climate projections are subject to uncertainty, because they are typically based on assumptions concerning future socio-economic and technological developments that may or may not be realized.

Climate Sensitivity: In IPCC Reports, equilibrium climate sensitivity refers to the equilibrium change in global mean surface temperature following a doubling of the atmospheric (equivalent) CO_2 concentration. More generally, equilibrium climate sensitivity refers to the equilibrium change in surface air temperature following a unit change in radiative forcing (degrees Celsius, per watts per square meter, °C/Wm-2). In practice, the evaluation of the equilibrium climate sensitivity requires very long simulations with Coupled General Circulation Models (Climate model). The effective climate sensitivity is a related measure that circumvents this requirement. It is evaluated from model output for evolving non-equilibrium conditions. It is a measure of the strengths of the feedbacks at a particular time and may vary with forcing history and climate state.

Climate Sensitivity: The equilibrium response of the *climate* to a change in *radiative forcing*; for example, a doubling of the *carbon dioxide* concentration.

Climate System: The five physical components (atmosphere, hydrosphere, cryosphere, lithosphere, and biosphere) that are responsible for the climate and its variations.

Climate Variability: Refers to changes in patterns, such as precipitation patterns, in the weather and climate.

Climate Smart 2050: The Queensland Government's strategy for contributing to a national greenhouse gas emissions target of 60% below 2000 levels by 2050.

The short, medium and long-term strategies include investments in clean coal and renewable technologies, energy efficiency incentives for industry and the community, and adaptation plans for vulnerable sectors.

Cloud Condensation Nuclei: Airborne particles that serve as an initial site for the condensation of liquid water and which can lead to the formation of cloud droplets.

CO_2 Fertilization: The enhancement of plant growth as a result of elevated atmospheric CO_2 concentrations.

CO_2-Equivalents: A metric measure used to compare the emissions of various greenhouse gases based upon their Global Warming Potential (GWP). The concentration of CO_2 that would give the same amount of radiative forcing as the given mixture of CO_2 and other greenhouse gases. Essentially, they sum the radiative forcing of all trace gases and treat the total forcing as if it comes from an equivalent CO_2 concentration.

Coal Mine Methane: Coal mine methane is the subset of CBM that is released from the coal seams during the process of coal mining. For more information, visit the Coalbed Methane Outreach program site.

Coal seam gas (CSG): Occurring naturally within coal deposits, this gas is largely composed of methane, which is also the principal component of natural gas.

Coalbed Methane: Coalbed methane is methane contained in coal seams, and is often referred to as virgin coalbed methane, or coal seam gas. For more information, visit the Coalbed Methane Outreach program site.

Co-Benefit: The benefits of policies that are implemented for various reasons at the same time – including climate change mitigation – acknowledging that most policies designed to address greenhouse gas mitigation also have other, often at least equally important, rationales (e.g., related to objectives of development, sustainability, and equity). The term co-impact is also used in a more generic sense to cover both the positive and negative side of the benefits.

Cogeneration: The process by which two different and useful forms of energy are produced at the same time. For example, while boiling water to generate electricity, the leftover steam can be sold for industrial processes or space heating.

Combustion: The process of burning.

Commitment Period: The period under the Kyoto Protocol during which Annex I Parties' GHG emissions, averaged over the period, must be within their emission targets. The first commitment period runs from January 1, 2008 to December 31, 2012.

Complementary measures: Establishing a carbon price alone via emissions trading will be an incomplete approach to mitigate climate change; additional, 'complementary' measures will be required. For example, the existence of market failures in the innovation system means that a price on emissions will not generate optimal levels of investment in technological change. A complementary measure, additional to emissions trading, is mandatory renewable energy targets.

Compliance: System for checking adherence to reduction obligations, including measures and sanctions to be implemented if a country does not fulfill its obligations to reduce emissions as laid down in the Kyoto Protocol.

Composting: The natural biological decomposition of solid organic wastes such as food scraps, paper and lawn clippings, creating a soil-like substance called humus, which can be used instead of artificial fertilizers. The decomposition process produces methane and carbon dioxide, therefore when composting occurs in controlled areas these greenhouse gases can be prevented from entering the atmosphere. Composting also reduces the amount of waste going to landfill.

Compound: A chemical combination of two or more elements, such as carbon and oxygen in carbon dioxide (CO_2)

Concentration Scenarios: Projections of greenhouse gas concentrations derived from emission scenarios and used as input into a climate model to compute climate projections.

Concentration: Amount of a chemical in a particular volume or weight of air, water, soil, or other medium. See parts per billion, parts per million.

Condensing boiler: A condensing boiler captures and uses energy contained in the water vapour given off when gas or oil is burned. In a non-condensing boiler this vapour leaves via a heat-resistant gas tube, or 'flue', and its energy is wasted. A condensing boiler cools the combustion gases sufficiently that the water vapour condenses back into liquid and its heat is recaptured. Approximately 10% of the energy value of the fuel is contained in this water vapour, so a condensing boiler converts far more of the fuel's energy into heat.

Conference of the Parties (COP): The supreme body of the United Nations Framework Convention on Climate Change (UNFCCC). It comprises more than 180 nations that have ratified the Convention. Its first session was held in Berlin, Germany, in 1995 and it is expected to continue meeting on a yearly basis. The COP's role is to promote and review the implementation of the Convention. It will periodically review existing commitments in light of the Convention's objective, new scientific findings, and the effectiveness of national climate change programs. See United Nations Framework Convention on Climate Change.

Conservation Easement: A legally enforceable transfer of usage rights for the purposes of conserving land and prohibiting real estate development.

Coriolis Effect: is caused by the rotation of the Earth (not curvature) and is responsible for the direction of rotation of cyclones. In general, the effect deflects objects moving along the surface of the Earth to the right in the Northern hemisphere and to the left in the Southern hemisphere. As a consequence, winds around the center of a cyclone rotate counterclockwise on the northern hemisphere and clockwise on the southern hemisphere.

Cost of carry: Generally, the costs associated with holding an investment over time - opportunity costs, fees and other expenses. In carbon markets, where the bulk of trade in emission allowances and credits is in forward contracts for their future delivery, it's largely the time value of money, or the interest rate. Here, 'cost of carry' represents the investment return foregone by the seller (who does not receive the sale proceeds until delivery) over the contract period, and is reflected in a premium built into forward carbon prices over spot prices.

Cost-effective: A criterion that specifies that a technology or measure deliver a good or service at equal or lower cost than current practice.

Credit for early action: A provision of an emissions trading scheme that allows crediting of emissions reductions achieved prior to the official start of the scheme. These credits can then be used to help achieve compliance under the scheme.

Credit: A term used in a number of contexts, most commonly in relation to emission reductions that have been achieved in excess of the required amount for (a) Joint Implementation, also known as Emission Reduction Units (ERUs) or Clean Development Mechanism projects, specifically known as Certified Emission Reductions (CERs).

CRS: Origin Energy's Carbon Reduction Scheme is an offsetting scheme that draws upon existing mandatory and voluntary frameworks.

Cryosphere: One of the interrelated components of the Earth's system, the cryosphere is frozen water in the form of snow, permanently frozen ground (permafrost), floating ice, and glaciers. Fluctuations in the volume of the cryosphere cause changes in ocean sea level, which directly impact the atmosphere and biosphere.

D

Dangerous climate change: Dangerous climate change is a term introduced by the 1992 UN Framework Convention on Climate Change - the treaty that led to the Kyoto protocol. In general terms, it refers to climate change of sufficient severity that it will have a major effect on societies, economies and the wider environment. The definition of what that level is varies, but Defra, for example, defines it as a rise in average global surface temperatures of 1-3 ºC (taking pre-industrial revolution average surface temperatures as the base).

Debundling (SSC - SSC A/R): Debundling is defined as the fragmentation of a large scale project activity into smaller parts. A small-scale Project activity that is part of a large scale project activity is not eligible to use the simplified modalities and procedures for small-scale CDM project activities. A large scale project activity or any component of a large scale project activity shall follow the regular CDM modalities and procedures. A proposed small-scale project activity shall be deemed to be a debundled component of a large scale project activity if there is a registered small-scale CDM project activity or a request for registration by another small-scale project activity:

Deforestation/reforestation: Deforestation is the conversion, long term, of areas of forested land to non-forested land through human activity. (Where cleared areas are quickly replanted, it is not regarded as deforestation.) Reforestation is the reversal of that process - i.e. planting trees on land that has historically been forested but has recently been used for other purposes. A less commonly seen term, afforestation, refers to the process of establishing forests on land that has never been forested, or not in recent history.

DEFRA: UK Government Department for Environment, Food and Rural Affairs.

Desertification: The progressive destruction or degradation of vegetative cover, especially in arid or semi-arid regions bordering existing deserts. Overgrazing of rangelands, large-scale cutting of forests and woodlands, drought, burning of extensive areas and climate changes all serve to destroy or degrade the vegetation cover.

Detection of Climate Change: The process of demonstrating that climate has changed in some defined statistical sense, without providing a reason for that change.

Diesel: Ordinary diesel, like regular petrol, is refined from oil but it is a thicker, heavier liquid with a higher 'energy density' - meaning it offers better fuel economy. On the down side, unless you buy an air filter, diesel exhaust is a significant source of particulates and other sources of air pollution. A type of diesel not derived from petroleum is increasingly widely available, commonly referred to as bio-diesel.

Dioxins: Dioxins are chlorine-containing chemical compounds formed and emitted into the atmosphere usually as by-products of human activity - waste incineration and fuel combustion being common examples. They are also formed by natural processes such as forest fires and volcanoes. Some dioxins have harmful properties and could, in sufficient concentration, be harmful to the environment and human health.

Direct emissions: Also known as Scope 1 emissions are greenhouse gas emissions from sources that are owned or controlled by an organization (e.g. emissions resulting from on-site combustion of fossil fuels). In contrast, Scope 2, indirect emissions result from an organization's activities but are not generated by sources the organization controls (e.g. power station emissions for electricity).Direct emissions also refer to those released from activities occurring in specified area - ex USA. Indirect emissions refer to those created elsewhere in order for the USA activities to occur — such as fires in Malaysia to make way for palm oil that is then imported as bio-fuel.

Double counting: Double counting can happen when two or more businesses claim the same emissions reduction. This can happen if an offset is sold to two or more entities, or when an entity upstream of the project unknowingly claims the reduction as its own. The establishment of protocols, and the use of an offsets registry can ensure offsets are adequately accounted for.

Double selling: Double selling occurs when a carbon offset is either sold by the producer and or retailer to more than one buyer. Therefore the emission reductions do not account for the total amount of emissions reductions the buyers have paid for.

Draught proofing: Draught proofing is the process of filling in unwanted gaps in the fabric of a building to reduce the heat loss and discomfort they cause. Common sources of draughts are gaps round window and door frames, places where pipes enter the building, ill-fitted floorboards, letter boxes and even keyholes. Materials used include foams, brushes and thin sections of rubber, plastic or metal.

Drought: A long period of unusually dry weather, long enough to cause serious shortages of water for agriculture and other needs.

Dual fuel: A dual-fuel vehicle is one designed to run on a combination of traditional fuels (petrol or diesel) and alternative fuels (such as liquid petroleum gas or bio diesel). The term is usually applied to cars with separate tanks for each fuel but is sometimes used to describe those diesel engines that accept a blend of traditional fuel and bio-diesel.

E

Eccentricity: The extent to which the Earth's orbit around the Sun departs from a perfect circle.

Eco-friendly: Eco-friendly, or environmentally friendly, is a term applied to goods, services, processes or people deemed to do minimal harm to the environment. The term is shorthand for 'ecologically friendly', ecology being the study of the relationships between living organisms and their environment.

Ecology: The study of the relationships that develop among living organisms and between these organisms and the environment. Ecology also means the amount of living matter in a given habitat.

Ecosystem: An ecosystem is the term applied to the interaction of a community of different living (organic) species - plants, animals and micro-organisms - with non-living (or inorganic) factors, such as atmospheric gases, temperature and light. When the balance of an ecosystem is changed - by the introduction of new elements or dramatic rises in one or more of them - the normal functioning of the ecosystem can be disrupted.

Efficiency ratings: All white goods and other household appliances sold in the UK are now required by law to carry a rating indicating their energy efficiency. Goods are graded A to G, with A the most efficient. Information on the EU-wide labelling scheme is available from the government's environment department, Defra.

EITs: Economies in Transition: Those nations in Annex I of the Kyoto Protocol considered developed but currently in transition to a market economy. Generally EIT means the nations and former republics of the old Soviet bloc.

El Niño - Southern Oscillation (ENSO): El Niño, in its original sense, is warm water current that periodically flows along the coast of Ecuador and Peru, disrupting the local fishery. This oceanic event is associated with a fluctuation of the inter-tropical surface pressure pattern and circulation in the Indian and Pacific Oceans, called the Southern Oscillation. This coupled atmosphere-ocean phenomenon is collectively known as El Niño-Southern Oscillation. During an El Niño event, the prevailing trade winds weaken and the equatorial countercurrent strengthens, causing warm surface waters in the Indonesian area to flow eastward to overlie the cold waters of the Peru Current. This event has great impact on the wind, sea surface temperature, and precipitation patterns in the tropical Pacific. It has climatic effects throughout the Pacific region and in many other parts of the world. The opposite of an El Niño event is called La Niña.

Embodied energy: Embodied energy - sometimes referred to as 'embedded energy' - is the energy used (and therefore the CO_2 emissions) in manufacturing, packaging and transporting a product, material or service. So when calculating the CO_2 savings from a new energy-efficient product such as a boiler or washing machine, its embodied energy needs to be taken into account. Usually the energy saved in use will quickly compensate for the embodied energy, but in some cases - such as small urban wind turbines - it may be that a product will use more energy in its manufacture than it will save across its lifetime.

Emission Reduction Purchase Agreement (ERPA): Binding purchase agreement signed between buyer (of CERs or ERUs) and seller.

Emission Reduction Units (ERU): Emission reductions achieved with JI projects receive credits referred to as emission reduction units (ERUs), whereby one ERU signifies an emission reduction of one tonne of CO_2 equivalent.

Emissions allowance: Permission to emit one ton of carbon dioxide or carbon dioxide equivalent in a specified period of time. Emissions allowances and the Kyoto units known as ERU (Emission Reduction Units) and CER (Certified Emission Reductions), which derive from the project-based mechanisms JI (Joint Implementation) and CDM (Clean Development Mechanism), are referred to jointly as "certificates." Other Kyoto units not relevant for the current 2005-2007 commitment period are AAU (Assigned Amount Units) and RMU (Removal Units).

Emissions Cap: A mandated restraint in a scheduled timeframe that puts a "ceiling" on the total amount of anthropogenic greenhouse gas emissions that can be released into the atmosphere. This can be measured as gross emissions or as net emissions (emissions minus gases that are sequestered).

Emissions conversion factor (ECF): When calculating emissions from energy use it is common to know what quantity of energy was used, either in kWh or by volume or mass of input material. Emissions factors enable a conversion to be made from the input measure of energy to the amount of carbon dioxide emissions that will result. UK conversion factors for energy to CO_2 are published by Defra and are also available from the Carbon Trust.

Emissions Factor: A unique value for scaling emissions to activity data in terms of a standard rate of emissions per unit of activity (e.g., grams of carbon dioxide emitted per barrel of fossil fuel consumed).

Emissions reduction obligations: Compulsory obligations on the industrial countries to limit and reduce their greenhouse gas emissions are laid down in the Kyoto Protocol for the first commitment period (2008-2012). In Annex B of the protocol, the following countries are listed as having undertaken to adjust their greenhouse gas emissions in relation to 1990 levels as follows: Bulgaria, all EU countries, Monaco, Romania, Switzerland: -8%, USA: -7%, Japan, Canada, Poland, Hungary: - 6%, Croatia: - 5%, New Zealand, Russia, Ukraine: +/- 0%, Norway: +1%, Australia: +8%, Iceland: +10%. This translates into an overall reduction of 5.2% in greenhouse gas emissions in the countries listed. The European Union countries have redistributed their reduction commitments in a so-called burden-sharing agreement.

Emissions Reductions: A measurable reduction in the level of greenhouse gases being emitted by a country, state, organization or individual.

Emissions Trading (ET): A market-based approach to achieving the environmental protection goals defined by the Kyoto Protocol. This approach allows countries that reduce their greenhouse gas emissions further than required to trade their excess certificates to offset emissions from other sources within or outside the country. Trading can take place at national or international level, or between companies. Emissions trading in Europe began on January 1, 2005, and are regulated by an EU directive. Companies receive a certain amount of certificates, which is reduced from period to period. Companies that have already made significant climate protection efforts or have shown themselves to be particularly innovative can sell their excess certificates, thus receiving an additional source of income. Heavy polluters must make extra efforts or buy in allowances to fulfill their obligations; those that do not fulfill them must pay a penalty. In this way, the reductions in emissions are made where the costs of reduction are lowest. The efficiency of climate protection is increased and the achievement of the ecological goal ensured.

Emissions Trading and Offsets - lists emissions trading, joint implementation and CDM solutions undertaken by members of The Pew Center's Business Environmental Leadership Council.

Emissions: Emissions are the CO_2 (and other greenhouse gases) produced by energy use, usually calculated and stated as an annual tally: also referred to as your carbon footprint. Your personal emissions can be direct - such as the gas you personally use to heat your home or the petrol you burn to power your car - or indirect - meaning the energy use that has gone into the products or services you buy. The latter, such as the emissions caused by the manufacture of your new TV, or the packaging your food comes in, are also referred to as embodied emissions.

Endangered Species: Animals, plants etc, that are in danger of becoming extinct, i.e. no longer alive in the world.

Energy conservation: Using less energy. For example, turning off stand-by power of electronic equipment

Energy Efficiencies Opportunities program (EEO program): An Australian government initiative that requires businesses with high energy usage to assess opportunities to improve energy efficiency and to report publicly on outcomes.

Energy efficiency (EE): Energy efficiency improvements refer to a reduction in the energy used for a given service (heating, lighting, etc.) or level of activity. Such savings are generally achieved by substituting technologically more advanced equipment to produce the same level of end-use services (e.g. lighting, heating, motor drive) with less electricity.

Energy Intensity: The ratio of energy consumption to a measure of the demand for services (e.g., number of buildings, total floorspace, floorspace-hours, number of employees, or constant dollar value of Gross Domestic Product for services).

Energy intensive: An energy-intensive process uses a great deal of energy - and therefore produces high emissions - relative to its useful output.

Energy Resources: The available supply and price of fossil and alternative resources will play a huge role in estimating how much a greenhouse gas constraint will cost. In the U.S. context, natural gas supply (and thus price) is particularly important, as it is expected to be a transition fuel to a lower carbon economy.

Energy Saver Incentive Scheme: Energy Saver Incentive is the public name for the Victorian Energy Efficiency Target (VEET)

Enhanced Greenhouse Effect: The natural *greenhouse effect* has been enhanced by anthropogenic *emissions* of *greenhouse gases*. Increased concentrations of *carbon dioxide, methane,* and *nitrous oxide,* CFCs, HFCs, PFCs, SF6, NF3, and other photo-chemically important gases caused by human activities such as fossil fuel consumption and adding waste to landfills, trap more *infra-red radiation,* thereby exerting a warming influence on the *climate.*

Entry Into Force: The point at which international climate change agreements become binding. The United Nations Framework Convention on Climate Change (UNFCCC) has entered into force. In order for the Kyoto Protocol to do so as well, 55 Parties to the Convention must ratify (approve, accept, or accede to) the Protocol, including Annex I Parties accounting for 55 percent of that group's carbon dioxide emissions in 1990. As of June 2003, 110 countries had ratified the Protocol, representing 43.9 percent of Annex I emissions.

Environmentally Friendly: Goods, services, processes or people causing minimal harm to the environment. Also know as eco-friendly.

Environment: A combination of the various physical and biological elements that affect the life of an organism, e.g, human beings. There are different environments e.g., aquatic or terrestrial, microscopic or global, all capable of change in time and place. All are closely linked and in combination they constitute the whole physical and atmospheric system of the earth.

Environmental Condition Indicators is measure the condition of land, air or water in the outside environment.

Environmental Refugees: People forced to leave their homes because of environmental factors such as drought, flooding and the rise of sea-levels.

Equipment performance (SSC): To determine equipment performance, project participants shall use:

ERPA (Emissions Reduction Purchase Agreement): Contracts governing the sale of CER carbon credits from UN CDM and JI projects. Heavily used for forward sales of CERs not yet issued, in projects under development, as a means of project financing. The price of such primary CERs is discounted in ERPAs to reflect the risks of non-delivery.

ERU-Emission Reduction Unit: Tradable credits generated from activities to reduce greenhouse emissions in industrialized countries, particularly those of the former Soviet-bloc, under the Kyoto Protocol's Joint Implementation (JI) mechanism.

ETS: An ETS is an organized system of emissions trading that can be applied within businesses, states, countries and also internationally. Through an ETS an organization is allocated an allowance for the amount of greenhouse gases it is permitted to produce. These systems allow those who reduce emissions beyond their obligations to sell their excess emission capacity to others within the ETS who are unable to meet their own emission reduction targets. There are two broad types of emissions trading schemes, cap and trade and baseline and credit.

EU Allowances (EUA): Emissions certificates as defined by the European emissions trading system. They are only traded by participants in the EU emissions trading system. Installation operators can use EUAs to fulfill their obligation to surrender emissions allowances.

EUA (European Union Allowances): Tradable emission credits from the EU Emissions Trading Scheme. Each allowance carries the right to emit one tonne of carbon dioxide.

European Community: As a regional economic integration organization, the European Community can be and is a Party to the UNFCCC; however, it does not have a separate vote from its members (Austria, Belgium, Denmark, Finland, France, Germany, Greece, Ireland, Italy, Luxemburg, the Netherlands, Portugal, Spain, Sweden, and the United Kingdom).

European Union Emissions Trading Scheme (EU ETS): This was launched in January 2005 by the European Commission to help achieve its Kyoto Protocol commitments. This scheme is a cap and trade system designed to make high emitting firms report their carbon emissions and by implementing a cap on emissions, they are encouraged to reduce them. The European Commission puts a cap on each country's emissions and issues a certain amount of allowances - measured in tons of carbon dioxide—and governments subsequently allocate credits/allowances to firms. Companies that do not have sufficient allowances to cover their emissions can buy more through the Emissions Trading Scheme (ETS). Companies with surplus allowances can sell them. Allowances can also be brought into the system from outside the EU through UN systems set up under Kyoto – for example Clean Development Mechanism (CDM).

Evaporation is the process by which a liquid becomes a gas.

Evapo-transpiration: The sum of evaporation and plant transpiration. Potential evapo-transpiration is the amount of water that could be evaporated or transpired at a given temperature and humidity, if there was plenty of water available. Actual evapo-transpiration can not be any greater than precipitation, and will usually be less because some water will run off in rivers and flow to the oceans. If potential evapo-transpiration is greater than actual precipitation, then soils are extremely dry during at least a major part of the year.

Exchange Forest Offset Credit (ZFO): A unit of carbon measure representing one hundred metric tons of CO_2 and identified by its annual vintage.

Extra tropical cyclones, sometimes called mid-latitude cyclones, are a group of cyclones defined as synoptic scale low pressure weather systems that occur in the middle latitudes of the Earth having neither tropical nor polar characteristics.

F

Feed-in tariff: A higher electricity tariff paid by an electricity retailer to customers who have solar systems and who sell the excess electricity back to the grid.

Flexible mechanisms: The Kyoto Protocol makes provision for three instruments that provide flexibility to its signatories in implementing their reduction goals: emissions trading, Joint Implementation (projects carried out jointly by industrial countries) and the Clean Development Mechanism (projects which reduce emissions in developing countries). The underlying philosophy of all three flexible mechanisms is that the Annex B countries can make some of the reductions to which they have committed themselves outside their own country.

Fluorocarbons: Carbon-fluorine compounds that often contain other elements such as hydrogen, chlorine, or bromine. Common fluorocarbons include chlorofluorocarbons (CFCs), hydrochlorofluorocarbons (HCFCs), hydrofluorocarbons (HFCs), and perfluorocarbons (PFCs).

Food miles: Refers to the distance foodstuffs travel through the various stages of production and processing to the point at which they reach the consumer. A measure of both distance traveled and mode of transportation allows comparisons of the energy use and the contribution to greenhouse emissions associated with various food products and their origin.

Forcing Mechanism: A process that alters the energy balance of the *climate system*, i.e. changes the relative balance between incoming *solar radiation* and outgoing *infrared radiation* from Earth. Such mechanisms include changes in solar irradiance, volcanic eruptions, and enhancement of the natural *greenhouse effect* by emission of carbon dioxide.

Forcing or **Radiative Forcing:** is a measure of how internal or external factors affect climate. Internal forcing is part of the natural chaos of the climate system, for example ENSO. External forcing may be natural (e.g. volcanic eruptions or solar fluctuations) or anthropogenic (e.g. increasing greenhouse gases or aerosols). External forcing can change the Earth's energy balance, and hence its climate patterns.

Forest (A/R - SSC A/R): "Forest" is a minimum area of land of 0.05-1.0 hectares with tree crown cover (or equivalent stocking level) of more than 10-30 per cent with trees with the potential to reach a minimum height of 2-5 metres at maturity *in situ*. A forest may consist either of closed forest formations where trees of various storeys and undergrowth cover a high proportion of the ground or open forest.

Forest Carbon Baseline: The quantity of carbon (expressed in metric tons CO_2 equivalent) stored on an individual property at the time it has completed all requirements necessary for enrollment in MACED's program.

Forestry/ Forestation: The science and art of cultivating, maintaining, and developing forests.

Forests: One of the major carbon sinks. There are a variety of definitions based on tree height.

Forward contract: A forward contract is an agreement between two parties to buy or sell an asset (e.g. carbon offsets) at a specified point of time in the future. The price of the abatement is paid before the asset changes hands.

Forward crediting: The buyer pays for and receives a certain number of offsets for emissions reductions or sequestration that will occur in the future. See also Forward sell.

Forward delivery: The buyer pays the purchase price for a certain number of offsets that have yet to be produced. The offsets will be delivered to the buyer once they have been realized and verified.

Forward price: The forward price (or sometimes forward rate) is the agreed upon price of an asset (e.g. carbon offsets) in a forward contract.

Forward purchases: The buyer invests the money upfront but does not get the credits until they are actually produced. These are long-term commitments that are predominantly done on a large scale.

Forward rate: The forward rate (or forward price) is the agreed upon price of an asset (e.g. carbon offsets) in a forward contract.

Forward sell: The sale of carbon offsets for emissions reductions or sequestration that have not yet occurred and therefore can not be precisely measured.

Fossil fuels: Fossil fuels are the deposits of crude oil, natural gas and coal formed by the decay, over millions of years, of organic material (plants, trees animals and bacteria). Because the combustion of fossil fuels releases carbon that has been out of the natural carbon cycle for so long (unlike with living or more recently dead organic matter, known as biomass) it affects the balance between stored carbon and carbon present in the atmosphere as CO_2, a greenhouse gas.

Fossil Fuel Combustion: Burning of coal, oil (including gasoline), or natural gas. This burning, usually to generate energy, releases *carbon dioxide*, as well as combustion by products that can include unburned *hydrocarbons*, *methane*, and carbon monoxide. Carbon monoxide, methane, and many of the unburned hydrocarbons slowly oxidize into carbon dioxide in the atmosphere. Common sources of fossil fuel combustion include cars and electric utilities.

Framework Convention on Climate Change: The landmark international treaty unveiled at the United Nations Conference on Environment and Development (UNCED, also known as the "Rio Summit"), in June 1992. The FCCC commits signatory countries to stabilize *anthropogenic (i.e., human-induced) greenhouse gas emissions* to 'levels that would prevent dangerous anthropogenic interference with the climate system'. The FCCC also requires that all signatory parties develop and update national inventories of anthropogenic emissions of all greenhouse gases not otherwise controlled by the *Montreal Protocol*. Out of 155 countries that have ratified this accord, the U.S. was the first industrialized nation to do so.

FT500: The 500 largest publicly traded companies in the world by market capitalization, according to the Financial Times.

FTSE 100: The UK's 100 largest companies based on market capitalization.

FTSE 350: The UK's 350 largest companies, based on market capitalization, with primary listings on the London Stock Exchange. The FTSE 350 represents a combination of the FTSE 100 Index of the largest 100 companies and the FTSE 250 Index of the next largest 250 companies.

Fuel duty: Fuel duty is the tax the government levies on fuel used for transportation, imposed at the point of sale.

Fuel economy: How many miles a vehicle will travel for each litre of fuel used.

Fuel switching: Reducing carbon dioxide emissions by switching to lower carbon-content fuels, such as from coal to natural gas can be a good first step for many companies to reduce their greenhouse gas emissions.

Fugitive emissions: Fugitive greenhouse gas emissions are waste or loss in the process of fuel production, storage, or transport, such as methane given off during oil and gas drilling and refining, or leakage of natural gas from pipelines.

Full fuel cycle emissions factors: Gives the quantity of emissions released per unit of energy for the entire fuel production and consumption chain. This term is used in the National Greenhouse Accounts Factors but is not used in the GHG Protocol.

Fungibility: In relation to carbon offsets, fungibility is the interchangeability of a carbon credit or permit between certification schemes or programs.

Futures Contract: A futures contract is a standardized contract, to buy or sell a specified commodity (e.g. carbon offsets) at a certain date in the future, at a market determined price (the futures price). A futures contract gives the holder the obligation to make or take delivery under the terms of the contract, whereas an option grants the buyer the right, but not the obligation.

Futures Exchange: A central financial exchange where people can trade standardized futures contracts; that is, a contract to buy specific quantities of a commodity (e.g. carbon offsets) at a specified price with delivery set at a specified time in the future.

G

General Circulation Model (GCM): A global, three-dimensional computer model of the climate system which can be used to simulate human-induced climate change. GCMs are highly complex and they represent the effects of such factors as reflective and absorptive properties of atmospheric water vapor, greenhouse gas concentrations, clouds, annual and daily solar heating, ocean temperatures and ice boundaries. The most recent GCMs include global representations of the atmosphere, oceans, and land surface.

Geo: sequestration: Also known as carbon capture and storage (CCS), geo-sequestration is the process of capture, transport, injection and long-term storage of CO_2 in underground geological formations for the primary purpose of mitigating greenhouse gas emissions.

Geo-sequestration: Involves the capture and long-term underground storage of carbon dioxide. Carbon dioxide produced by coal-fired electricity stations and other industrial sources is compressed to form a liquid and injected into deep underground geological formations. Possible storage sites include saline aquifers, coal seams, and used oil and gas reservoirs.

Geosphere: The soils, sediments, and rock layers of the Earth's crust, both continental and beneath the ocean floors.

Geothermal: Geothermal energy is energy from the inner core of the earth where molten rock heats the water just below the surface of the earth, or from the sun's energy trapped just below the earth's surface. This energy, hot water or steam can then be distributed to heat buildings. It is the reason we have natural spas. This natural heat has been used to heat buildings and water supplies since Roman times. See also ground source heat pump.

GHG (Greenhouse Gases): Greenhouse Gases in the earth's atmosphere absorb and re-emit infrared radiation. The Kyoto Protocol lists six major greenhouse gases, which vary in their relative warming effect. The six gases are: carbon dioxide (CO_2), methane (CH_4), nitrous oxide (N_2O), HFCs (hydrofluorocarbons), PFCs (perfluorocarbons) and sulphur hexafluoride (SF_6).

GIS-Green investment scheme: An arrangement whereby Annex 1 industrialized countries buy the surplus Kyoto carbon emissions credits, AAUs, of eastern European countries on the condition they invest the proceeds in low-emissions technology. GISs came about due to pressure on former Soviet bloc countries to use these surpluses, known as "hot air", responsibly, ie. to build environmentally-sustainable industry. Credits traded through GISs are termed 'greened AAUs'.

Glacier: A multi-year surplus accumulation of snowfall in excess of snowmelt on land and resulting in a mass of ice at least 0.1 km2 in area that shows some evidence of movement in response to gravity. A glacier may terminate on land or in water. Glacier ice is the largest reservoir of fresh water on Earth, and second only to the oceans as the largest reservoir of total water. Glaciers are found on every continent except Australia.

Global 500: The FTSE Global Equity Index Series. This lists the largest 500 global companies by market capitalization.

Global cooling: An average decrease in temperature of the Earth's atmosphere and oceans. Although not happening on a global scale, the IPCC concluded that relative cooling does occur near some industrial areas because of high aerosol concentrations.

Global dimming: A reduction in the amount of sunlight reaching the Earth. The cooling effect of global dimming is thought by some to have obscured the effects of global warming in recent decades. There was roughly a 5% reduction in sunlight from the 1960s to 1990s, but this trend seems to have reversed during the 1990s.

Global Surface Temperature: The average of the sea temperature in the first few metres of the oceans, and the temperature 1.5 metres above ground on land surfaces.

Global Warming Potential (GWP): The index used to translate the level of *emissions* of various gases into a common measure in order to compare the relative *radiative forcing* of different gases without directly calculating the changes in atmospheric concentrations. GWPs are calculated as the ratio of the *radiative forcing* that would result from the emissions of one kilogram of a *greenhouse gas* to that from emission of one kilogram of *carbon dioxide* over a period of time (usually 100 years). Gases involved in complex atmospheric chemical processes have not been assigned GWPs due to complications that arise. Greenhouse gases are expressed in terms of *Carbon Dioxide Equivalent*. The *International Panel on Climate Change (IPCC)* has presented these GWPs and regularly updates them in new assessments. The chart below shows the original GWPs (assigned in 1990) and the most recent GWPs (assigned in 1996) for the most important greenhouse gases.

Global Warming: An increase in the near surface temperature of the Earth. Global warming has occurred in the distant past as the result of natural influences, but the term is most often used to refer to the warming predicted to occur as a result of increased emissions of greenhouse gases. Scientists generally agree that the Earth's surface has warmed by about 1 degree Fahrenheit in the past 140 years. The Intergovernmental Panel on Climate Change (IPCC) recently concluded that increased concentrations of *greenhouse gases* are causing an increase in the Earth's surface temperature and that increased concentrations of sulfate aerosols have led to relative cooling in some regions, generally over and downwind of heavily industrialized areas.

Gold Standard VER: Is a certification standard for carbon offset projects, specifically aimed at small scale projects. (See Gold Standard above).

Gold Standard: A certification standard for carbon offset projects. Initiated by WWF, SSN and Helio International, the Gold Standard for CDM projects was launched in 2003 after wide-ranging stakeholder consultation among key actors of the carbon market as well as governments.

Green-e: An independent program run by a non-profit which certifies and verifies renewable energy certificates (RECs).

Greenhouse Effect: The effect produced as *greenhouse gases* allow incoming *solar radiation* to pass through the Earth's *atmosphere*, but prevent most of the outgoing *infra-red radiation* from the surface and lower atmosphere from escaping into outer space. This process occurs naturally and has kept the Earth's temperature about 59 degrees F warmer than it would otherwise be. Current life on Earth could not be sustained without the natural greenhouse effect.

Greenhouse gases: Greenhouse gases raise the earth's temperature through the greenhouse effect. There are six main examples. As well as carbon dioxide, they include: water vapour, methane, nitrous oxide, ozone and CFCs (which include sulphur hexafluoride, HFCs, and PFCs). Of these, carbon dioxide, methane, nitrous oxide, sulphur hexafluoride, HFCs, and PFCs are controlled under the Kyoto Protocol to limit their concentration in the atmosphere. The heat warming capacity of each gas is measured by its 'global warming potential': how much heat it traps depends on its chemical make-up and how long it stays in the atmosphere

Greenhouse Gas (GHG) Protocol: A widely used standard for emissions reporting. The protocol covers project emissions reporting and corporate emissions reporting. The corporate emissions reporting standard provides a methodology for calculation of a carbon footprint. The protocol was developed by the World Resources Institute and the World Business Council for Sustainable Development.

Greenhouse Gas Intensity: 'greenhouse gas intensity' refers to the volume of greenhouse gases emitted per unit of energy or economic output. It is a relative measure in that, if the economy is growing, greenhouse intensity per unit of economic output may be falling but greenhouse gas emissions may be increasing in absolute terms. Greenhouse gas intensity is to be contrasted with greenhouse gas emission reductions where the volume of gases emitted falls in absolute terms.

Greenhouse Pollution: Pollution of the Earth's atmosphere caused by excessive release of greenhouse gases by humans. This increases the volume of gases in the atmosphere, traps more solar radiation and leads to global warming

GreenPower standard: An accreditation standard for Australian renewable energy operated collaboratively by Australian governments. To be eligible for the GreenPower Standard a company must demonstrate that the renewable energy sources meet strict environmental standards and comes from a new renewable energy facility that was built since January 1997. Renewable resources eligible for the GreenPower Standard include solar power, wind, biomass, hydro-electric power, geothermal energy and wave and tidal power.

Grey Water: Grey water is waste water from baths, showers, washing machines, dishwashers and so on - in other words, water that has been used in the home but does not contain bodily waste - which can be reused for lawn and garden irrigation, say, or for flushing toilets.

Ground source heat pump: Electrically powered systems that tap the stable, residual heat stored immediately below the surface of the earth to provide heating and hot water for homes and offices.

Group of 77 and China, or G77/China: An international organization established in 1964 by 77 developing countries; membership has now increased to 133 countries. The group acts as a major negotiating bloc on some issues including climate change.

H

Habitat: The place or environment where a plant or animal naturally or normally lives and grows.

Halocarbons: Chemicals consisting of carbon, sometimes hydrogen, and either chlorine, fluorine bromine or iodine. Fossil fuels are made up of hydrocarbons. Some hydrocarbon compounds are major air *pollutants*.

Hardwoods and softwoods: Hardwoods come from broad-leaved trees, usually (but not always) deciduous. They are typically (but, again, there are exceptions) denser than softwoods, which come from coniferous trees. Because hardwoods are generally more resilient than softwoods, they are commonly used for making furniture and flooring, and in construction, particularly for exterior work. The use of tropical hardwoods is a contributor to deforestation, linked to climate change and reduced biodiversity.

Heat Stress: A variety of problems for humans and animals associated with very warm temperatures and high humidity. For humans heat exhaustion and heat stroke are examples of these problems.

Hedge: In finance, a hedge is a position established in one market in an attempt to balance exposure to the price risk another obligation.

Hedging: Hedging is a strategy designed to minimize exposure to such business risks as changing prices for commodities (e.g. carbon offsets).

Helium – (He) A light inert gas and the second most abundant element in the universe.

HGWP (High Global Warming Potential): Some industrially produced gases such as sulfur hexafluoride (SF_6), perfluorocarbons (PFCs), and hydrofluorocarbons (HFCs) have extremely high GWPs. Emissions of these gases have a much greater effect on global warming than an equal emission (by weight) of the naturally occurring gases. Most of these gases have GWPs of 1,300 - 23,900 times that of CO_2. These GWPs can be compared to the GWPs of CO_2, CH_4, and N_2O which are presently estimated to be 1, 23 and 296, respectively.

Hot Air: A situation in which emissions (of a country, sector, company or facility) are well below a target due to the target being above emissions that materialized under the normal course of events (i.e. without deliberate emission reduction efforts). Hot air can result from over-optimistic projections of growth. Emissions are often projected to grow roughly in proportion to GDP, and GDP is often projected to grow at historic rates. If a recession occurs and fuel use declines, emissions may be well below targets since targets are generally set in relation to emission projections. If emission trading is allowed, an emitter could sell the difference between actual emissions and emission targets. Such emissions are considered hot air because they do not represent reductions from what would have occurred in the normal course of events.

Hummer: Hummer is the brand name of a 4x4/off-road vehicle manufactured by General Motors and styled after the US military's high mobility multipurpose wheeled vehicle (HMMWV, or Humvee). The Hummer is widely criticised for its very poor fuel economy.

Hybrid: A hybrid car is one that has both a petrol engine and an electric motor and switches between them according to the driving conditions. It is different from a dual-fuel vehicle, which uses two types of liquid fuel.

Hydrocarbons: Hydrocarbons are compounds made up of hydrogen and carbon only. All fossil fuels - such as coal, oil and gas - are hydrocarbons, until they are burnt when they mix with oxygen, creating (amongst other gases) CO_2.

Hydrofluorocarbons (HFCs): These chemicals (along with *perfluorocarbons*) were introduced as alternatives to *ozone depleting substances* in serving many industrial, commercial, and personal needs. HFCs are emitted as by-products of industrial processes and are also used in manufacturing. They do not significantly deplete the stratospheric *ozone* layer, but they are powerful *greenhouse gases* with *global warming potentials* ranging from 140 (HFC-152a) to 12,100 (HFC-23). HFCs are synthetic industrial gases, primarily used in refrigeration and semi-conductor manufacturing as commercial substitutes for chlorofluorocarbons (CFCs). There are no natural sources of HFCs. The atmospheric lifetime of HFCs is decades to centuries, and they have 100-year "global warming potentials" thousands of times that of CO_2, depending on the gas. HFCs are among the six greenhouse gases to be curbed under the Kyoto Protocol.

Hydrogen – (H) At standard temperature and pressure it is a colorless, odorless, nonmetallic, univalent, tasteless, highly flammable diatomic gas. Having been used as an ingredient in some rocket fuels for several decades, hydrogen, or more specifically H_2, is now widely discussed in the context of energy. Hydrogen is not an energy source, since it is not an abundant natural resource and more energy is used to produce it than can be ultimately extracted from it.

Hydrologic Cycle: The process of evaporation, vertical and horizontal transport of vapor, condensation, precipitation, and the flow of water from continents to oceans. It is a major factor in determining climate through its influence on surface vegetation, the clouds, snow and ice, and soil moisture. The hydrologic cycle is responsible for 25 to 30 percent of the mid-latitudes' heat transport from the equatorial to Polar Regions.

Hydropower: Hydropower (or hydroelectric power) is the generation of electricity using the power of falling water.

Hydrosphere: The part of the Earth composed of water including clouds, oceans, seas, ice caps, glaciers, lakes, rivers, underground water supplies, and atmospheric water vapor.

Hypertext: On the web, text links move you easily from one location to another. For example, go to the site's navigation page.

I

Ice Cap: A permanent covering of ice over a large area, especially on the polar region of a planet

Icebergs: A large piece of ice floating on open water after breaking away from a glacier or ice shelf. More icebergs might suggest more melting.

Icecore: A cylindrical section of ice removed from a glacier or an ice sheet in order to study climate patterns of the past. By performing chemical analyses on the air trapped in the ice, scientists can estimate the percentage of carbon dioxide and other trace gases in the atmosphere at that time. Long tubes of ice extracted from an ice shelf or sheet and analyzed to determine past climate patterns. Some tests on the gas trapped within ice cores show that current carbon dioxide levels are higher than they have been for 440,000 years.

Icesheets: Large glaciers covering an area larger than 50,000 km^2 (so can also be applied to large enough ice caps). Currently, only two in the world (Antarctica and Greenland), both of which seem to be reducing in size.

IETA: International Emissions Trading Association IETA is an independent, non-profit organization dedicated to the establishment of effective systems for trading in greenhouse gas emissions by businesses.

IIED - International Institute for Environment and Development: IIED is an international policy research institute and non governmental body working towards sustainable and equitable global development.

Incentive-based Regulation: A regulation that uses the economic behavior of firms and households to attain desired environmental goals. Incentive-based programs involve taxes on emissions or tradable emission permits. The primary strength of incentive-based regulation is the flexibility it provides the polluter to find the least costly way to reduce emissions.

Indirect Emission: "Indirect" emissions are emissions that the Responding Company "caused" to occur, although the Responding Company did not own or control the facility that produced the emission. By way of example, a company's supply chain emissions would be categorized in this way.

Infrared Radiation: Radiation emitted by the Earth's surface, the atmosphere and the clouds. It is also known as terrestrial or long-wave radiation. Infrared radiation has a distinctive range of wavelengths ("spectrum") longer than the wavelength of the red color in the visible part of the spectrum. The spectrum of infrared radiation is practically distinct from that of solar or short-wave radiation because of the difference in temperature between the Sun and the Earth-atmosphere system.

Installation: Any stationary technical unit where one or more activities listed in Schedule I of S.I. 437 of 2004 are carried out. Also, any other directly associated activities that have a technical connection with the activities carried out on that site and which could have an effect on emissions and pollution.

Intergenerational Equity: The fairness of the distribution of the costs and benefits of a policy when costs and benefits are borne by different generations. In the case of a climate change policy the impacts of inaction in the present will be felt in future generations.

Intergovernmental Panel on Climate Change (IPCC): The IPCC was established jointly by the United Nations Environment Programme and the World Meteorological Organization in 1988. The purpose of the IPCC is to assess information in the scientific and technical literature related to all significant components of the issue of climate change. The IPCC draws upon hundreds of the world's expert scientists as authors and thousands as expert reviewers. Leading experts on climate change and environmental, social, and economic sciences from some 60 nations have helped the IPCC to prepare periodic assessments of the scientific underpinnings for understanding global climate change and its consequences. With its capacity for reporting on climate change, its consequences, and the viability of adaptation and mitigation measures, the IPCC is also looked to as the official advisory body to the world's governments on the state of the science of the climate change issue. For example, the IPCC organized the development of internationally accepted methods for conducting national greenhouse gas emission inventories.

Inventory: Quantitative method used to estimate the actual volume, composition, and market value of standing timber.

IPART: Independent Pricing and Regulatory Tribunal of New South Wales is an independent body that oversees regulation in the water, gas, electricity and public transport industries in NSW.

ISO 14000: ISO 14000 is a set of international standards, which provide a framework for the development of an environmental management system (EMS) and supporting audit programs. The ISO 14000 series are intended to help organizations comply with applicable laws, regulations and requirements and to continually improve on their environmental performance.

ISO 14064: A global GHG accounting, reporting and verification standard. The goal of the standard is to 'provide a set of unambiguous and verifiable requirements or specifications to support organizations and proponents of GHG emissions reductions projects.'

ISO 9001 and 14001: ISO 9001 (quality) and ISO 14001 (environment) are "generic management system standards". ISO 9001 contains a generic set of requirements for implementing a quality management system and ISO 14001 for an EMS. These two standards can be applied to any organization.

ISO14065: ISO 14065 defines requirements for companies performing greenhouse gas validation and verification. The standard provides assessing organizations with a basis for assessing the competence of validation and verification companies.

Issuance of Certified Emission Reductions (CERs) (P - SSC): Issuance of CERs refers to the instruction by the Executive Board to the CDM registry administrator to issue a specified quantity of CERs for a project activity into the pending account of the Executive Board in the CDM registry, in accordance with paragraph 66 and Appendix D of the CDM modalities and procedures. Upon issuance of CERs, the CDM registry administrator shall, in accordance with paragraph 66 of CDM modalities and procedures, promptly forward the CERs to the registry accounts of project participants involved, in accordance with their request, having deducted the quantity of CERs corresponding to the share of proceeds to cover administrative expenses for the Executive Board and to assist in meeting costs of adaptation for developing countries vulnerable to adverse impacts of climate change, respectively, in accordance with Article 12, paragraph 8, to the appropriate accounts in the CDM registry for the management of the share of proceeds.

Issuance of temporary certified emission reductions (tCERs) or of long-term certified emission reductions (lCERs) (SSC A/R): Issuance of lCERs or tCERs refers to the instruction by the Executive Board to the CDM registry administrator to issue a specified quantity of lCERs or tCERs for an A/R CDM project activity into the pending account of the Executive Board in the CDM registry, in accordance with paragraph 66 of the CDM modalities and procedures and sections J and K and appendix D of the A/R CDM modalities and procedures.

ITL-International Transaction Log: The means by which carbon allowances and credits generated under the mechanisms of the Kyoto Protocol - AAUs, CERs and ERUs - are traded between countries. An online IT platform that connects UN and national greenhouse emissions registries, facilitating the emerging global carbon market.

J

Joint Implementation: A Kyoto Protocol mechanism which allows developed countries, particularly those in transition to a market economy, to host carbon-reducing projects funded by another developed country. The arrangement sees the credits generated, called ERUs, go to the investor country while the emission allowances (AAUs) of the host country are reduced by the same anountAgreements made between two or more nations under the auspices of the Framework Convention on Climate Change to help reduce greenhouse gas emissions. Under this method, an Annex I Party may implement a project that reduces emissions (e.g. an energy efficiency scheme) or increases removals by sinks (e.g. a reforestation project) in the territory of another Annex I Party, and count the resulting Emission Reduction Units (ERUs) against its own target.

K

Kilowatt-hours (kWh): A kilowatt-hour is an amount of energy equivalent to a power of one kilowatt (a thousand watts) running for one hour. The unit is commonly used on electricity meters. If you know how many kilowatt-hours of energy your household uses, you can translate this into kilograms of CO_2 emitted by multiplying it by 0.527. A megawatt hour (MWh) is an amount of energy equivalent to the power of one megawatt (a million watts) running for one hour. Similarly, a gigawatt hour (GWh) is a billion watts for one hour, and a terawatt hour (TWh) is a trillion. While your domestic gas bill will be set out in kWh, the output for a power station, for example, will obviously be expressed in one of these much larger units.

King's Global Warming Collection (King's College Library, University of Cambridge) - "greenhouse gases, the ozone layer, carbon trading and carbon sequestration, fossil fuels, renewable, nuclear energy, energy storage and distribution, and the hydrogen economy".

Kitemark: The kitemark is a symbol signifying that a product or service has met a standard set by the British Standards Institution (BSI) Product Services division.

Kyoto Protocol Mechanisms: Three procedures established under the Kyoto Protocol to increase the flexibility and reduce the costs of making greenhouse-gas emissions cuts; they are the Clean Development Mechanism (CDM), Joint Implementation (JI) and Emissions Trading.

Kyoto Protocol: An international agreement adopted in December 1997 in Kyoto,

Japan. The protocol sets binding greenhouse gas emissions targets for developed countries that would reduce their emissions on average 5.2% below 1990 levels.The Kyoto Protocol is an addendum to the United Nations Framework Convention on Climate Change. An international agreement linked to the UNFCCC and sharing its aim of stabilizing atmospheric concentrations of greenhouse gases, but requiring separate ratification by governments. The Kyoto Protocol, among other things, sets binding targets for the reduction of greenhouse-gas emissions by industrialized countries. It entered into force for ratifying countries in February 2006 and commits developed nations to collectively cut their greenhouse gas emissions by 5.2 per cent of 1990 levels by 2012.

L

Lagging: Lagging is another term for insulating, or for the insulating material itself. To lag means to add material to spaces or containers liable to heat loss - typically hot water pipes and tanks, and loft spaces.

Landfill Gas: The gas that is generated by the decomposition of waste in landfills.

Landfill: Land waste disposal site in which waste is generally spread in thin layers, compacted, and covered with a fresh layer of soil each day.

Leakage (SSC): Leakage is defined as the net change of anthropogenic emissions by sources of greenhouse gases (GHG) which occurs outside the project boundary, and which is measurable and attributable to the CDM project activity. Reductions in anthropogenic emissions by sources shall be adjusted for leakage in accordance with the provisions of Appendix B for the relevant project categories. The Executive Board shall consider simplification of the leakage calculation for any other project categories added to Appendix B.

Leakage for A/R project activities (A/R - SSC A/R): Leakage is the increase in GHG emissions by sources which occurs outside the boundary of an A/R CDM project activity which is measurable and attributable to the A/R CDM project activity.

Leakage, Carbon leakage: Occurs when laws or activities designed to cut greenhouse gas emissions implemented in one jurisdiction or project area lead to the shifting of the targeted emitting activities elsewhere, thus undermining the attempt to reduce emissions.

LEDs: A LED is a light-emitting diode. Its light is produced by an electrical current passing through a semi-conductive material, so it does not require a filament like an incandescent bulb. They have a long life span, are durable, more energy efficient than CFLs, or incandescent bulbs cool to the touch and, while they have traditionally been used as small indicator lights, multiple LEDs are increasingly being used for household lighting.

Life Cycle Assessment (LCA): LCA is the investigation and valuation of the environmental, economic and social impacts of a product or service. A product's life cycle starts when the raw materials are extracted from the earth through to processing, transport, use, reuse, recycling or disposal. For each of these stages, the impact is measured in terms of the resources used and environmental impacts caused.

Lifetime (Atmospheric): The lifetime of a *greenhouse gas* refers to the approximate amount of time it would take for the anthropogenic increment to an atmospheric pollutant concentration to return to its natural level (assuming emissions cease) as a result of either being converted to another chemical compound or being taken out of the atmosphere via a *sink*. This time depends on the pollutant's sources and sinks as well as its reactivity. The lifetime of a pollutant is often considered in conjunction with the mixing of pollutants in the atmosphere; a long lifetime will allow the pollutant to mix throughout the atmosphere. Average lifetimes can vary from about a week (sulfate aerosols) to more than a century (CFCs, carbon dioxide).

Liquid petroleum gas (LPG): Liquid (or 'liquefied') petroleum gas (LPG) is a mixture of butane, propane and other light hydrocarbons produced either as a by-product of oil refining or from natural gas fields, and used as a vehicle fuel with lower CO_2 emissions than petrol. At standard temperature and pressure, it's a gas but the mixture can be cooled, or subjected to moderate pressure, and is then transformed into a liquid for ease of storage and transportation. LPG is used in dual-fuel vehicles, where it operates alongside a conventional petrol tank. It is also used in heating appliances.

Lithosphere: The outer part of the Earth, consisting of the crust and upper mantle, approximately 100 km thick.

Lobby Group: A group of people campaigning for a change.

Long-term Certified Emission Reductions (lCERs) (A/R - SSC A/R): A long-term certified emission reduction or lCER is a unit issued pursuant to Article 12 of the Kyoto Protocol for an A/R CDM project activity, which expires at the end of the crediting period of the A/R CDM project activity under the CDM for which it was issued. It is equal to one metric tonne of carbon dioxide equivalent. Where project participants have chosen the lCER approach to address non-permanence, a request to the Executive Board has to be made for issuance of lCERs equal to the verified amount of net anthropogenic GHG removals by sinks achieved by the A/R CDM project activity since the previous certification.

Long-Term Commitment: Evidence (e.g. conservation easement or forest certification) presented by all property owners enrolled in MACED's program indicating their commitment to maintaining their property's status as forested.

Long-wave Radiation: The radiation emitted in the spectral wavelength greater than 4 micrometers corresponding to the radiation emitted from the Earth and atmosphere. It is sometimes referred to as 'terrestrial radiation' or 'infrared radiation,' although somewhat imprecisely.

LPG: Liquefied Petroleum Gas is a mixture of hydrocarbon gases, used mostly in transport

LULUCF- Land use, Land Use Change and Forestry: A Kyoto Protocol acronym that stands for human-induced 'Land Use, Land-Use Change and Forestry'. Some of these activities result in greenhouse emissions and some lead to greenhouse gases being removed from the atmosphere. For example, land clearing generates greenhouse gas emissions. Planting trees can lead to greenhouse gases being absorbed from the atmosphere.

M

Management Performance Indicators (MPI) examine the management contribution to environmental performance

Marginal Abatement Cost (MAC): The cost of reducing emissions by one tonne of CO_2e. An aggregation of these costs against total tonnes abated creates a firm's marginal abatement cost curve. The lower the MAC curve, the more effective the firm's emission reduction strategies.

Market Benefits: Benefits of a climate policy that can be measured in terms of avoided market impacts such as changes in resource productivity (e.g., lower agricultural yields, scarcer water resources) and damages to human-built environment (e.g., coastal flooding due to sea-level rise).

Mauna Loa Record: The record of measurement of atmospheric CO_2 concentrations taken at Mauna Loa Observatory, Mauna Loa, Hawaii, since March 1958. This record shows the continuing increase in average annual atmospheric CO_2 concentrations.

Measurable and attributable (All types): In an operational context, the terms measurable and attributable in paragraph 51 (project boundary) of the CDM modalities and procedures should be read as "which can be measured" and "directly attributable", respectively.

Megatonnes: A megatonne is one million tonnes. It is commonly used to describe the large amounts of CO_2 emitted by power stations.

Methane (CH_4): A *hydrocarbon* that is a *greenhouse gas* with a *global warming potential* most recently estimated at 24.5. Methane is produced through anaerobic (without oxygen) decomposition of waste in landfills, animal digestion, decomposition of animal wastes, production and distribution of natural gas and oil, coal production , and incomplete *fossil fuel combustion*. The atmospheric concentration of methane has been shown to be increasing at a rate of about 0.6% per year and the concentration of about 1.7 parts per million by volume (ppmv) is more than twice its preindustrial value. However, the rate of increase of methane in the atmosphere may be stabilizing.

Methane capture: Methane is a greenhouse gas arising from sources such as landfills, animal waste and coal mines. Capturing it stops it entering the atmosphere. It is then burnt generate heat - which produces carbon dioxide, but with a much lower climate impact than the original methane

Micro-generation: Micro-generation is small-scale home production of electricity and heat using technology that produces little or no CO_2. Common examples include ground source heat pumps, solar panels and wind turbines. Where successful, people can meet their own energy needs and also sell excess energy back to the grid.

Micro-hydro: Hydroelectric power installations that typically produce up to 100 kW of power and are often installed in small dammed pools, at the top of a waterfall or on a river.

Microwave Sounding Units (MSU): Sensors carried aboard Earth orbiting satellites that have been used since 1979 to monitor troposphere temperatures.

Mineral wool: Mineral wool refers to any of several lightweight, fibrous materials used as insulation and created by blowing air or steam through molten blast-furnace slag (slag wool), molten rock (rock wool) or molten glass (glass wool). Definitions vary, but fibreglass is sometimes referred to as a mineral wool.

Mitigation: In the context of climate change is any action taken to permanently eliminate or reduce the long-term risk to human life, property, and function from the hazards of climate change. Also see the page explaining the relationship between adaptation and mitigation.

Mitigation of global warming: actions to reduce or avoid greenhouse gas emissions (in order to avoid global warming).

Modalities of communication of project participants with the Executive Board (All types): The modalities of communication between project participants and the Executive Board are indicated at the time of registration by submitting a statement signed by all project participants. All official communication from and to project participants, after a request for registration is submitted by a DOE, shall be handled in accordance with these modalities of communication. If these modalities have to be modified, the new statement shall be signed by all project participants and submitted in accordance with the modalities that are to be replaced.

Monitoring methodology - approved (P - A/R): A monitoring methodology approved by the Executive Board and made publicly available along with relevant guidance.

Monitoring methodology - new (P -A/R): Project participants may propose a new monitoring methodology. In developing a monitoring methodology, the first step is to identify the most appropriate methodology bearing in mind good monitoring practice in relevant sectors. Project participants shall submit a proposal for a new methodology to a designated operational entity by forwarding a completed "Proposed New Baseline and Monitoring methodologies (CDM-NM)" and the project design document (CDM-PDD) with sections A to C completed in order to demonstrate the application of the proposed new methodology to a proposed project activity.

Monitoring methodology (P - A/R - SSC): A monitoring methodology refers to the method used by project participants for the collection and archiving of all relevant data necessary for the implementation of the monitoring plan.

Monitoring methodology (SSC A/R): A monitoring methodology refers to the method used by project participants for the collection and archiving of all relevant data necessary for the implementation of the monitoring plan. A "simplified baseline and monitoring methodologies for selected small-scale afforestation and reforestation CDM project activities categories" is publicly available at the UNFCCC website (cdm.unfccc.int). Project participants may use the methodology for project categories, which fall under the applicability conditions of the methodology.

Monitoring methodology for small-scale CDM project activities - approved (SSC): A monitoring methodology approved by the Executive Board is included in the indicative list of simplified baseline and monitoring methodologies for selected small-scale CDM project activity categories (contained in the Appendix B to the simplified modalities and procedures for small-scale CDM project activities) and is publicly available along with relevant guidance on the UNFCCC- CDM.

Monitoring methodology for small-scale CDM project activities - new (SSC): In accordance with paragraphs 15 and 16 of the simplified modalities and procedures for small-scale CDM project activities, project participants may propose changes to the simplified baseline and monitoring methodologies or propose additional project categories for consideration by the Executive Board. Project participants who wish to submit a new small-scale project activity category or revisions to a methodology shall make a request in writing to be considered by the Executive Board through the working group to assist the Executive Board using a downloadable form 'F-CDM-SSC-Subm'. The Board may draw on expertise, as appropriate, in considering new project categories and/or revisions of and amendments to simplified methodologies. Once approved, the Executive Board shall amend the indicative list of simplified baseline and monitoring methodologies contained in Appendix B.

Monitoring of a CDM project activity (P - SSC): Monitoring refers to the collection and archiving of all relevant data necessary for determining the baseline, measuring anthropogenic emissions by sources of greenhouse gases (GHG) within the project boundary of a CDM project activity and leakage, as applicable.

Monitoring of an A/R CDM project activity (A/R - SSC A/R): Monitoring refers to the collection and archiving of all relevant data necessary for estimating or measuring the net anthropogenic GHG removals by sinks during the crediting period. For more information on the monitoring plan, please refer to paragraph 25 of the A/R CDM modalities and procedures.

Monoculture: Monoculture is the practice of planting a single crop, or genetically similar crops, over a wide area. Its advantages to growers are that the crop may thrive in conditions tailored to its success and without competition from other species. However, the practice will also reduce the region's biodiversity by limiting the number of other species for which it is a suitable habitat. Scientists warn that extensive forest clearance to make way for widespread growth of crops such as sugarcane to create biofuels is creating a monoculture harmful to biodiversity. The opposite of a monoculture is a polyculture.

Montreal Protocol: (on Substances that Deplete the Ozone Layer) An international agreement that entered into force in January 1989 to phase out the use of ozone-depleting compounds such as methyl chloroform, carbon tetrachloride, and CFCs. CFCs are potent greenhouse gases which are not regulated by the Kyoto Protocol since they are covered by the Montreal Protocol.

Moulin: A narrow, tubular chute, hole or crevasse worn in the ice by surface water, which carries water from the surface to the base far below. See the page on Moulin

Mount Pinatubo: A volcano in the Philippine Islands that erupted in 1991. The eruption of Mount Pinatubo ejected enough particulate and sulfate aerosol matter into the atmosphere to block some of the incoming solar radiation from reaching Earth's atmosphere. This effectively cooled the planet from 1992 to 1994, masking the warming that had been occurring for most of the 1980s and 1990s.

MRET (Mandatory Renewable Energy Target): a scheme to increase the amount of electricity generated from 'renewable energy' sources.

MRET: Mandatory Renewable Energy Target: The Australian Government has a policy commitment of a 20 percent share for renewable energy in Australia's electricity supply by 2020.

Municipal Solid Waste (MSW): Residential solid waste and some non-hazardous commercial, institutional, and industrial wastes. This material is generally sent to municipal landfills for disposal.

N

National Allocation Plans. These set out the overall emissions cap for countries in phases I and II of the EU Emissions Trading Scheme up to 2012, and the emissions allowances that each sector and individual installation within each country receives. Plans submitted to the Conference of the Parties (COP) by all Parties outlining the steps that they have adopted to limit their anthropogenic GHG emissions. Countries must submit these plans as a condition of participating in the UN Framework Convention on Climate Change and, subsequently, must communicate their progress to the COP regularly. National law that regulates the allocation of emissions reduction targets to the macro-sectors energy production; industry; trade, commerce and services; transport, and private households. It also regulates how emissions targets are allocated to the energy production and industry sectors, which participate in the trading system.

National Emissions Trading Taskforce (NETT): Established by Australian State and Territory governments to develop a multi-jurisdictional emissions trading scheme for consideration by State and Territory Governments. A key measure being investigated is a national emissions trading scheme (NETS) to provide a mechanism for the Australian economy to transition more smoothly to a carbon-constrained future, and potentially link Australia to international carbon markets.

National Greenhouse Accounts Factors: The National Greenhouse Accounts (NGA) Factors is an Australian guide to emission factors from a range of sectors that is used by companies to calculate greenhouse gases. It is prepared by the Department of Climate Change and replaces the AGO Factors & Methods Workbook.

National Greenhouse and Energy Reporting System (NGER System) - Under the Commonwealth's National Greenhouse and Energy Reporting System, registered corporations are to report annually to the Greenhouse and Energy Data Officer on the level of greenhouse gases emitted, energy produced and energy consumed or the reduction, removal or offset of greenhouse gas emissions achieved by its greenhouse gas projects. The NGER System will be used to underpin emissions trading.

National Grid: The grid, or National Grid, is a privately owned network of electricity supply lines that stretch from the power stations where the electricity is generated to the homes and businesses where it is used. In some parts of the UK, local or domestic micro-generation may make it feasible to live "off grid", while some homes in remote parts of the country have no choice because it does not reach them. In some cases, micro-generation means there is potential for people to sell surplus electricity back to the grid.

Natural Gas: Underground deposits of gases consisting of 50 to 90 percent methane (CH_4) and small amounts of heavier gaseous hydrocarbon compounds such as propane (C_3H_8) and butane (C_4H_{10}).

NCAS: The National Carbon Accounting System tracks emissions (sources) and removals (sinks) of greenhouse gases from Australian land based systems. It underpins National Greenhouse Gas Inventory reporting, and provides a basis for emissions projections to assess progress towards meeting Australia's emissions target.

NCAT: The National Carbon Accounting Toolbox provides the tools for tracking greenhouse gas emissions and carbon stock changes from land use and management.

Negative Feedback: A process that results in a reduction in the response of a system to an external influence. For example, increased plant productivity in response to global warming would be a negative feedback on warming, because the additional growth would act as a sink CO_2, reducing the atmospheric CO_2 concentration.

NEM: National Electricity Market. A wholesale market for the supply of electricity for most of Australia: Queensland, NSW, ACT, Victoria and South Australia.

Neon – (Ne): A colorless, nearly inert noble gas, neon gives a distinct reddish glow when used in vacuum discharge tubes and neon lamps and is found in air in trace amounts.

NGACs - **New South Wales Greenhouse Gas Abatement Scheme Credits:** NGACs are a particular kind of carbon credit issued by the New South Wales state Government of Australia. NGACs are generated, traded and regulated under NSW law. Each NGAC abates a single tonne of carbon for 100 years. **NGERS:** The National Greenhouse and Energy Reporting System is based on the National Greenhouse and Energy Reporting Act 2007, which was passed on 29 September 2007. The Act establishes a mandatory reporting system for corporate greenhouse gas emissions and energy production and consumption. The Act commences on 1 July 2008.

NGPAP: National GreenPower Accreditation Program. The system used to issue GreenPower Standards.

Nitrogen oxides: Nitrogen oxides are compounds of nitrogen and oxygen, two elements that do not normally react with each other but will do so during high temperature combustion – such as in a car engine. Examples include nitric oxide (NO) and nitrogen dioxide (NO2), which contributes to air pollution, and nitrous oxide (N2O), which is a major greenhouse gas. Although its warming effect is far less than CO_2, it persists in the atmosphere for far longer, so measured over 100 years its impact is 298 times greater.

Non-Annex B Parties: Countries that are not listed in Annex B of the Kyoto Protocol.

Non-Annex I Parties: Countries that have ratified or acceded to the UNFCCC that are not listed in Annex I of the UNFCCC.

Non-Compliance: Any landowner found to be noncompliant with the Terms and Conditions contained in MACED's XFO Sales Contract shall return a quantity of XFOs equal to the total quantity of XFOs found to be in non-compliance, or present payment in an amount equal to the cost of acquiring such XFO replacements.

Non-Market Benefits: Benefits of a climate policy that can be measured in terms of avoided non-market impacts such as human-health impacts (e.g., increased incidence of tropical diseases) and damages to ecosystems (e.g., loss of biodiversity).

Non-Methane Volatile Organic Compounds (NMVOCs): Organic compounds, other than methane, that participates in atmospheric photochemical reactions.

Non-Party: A state that has not ratified the UNFCCC. Non-parties may attend talks as observers.

Non-Renewable Energy Sources: Sources of energy such as fossil fuels that cannot be replaced.

NSW GGAS: New South Wales Greenhouse Gas Abatement Scheme commenced on 1 January 2003 and targets are set until 2012. It is one of the first mandatory greenhouse gas emissions trading schemes in the world. GGAS aims to reduce greenhouse gas emissions associated with the production and use of electricity.

O

OECD: The Organization for Economic Cooperation and Development

Offset: An emissions reduction, commonly resulting from a project undertaken in the developing world, which has been sold to compensate for emissions elsewhere. Offsets are commonly used to net off corporate emissions so that an organization can claim to be carbon neutral.

Offset Credits: These represent abatement and can be used to counterbalance emissions that are covered by the emissions trading scheme

Offset Insurance— CCX: eligible properties represented by an Aggregator will be issued XFOs (quantified as metric tons of CO_2 equivalence) on the basis of tonnage realized during the years 2006, 2007, 2008, 2009 and 2010.

Offsets, Carbon offsets, offset credits. Credits issued in return for a reduction of atmospheric carbon emissions through projects such as the provision of renewable energy to replace fossil fuel energy, or reforesting cleared land to create a carbon sink. By paying for such emission reducing activities, individuals and organizations can use the resulting credits to offset their own emissions, either voluntarily or under the rules of most emissions trading schemes. One offset credit equates to an emission reduction of one tonne of CO_2.

Offsetting: Offsetting, or carbon offsetting, refers to the (highly contentious) principle that a company or individual can mitigate or even entirely neutralise the effects of their CO_2 emissions by investing in programmes that promote energy-efficiency or CO_2 absorption, such as providing low-energy lightbulbs or planting trees.

Ofgem: The Office of Gas and Electricity Markets (Ofgem) is the government agency responsible for regulating the energy market in the UK. It describes its purpose as "protecting consumers... by promoting competition... and regulating the monopoly companies which run the gas and electricity networks". Its aims also include curbing climate change by helping energy providers meet environmental goals. There is also a non-governmental equivalent, the consumer watchdog Energywatch.

Oil refining: Oil refining is the process of taking crude oil and turning it into fuels such as petrol and diesel. Crude oil varies in its composition, consisting of hydrocarbons and other organic compounds containing nitrogen, oxygen and sulphur. Refining oil requires energy and releases air-polluting chemicals. However, the majority of the CO_2 produced by oil is emitted only once it has been turned into fuel and then burnt.

Operation Performance Indicators (OPIs) show a measure of how well the process may be performing.

Operational lifetime of a project activity (All types): It is defined as the period during which the project activity is in operation. No crediting period shall end after the end of the operational lifetime.

Opt-out: The option for member countries to exclude certain industries and installations from emissions trading until 2007 if they ensure comparable climate protection measures with respect to emissions limits, inspection, and sanctions.

ORER: The Office of the Renewable Energy Regulator is a statutory authority established to oversee the implementation of the Australian Government's mandatory renewable energy target (MRET).

Organic Farming: There is no universally accepted definition of organic farming, and each country regulates what can and cannot be labelled "organic". In general, it describes a form of agriculture that avoids use of synthetic fertilisers, pesticides, growth regulators, and livestock feed additives, and relies instead, as far as possible, on "natural" products and techniques - such as crop rotation and animal manures.

Organism: A form of life such as an animal or plant

OSCAR: The Australian Government's Online System for Comprehensive Activity Reporting. OSCAR is a web-based data collection tool for reporting energy and greenhouse data to the government.

Overall Monitoring Plan (SSC): If project activities are bundled, a separate monitoring plan shall apply for each of the constituent project activities in accordance with paragraphs 32 and 33 of the simplified modalities and procedures, or an overall monitoring plan shall apply for the bundled projects, as determined by the designated operational entity at validation to reflect good monitoring practice appropriate to the bundled project activities and to provide for collection and archiving of the data needed to calculate the emission reductions achieved by the bundled project activities.

Overgrazing: Excessive grazing (animals eating, for example grass, in a field) causing destruction of vegetation

Oxidize: To chemically transform a substance by combining it with oxygen.

Ozone (O_3): Ozone, the triatomic form of oxygen (O3), is a gaseous atmospheric constituent. In the troposphere, it is created both naturally and by photochemical reactions involving gases resulting from human activities (photochemical smog). In high concentrations, tropospheric ozone can be harmful to a wide range of living organisms. Tropospheric ozone acts as a greenhouse gas. In the stratosphere, ozone is created by the interaction between solar ultraviolet radiation and molecular oxygen (O2). Stratospheric ozone plays a decisive role in the stratospheric radiative balance. Depletion of stratospheric ozone, due to chemical reactions that may be enhanced by climate change, results in an increased ground-level flux of ultraviolet (UV-) B radiation.

Ozone Depleting Substance (ODS): A family of man-made compounds that includes, but is not limited to, chlorofluorocarbons (CFCs), bromofluorocarbons (halons), methyl chloroform, carbon tetrachloride, methyl bromide, and hydrochlorofluorocarbons (HCFCs). These compounds have been shown to deplete stratospheric ozone, and therefore are typically referred to as ODSs.

Ozone Depletion: Like CO_2, ozone is a greenhouse gas, and at low levels it is a part of smog. At ground level it can harm our health, but in the upper atmosphere it protects us from harmful UV radiation. Gases like CFCs break down ozone leading to depletion or holes in the protective ozone layer. Not to be confused with climate change

Ozone Hole: Any part of the ozone layer that has become thinner because of atmospheric pollution, resulting in excess ultraviolet radiation passing through the atmosphere.

Ozone Layer: The layer of ozone that begins approximately 15 km above Earth and thins to an almost negligible amount at about 50 km, shields the Earth from harmful ultraviolet radiation from the sun. The highest natural concentration of ozone (approximately 10 parts per million by volume) occurs in the stratosphere at approximately 25 km above Earth. The stratospheric ozone concentration changes throughout the year as stratospheric circulation changes with the seasons. Natural events such as volcanoes and solar flares can produce changes in ozone concentration, but man-made changes are of the greatest concern.

Ozone Precursors: Chemical compounds, such as carbon monoxide, methane, non-methane hydrocarbons, and nitrogen oxides, which in the presence of solar radiation react with other chemical compounds to form ozone, mainly in the troposphere.

P

Parabolic: A plane curve formed by the intersection of a right circular cone and a plane parallel to an element of the cone or by the locus of points equidistant from a fixed line and a fixed point not on the line. What's that? A bowl shaped trough that focuses sunlight into a narrow beam. The bowl reflects the incoming rays of the sun to a receiving element at the centre of the bowl or trough.

Particulate Matter (PM): Very small pieces of solid or liquid matter such as particles of soot, dust, fumes, mists or aerosols. The physical characteristics of particles, and how they combine with other particles, are part of the feedback mechanisms of the atmosphere.

Particulates: Particulates - sometimes referred to as particulate matter or just particles - are tiny pieces of dust, soot and other materials suspended in the atmosphere. These can be produced naturally, by volcanoes or forest fires for example, but are also caused by incomplete combustion of fossil fuels, particularly from diesel engines. Because diesel is a denser fuel than petrol, it needs more oxygen for all its hydrocarbons to react completely, and the unburnt carbon - soot or black smoke - is emitted through the exhaust pipe.

Parts Per Million (ppm): Number of parts of a chemical found in one million parts of a particular gas, liquid, or solid.

PCF (Prototype Carbon Fund): PCF was set up by the World Bank to facilitate CDM and JI projects from which it buys credits generate

Peak Oil refers to the maximum rate of the production of oil in any area under consideration, recognizing that it is a finite natural resource, subject to depletion.

Perfluorocarbons (PFCs): A group of human-made chemicals composed of carbon and fluorine only. These chemicals (predominantly CF4 and C2F6) were introduced as alternatives, along with hydrofluorocarbons, to the ozone depleting substances. In addition, PFCs are emitted as by-products of industrial processes and are also used in manufacturing. PFCs do not harm the stratospheric ozone layer, but they are powerful greenhouse gases: CF4 has a global warming potential (GWP) of 5,700 and C2F6 has a GWP of 11,900. The GWP is from the IPCC's Third Assessment Report (TAR).

Periodic Carbon Quantification: The process MACED will use to calculate additionality/year on each property enrolled in its program.

Permafrost: A thick layer of soil or earth under the surface that remains below freezing point throughout the year.

Permanence: A key pre-requisite for the credibility of any carbon sequestration activity, particularly tree planting; that it have in place safeguards to cover the possibility that carbon removed from the atmosphere may be released in the future, for example, due to fire, disease or logging. In practice, ongoing verification of planted trees must take place where carbon offset credits have been generated for those carbon reductions.

Permit: In relation to carbon offsets, a permit is a legal permission authorizing the holder to emit a defined quantity of greenhouse gases. Under an ETS a permit is equivalent to one tonne of CO_2e. If a company emits less greenhouse gases than authorized they can sell their permits within the trading scheme.

Personal Carbon Allowances (PCA): Based on the idea that we could all have a set amount of carbon to use or 'spend' each year. We would be able to trade allowances and limits would be reduced each year.

Petrochemicals: Chemicals derived from oil and natural gas.

Petroleum: Petroleum is the end-product of the partial decay of living organisms which once inhabited the world's oceans. As they died they sank to the bottom of the oceans, where they were preserved. It exists in the form of crude oil, natural gas or solid material.

Petrolhead: A car enthusiast; often used to describe those who are most reluctant to limit their car use or petrol consumption.

Photochemical Smog: Smog formed when gases from vehicle exhausts react in strong sunlight, producing substances that reduce visibility and make the air particularly difficult to breathe.

Photosynthesis: The process by which green plants use light to synthesize organic compounds from *carbon dioxide* and water. In the process oxygen and water are released. Increased levels of carbon dioxide can increase net photosynthesis in some plants. Plants create a very important reservoir for carbon dioxide. There are several pathways of photosynthesis with different responses to atmospheric CO_2 concentrations.

Photovoltaic cell: 'Photovoltaics' are materials that produce electricity from sunlight. A single photovoltaic cell can be used to power small devices such as calculators; combined in solar panels they can be used to power solar water heating. Most cells are manufactured from silicon.

Plan Vivo System: An offset project method for small-scale community-based LULUCF projects with a focus on promoting sustainable development and improving rural livelihoods and ecosystems.

Plankton: Free-floating, mostly microscopic, aquatic organisms. Plankton can be divided into phytoplankton (plants) and zooplankton (animals).

Planning permission: Planning permission is consent, usually from a local council, to build on land, or change the use of land or buildings. Significant structural change, such as that often associated with micro-generation projects, is likely to require planning permission. Changes to listed buildings require additional, separate consent.

Point of obligation: The point of obligation is the point in the supply chain - from those who produce goods and services that involve the release of greenhouse gases to the atmosphere, to those who consume those products - at which monitoring and reporting of emissions is required for the purposes of the emissions trading scheme.

Polar region: The area around the North or South Pole.

Pollutant: Strictly, too much of any substance in the wrong place or at the wrong time is a pollutant. More specifically, atmospheric pollution may be defined as the presence of substances in the *atmosphere*, resulting from man-made activities or from natural processes that cause adverse effects to human health, property, and the environment.

Polluter Pays Principle (PPP): The principle that countries should in some way compensate others for the effects of pollution that they (or their citizens) generate or have generated.

Pollution: The introduction of harmful substances or products into the environment

Polymer: A polymer is a giant molecule made up of thousands of atoms. It can be natural or man-made (synthetic). Many plastics in everyday use are polymers, such as polythene or PVC; but then so is cellulose, which makes up wood pulp from which paper is manufactured.

Pooled Projects: The total quantity of individual properties MACED will represent as a CCX-approved Aggregator.

Pooling: The opportunity for installations to join together in order to take part collectively in the emissions trading system. Pooling is included in the EU emissions trading guidelines after pressure from Germany.

Portfolio bundling (SSC): Refers to bundle of project activities of different categories.

Positive Feedback: A process that results in an amplification of the response of a system to an external influence. For example, increased atmospheric water vapor in response to global warming would be a positive feedback on warming, because water vapor is a GHG.

Precautionary Approach: The approach promoted under the *Framework Convention of Climate Change* to help achieve stabilization of *greenhouse gas* concentrations in the *atmosphere* at a level that would prevent dangerous interference with the *climate system.*

Precautionary Principle: The UNFCCC (Article 3.3) states: Parties should take precautionary measures to anticipate, prevent or minimize the causes of climate change and mitigate its adverse effects. Where there are threats of serious or irreversible damage, lack of full scientific certainty should not be used as a reason for postponing such measures taking into account that policies and measures to deal with climate change should be cost-effective so as to ensure global benefits at the lowest possible cost.

Precession: The comparatively slow torquing of the orbital planes of all satellites with respect to the Earth's axis, due to the bulge of the Earth at the equator which distorts the Earth's gravitational field. Precession is manifest by the slow rotation of the line of nodes of the orbit (westward for inclinations less than 90 degrees and eastward for inclinations greater than 90 degrees).

Products and services - In addition to reducing their direct emissions, many companies have introduced lower carbon products (e.g. energy efficient light bulbs and low emission vehicles) and services (e.g. energy audits, carbon offsets).

Project Activity (P - SSC): A project activity is a measure, operation or an action that aims at reducing greenhouse gases (GHG) emissions. The Kyoto Protocol and the CDM modalities and procedures use the term "project activity" as opposed to "project". A project activity could, therefore, be identical with or a component or aspect of a project undertaken or planned.

Project activity with more than one components (SSC): A single project activity composed of two or more distinct project activities being implemented by the same project participant, each applying an approved category/methodology separate from the other. Each component of a project activity should receive or provide an input from/to other components of the project activity.

Project boundary (P): The project boundary shall encompass all anthropogenic emissions by sources of greenhouse gases (GHG) under the control of the project participants that are significant and reasonably attributable to the CDM project activity.

Project boundary for A/R project activities (A/R - SSC A/R): The "project boundary" geographically delineates the A/R CDM project activity under the control of the project participants. An A/R CDM project activity may contain more than one discrete areas of land. If an A/R CDM project activity contains more than one discrete area of land:

Project boundary for small-scale CDM project activity (SSC): The project boundary shall encompass significant anthropogenic emissions by sources of greenhouse gases under the control of the project participants that are reasonably attributable to the small-scale CDM project activity, in accordance with provisions of Appendix B for the relevant project category. The project boundary shall be limited to the physical project activity. Project activities that displace energy supplied by external sources shall earn certified emission reductions (CERs) for the emission reductions associated with the reduced supply of energy by those external sources.

Project Design Document: The official application drawn up by an entity applying for project approval under the Clean Development Mechanism (CDM). PDDs must be validated by an independent third party, then approved and registered by the CDM Executive Board before a project qualifies as a CER carbon credit earner.

Project developer: The person or organization that set up an offset generating project for the purpose of selling carbon offsets and reducing greenhouse gas emissions.

Project participants (All types): In accordance with the use of the term project participant in the CDM modalities and procedures, a project participant is (a) a Party involved, which has indicated to be a project participant, or (b) a private and/or public entity authorized by a Party involved participating in a CDM project activity. In accordance with Appendix D of the CDM modalities and procedures, the decision on the distribution of CERs from a CDM project activity shall exclusively be taken by project participants. Project participants shall communicate with the Executive Board, through the secretariat, in writing in accordance with the "modalities of communication" as indicated at the time of registration or as subsequently altered If a project participant does not wish to be involved in taking decisions on the distribution of CERs, this shall be communicated to the Executive Board, through the secretariat, at the latest when the request regarding the distribution is made.

Prompt delivery: The buyer pays the agreed price for a specific number of offsets which have already been generated and are delivered to the buyer promptly.

Put option: A put option gives the buyer of the option the right, established through a contract to sell the offset/credit. For put options, the abatement is sold at the strike price.

R

Radiation: Energy transfer in the form of electromagnetic waves or particles that release energy when absorbed by an object.

Radiative Forcing: Radiative forcing is the change in the net vertical irradiance (expressed in Watts per square metre: Wm-2) at the tropopause due to an internal change or a change in the external forcing of the climate system, such as, for example, a change in the concentration of carbon dioxide or the output of the Sun. Usually radiative forcing is computed after allowing for stratospheric temperatures to readjust to radiative equilibrium, but with all tropospheric properties held fixed at their unperturbed values. Radiative forcing is called instantaneous if no change in stratospheric temperature is accounted for.

Radiosondes: Sensors carried aboard weather balloons that have been in continuous use since 1979 for the monitoring of troposphere temperatures.

Rainforest Alliance Certification: Rainforest Alliance Certification Programs cover agriculture, forestry, tourism, and community conservation enterprises. The Rainforest Alliance works to conserve biodiversity and ensure sustainable livelihoods by transforming land-use practices, business practices and consumer behaviour.

Ratification: After signing the UNFCCCC or the Kyoto Protocol, a country must ratify it, often with the approval of its parliament or other legislature. In the case of the Kyoto Protocol, a Party must deposit its instrument of ratification with the UN Secretary General in New York.

REC: Renewable Energy Certifications in Australia are issued by the Australian Government's Office of Renewable Energy Regulator. They are equivalent to one-megawatt hour of renewable electricity. RECs can be bought and sold both by electricity retailers and by other businesses in order to meet the legal requirements of MRET.

Recycling: Collecting and reprocessing already manufactured materials for remanufacture, either as the same thing or as part of a different product. An example is collecting aluminium cans, melting them down, and using the aluminum to make new cans or other aluminium products.

REDD: Reduced Emissions from Deforestation and Degradation. An initiative to cut greenhouse gas emissions associated with forest clearing by the inclusion of "avoided deforestation" in carbon market mechanisms.

RMUs: Removal Units: Credits earned from land use, land-use change and forestry projects (LULUCF) in industrialized countries, including such projects under the Kyoto Protocol's JI mechanism. Environmental and economic benefits from the trading system, commonly called the 'double dividend.'

Reduced Emissions from Deforestation and Degradation: An initiative to cut greenhouse gas emissions associated with forest clearing by the inclusion of "avoided deforestation" in carbon market mechanisms.

Reforestation: Planting of forests on lands that have previously contained forests but have then been used for something different.

Regional Groups: The five regional groups meet privately to discuss issues and nominate bureau members and other officials. They are Africa, Asia, Central and Eastern Europe (CEE), Latin America and the Caribbean (GRULAC), and the Western Europe and Others Group (WEOG).

Register: The Register is a database which shows who own what emissions allowances. Account balances can be viewed and transactions initiated online. It combines the functionality of a land registry with that of online banking. The Register is not a trading platform; it does not support the statement of sale and purchase orders or prices.

Removal Monitoring Units (RMU): Credits arising from emissions reductions created by the respective countries by means of projects that reduce emissions (land-use change and forestry) according to Article 3 of the Kyoto Protocol. RMUs are only traded at country level, and can be used towards fulfilling national obligations as of 2008.

Renewable energy: Renewable energy comes from natural sources that can be replenished and not permanently depleted - such as biomass, hydro-power, geothermal heat, solar power, wind power, and wave and tidal power - and most of which do not produce CO_2 emissions. They are unlike fossil fuels, which took millennia to form and cannot be replenished.

Renewables obligation: A renewables obligation is the legal requirement for UK energy providers to source a certain percentage (currently 8%) of their energy from renewable sources. They are awarded renewable obligation certificates (ROCs) to demonstrate that they have done so.

Reserve funds: Part of the overall budget of emission certificates reserved for installations entering the market for the first time.

Residence Time: The average time spent in a reservoir by an individual atom or molecule. With respect to greenhouse gases, residence time usually refers to how long a particular molecule remains in the atmosphere. See atmospheric lifetime.

Respiration: Cellular respiration is the process by which organisms release energy from complex organic molecules, typically sugars. All living things, including plants, respire. Most use oxygen (aerobic respiration) and release carbon dioxide. The balanced chemical equation for aerobic respiration is

Retailer: Carbon offset retailers either funds or purchase carbon offsets in large quantities and then on sells them to individual consumers in smaller quantities.

Retire: In the context of carbon offsets, this means to remove a carbon credit or permit from the market. As a result of retirement those offsets cannot be traded any further. Retiring offsets is a means of regulating offsetting and preventing companies and individuals from selling or buying offsets that have already been used.

Revenue Recycling: If permits are auctioned, this gives considerable sums of money to be recycled back into the economy, either through a lump sum payment of offsetting other taxes. If the existing taxes that are correspondingly reduced were very inefficient, this allows this allows the possibility of both environmental and economic benefits from the trading system, commonly called the 'double dividend.'

RFI: The Radiative Forcing Index calculates the impact non-CO_2 emissions have on the earth in terms of CO_2. RFI is usually used in relation to aviation.

S

S&P 500: Standard & Poor's 500 Index. The 500 largest companies in the US based on market capitalization.

SBF 120: Société des Bourses Françaises 120 Index is a French stock market index based on the 120 most actively traded stocks listed in Paris.

Scenarios: There are two main scenarios used to predict climate change. Low scenario consists of "convergent world with the same global population, that peaks in mid-century and declines thereafter, ... but with rapid change in economic structures toward a service and information economy, with reductions in material intensity and the introduction of clean and resource efficient technologies." High scenario (= 'business as usual') consists of "a world of very rapid economic growth, global population that peaks in mid-century and declines thereafter, and the rapid introduction of new and more efficient technologies." IPCC predicts that, surface air warming during the 21st century for a "low scenario" is 1.8°C (range of 1.1 to 2.9°C to 3.2°F (range of 2.0 to 5.2°F). For a high scenario the best estimate is 4.0°C (range of 2.4 to 6.4°C) to 7.2°F (range of 4.3 to 11.5°F)

Second Assessment Report (SAR): The Second Assessment Report, prepared by the Intergovernmental Panel on Climate Change, reviewed the existing scientific literature on climate change. Finalized in 1995, it is comprised of three volumes: Science; Impacts, Adaptations and Mitigation; and Economic and Social Dimensions of Climate Change.

Secondary glazing: Secondary glazing is a less effective form of double glazing, used where the latter is unaffordable or undesirable in some way (such as in listed buildings). A separate layer of glass or plastic is fitted to the inside of the existing window frame. The gap between the two panes determines its effectiveness in insulating against heat loss and noise.

Secretariat of the UN Framework Convention: The United Nations staff assigned the responsibility of conducting the affairs of the UNFCCC. In 1996 the Secretariat moved from Geneva, Switzerland, to Bonn, Germany.

Semiconductor: A semiconductor is a solid material that is able to conduct electricity, but with conductivity less than that of a good conductor and greater than an insulator. Semiconductors include silicon, and are used in the manufacture of photovoltaic cells, as well as computer chips and other electronic devices.

Sequester: The uptake and storage of carbon from the atmosphere. For example trees and other plants sequester carbon dioxide from the atmosphere as they grow, through the process of photosynthesis.

Sequestrate originally meant to seize assets (as UK government did with trade union funds in 1980's), but now means to capture gaseous carbon and keep in solid form in biomass, soils or oceans.

Sequestration: Opportunities to remove atmospheric CO_2, either through biological processes (e.g. plants and trees), or geological processes through storage of CO_2 in underground reservoirs.

Short Ton: Common measurement for a ton in the United States. A short ton is equal to 2,000 lbs or 0.907 metric tons.

Silicon: Silicon is a chemical element known as a 'metalloid' (because it has intermediate properties between a metal and a non-metal). It is used as a semiconductor, for example in microchips in computers, and to make photovoltaic cells for solar panels.

Sinks: Any process, activity or mechanism that results in the net removal of greenhouse gases, aerosols, or precursors of greenhouse gases from the atmosphere.

Smart meter: A smart meter is a more sophisticated version of your normal gas or electricity meter (in fact, a single smart meter may monitor both) that is linked to your energy provider and can also be linked to a computer in your home. Not yet available, it will eliminate the need for estimated bills, and allow you to monitor your own usage: which devices are using the most energy and when. Smart meters will generally also store up to a month of usage data.

SMI: Swiss Market Index: This is the blue-chip Index in Switzerland and lists 30 of the country's largest companies. The CDP sample is extended to include an additional 20 companies.

Smog: Thick, dark, unpleasant mist formed of smoke and waste gases.

Smoke-free Zone: A smoke-free zone (or smokeless zone, or smoke control area) is not, in this context, anything to do with cigarettes. It refers to an area, designated by a local authority, where only authorised fuels - for example smokeless fuel, coke, and gas - may be burned. Other fuels may sometimes be permitted when used on specified authorised appliances. Ordinary coal and wood (other than small amounts of kindling) are prohibited. The zones originated in London after legislation in 1946 in response to terrible smog but are now very common in towns and cities nationwide.

Soil Carbon: A major component of the terrestrial biosphere pool in the carbon cycle. The amount of carbon in the soil is a function of the historical vegetative cover and productivity, which in turn is dependent in part upon climatic variables.

Solar energy: Energy derived from the sun's rays.

Solar Panels: Solar panels, usually roof-mounted, use the energy of the sun to generate electricity for home use, with the potential to sell surplus back to the grid. Light shining on a panel of photovoltaic cells creates an electric field across layers of semi conductive material, causing electricity to flow.

Solar Power Stations: Stations that use thousands of wide mirrors to collect and concentrate sunlight to be used as energy.

Solar Power: Sun's energy collected by using solar power stations .

Solar Radiation: Radiation emitted by the Sun. It is also referred to as short-wave radiation. Solar radiation has a distinctive range of wavelengths (spectrum) determined by the temperature of the Sun. See ultraviolet radiation, infrared radiation, radiation.

Solar water heater: A solar water heater uses the energy of the sun directly to heat a fluid that is fed into your hot water tank, meaning your boiler has less work to do (and requires less energy) to get your hot water to the required temperature.

Source: Any process or activity that results in the net release of greenhouse gases, aerosols, or precursors of greenhouse gases into the atmosphere.

Sources and sinks: A source is the place or compartment from which a substance comes while a sink is the place or process that takes the substance out of circulation. For example, young growing forests are a carbon sink, while the burning of fossil fuels is a carbon source

Spot price: The spot price or spot rate of an offset product is the price that is quoted for immediate (spot) settlement (payment and delivery) for commodity (e.g. carbon offsets). Spot settlement is normally one or two business days from trade date. This is in contrast with the forward price established in a forward contract or futures contract, where contract terms (price) are set now, but delivery and payment will occur at a future date.

Spot trades: The purchase or sale of abatement (e.g. carbon offsets) for immediate delivery. Spot trades are settled "on the spot" (usually within one or two business days), as opposed to at a set date in the future. Futures transactions that expire in the current month are also considered spot trades. Spot trades are also known as "cash trades". Spot trades are the opposite of forward contracts.

SRES Scenarios: A suite of emissions scenarios developed by the Intergovernmental Panel on Climate Change in its Special Report on Emissions Scenarios (SRES). These scenarios were developed to explore a range of potential future greenhouse gas emissions pathways over the 21st century and their subsequent implications for global climate change.

Standby: Standby, or 'sleep mode', is a mode in which electronic appliances are turned off but still drawing current and ready to activate on command. Although legislation has limited the energy new appliances can use in standby mode, they still use more energy than if they are switched off at the wall.

Stationary Energy Emissions: Includes emissions from electricity generation and from fuels consumed in the manufacturing, construction and commercial sectors, and emissions from other sources like domestic heating.

Stern Report: The Stern report - or the Stern Review on the Economics of Climate Change - is a 700-page report, written by economist Sir Nicholas Stern for the UK government, and published in October 2006. It discusses the effect of climate change and global warming on the world economy and concludes that global warming could shrink the global economy by 20%, but that immediate action to combat it effectively would cost just 1% of global gross domestic product.

Stern Review: The Stern Review on the Economics of Climate Change was released in October 2006 by economist Sir Nicholas Stern for the British government. The report discusses the effect of climate change and global warming on the world economy, the principal message being that the world must act now on climate change or face devastating economic consequences. Some of the main points are:

Stratosphere: Region of the atmosphere between the troposphere and mesosphere, having a lower boundary of approximately 8 km at the poles to 15 km at the equator and an upper boundary of approximately 50 km. Depending upon latitude and season, the temperature in the lower stratosphere can increase, be isothermal, or even decrease with altitude, but the temperature in the upper stratosphere generally increases with height due to absorption of solar radiation by ozone.

Stream flow: The volume of water that moves over a designated point over a fixed period of time. It is often expressed as cubic feet per second (ft^3/sec).

Strike price: Generally refers to a fixed price in a contract between two parties. Where the contract requires delivery of the abatement, the trade will be at the strike price, regardless of the spot price (market price) at that time. See also exercise price.

Subsidiary Body for Implementation (SBI): A permanent body established by the UNFCCC that makes recommendations to the COP on policy and implementation issues. It is open to participation by all Parties and is composed of government representatives.

Subsidiary Body for Scientific & Technical Advice: (SBSTA): A permanent body established by the UNFCCC that serves as a link between expert information sources such as the IPCC and the COP.

Substitution: The economic process of trading off inputs and consumption due to changes in prices arising from a constraint on greenhouse gas emissions. How the extremely flexible U.S. economy adapts to available substitutes and/or finds new methods of production under a greenhouse gas constraint will be critical in minimizing overall costs of reducing emissions.

Sulfate Aerosol: *Particulate* matter that consists of compounds of sulfur formed by the interaction of *sulfur dioxide* and sulfur trioxide with other compounds in the *atmosphere*. Sulfate aerosols are injected into the atmosphere from the combustion of *fossil fuels* and the eruption of volcanoes like *Mt. Pinatubo*. Recent theory suggests that sulfate aerosols may lower the earth's temperature by reflecting away solar *radiation* (negative *radiative forcing*). *Global Climate Models* which incorporate the effects of sulfate aerosols more accurately predict global temperature variations.

Sulfur Dioxide (SO_2): A compound composed of one sulfur and two oxygen molecules. Sulfur dioxide emitted into the *atmosphere* through natural and *anthropogenic* processes is changed in a complex series of chemical reactions in the atmosphere to sulfate *aerosols*. These aerosols result in negative *radiative forcing* (i.e., tending to cool the Earth's surface).

Sulfur Hexafluoride (SF_6): SF_6 is among the six types of greenhouse gases to be curbed under the Kyoto Protocol. SF_6 is a synthetic industrial gas largely used in heavy industry to insulate high-voltage equipment and to assist in the manufacturing of cable-cooling systems. There are no natural sources of SF_6. SF_6 has an atmospheric lifetime of 3,200 years. Its 100-year GWP is currently estimated to be 22,200 times that of CO_2.

Supplementarity: The Protocol does not allow Annex I parties to meet their emission targets entirely through use of emissions trading and the other Kyoto Mechanisms; use of the mechanisms must be supplemental to domestic actions to limit or reduce their emissions.

Supply chain: A network of retailers, distributors, transporters, storage facilities, and suppliers that participate in the production, delivery, and sale of a product to the consumer.

Surrender: In an ETS, countries or businesses are allocated carbon credits or permits, usually for a designated time frame. They must then surrender one carbon credit for each tonne of carbon dioxide equivalent (CO_2e) emitted within the relevant time frame. This ensures the credits are not double-counted.

Sustainability: Sustainability - whether applied to energy, technology, industry, agriculture or just consumption of resources in general - refers to the concept of using things at a rate that, while meeting our own needs, does not compromise future generations' ability to meet theirs. In environmental terms, a process or industry is unsustainable when it requires natural resources to be used up faster than they can be replenished.

SUV: An SUV is a sports utility vehicle, an American term for what in the UK tends to be called a 4x4 or four-wheel drive, although in fact not all SUVs are actually four-wheel drive.

T

Targets and Timetables: Targets refer to the emission levels or emission rates set as goals for countries, sectors, companies, or facilities. When these goals are to be reached by specified years, the years at which goals are to be met are referred to as the timetables. In the Kyoto Protocol, a target is the percent reduction from the 1990 emissions baseline that the country has agreed to. On average, developed countries agreed to reduce emissions by 5.2% below 1990 emissions during the period 2008-12, the first commitment period.

tCO_2e, $MtCO_2e$: Tonnes of carbon dioxide equivalent, and millions of tonnes of carbon dioxide equivalent. This is the metric measurement unit for greenhouse emissions. The global warming impact of all greenhouse gases is measured in terms of equivalency to the impact of carbon dioxide (CO_2). For example, one million tonnes of emitted methane, a far more potent greenhouse gas than carbon dioxide, is measured as 23 million tonnes of CO_2-equivalent, or 23 $MtCO_2e$.

Technological Change: How much technological change will be additionally induced by climate policies is a crucial, but not well quantified, factor in assessing the costs of long-term mitigation of greenhouse gas emissions.

Textile banks: A textile bank is a place to take old clothes for recycling. Compared with other materials such as glass, there is, as yet, very little textile recycling in the UK.

Thaw: To become liquid or soft, after being frozen.

Thermal Expansion: Expansion of a substance as a result of the addition of heat. In climate change, thermal expansion of the world's oceans as a result of global warming is considered the main source of current and future sea-level rise.

Thermohaline Circulation (THC): A three-dimensional pattern of ocean circulation driven by wind, heat and salinity that is an important component of the ocean-atmosphere climate system. In the Atlantic, winds transport warm tropical surface water northward where it cools, becomes denser, and sinks into the deep ocean, at which point it reverses direction and migrates back to the tropics, where it eventually warms and returns to the surface. This cycle or "conveyor belt" is a major mechanism for the global transport of heat, and thus has an important influence on the climate. Global warming is projected to increase sea-surface temperatures, which may slow the THC by reducing the sinking of cold water in the North Atlantic. In addition, ocean salinity also influences water density, and thus decreases in sea-surface salinity from the melting of ice caps and glaciers may also slow the THC.

Thermosphere: The upper region of the atmosphere above the mesosphere. Also called ionosphere, this part of the atmosphere gradually lessens and forms a fuzzy, or indistinct, border with outer space.

Thermostat: A thermostat maintains the temperature of a system at or near a stipulated level using sensors that tell it when to turn off heating devices. In most household central heating systems, a wall-mounted thermostat is used to set a maximum temperature for the house, while thermostatic radiator valves can be used to set a desired level for individual radiators.

Thermostatic radiator valves: Thermostatic radiator valves (TRVs) automatically open and close as necessary to control the flow of hot water into a radiator and so achieve the pre-set air temperature for a particular room, rather than for the whole house. The level is set by turning a dial on the radiator.

Third Assessment Report (TAR): The most recent Assessment Report prepared by the Intergovernmental Panel on Climate Change, which reviewed the existing scientific literature on climate change, including new information acquired since the completion of the Second Assessment report (SAR). Finalized in 2001, it is comprised of three volumes: Science; Impacts and Adaptation; and Mitigation.

Tonne-tCO_2e, $MtCO_2e$: Tonnes of CO_2e, and millions of tonnes of carbon dioxide equivalent. This is the metric measurement unit for greenhouse gas emissions.

Toxins: A toxin is a chemical compound from one organism that is harmful to another organism.

Trace Gas: A term used to refer to gases found in the Earth's atmosphere other than nitrogen, oxygen, argon and water vapor. When this terminology is used, carbon dioxide, methane, and nitrous oxide are classified as trace gases. Although trace gases taken together make up less than one percent of the atmosphere, carbon dioxide, methane and nitrous oxide are important in the climate system. Water vapor also plays an important role in the climate system; its concentrations in the lower atmosphere vary considerably from essentially zero in cold dry air masses to perhaps 4 percent by volume in humid tropical air masses.

Trader: Carbon offset traders purchase carbon offsets in bulk from project developers with the plan to sell

Trading Authority: MACED shall have sole authority to: 1) access the CCX Trading Platform and the Registry Account holding the XFOs it represents; 2) execute sales on the CCX electronic trading platform on behalf of property owners; and 3) distribute XFO sales proceeds to individual property owners.

Trading period: Period of time for which ETS emissions certificates are issued. Initially, two trading periods have been defined: 2005-07 and 2008-12.

Transactions: Transfer of emissions allowances from one Register account to another. It is possible to have internal transactions between two accounts on the same Register as well as external transactions between two accounts that exist on different national Registers (e.g., Germany and Italy). The deleting and surrender of emissions allowances are also transactions.

Transfer rule: Recommendation of the BMU (German Ministry for the Environment, Nature Conservation and Nuclear Safety) on emissions trading. When replacing an old installation with a new one, the certificates allocated to the old installation may be transferred in their entirety to the new one. It is expected that most operators will make use of this transfer rule. As new technology generally emits less carbon dioxide, there will be an excess of certificates that can be sold on the market.

Troposphere: The lowest part of the atmosphere from the surface to about 10 km in altitude in mid-latitudes (ranging from 9 km in high latitudes to 16 km in the tropics on average) where clouds and "weather" phenomena occur. In the troposphere temperatures generally decrease with height.

Tropospheric Ozone (O_3): Ozone that is located in the *troposphere* and plays a significant role in the *greenhouse gas effect* and urban smog. See *Ozone* for more details.

Tropospheric Ozone Precursor: Gases that influence the rate at which *ozone* is created and destroyed in the atmosphere. Such gases include: carbon monoxide (CO), *nitrogen oxides (NOx)*, and non-methane volatile organic compounds (NMVOCs).

Turbulence: Turbulence describes irregular eddies of air within the general air current. It can be caused when wind flows over obstacles such as trees or buildings and gets 'churned up', so it no longer has a smooth flow. Wind turbines do not handle turbulence well and power production can be dramatically decreased.

U

UK CO_2 targets: UK CO_2 targets are this country's self-imposed goals for reducing carbon dioxide emissions, and go beyond the targets set out in the Kyoto protocol. The Climate Change Bill, which has been passed with amendments by the House of Lords but, as of summer 2008, is not yet law, stipulates a 60% reduction in emissions of CO_2 by 2050 and an interim reduction of 26-32% by 2020. The figures are compared to a 1990 baseline. Once the Climate Change Bill is enacted, those targets (or any amendments to them) will become legally binding.

Ultraviolet Radiation/Ultraviolet Rays: High energy sun radiation, transmitted through rays. At normal levels it is an important for the human body. At high levels, for humans it causes sunburn and skin cancer, and can produce changes in the genetic make-up of an organism. It also has a role in the formation of photochemical smog. Most of the UV radiation which reaches the Earth from the Sun is absorbed by the ozone layer in the stratosphere. Thinning of the ozone layer, however, has increased the amount of ultraviolet radiation reaching the Earth's surface, which might increase the incidence of radiation-related problems.

Umbrella Group: Negotiating group within the UNFCCC process comprising the United States, Canada, Japan, Australia, New Zealand, Norway, Iceland, Russia, and Ukraine.

UN Framework Convention on Climate Change: (UNFCCC) A treaty signed at the 1992 Earth Summit in Rio de Janeiro that calls for the "stabilization of greenhouse gas concentrations in the atmosphere at a level that would prevent dangerous anthropogenic interference with the climate system." The treaty includes a non-binding call for developed countries to return their emissions to 1990 levels by the year 2000. The treaty took effect in March 1994 upon ratification by more than 50 countries. The United States was the first industrialized nation to ratify the Convention.

Uncertainty: Uncertainty is a prominent feature of the benefits and costs of climate change. Decision makers need to compare risk of premature or unnecessary actions with risk of failing to take actions that subsequently prove to be warranted. This is complicated by potential irreversibilities in climate impacts and long term investments.

Upper Mantle: The upper mantle is the upper part of the Earth, from the base of the crust downward to a depth of about 670 kilometres.

Urban Heat Island (UHI): The tendency for urban, inner-city, areas to have warmer air temperatures than the surrounding rural landscape, due to the low albedo of streets, pavements, car parks and buildings. These surfaces absorb solar radiation during the day and release it at night, resulting in higher night temperatures.

Urban Heat Island (UHI): Refers to the tendency for urban areas to have warmer air temperatures than the surrounding rural landscape, due to the low albedo of streets, sidewalks, parking lots, and buildings. These surfaces absorb solar radiation during the day and release it at night, resulting in higher night temperatures.

V

Validation (SSC-SSC A/R): Validation is the process of independent evaluation of a proposed SSC/SSC A/R CDM project activity under the CDM by a designated operational entity (DOE) against the requirements of small-scale/ small-scale afforestation and reforestation project activities under the CDM as set out in decision 3/CMP.1, 4/CMP.1, 5/CMP.1 and 6/CMP.1, its annex and relevant decisions of the COP/MOP, on the basis of the project design document. A single designated operational entity (DOE) may perform validation as well as verification and certification for a small-scale/small-scale afforestation or reforestation project activity under the CDM or for bundled small-scale/small-scale afforestation and reforestation project activities under the CDM.

Vapour: Very small drops of water or other liquids in the air, often the result of the heating of a liquid

VCS (Voluntary Carbon Standard): The Voluntary Carbon Standard (VCS) has been developed by the International Emission Trading Association, the Climate Change Group and the World Economic Forum. The VCS has been designed to be a global benchmark standard for project-based voluntary emission reductions and provides a degree of standardization to the Voluntary Carbon Market. It creates a voluntary emission reduction credit, the VCU that can be, traded and used by VCM participants.

VCS 2007: Is a certification standard for carbon offset projects. It provides a global standard for voluntary GHG emission reduction and removal projects and their validation and verification.

VCU: Voluntary Carbon Unit: The name of carbon offset credits specifically verified to the Voluntary Carbon Standard, one of the leading independent standards established to demonstrate integrity in project-based emission reductions in the unregulated voluntary carbon market.

VERs (Verified Emission Reductions): The general name given to carbon offset credits in the voluntary carbon market. These are tradable credits for greenhouse emission reductions generated to meet voluntary demand for carbon credits by organizations and individuals wanting to offset their own emissions. See VER market reports

Vector-borne disease: Disease that results from an infection transmitted to humans and other animals by blood-feeding arthropods, such as mosquitoes, ticks, and fleas. Examples of vector-borne diseases include Dengue fever, viral encephalitis, Lyme disease, and malaria.

VEEC: Victorian Energy Efficiency Certificates are created under the Victorian Energy Efficiency Target (VEET) scheme. One VEECis equal to 0.1 tonnes of CO_2e of energy savings, which is accredited via a certificate system. Certificates are then surrendered to the scheme administrator to achieve compliance. Retailers will need to surrender a specific amount of certificates to ensure they are meeting their responsibilities for the overall reduction of greenhouse gas emissions.

VEET: The Victorian Energy Efficiency Target (VEET) is the target within the Energy Saver Incentive, the mandatory energy efficiency target scheme in Victoria that commenced on 1 January 2009. The legislative requirement is administered by the Essential Services Commission, while compliance is placed on energy retailers through the Victorian Energy Efficiency Target Act 2007.

VER+ Standard: The main criteria for VER+ are in line with those for the Kyoto Protocol project-based mechanisms (JI and CDM), including requirements for project additionality. VER+ projects are not brought to registration with UNFCCC and therefore will not be accounted on any Annex I country's Kyoto balance. For projects in developing countries larger flexibility is provided on the choice of the applied methodologies, which may be composed according to the guidelines applied for JI projects.

VER+: The VER plus (VER+) is a carbon offset standard and closely follows the Kyoto Protocol's project-based mechanisms (CDM and JI). It does not focus on co-benefits. The VER+ standard was developed by TÜV SÜD, a Designated Operational Entity (DOE) for the validation and verification of CDM projects. It was designed for project developers who have projects that cannot be implemented under CDM yet who want to use very similar procedures as the CDM. The VER Plus was launched in mid 2007.

Verification (A/R -SSC A/R): Verification is the periodic independent review and ex post determination by the DOE of the net anthropogenic GHG removals by sinks achieved, since the start of the project, by an A/R CDM project activity under the CDM. Certification is the written assurance by a DOE that an A/R CDM project activity under the CDM achieved the net anthropogenic GHG removals by sinks since the start of the project, as verified.

Verification (P - SSC): Verification is the periodic independent review and ex post determination by a designated operational entity of monitored reductions in anthropogenic emissions by sources of greenhouse gases (GHG) that have occurred as a result of a registered CDM project activity during the verification period. There is no prescribed length of the verification period. It shall, however, not be longer than the crediting period. *To be applied mutatis mutandis to SSC - Verification.*

Verification Report: A report prepared by an Operational Entity, or by another independent third party, which reports the findings of the Verification process of offset projects, including the amount of reductions in emission of greenhouse gases that have been found to have been generated

Verifier: A technical expert approved by CCX who verifies the amount of XFOs an Aggregator calculates is present on an individual property.

VERs (Verified Emission Reductions): Credits generated by small scale projects which are assessed and verified by third party rather than through the UNFCCC. Projects can be relatively cheap and based anywhere, unlike CDM projects which must be hosted in developing countries.

Vintage: The vintage of an XFO is defined as the first year the designated XFO may be used for compliance with the CCX emission reduction schedule.

Voluntary Carbon Market (VCM): Voluntary buying and selling of greenhouse gas abatements, normally in the form of carbon offset credits - in order to reduce or completely offset the carbon emissions of an organization or activity. By contrast, participants in an emissions trading scheme will be legally obliged to acquire permits or offset credits to cover their emissions.

Voluntary Carbon Standard has been proposed by The Climate Group and the

International Emissions Trading Association for carbon offsets bought and sold. These are standards, including gold standard, under which projects selling credits in the voluntary market can receive accreditation that they have met certain stringent criteria.

Voluntary Carbon Standard Program: The VCS Program includes the standard (VCS 2007) and the Program Guidelines 2007. The VCS Program provides a new global standard and criteria for validating, measuring, and monitoring voluntary carbon offset projects. For more information,

Voluntary market: unregulated market for carbon credits, outside the Kyoto protocol and the EUETS.

Voluntary offset: A carbon offset which is purchased by a concerned individual or business out of choice, not as a result of regulation

Vulnerability is the degree to which a system is susceptible to, or unable to cope with, adverse effects of climate change, including climate variability and extremes. Vulnerability is a function of the character, magnitude, and rate of climate change and variation to which a system is exposed, its sensitivity, and its adaptive capacity.

W

Waste management: Waste management policies which reduce greenhouse gas emissions include landfill gas capture and use and 'reduce, reuse, recycle'. Reducing refuse production or using alternative means of disposal that reduce greenhouse gas emissions from waste decomposition and processing can have a large impact.

Wastewater: Water that has been used and contains dissolved or suspended waste materials.

Water pressure: Water pressure is the force that pushes water through pipes and determines the rate of flow from your taps. Water companies are required to provide water at a pressure that will reach the upper floors of houses, but where this is a problem, pressure can be improved using pumps. However, devices such as power showers, which use a pump to boost flow for a more powerful shower spray, also greatly increase the volume of water used and the energy required to heat it.

Water Vapor: The most abundant *greenhouse gas*, it is the water present in the *atmosphere* in gaseous form. Water vapor is an important part of the natural *greenhouse effect*. While humans are not significantly increasing its concentration, it contributes to the *enhanced greenhouse effect* because the warming influence of greenhouse gases leads to a positive water vapor *feedback*. In addition to its role as a natural greenhouse gas, water vapor plays an important role in regulating the temperature of the planet because clouds form when excess water vapor in the atmosphere condenses to form ice and water droplets and precipitation.

Wave and Tidal Power: Power or energy that comes from movement of the sea.

Weather: Atmospheric condition at any given time or place. It is measured in terms of such things as wind, temperature, humidity, atmospheric pressure, cloudiness, and precipitation. In most places, weather can change from hour-to-hour, day-to-day, and season-to-season. Climate in a narrow sense is usually defined as the "average weather", or more rigorously, as the statistical description in terms of the mean and variability of relevant quantities over a period of time ranging from months to thousands or millions of years. The classical period is 30 years, as defined by the World Meteorological Organization (WMO). These quantities are most often surface variables such as temperature, precipitation, and wind. Climate in a wider sense is the state, including a statistical description, of the climate system.

Weather Satellite: A satellite that orbits Earth taking pictures, for example monitoring the temperature and measuring the height of waves

Weather Stations: A network of observing stations on land and at sea that send reports to a weather forecast centre, giving details of, for example, cloud types, wind, speed, temperature and pressure.

Wind Farm: An area with windmills that turn the wind's energy into electricity.

Wind turbine: All turbines use kinetic (movement) energy to cause a bladed rotor to turn and so generate electricity. A wind turbine is a machine that captures the force of the wind. Often seen in remote, exposed coastal regions or offshore where wind speeds are greatest and most consistent, smaller versions can be installed on houses, but research suggests they may be useless, or even counter-productive, in urban settings.

Working Forests: Forested land harvested in accordance with a MACED-approved forest stewardship plan, forest certification, and a current forest inventory.

Z

Zero Carbon is any activity (whether an operation, plan or policy) where absolute carbon emissions are zero.

Index